Sixth Edition

INTRODUCTION TO GEOGRAPHIC INFORMATION SYSTEMS

Kang-tsung Chang

Mc
Graw
Hill

*Connect
Learn
Succeed*™

Connect
Learn
Succeed™

INTRODUCTION TO GEOGRAPHIC INFORMATION SYSTEMS, SIXTH EDITION
International Edition 2012

Exclusive rights by McGraw-Hill Education (Asia), for manufacture and export. This book cannot be re-exported from the country to which it is sold by McGraw-Hill. This International Edition is not to be sold or purchased in North America and contains content that is different from its North American version.

Published by McGraw-Hill, a business unit of The McGraw-Hill Companies, Inc., 1221 Avenue of the Americas, New York, NY 10020. Copyright © 2012 by The McGraw-Hill Companies, Inc. All rights reserved. Previous editions © 2010, 2008, and 2006. No part of this publication may be reproduced or distributed in any form or by any means, or stored in a database or retrieval system, without the prior written consent of The McGraw-Hill Companies, Inc., including, but not limited to, in any network or other electronic storage or transmission, or broadcast for distance learning.
Some ancillaries, including electronic and print components, may not be available to customers outside the United States.

10 09 08 07 06 05 04 03 02 01
20 15 14 13 12 11
CTP   COS

When ordering this title, use ISBN 978-007-108616-5 or MHID 007-108616-1

Printed in Singapore

www.mhhe.com

BRIEF CONTENTS

CONTENTS

CHAPTER 3

VECTOR DATA MODEL   44

CHAPTER 4

RASTER DATA MODEL   68

CHAPTER 5

GIS DATA ACQUISITION   *90*

CHAPTER 6

GEOMETRIC TRANSFORMATION   *111*

CHAPTER 9

DATA DISPLAY AND CARTOGRAPHY   *168*

CHAPTER 10

DATA EXPLORATION   *198*

CHAPTER 11

VECTOR DATA ANALYSIS   *222*

CHAPTER 12

RASTER DATA ANALYSIS   *248*

CHAPTER 13

TERRAIN MAPPING AND ANALYSIS   *268*

CHAPTER 14

VIEWSHEDS AND WATERSHEDS   *291*

CHAPTER 15

SPATIAL INTERPOLATION   *314*

CHAPTER 16

GEOCODING AND DYNAMIC SEGMENTATION 344

CHAPTER 17

LEAST-COST PATH ANALYSIS AND NETWORK ANALYSIS 364

CHAPTER **18**

GIS MODELS AND MODELING   *389*

PREFACE

ABOUT GIS

A geographic information system (GIS) is a computer system for storing, managing, and displaying geospatial data. Since the 1970s GIS has been important for researchers in natural resource management, crime analysis, emergency planning, land records management, market analysis, and transportation planning. It has also become a necessary tool for government agencies of all levels for routine operations. More recent integration of GIS with the Internet, GPS (global positioning systems), wireless technology, and web service has found commercial applications in location-based services, interactive mapping, in-vehicle navigation systems, and precision farming. It is therefore no surprise that, for the past several years, the U.S. Department of Labor has listed geospatial technology as an emerging field for career development. Geospatial technology centers on GIS and uses GIS to integrate data from remote sensing, GPS, cartography, and surveying to produce useful geographic information.

Many of us actually use geospatial technology on a daily basis. To locate a restaurant, we go online, type the name of the restaurant, and find it on a location map. To make a map for a project, we go to Google Maps, locate a reference map, and superimpose our own contents and symbols to complete the map. To find the shortest route for driving, we use an in-vehicle navigation system to get the directions. And, to record places we have visited, we use geotagged photographs. All of these activities involve the use of geospatial technology, even though we may not be aware of it.

It is, however, easier to be GIS users than GIS professionals. To become GIS professionals, we must be familiar with the technology as well as the basic concepts that drive the technology. Otherwise, it can easily lead to the misuse or misinterpretation of geospatial information. This book is designed to provide students with a solid foundation in GIS concepts and practice.

UPDATES TO THE SIXTH EDITION

This book can be used in a first or second GIS course. The sixth edition has 18 chapters. Chapters 1 to 4 explain GIS concepts and data models. Chapters 5 to 8 cover data input, editing, and management. Chapters 9 and 10 include data display and exploration. Chapters 11 and 12 provide an overview of core data analysis. Chapters 13 to 15 focus on surface mapping and analysis. Chapters 16 and 17 examine linear features and movement. Chapter 18 presents GIS models and modeling. This book covers a large variety of topics in GIS to meet the needs of students from different disciplines.

In this edition, I have introduced a number of new topics, including Web mapping (Chapter 1), Web Mercator (Chapter 2), map algebra (Chapter 12), vehicle routing problems (Chapter 17), and local regression analysis (Chapter 18). I have also completely re-written Chapter 16 on geocoding and dynamic segmentation and Chapter 18 on GIS models and modeling. This edition retains boxes, review questions, task-related questions, and challenge tasks, which have proven to be useful to readers of the earlier editions. At the same time, references have been updated and new figures have been added.

This sixth edition continues to emphasize the practice of GIS. Each chapter has problem-solving tasks in the applications section, complete with data sets and instructions. The number of tasks totals 77, with two to seven tasks in each chapter. The instructions for performing the tasks correlate to ArcGIS 10.0. All tasks in this edition use ArcGIS Desktop and its extensions of Spatial Analyst, 3-D Analyst, Geostatistical Analyst, Network Analyst, and ArcScan. A challenge task is found at the end of each applications section, challenging students to complete the task without given instructions.

The website for the sixth edition, located at **www.mhhe.com/changgis6e,** contains a password-protected instructor's manual. Contact your McGraw-Hill sales representative for a user ID and password.

CREDITS

Data sets downloaded from the following websites are used for some tasks in this book:

Montana GIS Data Clearinghouse
http://www.nris.state.mt.us/

Northern California Earthquake Data
http://quake.geo.berkeley.edu/

University of Idaho Library
http://inside.uidaho.edu

Washington State Department of Transportation GIS Data
http://www.wsdot.wa.gov/mapsdata/ geodatacatalog/default.htm

Wyoming Geographic Information Advisory Council
http://wgiac2.state.wy.us/html/

ACKNOWLEDGMENTS

I would like to thank the following reviewers who provided many helpful comments on the fifth edition:

Kirk Goldsbury
Michigan State University

Daniel Sui
Texas A&M University

Adrienne Domas
Michigan State University

Jennifer Miller
University of Texas at Austin

Wing Cheung
Palomar College

Paul Sutton
University of Denver

Danielson Kisanga
Miami University

I have read their reviews carefully and incorporated many comments into the revision. Of course, I take full responsibility for the book.

I wish to thank Todd Turner, Lisa Nicks, Robin Reed, Brenda Rolwes, and Sue Culbertson at McGraw-Hill and Lori Bradshaw at S4Carlisle Publishing for their guidance and assistance during various stages of this project. Finally, this book is dedicated to Gary and Mark.

Kang-tsung Chang

CHAPTER 1

INTRODUCTION

Are you looking for a twenty-first century career that combines interests in the Earth, space, and high technology? If you are, you should look into the geospatial industry according to the U.S. Department of Labor **(http://www.doleta.gov/ brg/jobtraininitiative/).** Since 2004, the U.S. Department of Labor has listed geospatial technology, along with biotechnology and information technology, as a high-growth industry. What can one do with geospatial technology? Geospatial

technology can help create large databases of geographic information and turn the information into maps and decision-making tools for problem solving. For example, this kind of geographic information has proved to be critical for identifying priorities, planning logistics, and mapping access routes for relief operations in the aftermath of major disasters such as Hurricane Katrina (Nourbakhsh et al. 2006).

Geographic information is equally important in the business sector. As an example, a recent article published in *Advertising Age,* a magazine that delivers news, analysis, and data on marketing and media, suggests that adding geographic information to people's mobile habits will "spawn" a flood of useful on-the-go brand experiences.

Geospatial technology covers a number of fields, including remote sensing, cartography, surveying, and photogrammetry; however, to integrate data from these different fields to produce useful geographic information we rely on geographic information systems.

1.1 GIS

A **geographic information system (GIS)** is a computer system for capturing, storing, querying, analyzing, and displaying geospatial data. **Geospatial data** describe both the locations and characteristics of spatial features. To describe a road, for example, we refer to its location (i.e., where it is) and its characteristics (e.g., length, name, speed limit, and direction), as shown in Figure 1.1. The ability of a GIS to handle and process geospatial data distinguishes GIS from other information systems and allows GIS to be used for integration of geospatial data and other data. It also establishes GIS as a technology important to geoscientists, cartographers, photogrammetrists, environmental engineers, and urban and regional

planners, some of the occupations listed on the U.S. Department of Labor's website.

1.1.1 Components of a GIS

GIS operations, similar to other information technologies, require the following components besides geospatial data:

- Hardware. The hardware includes computers such as PCs and workstations, and operating systems such as Windows, Linux, and UNIX. Additional equipment may include monitors for display, digitizers and scanners for digitization of spatial data, GPS (global positioning systems) receivers and mobile devices for fieldwork, and printers and plotters for hard-copy data display.

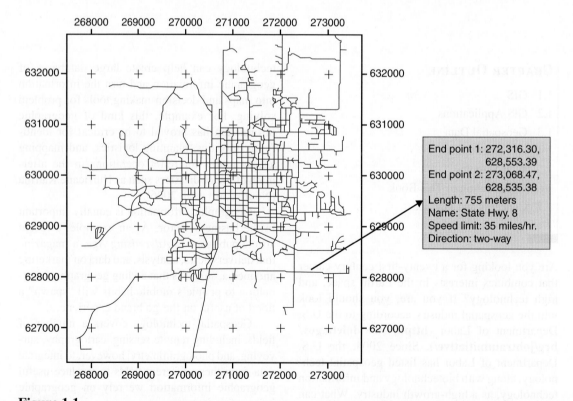

Figure 1.1

An example of geospatial data. The street network is based on a plane coordinate system. The box on the right lists the *x*- and *y*-coordinates of the end points and other attributes of a street segment.

- Software. The software includes the source code and the user interface. The code may be written in C++, Visual Basic, or Python. Common user interfaces are menus, graphical icons, command lines, and scripts.
- People. GIS professionals define the purpose and objectives and provide the reason and justification for using GIS.
- Infrastructure. The infrastructure refers to the necessary physical, organizational, administrative, and cultural environments that support GIS operations. The infrastructure includes requisite skills, data standards, data clearinghouses, and general organizational patterns.

1.1.2 A Brief History of GIS

GIS is not new. The first operational GIS is reported to be developed by Tomlinson in the early 1960s for storing, manipulating, and analyzing data collected for the Canada Land Inventory (Tomlinson 1984). In 1964, Fisher founded the Harvard Laboratory for Computer Graphics, where several well-known computer programs such as SYMAP, SYMVU, GRID, and ODESSEY were developed and distributed throughout the 1970s (Chrisman 1988). These earlier programs were run on mainframe- and mini-computers, and maps were made on line printers and pen plotters. In the United Kingdom, computer mapping and spatial analysis were also introduced at the University of Edinburgh and the Experimental Cartography Unit (Coppock 1988; Rhind 1988). Two other events must also be noted about the early development of GIS: publication of Ian McHarg's *Design with Nature* and its inclusion of the map overlay method for suitability analysis (McHarg 1969), and introduction of an urban street network with topology in the U.S. Census Bureau's DIME (Dual Independent Map Encoding) system (Broome and Meixler 1990).

For many years, though, GIS was considered to be too difficult, expensive, and proprietary. The advent of the graphical user interface (GUI), powerful and affordable hardware and software,

and public digital data broadened the range of GIS applications and brought GIS to mainstream use in the 1990s.

Along with the proliferation of GIS activities, numerous GIS textbooks have been published, and several journals and trade magazines are now devoted to GIS and GIS applications. A GIS certification program, sponsored by several nonprofit associations, is also available to those who want to become certified GIS professionals. The certification uses a point system that is based on educational achievement, professional experience, and contribution to the profession (**http://www.gisci.org/**).

1.1.3 GIS Software Products

Box 1.1 lists GIS software producers and their main products. According to the 2007 salary survey by the Urban and Regional Information Systems Association (URISA), Esri software products were the most popular among GIS professionals who participated in the survey (**http://www.urisa.org/2007_salary_survey**). Various other trade reports also suggest that Esri and Intergraph lead the GIS industry in terms of the software market and software revenues.

The main software product from Esri is ArcGIS, a scalable system with ArcView, ArcEditor, and ArcInfo. All three versions of the system operate on the Windows platforms and share the same applications and extensions, but they differ in their capabilities: ArcView has data integration, query, display, and analysis capabilities; ArcEditor has additional functionalities for data editing; and ArcInfo has more data conversion and analysis capabilities than ArcView and ArcEditor. Esri, delivers these three versions of ArcGIS with separate licenses. Intergraph has two main products: GeoMedia and MGE. GeoMedia is a desktop GIS package for data integration and visualization and is compatible with standard Windows development tools. MGE, which operates on the Windows operating systems or on UNIX, consists of a series of products for data production and analysis. Data

Box 1.1   A List of GIS Software Producers and Their Main Products

The following is a list of GIS software producers and their main products:

- Autodesk Inc. (**http://www3.autodesk.com/**): **Autodesk Map**
- Bentley Systems, Inc. (**http://www2.bentley.com/**): **Microstation**
- Cadcorp (**http://www.cadcorp.com/**): **Cadcorp SIS—Spatial Information System**
- Caliper Corporation (**http://www.caliper.com/**): **TransCAD, Maptitude**
- CARIS (**http://www.caris.com/**): **CARIS system**
- Clark Labs (**http://www.clarklabs.org/**): **IDRISI**
- Environmental Systems Research Institute (Esri) (**http://www.esri.com/**): **ArcGIS**
- Intergraph Corporation (**http://www.intergraph.com/**): **MGE, GeoMedia**

- International Institute for Aerospace Survey and Earth Sciences, the Netherlands (**http://www.itc.nl/ilwis/**): **ILWIS**
- Land Management Information Center at Minnesota Planning (**http://www.lmic.state.mn.us/EPPL7/**): **EPPL7**
- Manifold.net (**http://www.manifold.net/**): **Manifold System**
- MapInfo Corporation (**http://www.mapinfo.com/**): **MapInfo**
- Open Source Geospatial Foundation (**http://grass.osgeo.org/**): **GRASS**
- PCI Geomatics (**http://www.pcigeomatics.com/**): **Geomatica**
- SAGA User Group (**http://www.saga-gis.uni-goettingen.de/html/index.php**): **SAGA GIS**
- Terralink International (**http://www.terralink.co.nz/**): **Terraview**

can be migrated from GeoMedia to MGE, and vice versa.

The Geographic Resources Analysis Support System (GRASS) is an open-source GIS software package. Originally developed by the U.S. Army Construction Engineering Research Laboratories, GRASS is currently maintained and developed by a worldwide network of users. DIVA-GIS is also a free GIS program useful for mapping and analyzing biodiversity data (**http://www.diva-gis.org/**). Some GIS packages are targeted at certain user groups. TransCAD, for example, is a package designed for use by transportation professionals. Oracle and IBM have also entered the GIS database industry. Oracle Spatial can store, access, and manage geospatial data in Oracle's relational database management system (**http://www.oracle.com/**). IBM offers Spatial DataBlade, an extension that allows location-based data to be stored in IBM's relational database (**http://www-306.ibm.com/software/data/informix/blades/spatial/**).

1.2 GIS APPLICATIONS

Since its inception, GIS has been important in natural resource management including land-use planning, natural hazard assessment, wildlife habitat analysis, riparian zone monitoring, and timber management. The U.S. Geological Survey (USGS), a leading agency in the early development and promotion of GIS, provides nationwide geospatial data for applications in natural hazards, risk assessment, homeland security, and many other areas through its National Map program (**http://nationalmap.usgs.gov**). For example, the USGS created and posted mapping products in support of Hurricane Katrina relief efforts. The following are other examples of GIS applications in natural resource management on the Internet:

- The Incident Information System, an interagency system, catalogs wildland fire incidents and provides information on active fires (**http://www.inciweb.org/**).

- The National Integrated Land System, maintained by the Bureau of Land Management and the U.S. Forest Service, has a publication website for the distribution of data on land parcels, minerals, and mining claims (**http://www.geocommunicator.gov/ GeoComm/index.shtm**).
- The National Weather Service offers weather data such as precipitation estimates, hydrometeorological data, and radar imagery in GIS-compatible format on its website (**http://www.weather.gov/gis/**) and delivers tropical cyclone wind-speed probabilities and historical track data through its Hurricane Center (**http://www .nhc.noaa.gov/**).

In more recent years GIS has been used for emergency planning, crime analysis, public health, land records management, transportation applications, precision farming, and economic recovery. For example, the U.S. Census Bureau maintains the Maps and Cartographic Resources website, where Internet users can choose American Fact Finder or the Topologically Integrated Geographic Encoding and Referencing (TIGER) Map Server (**http://www.census.gov/geo/www/maps/**) for online mapping. Users of American Fact Finder can access and map Census 2000 and other census data. TIGER Map Server lets users map cities, boundaries, transportation, hydrography, and ZIP code points from the TIGER database for anywhere in the United States. The following are other examples of more recent GIS applications on the Internet:

- The U.S. Department of Housing and Urban Development has an online "map your community" service that lets users map housing development information with environmental data, flood hazards, census geographies, and other data (**http://egis.hud .gov/egis/**).
- The National Institute of Justice uses GIS to map crime records and to analyze their spatial patterns by location and time (**http:// www.ojp.usdoj.gov/nij/maps/**).

- The U.S. Department of Health and Human Services' warehouse provides access to information about health resources including community health centers (**http:// datawarehouse.hrsa.gov/**).
- The Federal Highway Administration offers a census transportation planning package, a national highway planning network, and other data resources for GIS in transportation (**http://www.gis.fhwa.dot.gov/gisData.asp**).
- The U.S. Department of Agriculture promotes precision farming by linking GIS to site-specific farming activities such as applications of herbicides, pesticides, and fertilizers (**http://www.csrees.usda.gov/nea/ag_ systems/in_focus/precision_if_crop.html**).
- The Recovery Accountability and Transparency Board has a website that uses maps, charts, and graphs to track the use of funds made available in the American Recovery and Reinvestment Act of 2009 (**http://www.recovery.gov**).

Integration of GIS with the Internet, GPS, wireless technology, and Web services has also introduced exciting commercial applications involving a number of well-known companies such as Microsoft, Yahoo!, and Google. These types of applications can generally be grouped into the following categories:

- Online mapping websites offer locators for finding real estate listings, credit union service centers, restaurants, coffee shops, and hotels.
- Location-based services allow mobile phone users to search for location information such as nearby banks and restaurants, and to track friends, dates, children, and the elderly.
- Mobile GIS allows field-workers to collect and access geospatial data in the field.
- Mobile resource management tools track and manage the location of field crews and mobile assets in real time (Box 1.2).
- Automotive navigation systems provide turn-by-turn guidance, optimal routes, and live traffic updates to drivers (Box 1.3).

Box 1.2   Monitoring by GPS

Monitoring by GPS can be a source of conflict. In September 2007, New York City cab drivers called for a strike to protest the installation of a high-tech video-and-fare system in their taxis. The system includes a video screen in the back of the taxi where passengers can see a map of their route. It also allows taxi drivers to receive text messages from the taxi commission directing them to places where they are needed. Besides complaining about the cost of installation and noises from the system, taxi drivers also argued that the city might use the information to send speeding tickets to drivers who drove to a destination, such as one of the airports, in less time than the trip should take.

Box 1.3   Smartphones

Automotive navigation systems use stand-alone GPS units. But because GPS has become a popular feature on smartphones such as the BlackBerry and iPhone, many people are now getting turn-by-turn driving directions from their smartphones. This has forced traditional GPS companies such as TomTom and Garmin to introduce portable navigation applications for cell phones. Google has also announced the beta release of Google Maps Navigation, a voice-guided, turn-by-turn GPS navigation application for cell phone users, which gathers mapping information from Google Maps.

Besides cell phones, other GPS-enabled devices include digital cameras and sport watches. Geotagged photographs are now found on popular websites such as flickr and Facebook. The next new tool in the integration of geospatial and mobile technologies is probably "augmented reality," which allows mobile device the ability of "point and discover" **(http://p2d.ftw.at/),** getting information of what they are looking at.

As GIS becomes better known, new applications will arise, both conventional and imaginative.

1.3 GEOSPATIAL DATA

To use geospatial data properly in a GIS, we must understand coordinate systems and spatial data models that distinguish geospatial data from other types of data.

1.3.1 Coordinate Systems

Geospatial data are geographically referenced. Spatial features on Earth's surface are referenced to a geographic coordinate system in terms of longitude and latitude values. However, when features are displayed on maps, they are typically based on a projected coordinate system in terms of x-, y-coordinates. These two spatial reference systems are connected by the process of **projection,** which displays Earth's spherical surface as a plane surface. Thousands of geographic and projected coordinate systems are in use. Therefore, a basic understanding of projection and coordinate systems is a first step toward the intelligent use of GIS.

1.3.2 Vector Data Model

A data model defines how spatial features are represented in a GIS. There are vector and raster data models (Figure 1.2). The **vector data model** uses points and their x-, y-coordinates to represent discrete features with a clear spatial location and boundary. Examples of vector data are point features such as wells, line features such as streams, and polygon features such as land parcels (Figure 1.3). Depending on the data structure, a vector data model can be georelational or object-based, with or without topology, and simple or composite.

The **georelational data model** stores geometries and attributes of spatial features separately and uses the feature IDs to link them. A more recent development, the **object-based data model** stores geometries and attributes in a single system and treats spatial features as objects with associated properties and methods.

Topology explicitly expresses the spatial relationships between features, such as two lines meeting perfectly at a point and a directed line having an explicit left and right side. Topology ensures the integrity of vector data and is necessary for some GIS analyses, but data without topology are interoperable and can display faster on the computer monitor. Composite features are built on simple features of points, lines, and polygons and are designed for handling complex spatial relationships. Examples of composite features include the **triangulated irregular network (TIN)** (Figure 1.4), which approximates the terrain with a set of nonoverlapping triangles;

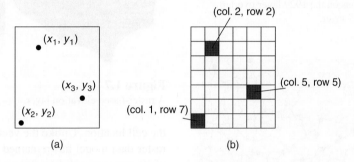

(a)   (b)

Figure 1.2
The vector data model uses x-, y-coordinates to represent point features (a), and the raster data model uses cells in a grid to represent point features (b).

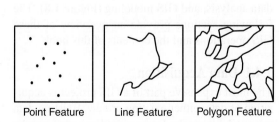

Point Feature   Line Feature   Polygon Feature

Figure 1.3
Point, line, and polygon features.

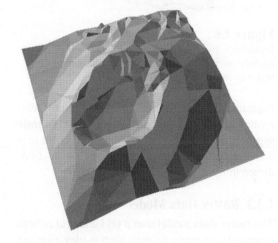

Figure 1.4
An example of the TIN model.

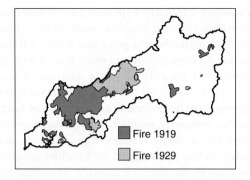

Figure 1.5
The map shows two regions layers, one for burned areas in 1919 and the other for burned areas in 1929. Both layers consist of spatially disjoint polygons. Additionally, polygons on the 1929 layer overlap polygons on the 1919 layer.

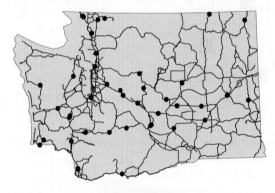

Figure 1.6
Dynamic segmentation allows rest areas, which are linearly referenced, to be plotted as point features on highway routes in Washington State.

regions (Figure 1.5), which allow spatially disjoint and overlapping polygons; and **dynamic segmentation** (Figure 1.6), which combines one-dimensional linear measures with two-dimensional projected coordinates.

1.3.3 Raster Data Model

The **raster data model** uses a grid and grid cells to represent continuous features such as elevation and precipitation (Figure 1.7). Each cell has a value that captures the magnitude of the continuous surface at

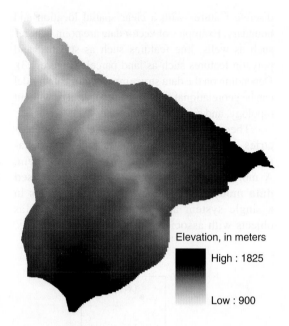

Figure 1.7
A raster-based elevation layer.

the cell location. Unlike the vector data model, the raster data model has remained the same in terms of its concept and data structure since the beginning of GIS, but methods for storing and compressing raster data have continually changed over the past three decades.

1.4 GIS OPERATIONS

Although GIS activities no longer follow a set sequence, to explain what we do in GIS, we can group GIS activities into data acquisition, attribute data management, data display, data exploration, data analysis, and GIS modeling (Figure 1.8). The following sections provide an overview of these GIS operations and the contents of this book.

1.4.1 Data Acquisition

The most expensive part of a GIS project is acquisition of data, either existing or new. Existing data are available on the Internet, and new data can be digitized from paper maps, satellite images, GPS data, field surveys, street addresses, and text files

Spatial data input	1. Data entry: use existing data, create new data 2. Data editing 3. Geometric transformation 4. Projection and reprojection
Attribute data management	1. Data entry and verification 2. Database management 3. Attribute data manipulation
Data display	1. Cartographic symbolization 2. Map design
Data exploration	1. Attribute data query 2. Spatial data query 3. Geographic visualization
Data analysis	1. Vector data analysis: buffering, overlay, distance measurement, spatial statistics, map manipulation 2. Raster data analysis: local, neighborhood, zonal, global, raster data manipulation 3. Terrain mapping and analysis 4. Viewshed and watershed 5. Spatial interpolation 6. Geocoding and dynamic segmentation 7. Path analysis and network applications
GIS modeling	1. Binary models 2. Index models 3. Regression models 4. Process models

Figure 1.8
A classification of GIS operations.

with *x*-, *y*-coordinates. A newly digitized map typically requires editing and geometric transformation. Editing removes digitizing errors from the map, such as missing polygons, or topological errors, such as unclosed polygons. Geometric transformation converts the digitized map into a projected coordinate system.

1.4.2 Attribute Data Management

Attribute data describe the characteristics of spatial features. Such data are entered and verified through digitizing and editing. The relational database model is the norm for managing attribute data in GIS. A **relational database** is a collection of

tables (relations), each of which can be prepared, maintained, and edited separately from other tables; but these tables can be joined or related for data search and retrieval. Two basic elements in designing a relational database are therefore the key and the type of data relationship: the former establishes a connection between corresponding records in two tables, and the latter dictates how the tables are actually linked.

1.4.3 Data Display

Because maps are most effective in communicating spatial information, mapmaking is a routine GIS operation. Maps are derived from data query and analysis and prepared for data visualization and presentation. A map for presentation usually has a number of elements: title, subtitle, body, legend, north arrow, scale bar, acknowledgment, neatline, and border. These elements are combined to convey spatial information to the map reader.

The first step in mapmaking is to assemble map body and other elements, using choices from menus and palettes in a Windows-based GIS package. However, caution is required with "default options"; without a basic understanding of map symbols, colors, and typology, it is easy to produce a bad map using these options. The second step is map design, a creative process that cannot easily be replaced by default templates and computer code. A well-designed map can enhance map communication, whereas a poorly designed map can confuse the map reader and even distort the information intended by the mapmaker.

1.4.4 Data Exploration

Usually a precursor to data analysis, **data exploration** involves the activities of exploring the general trends in the data, taking a close look at data subsets, and focusing on possible relationships among data sets. A Windows-based GIS allows maps, graphs, and tables to be displayed in multiple but dynamically linked windows, so that when a data subset is selected from a table, it automatically highlights the corresponding features in a graph and a map. This type of interactivity increases the capacity for information processing and synthesis. Data exploration in GIS can be approached from spatial data or attribute data, or from both. Additionally, it can employ map-based tools such as data classification, data aggregation, and map comparison.

1.4.5 Data Analysis and GIS Modeling

Figure 1.8 classifies data analysis into seven groups. The first two include basic analytical tools. For vector data, these basic tools include buffering, overlay, distance measurement, spatial statistics, and feature manipulation. Buffering creates buffer zones by measuring straight-line distances from select features. Overlay creates output by combining geometries and attributes from different layers (Figure 1.9). Distance measurement calculates distances between spatial features. Spatial statistics detect spatial dependence and patterns of concentration among features, and feature manipulation tools manage and alter spatial features in a layer.

Common tools for analyzing raster data are traditionally grouped into local, neighborhood, zonal, and global operations. A local operation operates on individual cells (Figure 1.10); a neighborhood operation, on a specified neighborhood such as a 3-by-3 window; a zonal operation, on a group of cells with same values or like features; and a global operation, on the entire raster. These operations create new raster data by using algebraic functions to relate the input and the output. Other analysis tools can perform data extraction and generalization.

The terrain has been the object for mapping and analysis for hundreds of years. Mapping techniques such as contouring, profiling, hill shading, hypsometric tinting, and 3-D views are useful for visualizing the land surface. Topographic measures including slope, aspect, and surface curvature are important for studies of timber management, soil erosion, hydrologic modeling, wildlife habitat suitability, and other fields. Terrain analysis also includes viewshed and watershed. A viewshed analysis determines areas of the land surface that are visible from one or more observation points. A watershed analysis

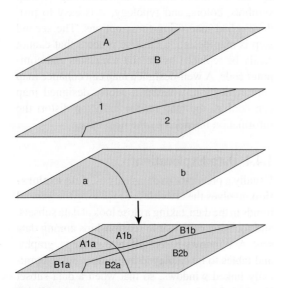

Figure 1.9
A vector-based overlay operation combines geometries and attributes from different layers to create the output.

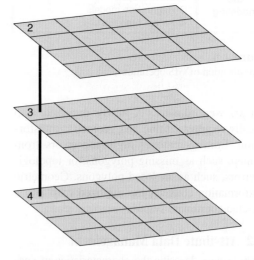

Figure 1.10
A raster data operation with multiple rasters can take advantage of the fixed cell locations. For example, a local average can easily be computed by dividing the sum of 2, 3, and 4 (9) by 3.

can derive topographical features such as flow direction, stream networks, and watershed boundaries for hydrologic applications.

Precipitation, snow accumulation, water table, and many other spatial phenomena are similar to the terrain in visualization and analysis; but unlike the terrain, data for these phenomena are available at only a number of sample points. To construct a surface from these sample points, values must be interpolated at unknown points with methods such as trend surface, Thiessen polygons, kernel density estimation, inverse distance weighted, splines, and kriging. Kriging is also known as a geostatistical method because it can not only predict unknown values but also estimate prediction errors.

Geocoding converts street addresses and intersections into point features. Dynamic segmentation plots linearly referenced data on a coordinate system. Both techniques can locate data from a data source that lacks x-, y-coordinates, and both use linear features (e.g., streets, highways) as reference data. Data generated from these two techniques are necessary inputs for certain data analyses: geocoded data for crime mapping and analysis, and dynamic segmentation layers for managing and analyzing highway-related data.

Path analysis finds the least-cost path between cells by using a cost raster that defines the cost to move through each cell. Network analysis solves for the shortest path between stops on a topological network that also has the appropriate attributes for the flow of objects. The two analyses share the same problem-solving algorithm but differ in applications. Path analysis is raster-based and works with "virtual" paths, whereas network analysis is vector-based and works with an existing network.

A *model* is a simplified representation of a phenomenon or a system. A GIS and its functionalities can be used to build, or help build, a spatially explicit model with geospatial data. GIS models can be grouped into four general types: binary, index, regression, and process models.

Both binary and index models rely on overlay operations, either vector- or raster-based, to combine geospatial data from different layers for multicriteria evaluation. A binary model queries the overlay output to separate areas that satisfy the criteria from those that do not. An index model calculates the index value from the overlay output and ranks areas by the index value. A regression model also relies on overlay operations to combine data for statistical analysis between a dependent variable and independent variables. If it is statistically significant, a regression model can be used to predict the dependent variable. A process model, such as a soil erosion model, integrates existing knowledge about the environmental processes in the real world and quantifies the processes with a set of relationships and equations. A GIS can assist environmental modelers in data visualization, database management, and data exploration.

1.5 WEB MAPPING

1.5.1 Online Mapping

In 1996, MapQuest offered the first online mapping services, including address matching and travel planning with the map output (**http://www .mapquest.com/**). This was followed by other mapping services, including some maintained by government agencies. In 1997, the USGS received the mandate to coordinate and create a national atlas, including electronic maps and services to be delivered online. In 2001, the U.S. Census Bureau began an online mapping service using census and TIGER data. And in 2004, the U.S. National Oceanic and Atmospheric Administration (NOAA) introduced World Wind, a free, open-source program that allows users to overlay satellite imagery, aerial photographs, topographic maps, and GIS data on 3-D models of the Earth.

Although Web mapping had become common by the early 2000s, it was not until 2005, when Google introduced Google Maps and Google Earth, that Web mapping became popular with Web users. Google Maps is a free mapping service that lets users search for an address or a business, find the location on a reference map, a satellite image, or both, and get travel directions to the location. Google Earth, a free browser, uses digital elevation models (DEMs), satellite imagery, and aerial photographs to display the 3-D maps of the Earth's surface. It was an instant success primarily because of its ease with which the user can zoom in from space down to street level (Butler 2006). To capitalize the success of Google Maps, Microsoft and Yahoo! also came up with their online map services: Microsoft Virtual Earth and Yahoo! Maps.

1.5.2 Collaborative Web Mapping

Since 2005, Web mapping has undergone major changes in terms of map contents and services. In April 2006 Google Maps introduced a free Application Programming Interface (API) for users to combine their own contents (e.g., text, photos, and videos) with Web-based maps to make "Google Maps mashups," thus allowing Google Maps users to become instant new cartographers (Liu and Palen 2010). An assortment of such mashups, many offbeat, can be viewed at Google Maps Mania (**http://www.googlemapsmania .blogspot.com/**). Wikimapia was also launched in 2006. Wikimapia combines Google Maps with a wiki system and allows users to add information, typically in the form of a note, to any point on the Earth surface (**http://wikimapia.org**). Both Google Maps mashups and Wikimapia represent Web 2.0, a term describing Web applications that facilitate user-centered design and collaboration on the World Wide Web.

Then in April 2007 Google My Maps was introduced, allowing users to add maps and tools created by others and to create personalized, annotated maps. Like Google My Maps, Microsoft Popfly (discontinued in August 2009) and Yahoo! Pipes also let people with limited to almost no programming skills to integrate maps, 3-D imagery, text, photos, and videos. As another example of advances in Web services, Microsoft Virtual Earth, when integrated with MapPoint Web Service, can take requests online, process them, and render the information as a Web page to the customer.

One of the latest developments in collaborative Web mapping is integrating maps generated from a GIS package with Google Earth or Microsoft Virtual Earth. ArcGIS users, for example, can convert a shapefile to kml (Keyhole Markup Language) and import the kml file into Google Earth. With a proper license file, ArcGIS users can also open Microsoft Virtual Earth in ArcMap and superimpose maps on Microsoft Virtual Earth's aerial or road layers.

1.5.3 Web Mapping and GIS

Web mapping has attracted a lot of users, who probably have not heard of GIS before. These users create *volunteered geographic information*, a term coined by Goodchild (2007), which can be quite useful for national security, disaster response, and emergency relief (Shneiderman and Preece 2007).

Will Web mapping services replace GIS? They are certainly useful for location-based searches and queries and for personalized mapping purposes. Some basic analytical capabilities such as proximity analysis are available. And, with the data sharing capability of Google Earth, Microsoft Virtual Earth, and ArcGIS, the boundary between Web mapping and GIS becomes blurry at least in the area of mapping. Given the current condition, GIS professionals can perhaps integrate these popular Web services into their GIS projects for such tasks as data editing, data delivery, and display and to perform data management, data analysis, and other tasks using GIS packages.

1.6 ORGANIZATION OF THIS BOOK

Based on the topics outlined in Sections 1.3 and 1.4, this book is organized into seven sections: GIS data and data models (Chapters 2–4), data acquisition and management (Chapters 5–8), data display and exploration (Chapters 9–10), core data analysis (Chapters 11–12), surface analysis (Chapters 13–15), line analysis (Chapters 16–17), and GIS models and modeling (Chapter 18).

Chapter 2 discusses coordinate systems, a topic essential to geospatial data. Common coordinate systems including the transverse Mercator projection, the Lambert conformal conic projection, Web Mercator, the Universal Transverse Mercator

(UTM) grid system, and the State Plane Coordinate (SPC) system are covered in Chapter 2 along with their parameters. Chapter 3 examines the vector data model, both georelational and object-based. Chapter 3 also deals with topology and the representation of composite features. Chapter 4 focuses on the raster data model and discusses different types of raster data, raster data structure, data compression, and conversion between raster and vector data.

Chapter 5 describes online digital data, metadata (information about data), data conversion, and creation of new data from satellite images, field data, and paper maps. Chapter 6 concentrates on the geometric transformation of newly digitized maps and satellite images and the interpretation of transformation results. Chapter 7 covers spatial data editing related to location and topological errors. Chapter 7 also discusses spatial data accuracy standards, edgematching, line simplification, and line smoothing. Chapter 8 examines attribute data input and management, including the relational model and its design, relational database examples, and attribute data entry and manipulation.

Chapter 9 deals with data display and cartography. Starting with the elements of cartographic symbolization, Chapter 9 proceeds to discuss types of maps, text and text placement, map design, and map production. Chapter 10 covers data exploration. The topics include exploratory data analysis, map-based data manipulation, attribute data query, spatial data query, and raster data query.

Chapters 11 and 12 provide the basic tools for GIS analysis and their applications. Chapter 11 includes vector-based tools of buffering, overlay, distance measurement, pattern analysis, and feature manipulation. Chapter 12 covers local, neighborhood, zonal, global, and other raster data operations. Chapter 12 also uses overlay and buffering as examples to compare vector- and raster-based data analysis.

Chapter 13 examines terrain mapping and analysis. Terrain mapping includes contouring, profiling, hill shading, hypsometric tinting, and perspective view. Terrain analysis includes slope, aspect, and surface curvature. Worked examples are provided for the computing algorithms whenever appropriate,

and a comparison is made between elevation raster and TIN for terrain analysis. Chapter 14 focuses on viewshed and watershed analysis and their applications. Chapter 14 also discusses the parameters and options for both analyses. Chapter 15 covers spatial interpolation including the basic elements of spatial interpolation, global methods, local methods, kriging, and cross-validation statistics. Throughout Chapter 15, a small data set and worked examples illustrate how the interpolation algorithms work.

Chapter 16 discusses geocoding and dynamic segmentation. Both techniques are popular in GIS because they can convert data that lack x-, y-coordinates into point and line features. Chapter 17 examines least cost path analysis and network applications. Worked examples illustrate the same shortest path algorithm, first in raster format and then in vector format. Chapters 16 and 17 are linked by their focus on linear features, which serve either as the reference data or as inputs in the operations.

Chapter 18 gives an overview of GIS models and modeling. Following an introduction to the classification of models, the modeling process, and the role of GIS in modeling, Chapter 18 describes the basics of building binary, index, regression, and process models, with numerous examples from a variety of disciplines for each type of model.

1.7 CONCEPTS AND PRACTICE

Each chapter in this book has two main sections. The first section covers a set of topics and concepts, and the second section covers applications with three to seven problem-solving tasks. Additional materials in the first section include information boxes, websites, key concepts and terms, and review questions. The applications section provides step-by-step instructions as well as questions to reinforce the learning process. Persons do not learn well if they merely follow the instructions to complete a task without pausing and thinking about the process. A challenge question is also included at the end of each applications section to further develop the necessary skills for problem solving. Each chapter concludes with an extensive, updated bibliography.

This book stresses both concept and practice. GIS concepts explain the purpose and objectives of GIS operations and the interrelationship among GIS operations. A basic understanding of map projection, for example, explains why map layers must be projected into a common coordinate system before being used together and why numerous projection parameters must be inputted. Knowledge of map projection is long lasting, because the knowledge will neither change with the technology nor become outdated with new versions of a GIS package.

GIS is a science as well as a problem-solving tool (Wright, Goodchild, and Proctor 1997; Goodchild 2003; Goodchild and Haining 2004; Gold 2006; DiBiase 2007). To apply the tool correctly and efficiently, one must become proficient in using the tool. Although a Windows-based GIS package improves the human–computer interaction over a command-driven package, it still requires sorting out a multitude of menus, buttons, and tools and knowing how and when to use them. Practice, which is a regular feature in mathematics and statistics textbooks, is really the only way to become proficient in using GIS. Practice can also help one grasp GIS concepts. For instance, the root

mean square (RMS) error, an error measure for geometric transformation and spatial interpolation, may be difficult to comprehend mathematically; but after a couple of geometric transformations, the RMS error starts to make more sense because the user can see how the error measure changes each time with a different set of control points.

Practice sections in a GIS textbook require data sets and GIS software. Many data sets used in this book are from GIS classes taught at the University of Idaho and National Taiwan University over a period of more than 20 years, while some are downloaded from the Internet. Instructions accompanying the exercises correlate to ArcGIS 10.0. Most tasks use the ArcView version of ArcGIS Desktop and the extensions of Spatial Analyst, 3D Analyst, Geostatistical Analyst, Network Analyst, and ArcScan. Like other software packages, ArcGIS can have bugs. This is why Esri offers patches (latest bug fixes) and service packs (compilations of bug fixes) for downloads at their website (**http://www.esri.com/**). Esri also has a user support website called Knowledge Base, where one can search for known bugs or report a problem or bug (**http://support.esri.com/index .cfm?fa=knowledgeBase.gateway**).

KEY CONCEPTS AND TERMS

Data exploration: Data-centered query and analysis.

Dynamic segmentation: A data model that allows the use of linearly measured data on a coordinate system.

Geographic information system (GIS): A computer system for capturing, storing, querying, analyzing, and displaying geospatial data.

Georelational data model: A vector data model that uses a split system to store geometries and attributes.

Geospatial data: Data that describe both the locations and characteristics of spatial features on the Earth's surface.

Object-based data model: A data model that uses objects to organize spatial data and stores geometries and attributes in a single system.

Projection: The process of transforming from a geographic coordinate system to a projected coordinate system.

Raster data model: A data model that uses a grid and cells to represent the spatial variation of a feature.

Regions: Composite vector data that can have spatially disjoint and overlapping polygons.

Relational database: A collection of tables in which tables are connected to one another by keys.

Topology: A subfield of mathematics that, when applied to GIS, ensures that the spatial relationships between features are expressed explicitly.

Triangulated irregular network (TIN):
Composite vector data that approximate the terrain with a set of nonoverlapping triangles.

Vector data model: A spatial data model that uses points and their x-, y-coordinates to construct spatial features of points, lines, and polygons.

REVIEW QUESTIONS

1. Define geospatial data.

2. Describe an example of GIS application from your discipline.

3. Go to the U.S. Department of Housing and Urban Development website **(http://egis.hud.gov/egis)** and plot a map of your community. Are there any hazardous sites?

4. Go to the National Institute of Justice website **(http://www.ojp.usdoj.gov/nij/maps/).** What does it mean by "hot spot" analysis?

5. Location-based services are probably the most commercialized GIS-related field. Search for "location-based service" on Wikipedia **(http://www.wikipedia.org/)** and read what has been posted on the topic.

6. What types of software and hardware are you currently using for GIS classes and projects?

7. Try the map locators offered by Microsoft Virtual Earth, Yahoo! Maps, and Google

Maps, respectively. State the major differences among these three systems.

8. Define geometries and attributes as the two components of GIS data.

9. Explain the difference between vector data and raster data.

10. Explain the difference between the georelational data model and the object-based data model.

11. How does data exploration differ from data analysis?

12. Suppose you are required to do a GIS project for a class. What types of activities or operations do you have to perform to complete the project?

13. Name *two* examples of vector data analysis.

14. Name *two* examples of raster data analysis.

15. Describe an example from your discipline in which a GIS can provide useful tools for building a model.

APPLICATIONS: INTRODUCTION

ArcGIS uses a single scalable architecture and user interface. ArcGIS has three versions: ArcView, ArcEditor, and ArcInfo. ArcView is the simplest version and has fewer capabilities than the other two versions. All three versions use the same applications of ArcCatalog and ArcMap and share the same extensions such as Spatial Analyst, 3D Analyst, Network Analyst, and Geostatistical Analyst.

You can tell which version of ArcGIS you are using by looking at the title of an application. For example, the title of ArcCatalog may appear as

ArcCatalog—ArcView or ArcCatalog—ArcInfo, depending on if you are using ArcView or ArcInfo. Both ArcCatalog and ArcMap have the Customize menu. When you click on Extensions on the Customize menu, it displays a list of extensions and allows you to select the extensions to use. If the controls of an extension (e.g., Geostatistical Analyst) are on a toolbar, you must also check its toolbar (e.g., Geostatistical Analyst) from the Toolbars pullright in the Customize menu.

This applications section covers two tasks. Task 1 introduces ArcCatalog and ArcToolbox, and

Task 2 ArcMap and the Spatial Analyst extension. Typographic conventions used in the instructions include italic typeface for data sets (e.g., *emidalat*) and boldface for questions (e.g., **Q1**).

Task 1 Introduction to ArcCatalog

What you need: *emidalat*, an elevation raster; and *emidastrm.shp*, a stream shapefile.

Task 1 introduces ArcCatalog, an application for managing data sets.

1. Start ArcCatalog. ArcCatalog lets you set up connections to your data sources, which may reside in a folder on a local disk or on a database on the network. For Task 1, you will first connect to the folder containing the Chapter 1 database (e.g., chap1). Click the Connect to Folder button. Navigate to the chap1 folder and click OK. The chap1 folder now appears in the Catalog tree under Folder Connections. Expand the folder to view the data sets.

2. Click *emidalat* in the Catalog tree. Click the Preview tab to view the elevation raster. Click *emidastrm.shp* in the Catalog tree. On the Preview tab, you can preview the geography or table of *emidastrm.shp*.

3. ArcCatalog has tools for various data management tasks. You can access these tools by right-clicking a data set to open its context menu. Right-click *emidastrm.shp*, and the menu shows Copy, Delete, Rename, Create Layer, Export, and Properties. Using the context menu, you can copy *emidastrm.shp* and paste it to a different folder, rename it, or delete it. A layer file is a visual representation of a data set. The export tool can export a shapefile to a geodatabase, a coverage, and other formats. The properties dialog shows the data set information. Right-click *emidalat* and select Properties. The Raster Dataset Properties dialog shows that *emidalat* is a raster dataset projected onto the Universal Transverse Mercator (UTM) coordinate system.

4. This step lets you create a personal geodatabase and then import *emidalat* and *emidastrm.shp* to the geodatabase. Right-click the Chapter 1 database in the Catalog tree, point to New, and select Personal Geodatabase. Click the new geodatabase and rename it *Task1.mdb*. If the extension .mdb does not appear, select ArcCatalog Options from the Customize menu and on the General tab uncheck the box to hide file extensions.

5. There are two options for importing *emidalat* and *emidastrm.shp* to *Task1.mdb*. For the first option, right-click *Task1.mdb*, point to Import, and select Raster Datasets. In the next dialog, navigate to *emidalat*, add it for the input raster, and click OK to import.

6. Now you will use the second option, ArcToolbox, to import *emidastrm.shp* to *Task1.mdb*. ArcCatalog's standard toolbar has a button called ArcToolbox window. Click the button to open ArcToolbox. Dock the ArcToolbox window so that you can see both the window and the Catalog tree. Right-click ArcToolbox, and select Environments. The Environment Settings dialog can let you set the workspace, which is important for most operations. Click the dropdown arrow for Workspace. Navigate to the Chapter 1 database and set it to be the current workspace. Close the Environment Settings window. Tools in ArcToolbox are organized into a hierarchy. The tool you need for importing *emidastrm.shp* resides in the Conversion Tools/To Geodatabase toolset. Double-click Feature Class to Feature Class to open the tool. Select *emidastrm.shp* for the input features, select *Task1.mdb* for the output location, specify *emidastrm* for the output feature class name, and click OK. When the import operation is completed, you will see a message at the bottom of the screen. (You will also see a message with a red X if the operation fails.) Right-click *Task1.mdb* and select Refresh. Make sure that the import operations have been completed.

Q1. The number of usable tools in ArcToolbox varies depending on which version of ArcGIS you are using. Go to ArcGIS Desktop Help/ArcGIS Desktop 10 Help/Professional Library/Geoprocessing/Geoprocessing tool reference. The licensing requirement is explained for each toolset. You will see the file "Conversion toolbox licensing" by opening the Conversion toolbox.

Task 2 Introduction to ArcMap

What you need: *emidalat* and *emidastrm.shp*, same as Task 1.

In Task 2, you will learn the basics of working with ArcMap. Starting in ArcGIS 10.0, ArcMap has the Catalog window button that lets you open Catalog directly in ArcMap. Catalog allows you to perform many of the same functions and tasks such as copy and delete as in ArcCatalog.

1. ArcMap is the main application for data display, data query, data analysis, and data output. You can start ArcMap by clicking the Launch ArcMap button in ArcCatalog or from the Programs menu. Start with a blank map document. ArcMap organizes data sets into data frames (also called *maps*). You open a new data frame called Layers when you launch ArcMap. Right-click Layers, and select Properties. On the General tab, change the name Layers to Task 2 and click OK.

2. Next, add *emidalat* and *emidastrm.shp* to Task 2. Click the Add Data button in ArcMap, navigate to the Chapter 1 database, and select *emidalat* and *emidastrm.shp*. To select more than one data set to add, click the data sets while holding down the Ctrl key. An alternative to using the Add Data button is to use the drag-and-drop method. ArcMap has the Catalog window button that lets you open Catalog directly in ArcMap. You can add a data set in ArcMap by dragging it from the Catalog tree and dropping it in ArcMap's view window.

3. A warning message states that one or more layers are missing spatial reference information. Click OK to dismiss the dialog; *emidastrm.shp* does not have the projection information, although it is based on the UTM coordinate system, as is *emidalat*. You will learn in Chapter 2 how to define a coordinate system.

4. Both *emidastrm* and *emidalat* are highlighted in the table of contents, meaning that they are both active. You can deactivate them by clicking on the empty space. The table of contents has four tabs: List by Drawing Order, List by Source, List by Visibility, and List by Selection. On the List by Drawing Order tab, you can change the drawing order of the layers by dragging and dropping a layer up or down. You can also rename or remove layers on the List by Drawing Order tab. The List by Source tab shows the data source of each layer. The List by Visibility tab shows the list of layers that are currently displayed in the active data frame. The List by Selection tab lists the selectable layer(s) and selected features. The Options button offers preferences and patches.

Q2. Does ArcMap draw the top layer in the table of contents first?

5. The Standard toolbar in ArcMap has such tools as Zoom In, Zoom Out, Pan, Global, Select Elements, and Identify. When you hold the mouse point over a tool, a tooltip appears in a floating box giving the name of the tool, and a short message about the use of the tool appears in the status bar at the bottom of the ArcMap window.

6. ArcMap has two views: Data View and Layout View. (The buttons for the two views are located at the bottom of the view window.) Data View is for viewing data, whereas Layout View is for viewing the map product for printing and plotting. For this task, you will stay with Data View.

7. This step is to change the symbol for *emidastrm*. Click the symbol for *emidastrm*

in the table of contents to open the Symbol Selector dialog. You can either select a preset symbol (e.g., river) or make up your own symbol for *emidastrm* by specifying the color, width, and properties of the symbol. Choose the preset symbol for river.

8. Next, classify *emidalat* into the elevation zones <900, 900−1000, 1000−1100, 1100−1200, 1200−1300, and >1300 meters. Right-click *emidalat*, and select Properties. Click the Symbology tab. Click Classified in the Show frame. Click Yes in the Compute Unique Values dialog. Change the number of classes to 6, and click the Classify button. The Method dropdown list shows seven methods. Select Manual. There are two ways to set the break values for the elevation zones manually. To use the first method, click the first break line and drag it to a data value near 900. Then, set the other break lines near 1000, 1100, 1200, 1300, and 1337. To use the second method, which is normally preferred, click the first cell in the Break Values frame and enter 900. Then enter 1000, 1100, 1200, and 1300 for the next four cells. (If the break value you entered is changed to a different value, reenter it.) Use the second method to set the break values, and click OK to dismiss the Classification dialog.

Q3. List the other classification methods besides Manual that are available in ArcMap.

9. You can change the color scheme for *emidalat* by using the Color Ramp dropdown list in the Layer Properties dialog box. Sometimes it is easier to select a color scheme using words instead of graphic views. In that case, you can right-click inside the Color Ramp box and uncheck Graphic View. The Color Ramp dropdown list now shows White to Black, Yellow to Red, etc. Select Elevation #1. Click OK to dismiss the dialog.

10. This step lets you derive a slope layer from *emidalat*. Select Extensions from the Customize menu and check Spatial Analyst. Then click the ArcToolbox window button to open ArcToolbox. Right-click ArcToolbox to select Environments. Change the Workspace (both Current and Scratch) to the Chapter 1 database. The Slope tool resides in the Spatial Analyst Tools/Surface toolset. Double-click the Slope tool. In the Slope dialog, select *emidalat* for the input raster, save the output raster as *slope*, and click OK. *slope* is added to Task 2.

11. You can save Task 2 as a map document before exiting ArcMap. Select Save from the File menu in ArcMap. Navigate to the Chapter 1 database, enter *chap1* for the file name, and click Save. ArcMap automatically adds the extension .mxd to chap1. Data sets displayed in Task 2 are now saved with *chap1.mxd*. To re-open *chap1.mxd*, *chap1.mxd* must reside in the same folder as the data sets it references. You can save a map document with the relative path name option (e.g., without the drive name). Select Map Document Properties from ArcMap's File menu. In the next dialog, check the box to store relative path names to data sources.

12. To make sure that *chap1.mxd* is saved correctly, first select Exit from ArcMap's File menu. Then launch ArcMap again. Click on chap1 or select *chap1.mxd* from the File menu.

Challenge Task

What you need: *menan-buttes*, an elevation raster.

This challenge question asks you to display *menan-buttes* in 10 elevation zones and save the map along with Task 2 in *chap1.mxd*.

1. Open *chap1.mxd*. Select Data Frame from ArcMap's Insert menu. Rename the new data frame Challenge, and add *menan-buttes* to Challenge.

2. Display *menan-buttes* in 10 elevation zones by using the elevation #2 color ramp and the following break values: 4800, 4900, 5000, 5100, 5200, 5300, 5400, 5500, 5600, and 5619 (feet).

3. Save Challenge with Task 2 in *chap1.mxd*.

REFERENCES

Broome, F. R., and D. B. Meixler. 1990. The TIGER Data Base Structure. *Cartography and Geographic Information Systems* 17: 39–47.

Butler, D. 2006. Virtual Globes: The Web-Wide World. *Nature* 439: 776–778.

Chrisman, N. 1988. The Risks of Software Innovation: A Case Study of the Harvard Lab. *The American Cartographer* 15: 291–300.

Coppock, J. T. 1988. The Analogue to Digital Revolution: A View from an Unreconstructed Geographer. *The American Cartographer* 15: 263–75.

DiBiase, D. W. 2007. Is GIS a Wampeter? *Transactions in GIS* 11: 1–8.

Gold, C. M. 2006. What Is GIS and What Is Not? *Transactions in GIS* 10: 505–19.

Goodchild, M. F. 2003. Geographic Information Science and Systems for Environmental Management. *Annual Review of Environment & Resources* 28: 493–519.

Goodchild, M. F. 2007. Citizens as Sensors: The World of Volunteered Geography. *GeoJournal* 69: 211–21.

Goodchild, M. F., and R. P. Haining. 2004. GIS and Spatial Data Analysis: Converging Perspectives. *Papers in Regional Science* 83: 363–85.

Liu, S. B., and L. Palen. 2010. The new cartographers: Crisis map mashups and the emergence of neogeographic practice. *Cartography and Geographic Information Science* 37: 69–90.

McHarg, I. L. 1969. *Design with Nature.* New York: Natural History Press.

Nourbakhsh, I., R. Sargent, A. Wright, K. Cramer, B. McClendon, and M. Jones. 2006. Mapping Disaster Zones. *Nature* 439: 787–88.

Rhind, D. 1988. Personality as a Factor in the Development of a Discipline: The Example of Computer-Assisted Cartography. *The American Cartographer* 15: 277–89.

Shneiderman, B., and J. Preece. 2007. 911.gov. *Science* 315: 944.

Tomlinson, R. F. 1984. Geographic Information Systems: The New Frontier. *The Operational Geographer* 5:31–35.

Wright, D. J., M. F. Goodchild, and J. D. Proctor. 1997. Demystifying the Persistent Ambiguity of GIS as "Tool" versus "Science." *Annals of the Association of American Geographers* 87: 346–62.

CHAPTER 2

COORDINATE SYSTEMS

CHAPTER OUTLINE

2.1 Geographic Coordinate System
2.2 Map Projections
2.3 Commonly Used Map Projections
2.4 Projected Coordinate Systems
2.5 Working with Coordinate Systems in GIS

A basic principle in GIS is that map layers to be used together must align spatially. Obvious mistakes can occur if they do not. For example, Figure 2.1 shows the interstate highway maps of Idaho and Montana downloaded separately from the Internet. Obviously, the two maps do not register spatially. To connect the highway networks across the shared state border, we must convert them to a common spatial reference system. Chapter 2 deals with coordinate systems, which provide the spatial reference.

GIS users typically work with map features on a plane. These map features represent spatial features on the Earth's surface. The locations of map features are based on a plane coordinate system expressed in x- and y-coordinates, whereas the locations of spatial features on the Earth's surface are based on a geographic coordinate system expressed in longitude and latitude values. A map projection bridges the two types of coordinate systems. The process of projection transforms the Earth's surface to a plane, and the outcome is a map projection, ready to be used for a projected coordinate system.

We regularly download data sets from the Internet or get them from government agencies for GIS projects. Some digital data sets are measured in longitude and latitude values, whereas others are in different projected coordinate systems. Invariably, these data sets must be processed before they can be used together. Processing in this case means projection and reprojection. **Projection** converts data sets from geographic coordinates to projected coordinates, and **reprojection** converts from one system of projected coordinates to another system. Typically, projection and reprojection are among the initial tasks performed in a GIS project.

Chapter 2 is divided into the following five sections. Section 2.1 describes the geographic coordinate system. Section 2.2 discusses projection,

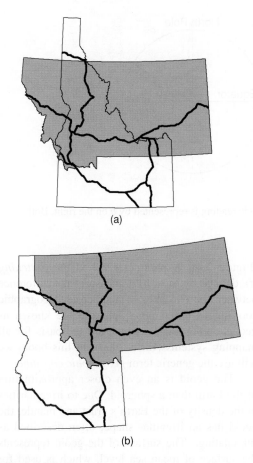

(a)

Figure 2.2
The geographic coordinate system.

(b)

Figure 2.1
The top map shows the interstate highways in Idaho
and Montana based on different coordinate systems.
The bottom map shows the connected interstate
networks based on the same coordinate system.

types of map projections, and map projection
parameters. Sections 2.3 and 2.4 cover commonly
used map projections and coordinate systems,
respectively. Section 2.5 discusses how to work
with coordinate systems in a GIS package.

2.1 GEOGRAPHIC COORDINATE SYSTEM

The **geographic coordinate system** is the location
reference system for spatial features on the Earth's
surface (Figure 2.2). The geographic coordinate

system is defined by **longitude** and **latitude.** Both
longitude and latitude are angular measures: longitude measures the angle east or west from the
prime meridian, and latitude measures the angle
north or south of the equatorial plane (Figure 2.3).

Meridians are lines of equal longitude. The
prime meridian passes through Greenwich, England,
and has the reading of 0°. Using the prime meridian
as a reference, we can measure the longitude value
of a point on the Earth's surface as 0° to 180° east or
west of the prime meridian. Meridians are therefore
used for measuring location in the E–W direction.
Parallels are lines of equal latitude. Using the equator as 0° latitude, we can measure the latitude value
of a point as 0° to 90° north or south of the equator.
Parallels are therefore used for measuring location
in the N–S direction. A point location denoted by
(120° W, 60° N) means that it is 120° west of the
prime meridian and 60° north of the equator.

The prime meridian and the equator serve as
the baselines of the geographic coordinate system.
The notation of geographic coordinates is therefore
like plane coordinates: longitude values are equivalent to *x* values and latitude values are equivalent

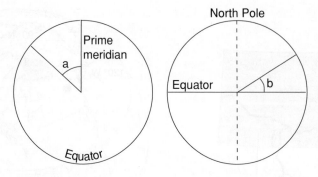

Figure 2.3

A longitude reading is represented by *a* on the left, and a latitude reading is represented by *b* on the right. Both longitude and latitude readings are angular measures.

to *y* values. And, as with *x*-, *y*-coordinates, it is conventional in GIS to enter longitude and latitude values with positive or negative signs. Longitude values are positive in the eastern hemisphere and negative in the western hemisphere. Latitude values are positive if north of the equator, and negative if south of the equator.

The angular measures of longitude and latitude may be expressed in **degrees-minutes-seconds (DMS), decimal degrees (DD),** or radians (rad). Given that 1 degree equals 60 minutes and 1 minute equals 60 seconds, we can easily convert between DMS and DD. For example, a latitude value of 45°52′30″ would be equal to 45.875° (45 + 52/60 + 30/3600). Radians are typically used in computer programs. One radian equals 57.2958°, and one degree equals 0.01745 rad.

2.1.1 Approximation of the Earth

The first step to map spatial features on the Earth's surface is to select a model that approximates the shape and size of the Earth. The simplest model is a sphere, which is typically used in discussing map projections (Section 2.3). But the Earth is not a perfect sphere: the Earth is wider along the equator than between the poles. Therefore a better approximation of the shape of the Earth is a **spheroid,** also called **ellipsoid,** an ellipse rotated about its minor axis.

A spheroid has its major axis *(a)* along the equator and its minor axis *(b)* connecting the poles

(Figure 2.4). A parameter called the *flattening (f),* defined by *(a − b)/a,* measures the difference between the two axes of a spheroid. Geographic coordinates based on a spheroid are known as geodetic coordinates, which are the basis for all mapping systems (Iliffe 2000). In this book, we will use the generic term *geographic coordinates.*

The geoid is an even closer approximation of the Earth than a spheroid. Due to irregularities in the density of the Earth's crust and mantle, the geoid has an irregular shape, often described as 'undulating.' The surface of the geoid represents the surface of mean sea level, which is used for measuring the elevation, or height, of a geographic

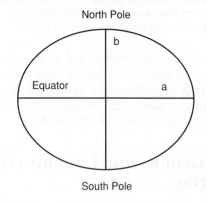

Figure 2.4

The flattening is based on the difference between the semimajor axis *a* and the semiminor axis *b*.

location. When heights are obtained from a GPS (global positioning system) receiver, which is based on a spheroid, they must be transformed so that they are measured from the surface of the geoid. More information on this transformation is included in Chapter 5.

2.1.2 Datum

A **datum** is a mathematical model of the Earth, which serves as the reference or base for calculating the geographic coordinates of a location (Burkard 1984; Moffitt and Bossler 1998). The definition of a datum consists of an origin, a spheroid, and the separation of the spheroid and the Earth at the origin. Because of this definition, datum and spheroid are often interchangeable to GIS users.

To attain a better fit of the geoid locally, many countries have developed their own datums in the past. Among these local datums are the European Datum, the Australian Geodetic Datum, the Tokyo Datum, and the Indian Datum (for India and several adjacent countries). A recent trend, however, is to adopt an Earth-centered (also called geocentric) datum such as the **GRS80** (Geodetic Reference System 1980) spheroid. A geocentric datum has the advantage of being compatible with the GPS. Chapter 5 has more detailed information about GPS measurements.

In the United States, **Clarke 1866,** a ground-measured spheroid, was the standard spheroid for mapping until the late 1980s. Clarke 1866's semimajor axis (equatorial radius) and semiminor axis (polar radius) measure 6,378,206.4 meters

(3962.96 miles) and 6,356,583.8 meters (3949.21 miles), respectively, with the flattening of 1/294.979. **NAD27** (North American Datum of 1927) is a local datum based on the Clarke 1866 spheroid, with its origin at Meades Ranch in Kansas. Hawaii was the only state that did not actually adopt NAD27; Hawaii used the Old Hawaiian Datum, an independent datum derived from NAD27.

In 1986 the National Geodetic Survey (NGS) introduced **NAD83** (North American Datum of 1983) based on the **GRS80.** GRS80's semimajor axis and semiminor axis measure 6,378,137.0 meters (3962.94 miles) and 6,356,752.3 meters (3949.65 miles), respectively, with the flattening of 1/298.257. In the case of GRS80, the shape and size of the Earth were determined through measurements made by Doppler satellite observations. The shift from NAD27 to NAD83 represents a shift from a local to a geocentric datum. Datum shift is taking place in other countries as well. Box 2.1 describes an example from Australia.

Datum shift from NAD27 to NAD83 can result in substantial shifts of positions of points. As shown in Figure 2.5, horizontal shifts are between 10 and 100 meters in the conterminous United States. (The shifts are more than 200 meters in Alaska and in excess of 400 meters in Hawaii.) For example, for the Ozette quadrangle map from the Olympic Peninsula in Washington, the shift is 98 meters to the east and 26 meters to the north. The horizontal shift is therefore 101.4 meters ($\sqrt{98^2 + 26^2}$).

Many GIS users in the United States have migrated from NAD27 to NAD83, whereas others are still in the process of adopting NAD83. The same is

---

Box 2.1 **Australian Datums**

As GIS users in the United States have migrated from NAD27 to NAD83, GIS users in Australia have migrated from the Australian Geodetic Datum (AGD) to the Geocentric Datum of Australia (GDA). The AGD introduced in 1966 is based on the Australian

National Spheroid, a spheroid that best estimates the Earth's shape around the Australian continent. AGD is therefore a local datum like NAD27. The new datum GDA, similar to NAD83, is a geocentric datum based on the GRS80.

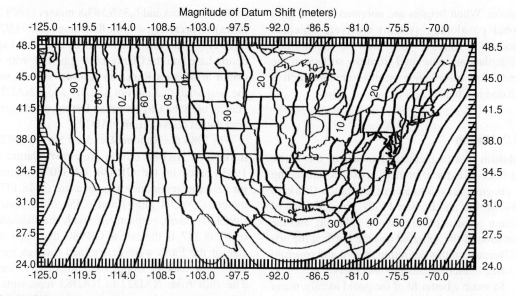

Figure 2.5

The isolines show the magnitudes of the horizontal shift from NAD27 to NAD83 in meters. See Section 2.1.2 for the definition of the horizontal shift. (By permission of the National Geodetic Survey.)

true with data sets downloadable from GIS data clearinghouses: some are based on NAD83, and others on NAD27. Until the switch from NAD27 to NAD83 is complete, we must keep watchful eyes on the datum because digital layers based on the same projection but different datums will not register correctly.

WGS84 (World Geodetic System 1984) is a datum established by the National Imagery and Mapping Agency (NIMA, now the National Geospatial-Intelligence Agency or NGA) of the U.S. Department of Defense (Kumar 1993). WGS84 agrees with GRS80 in terms of measures of the semimajor and semiminor axes. But WGS84 has a set of primary and secondary parameters. The primary parameters define the shape and size of the Earth, whereas the secondary parameters refer to local datums used in different countries (National Geospatial-Intelligence Agency 2000; Iliffe 2000). WGS84 is the datum for GPS readings. The satellites used by GPS send their positions in WGS84 coordinates and all calculations internal to GPS receivers are also based on WGS84.

Datum shift, such as from NAD27 to NAD83 or from NAD27 to WGS84 requires a datum transformation, which recomputes longitude and latitude values from one geographic coordinate system to another. A commercial GIS package may offer several transformation methods such as three-parameter, seven-parameter, Molodensky, and abridged Molodensky. A good reference on datum transformation and its mathematical methods is available online from the NGA (2000). Free software packages for data conversion are also available online; for example, Nadcon is a software package that can be downloaded at the NGS website for conversion between NAD27 and NAD83 **(http://www.ngs .noaa.gov/ TOOLS/Nadcon/Nadcon.html).**

Although the migration from NAD27 to NAD83 is not yet complete, new developments on datums continue in the United States for local surveys (Kavanagh 2003). In the late 1980s, the NGS began a program of using GPS technology to establish the High Accuracy Reference

Network (HARN) on a state-by-state basis. In 1994, the NGS started the Continuously Operating Reference Stations (CORS) network, a network of over 200 stations that provide measurements for the postprocessing of GPS data. The positional difference of a control point may be up to a meter between NAD83 and HARN but less than 10 centimeters between HARN and CORS (Snay and Soler 2000).

HARN and CORS networks can provide data for refining NAD83. NAD83 (HARN) is a refined NAD83 based on HARN data, and NAD83 (CORS96) is a refined NAD83 based on CORS data. Both refined NAD83 datums are more accurate than the original NAD83 and are thus important to surveyors and GPS users, who require highly accurate data (e.g., centimeter-level accuracy) for their work.

2.2 MAP PROJECTIONS

A map projection transforms the Earth's surface, typically based on a spheroid in GIS, to a plane (Robinson et al. 1995; Dent 1999; Slocum et al. 2005).

The outcome of this transformation process is a **map projection:** a systematic arrangement of parallels and meridians on a plane surface representing the geographic coordinate system.

We can use data sets based on geographic coordinates directly in a GIS, and in fact, we are seeing more maps made with such data sets. But a map projection provides a couple of distinctive advantages. First, a map projection allows us to use two-dimensional maps, either paper or digital, instead of a globe. Second, a map projection allows us to work with plane coordinates rather than longitude and latitude values. Computations with geographic coordinates are more complex and can yield less accurate distance measurements (Box 2.2).

But the transformation from the Earth's surface to a flat surface always involves distortion, and no map projection is perfect. This is why hundreds of map projections have been developed for mapmaking (Maling 1992; Snyder 1993). Every map projection preserves certain spatial properties while sacrificing other properties.

| *Box 2.2* | **How to Measure Distances on the Earth's Surface** |

The equation for measuring distances on a plane coordinate system is:

$$D = \sqrt{(x_1 - x_2)^2 + (y_1 - y_2)^2}$$

where x_i and y_i are the coordinates of point i.

This equation, however, cannot be used for measuring distances on the Earth's surface. Because meridians converge at the poles, the length of 1-degree latitude does not remain constant but gradually decreases from the equator to 0 at the pole. The standard and simplest method for calculating the shortest distance between two points on the Earth's surface uses the equation:

$$\cos (d) = \sin (a) \sin (b) + \cos (b) \cos (c)$$

where d is the angular distance between points A and B in degrees, a is the latitude of A, b is the latitude of B, and c is the difference in longitude between A and B. To convert d to a linear distance measure, one can multiply d by the length of 1 degree at the equator, which is 111.32 kilometers or 69.17 miles. This method is accurate unless d is very close to zero (Snyder 1987).

Most commercial data producers deliver spatial data in geographic coordinates to be used with any projected coordinate system the end user needs to work with. But more GIS users are using spatial data in geographic coordinates directly for data display and even simple analysis. Distance measurements from such spatial data are usually derived from the shortest spherical distance between points.

2.2.1 Types of Map Projections

Map projections can be grouped by either the preserved property or the projection surface. Cartographers group map projections by the preserved property into the following four classes: conformal, equal area or equivalent, equidistant, and azimuthal or true direction. A **conformal projection** preserves local angles and shapes. An **equivalent projection** represents areas in correct relative size. An **equidistant projection** maintains consistency of scale along certain lines. And an **azimuthal projection** retains certain accurate directions. The preserved property of a map projection is often included in its name, such as the Lambert conformal conic projection or the Albers equal-area conic projection.

The conformal and equivalent properties are mutually exclusive. Otherwise a map projection can have more than one preserved property, such as conformal and azimuthal. The conformal and equivalent properties are global properties, meaning that they apply to the entire map projection. The equidistant and azimuthal properties are local properties and may be true only from or to the center of the map projection.

The preserved property is important for selecting an appropriate map projection for thematic mapping. For example, a population map of the world should be based on an equivalent projection. By representing areas in correct size, the population map can create a correct impression of population densities. In contrast, an equidistant projection would be better for mapping the distance ranges from a missile site.

Cartographers often use a geometric object and a globe (i.e., a sphere) to illustrate how to construct a map projection. For example, by placing a cylinder tangent to a lighted globe, one can draw a projection by tracing the lines of longitude and latitude onto the cylinder. The cylinder is called the projection surface or the developable surface, and the globe is called the **reference globe.** Other common projection surfaces include a cone and a plane. Therefore, map projections can be grouped by their projection surfaces into cylindrical, conic, and azimuthal. A map projection is called a **cylindrical projection** if it can be constructed using a cylinder, a **conic projection** if using a cone, and an **azimuthal projection** if using a plane.

The use of a geometric object helps explain two other projection concepts: case and aspect. For a conic projection, the cone can be placed so that it is tangent to the globe or intersects the globe (Figure 2.6). The first is the simple case, which results in one line of tangency, and the second is the secant case, which results in two lines of tangency. A cylindrical projection behaves the same way as a conic projection in terms of case. An azimuthal projection, on the other hand, has a point of tangency in the simple case and a line of tangency in the secant case. Aspect describes the placement of a geometric object relative to a globe. A plane, for example, may be tangent at any point on a globe. A polar aspect refers to tangency at the pole, an equatorial aspect at the equator, and an oblique aspect anywhere between the equator and the pole (Figure 2.7).

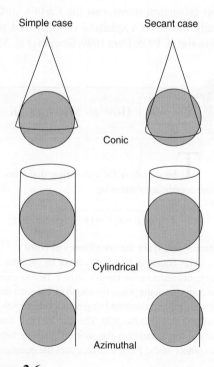

Figure 2.6
Case and projection.

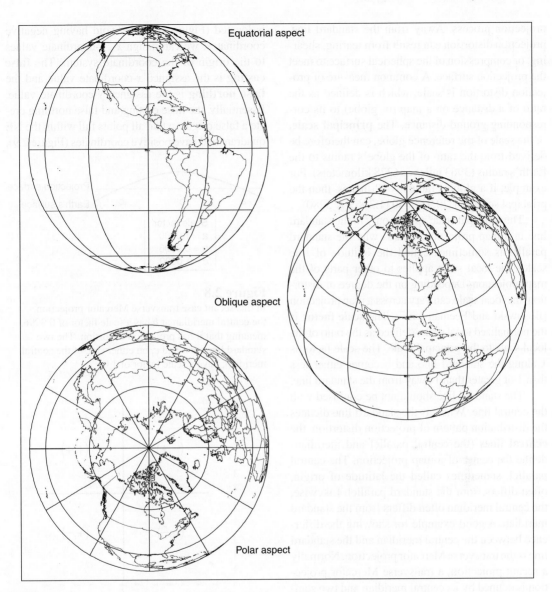

Equatorial aspect

Oblique aspect

Polar aspect

Figure 2.7
Aspect and projection.

2.2.2 Map Projection Parameters

The concept of case relates directly to the standard line, a common parameter in defining a map projection. A **standard line** refers to the line of tangency between the projection surface and the reference globe. For cylindrical and conic projections the simple case has one standard line, whereas the secant case has two standard lines. The standard line is called the **standard parallel** if it follows a parallel, and the **standard meridian** if it follows a meridian.

Because the standard line is the same as on the reference globe, it has no distortion from the projection process. Away from the standard line, projection distortion can result from tearing, shearing, or compression of the spherical surface to meet the projection surface. A common measure of projection distortion is scale, which is defined as the ratio of a distance on a map (or globe) to its corresponding ground distance. The **principal scale,** or the scale of the reference globe, can therefore be derived from the ratio of the globe's radius to the Earth's radius (3963 miles or 6378 kilometers). For example, if a globe's radius is 12 inches, then the principal scale is 1:20,924,640 (1:3963 × 5280).

The principal scale applies only to the standard line in a map projection. This is why the standard parallel is sometimes called the latitude of true scale. The local scale applies to other parts of the map projection. Depending on the degree of distortion, the local scale can vary across a map projection (Bosowski and Feeman 1997). The **scale factor** is the normalized local scale, defined as the ratio of the local scale to the principal scale. The scale factor is 1 along the standard line and becomes either less than 1 or greater than 1 away from the standard line.

The standard line should not be confused with the central line. Whereas the standard line dictates the distribution pattern of projection distortion, the **central lines** (the central parallel and meridian) define the center of a map projection. The central parallel, sometimes called the latitude of origin, often differs from the standard parallel. Likewise, the central meridian often differs from the standard meridian. A good example for showing the difference between the central meridian and the standard line is the transverse Mercator projection. Normally a secant projection, a transverse Mercator projection is defined by its central meridian and two standard lines on either side. The standard line has a scale factor of 1, and the central meridian has a scale factor of less than 1 (Figure 2.8).

When a map projection is used as the basis of a coordinate system, the center of the map projection, as defined by the central parallel and the central meridian, becomes the origin of the coordinate system and divides the coordinate system into four quadrants. The x-, y-coordinates of a point are either positive or negative, depending on where the point

is located (Figure 2.9). To avoid having negative coordinates, we can assign x-, y-coordinate values to the origin of the coordinate system. The **false easting** is the assigned x-coordinate value and the **false northing** is the assigned y-coordinate value. Essentially, the false easting and false northing create a false origin so that all points fall within the NE quadrant and have positive coordinates (Figure 2.9).

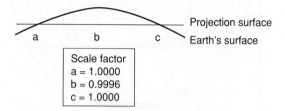

Figure 2.8
In this secant case transverse Mercator projection, the central meridian at *b* has a scale factor of 0.9996, meaning that it has projection distortion. The two standard lines at *a* and *c*, on either side of the central meridian, have a scale factor of 1.0.

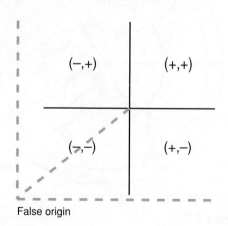

False origin

Figure 2.9
The central parallel and the central meridian divide a map projection into four quadrants. Points within the NE quadrant have positive x- and y-coordinates, points within the NW quadrant have negative x-coordinates and positive y-coordinates, points within the SE quadrant have positive x-coordinates and negative y-coordinates, and points within the SW quadrant have negative x- and y-coordinates. The purpose of having a false origin is to place all points within the NE quadrant.

2.3 COMMONLY USED MAP PROJECTIONS

Hundreds of map projections are in use. Commonly used map projections in GIS are not necessarily the same as those we see in classrooms or in magazines. For example, the Robinson projection is a popular projection for general mapping at the global scale because it is aesthetically pleasing (Dent 1999). But the Robinson projection is not suitable for GIS applications. A map projection for GIS applications usually has one of the preserved properties mentioned earlier, especially the conformal property. Because it preserves local shapes and angles, a conformal projection allows adjacent maps to join correctly at the corners. This is important in developing a

map series such as the U.S. Geological Survey (USGS) quadrangle maps.

2.3.1 Transverse Mercator

The **transverse Mercator projection,** also known as the Gauss-Kruger, is a variation of the Mercator projection, probably the best-known projection for mapping the world (Figure 2.10). The Mercator projection uses the standard parallel, whereas the transverse Mercator projection uses the standard meridian. Both projections are conformal.

The transverse Mercator is the basis for two common coordinate systems to be discussed in Section 2.4. The definition of the projection requires the following parameters: scale factor at central meridian, longitude of central meridian,

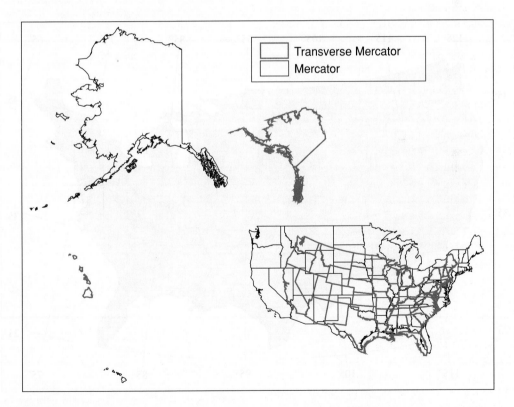

Figure 2.10
The Mercator and the transverse Mercator projection of the United States. For both projections, the central meridian is 90° W and the latitude of true scale is the equator.

latitude of origin (or central parallel), false easting, and false northing.

2.3.2 Lambert Conformal Conic

The **Lambert conformal conic projection** is a standard choice for mapping a midlatitude area of greater east–west than north–south extent, such as the state of Montana or the conterminous United States (Figure 2.11). The USGS has used the Lambert conformal conic for many topographic maps since 1957.

Typically used as a secant projection, the Lambert conformal conic is defined by the following parameters: first and second standard parallels, central meridian, latitude of projection's origin, false easting, and false northing.

2.3.3 Albers Equal-Area Conic

The Albers equal-area conic projection has the same parameters as the Lambert conformal conic projection. In fact, the two projections are quite similar except that one is equal area and the other is conformal. The Albers equal-area conic is the projection for the National Land Cover Data 1992 and the National Land Cover Database 2001 for the conterminous United States.

2.3.4 Equidistant Conic

The equidistant conic projection is also called the simple conic projection. The projection preserves the distance property along all meridians and one or two standard parallels. It uses the same parameters as the Lambert conformal conic.

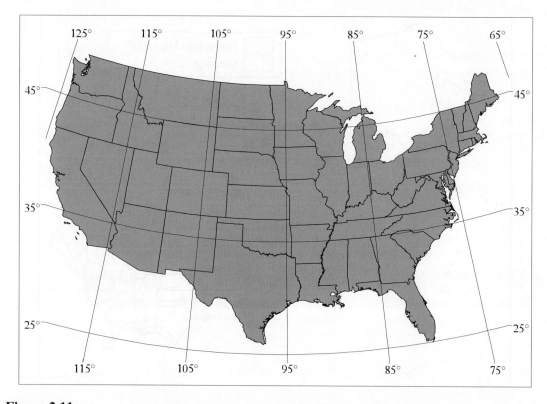

Figure 2.11

The Lambert conformal conic projection of the conterminous United States. The central meridian is 96° W, the two standard parallels are 33° N and 45° N, and the latitude of projection's origin is 39° N.

2.3.5 Web Mercator

Both Google Earth and Microsoft Virtual Earth (Section 1.5 in Chapter 1) use Web Mercator. What is Web Mercator? It is the Mercator projection based on a sphere instead of a spheroid. This simplifies the calculations. Because Google Earth and Microsoft Virtual Earth are primarily used for map display rather than numerical analysis, the loss of accuracy in projection by using a sphere is not deemed to be important. With Web Mercator, GIS users must consider reprojection when they want to overlay GIS layers on Google Earth and Microsoft Virtual Earth for data analysis.

2.4 PROJECTED COORDINATE SYSTEMS

A **projected coordinate system** is built on a map projection. Projected coordinate systems and map projections are often used interchangeably. For example, the Lambert conformal conic is a map projection but it can also refer to a coordinate system. In practice, however, projected coordinate systems are designed for detailed calculations and positioning, and are typically used in large-scale mapping such as at a scale of 1:24,000 or larger

(Box 2.3). Accuracy in a feature's location and its position relative to other features is therefore a key consideration in the design of a projected coordinate system.

To maintain the level of accuracy desired for measurements, a projected coordinate system is often divided into different zones, with each zone defined by a different projection center. Moreover, a projected coordinate system is defined not only by the parameters of the map projection it is based on but also the parameters of the geographic coordinate system (e.g., datum) that the map projection is derived from. All mapping systems are based on a spheroid rather than a sphere. The difference between a spheroid and a sphere may not be a concern for general mapping at small map scales but can be a matter of importance in the detailed mapping of land parcels, soil polygons, or vegetation stands.

Three coordinate systems are commonly used in the United States: the Universal Transverse Mercator (UTM) grid system, the Universal Polar Stereographic (UPS) grid system, and the State Plane Coordinate (SPC) system. As a group, coordinates of these common systems are sometimes called real-world coordinates. This section also includes the Public Land Survey System (PLSS),

---

Box 2.3 **Map Scale**

Map scale is the ratio of the map distance to the corresponding ground distance. This definition applies to different measurement units. A 1:24,000 scale map can mean that a map distance of 1 centimeter represents 24,000 centimeters (240 meters) on the ground. A 1:24,000 scale map can also mean that a map distance of 1 inch represents 24,000 inches (2000 feet) on the ground. Regardless of its measurement unit, 1:24,000 is a larger map scale than 1:100,000 and the same spatial feature (e.g., a town) appears larger on a 1:24,000 scale map than on a

1:100,000 scale map. Some cartographers consider maps with a scale of 1:24,000 or larger to be large-scale maps.

Map scale should not be confused with spatial scale, a term commonly used in natural resource management. Spatial scale refers to the size of area or extent. Unlike map scale, spatial scale is not rigidly defined. A large spatial scale simply means that it covers a larger area than a small spatial scale. A large spatial scale to an ecologist is therefore a small map scale to a cartographer.

a land partitioning system used in the United States for land parcel mapping. Although it is not a coordinate system, the PLSS is covered here as an example of a locational reference system that can be used for the same purpose as a coordinate system. Additional readings on these systems can be found in Robinson et al. (1995), Muehrcke et al. (2001), and Slocum et al. (2005).

2.4.1 The Universal Transverse Mercator (UTM) Grid System

Used worldwide, the **UTM grid system** divides the Earth's surface between 84° N and 80° S into 60 zones. Each zone covers 6° of longitude and is numbered sequentially with zone 1 beginning at 180° W. Each zone is further divided into the northern and southern hemispheres. The designation of a UTM zone therefore carries a number and a letter. For example, UTM Zone 10N refers to the zone between 126° W and 120° W in the northern hemisphere. The inside of this book's front cover has a list of the UTM zone numbers and their longitude ranges. Figure 2.12 shows the UTM zones in the conterminous United States.

Because datum is part of the definition of a projected coordinate system, the UTM grid system may be based on NAD27, NAD83, or WGS84. Thus, if UTM Zone 10N is based on NAD83, then its full designation reads NAD 1983 UTM Zone 10N.

Each UTM zone is mapped onto a secant case transverse Mercator projection, with a scale factor of 0.9996 at the central meridian and the equator as the latitude of origin. The standard meridians are 180 kilometers to the east and the west of the central meridian (Figure 2.13). The use of a projection per UTM zone is designed to maintain the accuracy of at least one part in 2500 (i.e., distance measured over a 2500-meter course on the UTM grid system would be accurate within a meter of the true measure) (Muehrcke, Muehrcke, and Kimerling 2001).

In the northern hemisphere, UTM coordinates are measured from a false origin located at the equator and 500,000 meters west of the UTM zone's central meridian. In the southern hemisphere, UTM coordinates are measured from a false origin located at 10,000,000 meters south of the equator and 500,000 meters west of the UTM zone's central meridian.

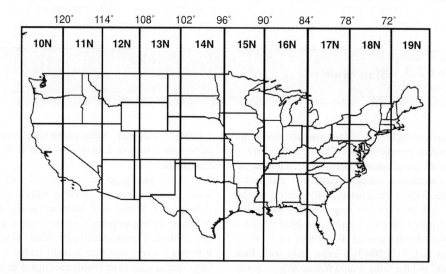

Figure 2.12
UTM zones range from zone 10N to 19N in the conterminous United States.

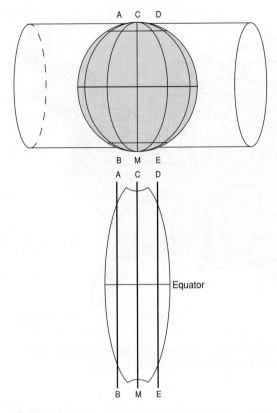

Figure 2.13

A UTM zone represents a secant case transverse Mercator projection. CM is the central meridian, and AB and DE are the standard meridians. The standard meridians are placed 180 kilometers west and east of the central meridian. Each UTM zone covers 6° of longitude and extends from 84° N to 80° S. The size and shape of the UTM zone are exaggerated for illustration purposes.

The use of a false origin means that UTM coordinates can be very large numbers. For example, the NW corner of the Moscow East, Idaho, quadrangle map has the UTM coordinates of 500,000 and 5,177,164 meters. To preserve data precision for computations with coordinates, we can apply **x-shift** and **y-shift** values to all coordinate readings to reduce the number of digits. For example, if the x-shift value is set as −500,000 meters and the y-shift value as −5,170,000 meters for the previous quadrangle map, the coordinates for its NW corner

become 0 and 7164 meters. Small numbers such as 0 and 7164 reduce the chance of having truncated computational results. The x-shift and y-shift are therefore important if coordinates are stored in single precision (i.e., up to seven significant digits). Like false easting and false northing, x-shift and y-shift change the values of x-, y-coordinates in a data set. They must be documented along with the projection parameters in the metadata (information about data, Chapter 5), especially if the map is to be shared with other users.

2.4.2 The Universal Polar Stereographic (UPS) Grid System

The **UPS grid system** covers the polar areas. The stereographic projection is centered on the pole and is used for dividing the polar area into a series of 100,000-meter squares, similar to the UTM grid system.

2.4.3 The State Plane Coordinate (SPC) System

The **SPC system** was developed in the 1930s to permanently record original land survey monument locations in the United States. To maintain the required accuracy of one part in 10,000 or less, a state may have two or more SPC zones. As examples, Oregon has the North and South SPC zones and Idaho has the West, Central, and East SPC zones (Figure 2.14). Each SPC zone is mapped onto a map projection. Zones that are elongated in the north-south direction (e.g., Idaho's SPC zones) use the transverse Mercator and zones that are elongated in the east-west direction (e.g., Oregon's SPC zones) use the Lambert conformal conic. (The only exception is zone 1 of Alaska, which uses the oblique Mercator to cover the panhandle of Alaska.) Point locations within each SPC zone are measured from a false origin located to the southwest of the zone.

Because of the switch from NAD27 to NAD83, there are SPC27 and SPC83. Besides the change of the datum, SPC83 has a few other changes. SPC83 coordinates are published in meters instead of feet. The states of Montana,

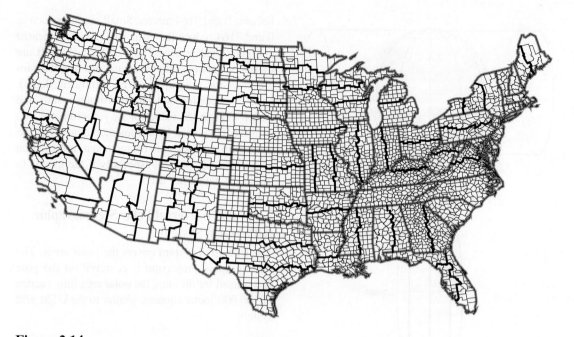

Figure 2.14
SPC83 zones in the conterminous United States. The thin lines are county boundaries, and the bold lines are SPC zone boundaries. This map corresponds to the SPC83 table on the inside of this book's back cover.

Nebraska, and South Carolina have each replaced multiple zones with a single SPC zone. California has reduced SPC zones from seven to six. And Michigan has changed from transverse Mercator to Lambert conformal conic projections. A list of SPC83 is available on the inside of this book's back cover.

Some states in the United States have developed their own statewide coordinate system. Montana, Nebraska, and South Carolina all have a single SPC zone, which can serve as the statewide coordinate system. Idaho is another example. Idaho is divided into two UTM zones (11 and 12) and three SPC zones (West, Central, and East). These zones work well as long as the study area is within a single zone. When a study area covers two or more zones, the data sets must be converted to a single zone for spatial registration. But the conversion to a single zone also means that the data sets can no longer maintain the accuracy level designed for the UTM or the SPC coordinate system. The

Idaho statewide coordinate system, adopted in 1994 and modified in 2003, is still based on a transverse Mercator projection but its central meridian passes through the center of the state (114° W). (A complete list of parameters of the Idaho statewide coordinate system is included in Task 1 of the applications section.) Changing the location of the central meridian results in one zone for the entire state.

2.4.4 The Public Land Survey System (PLSS)

The **PLSS** is a land partitioning system (Figure 2.15). Using the intersecting township and range lines, the system divides the lands mainly in the central and western states into 6 × 6 mile squares or townships. Each township is further partitioned into 36 square-mile parcels of 640 acres, called sections. (In reality, many sections are not exactly 1 mile by 1 mile in size.)

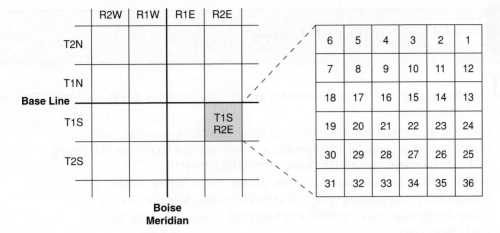

Figure 2.15

The shaded survey township has the designation of T1S, R2E. T1S means that the survey township is south of the base line by one unit. R2E means that the survey township is east of the Boise (principal) meridian by two units. Each survey township is divided into 36 sections. Each section measures 1 mile by 1 mile and has a numeric designation.

Land parcel layers are typically based on the PLSS. The Bureau of Land Management (BLM) has been working on a **Geographic Coordinate Data Base (GCDB)** of the PLSS for the western United States **(http://www.blm.gov/gcdb/).** Generated from BLM survey records, the GCDB contains coordinates and other descriptive information for section corners and monuments recorded in the PLSS. Legal descriptions of a parcel layer can then be entered using, for example, bearing and distance readings originating from section corners.

2.5 WORKING WITH COORDINATE SYSTEMS IN GIS

Basic GIS tasks with coordinate systems involve defining a coordinate system, projecting geographic coordinates to projected coordinates, and reprojecting projected coordinates from one system to another.

A GIS package typically has many options of datums, spheroids, and coordinate systems. For example, Autodesk Map offers 3000 global systems, presumably 3000 combinations of coordinate system, datum, and spheroid. A constant challenge for GIS users is how to work with this large number of coordinate systems. Recent trends suggest that commercial GIS companies have tried to provide assistance in the following three areas: projection file, predefined coordinate systems, and on-the-fly projection.

2.5.1 Projection File

A projection file is a text file that stores information on the coordinate system that a data set is based on. Box 2.4, for example, shows a projection file for the NAD 1983 UTM Zone 11N coordinate system. The projection file contains information on the geographic coordinate system, the map projection parameters, and the linear unit.

Besides identifying a data set's coordinate system, a projection file serves at least two other purposes: it can be used as an input for projecting or reprojecting the data set, and it can be exported to other data sets that are based on the same coordinate system.

Box 2.4   **A Projection File Example**

The following projection file example is used by ArcGIS to store information on the NAD 1983 UTM Zone 11N coordinate system:

PROJCS ["NAD_1983_UTM_Zone_11N", GEOGCS["GCS_North_American_1983",
DATUM["D_North_American_1983",SPHEROID["GRS_1980",6378137.0,298.257222101]],
PRIMEM["Greenwich",0.0], UNIT["Degree",0.0174532925199433]],
PROJECTION ["Transverse_Mercator"], PARAMETER["False_Easting",500000.0],
PARAMETER["False_Northing",0.0], PARAMETER["Central_Meridian",−117.0],
PARAMETER["Scale_Factor",0.9996], PARAMETER["Latitude_Of_Origin",0.0],
UNIT["Meter",1.0]]

The information comes in three parts. The first part defines the geographic coordinate system: NAD83 for the datum, GRS80 for the spheroid, the prime meridian of 0° at Greenwich, and units of degrees. The file also lists the major axis (6378137.0) and the denominator of the flattening (298.257222101) for the spheroid. The number of 0.0174532925199433 is the conversion factor from degree to radian (an angular unit typically used in computer programming). The second part defines the map projection parameters of name, false easting, false northing, central meridian, scale factor, and latitude of origin. And the third part defines the linear unit in meters.

2.5.2 Predefined Coordinate Systems

A GIS package typically groups coordinate systems into predefined and custom (Table 2.1). A predefined coordinate system, either geographic or projected, means that its parameter values are known and are already coded in the GIS package. The user can therefore select a predefined coordinate system without defining its parameters. Examples of predefined coordinate systems include NAD27 (based on Clarke 1866) and Minnesota SPC83, North (based on a Lambert conformal conic projection). In contrast, a custom coordinate system requires its parameter values to be specified by the user. The Idaho statewide coordinate system (IDTM) is an example of a custom coordinate system.

2.5.3 On-the-Fly Projection

On-the-fly projection is a utility that has been heavily advertised by GIS vendors. On-the-fly projection is designed for displaying data sets that are

TABLE 2.1   **A Classification of Coordinate Systems in GIS Packages**

	Predefined	Custom
Geographic	NAD27, NAD83	Undefined local datum
Projected	UTM, State Plane	IDTM

based on different coordinate systems. The software package uses the projection files available and automatically converts the data sets to a common coordinate system. This common coordinate system is by default the coordinate system of the first data set in display. If a data set has an unknown coordinate system, the GIS package may use an assumed coordinate system such as NAD27 as the assumed geographic coordinate system.

On-the-fly projection does not actually change the coordinate system of a data set. Thus it cannot replace the task of projecting and reprojecting data sets in a GIS project. If a data set is to be used frequently in a different coordinate system, we should reproject the data set. And if the data sets to be used in spatial analysis have different coordinate systems, we should convert them to the same coordinate system to obtain the most accurate results.

Like other GIS packages, ArcGIS provides a suite of tools to work with coordinate systems. Box 2.5 is a summary of how to work with coordinate systems in ArcGIS.

Box 2.5 Coordinate Systems in ArcGIS

ArcGIS divides coordinate systems into geographic and projected. The user can define a coordinate system by selecting a predefined coordinate system, importing a coordinate system from an existing data set, or creating a new (custom) coordinate system. The parameters that are used to define a coordinate system are stored in a projection file. A projection file is provided for a predefined coordinate system. For a new coordinate system, a projection file can be named and saved for future use or for projecting other data sets.

The predefined geographic coordinate systems in ArcGIS have the main options of world, continent, and spheroid-based. WGS84 is one of the world files. Local datums are used for the continental files. For example, the Indian Datum and Tokyo Datum are available for the Asian continent. The spheroid-based options include Clarke 1866 and GRS80. The predefined projected coordinate systems have the main options of world, continent, polar, national grids, UTM, State Plane, and Gauss Kruger (one type of the transverse Mercator projection mainly used in Russia and China). For example, the Mercator is one of the world projections; the Lambert conformal conic and Albers equal-area are among the continental projections; and the UPS is one of the polar projections.

A new coordinate system, either geographic or projected, is user-defined. The definition of a new geographic coordinate system requires a datum including a selected spheroid and its major and minor axes. The definition of a new projected coordinate system must include a datum and the parameters of the projection such as the standard parallels and the central meridian.

KEY CONCEPTS AND TERMS

Azimuthal projection: One type of map projection that retains certain accurate directions. Azimuthal also refers to one type of map projection that uses a plane as the projection surface.

Central lines: The central parallel and the central meridian. Together, they define the center or the origin of a map projection.

Clarke 1866: A ground-measured spheroid, which is the basis for the North American Datum of 1927 (NAD27).

Conformal projection: One type of map projection that preserves local shapes.

Conic projection: One type of map projection that uses a cone as the projection surface.

Cylindrical projection: One type of map projection that uses a cylinder as the projection surface.

Datum: The basis for calculating the geographic coordinates of a location. A spheroid is a required input to the derivation of a datum.

Datum shift: A change from one datum to another, such as from NAD27 to NAD83, which can result in substantial horizontal shifts of point positions.

Decimal degrees (DD) system: A measurement system for longitude and latitude values such as 42.5°.

Degrees-minutes-seconds (DMS) system: A measuring system for longitude and latitude values such as 42°30′00″, in which 1 degree equals 60 minutes and 1 minute equals 60 seconds.

Ellipsoid: A model that approximates the Earth. Also called *spheroid*.

Equidistant projection: One type of map projection that maintains consistency of scale for certain distances.

Equivalent projection: One type of map projection that represents areas in correct relative size.

False easting: A value applied to the origin of a coordinate system to change the *x*-coordinate readings.

False northing: A value applied to the origin of a coordinate system to change the *y*-coordinate readings.

Geographic Coordinate Data Base (GCDB):   A database developed by the U.S. Bureau of Land Management (BLM) to include longitude and latitude values and other descriptive information for section corners and monuments recorded in the PLSS.

Geographic coordinate system: A location reference system for spatial features on the Earth's surface.

GRS80:   A satellite-determined spheroid for the Geodetic Reference System 1980.

Lambert conformal conic projection: A common map projection, which is the basis for the SPC system for many states.

Latitude: The angle north or south of the equatorial plane.

Longitude: The angle east or west from the prime meridian.

Map projection: A systematic arrangement of parallels and meridians on a plane surface.

Meridians: Lines of longitude that measure locations in the E–W direction on the geographic coordinate system.

NAD27:   North American Datum of 1927, which is based on the Clarke 1866 spheroid and has its center at Meades Ranch, Kansas.

NAD83:   North American Datum of 1983, which is based on the GRS80 spheroid and has its origin at the center of the spheroid.

Parallels: Lines of latitude that measure locations in the N–S direction on the geographic coordinate system.

Principal scale: Same as the scale of the reference globe.

Projected coordinate system: A plane coordinate system that is based on a map projection.

Projection: The process of transforming the spatial relationship of features on the Earth's surface to a flat map.

Public Land Survey System (PLSS):   A land partitioning system used in the United States.

Reference globe: A reduced model of the Earth from which map projections are made. Also called a *nominal* or *generating globe*.

Reprojection: Projection of spatial data from one projected coordinate system to another.

Scale factor: Ratio of the local scale to the scale of the reference globe. The scale factor is 1.0 along a standard line.

Spheroid: A model that approximates the Earth. Also called *ellipsoid*.

Standard line: Line of tangency between the projection surface and the reference globe. A standard line has no projection distortion and has the same scale as that of the reference globe.

Standard meridian: A standard line that follows a meridian.

Standard parallel: A standard line that follows a parallel.

State Plane Coordinate (SPC) system: A coordinate system developed in the 1930s to permanently record original land survey monument locations in the United States. Most states have more than one zone based on the SPC27 or SPC83 system.

Transverse Mercator projection: A common map projection, which is the basis for the UTM grid system and the SPC system.

Universal Polar Stereographic (UPS) grid system: A grid system that divides the polar area into a series of 100,000-meter squares, similar to the UTM grid system.

Universal Transverse Mercator (UTM) grid system: A coordinate system that divides the Earth's surface between 84° N and 80° S into 60 zones, with each zone further divided into the northern hemisphere and the southern hemisphere.

WGS84: A satellite-determined spheroid for the World Geodetic System 1984.

x-**shift:** A value applied to *x*-coordinate readings to reduce the number of digits.

y-**shift:** A value applied to *y*-coordinate readings to reduce the number of digits.

REVIEW QUESTIONS

1. Describe the three levels of approximation of the shape and size of the Earth for GIS applications.

2. Why is the datum important in GIS?

3. Describe two common datums used in the United States.

4. Pick up a USGS quadrangle map of your area. Examine the information on the map margin. If the datum is changed from NAD27 to NAD83, what is the expected horizontal shift?

5. Go to the NGS-CORS website (**http://www.ngs.noaa.gov/CORS/corsdata.html**). How many continuously operating reference stations do you have in your state? Use the links at the website to learn more about CORS.

6. Explain the importance of map projection.

7. Describe the four types of map projections by the preserved property.

8. Describe the three types of map projections by the projection or developable surface.

9. Explain the difference between the standard line and the central line.

10. How is the scale factor related to the principal scale?

11. Name two commonly used projected coordinate systems that are based on the transverse Mercator projection.

12. Find the GIS data clearinghouse for your state at the Geospatial One-Stop website (**http://gos2.geodata.gov/wps/portal/gos**). Go to the clearinghouse website. Does the website use a common coordinate system for the statewide data sets? If so, what is the coordinate system? What are the parameter values for the coordinate system? Is the coordinate system based on NAD27 or NAD83?

13. Explain how a UTM zone is defined in terms of its central meridian, standard meridian, and scale factor.

14. Which UTM zone are you in? Where is the central meridian of the UTM zone?

15. How many SPC zones does your state have? What map projections are the SPC zones based on?

16. Describe how on-the-fly projection works.

Applications: Coordinate Systems

This applications section has four tasks. Task 1 shows you how to project a shapefile from a geographic coordinate system to a custom projected coordinate system. In Task 2, you will also project a shapefile from a geographic to a projected coordinate system but use the coordinate systems already defined in Task 1. In Task 3, you will create a shapefile from a text file that contains point locations in geographic coordinates and project the shapefile onto a predefined projected coordinate system. In Task 4, you will see how on-the-fly projection works and then reproject a shapefile onto a different projected coordinate system. Designed for data display, on-the-fly projection does not change the spatial reference of a data set. To change the spatial reference of a data set, you must reproject the data set.

All four tasks use the Define Projection and Project tools in ArcToolbox, which are available in ArcCatalog as well as ArcMap. The Define Projection tool defines a coordinate system. The Project tool projects a geographic or projected coordinate system. ArcToolbox has three options for defining a coordinate system: selecting a predefined coordinate system, importing a coordinate system from an existing data set, or creating a new (custom) coordinate system. A predefined coordinate system already has a projection file. A new coordinate system can be saved into a projection file, which can then be used to define or project other data sets.

This applications section uses shapefiles or vector data for all four tasks. ArcToolbox has a separate tool in the Data Management Tools/Projections and Transformations/Raster toolset for projecting rasters or raster data.

Task 1 Project a Shapefile from a Geographic to a Projected Coordinate System

What you need: *idll.shp,* a shapefile measured in geographic coordinates and in decimal degrees. *idll.shp* is an outline layer of Idaho.

For Task 1, you will first define *idll.shp* by selecting a predefined geographic coordinate system and then project the shapefile onto the Idaho transverse Mercator coordinate system (IDTM). A custom coordinate system, IDTM has the following parameter values:

Projection Transverse Mercator
Datum NAD83
Units meters
Parameters
    scale factor: 0.9996
    central meridian: −114.0
    reference latitude: 42.0
    false easting: 2,500,000
    false northing: 1,200,000

1. Start ArcCatalog, and make connection to the Chapter 2 database. Right-click *idll.shp* in the Catalog tree and select Properties. The XY Coordinate System tab in the Shapefile Properties dialog shows the name to be unknown. Close the dialog.

2. First define the coordinate system for *idll.shp.* Click ArcToolbox window to open it in ArcCatalog. Right-click ArcToolbox and select Environments. In the Environment Settings dialog, select the Chapter 2 database for the current workspace. Double-click the Define Projection tool in the Data Management Tools/Projections and Transformations toolset. Select *idll.shp* for the input feature class. The dialog shows that *idll.shp* has an unknown coordinate system. Click the button for the coordinate system to open the Spatial Reference Properties dialog. Click Select. Double-click Geographic Coordinate Systems, North America, and NAD 1927 .prj. Click OK to dismiss the dialogs. Check the properties of *idll.shp* again. The XY Coordinate System tab should show GCS_ North_American_1927.

3. Next project *idll.shp* to the IDTM coordinate system. Double-click the Project tool in the Data Management Tools/Projections and Transformations/Feature toolset. In the Project dialog, select *idll.shp* for the input feature class, specify *idtm.shp* for the output feature class, and click the button for the output coordinate system to open the Spatial Reference Properties dialog. Click the New dropdown arrow and select Projected. In the New Projected Coordinate System dialog, first enter idtm for the Name. Then you need to provide projection information in the Projection frame and for the Geographic Coordinate System. In the Projection frame, select Transverse_Mercator from the Name dropdown list. Enter the following parameter values: 2500000 for False_Easting, 1200000 for False_Northing, −114 for Central_Meridian, 0.9996 for Scale_Factor, and 42 for Latitude_ Of_ Origin. Make sure that the Linear Unit is Meter. Click Select for the Geographic Coordinate System. Double-click North America, and NAD 1983.prj. Click Finish to dismiss the New Projected Coordinate System dialog. Click Save As in the Spatial Reference Properties dialog, and save the projection file as *idtm83.prj* in the Chapter 2 workspace. Dismiss the Spatial Reference Properties dialog.

4. A green dot appears next to Geographic Transformation in the Project dialog. This is because *idll.shp* is based on NAD27 and IDTM is based on NAD83. The green dot indicates that the projection requires a geographic transformation. Click Geographic Transformation's dropdown arrow and select NAD_1927_To_NAD_1983_NADCON. Click OK to run the command.

5. You can verify if *idll.shp* has been successfully projected to *idtm.shp* by checking the properties of *idtm.shp*.

Q1. Summarize in your own words the steps you have followed to complete Task 1.

Task 2 Import a Coordinate System

What you need: *stationsll.shp,* a shapefile measured in longitude and latitude values and in decimal degrees. *stationsll.shp* contains snow courses in Idaho.

In Task 2, you will complete the projection of *stationsll.shp* by importing the projection information on *idll.shp* and *idtm.shp* from Task 1.

1. Verify that *stationsll.shp* has an unknown geographic coordinate system by checking its properties. Double-click the Define Projection tool. Select *stationsll.shp* for the input feature class. Click the button for the coordinate system. Click Import in the Spatial Reference Properties dialog. Double-click *idll.shp* to add. Dismiss the dialogs.

Q2. Describe in your own words what you have done in Step 1.

2. Double-click the Project tool. Select *stationsll.shp* for the input feature class, specify *stationstm.shp* for the output feature class, and click the button for the output coordinate system. Click Import in the Spatial Reference Properties dialog. Double-click *idtm.shp* to add. Dismiss the Spatial Reference Properties dialog. Click the Geographic Transformation's dropdown arrow and select NAD_1927_To_NAD_ 1983_NADCON. Click OK to complete the operation. *stationstm.shp* is now projected onto the same (IDTM) coordinate system as *idtm.shp.*

Task 3 Project a Shapefile by Using a Predefined Coordinate System

What you need: *snow.txt,* a text file containing the geographic coordinates of 40 snow courses in Idaho.

In Task 3, you will first create an event layer from *snow.txt*. Then you will project the event layer, which is still measured in longitude and latitude values, to a predefined projected (UTM) coordinate system and save the output into a shapefile.

1. Launch ArcMap. Rename the new data frame Tasks 3&4 and add *snow.txt* to Tasks 3&4. (Notice that the table of contents is on the Source tab.) Right-click *snow.txt* and select Display XY Data. In the next dialog, make sure that *snow.txt* is the input table, longitude is the X field, and latitude is the Y field. The dialog shows that the spatial reference of the input coordinates is an unknown coordinate system. Click the Edit button to open the Spatial Reference Properties dialog. Click Select. Double-click Geographic Coordinate Systems, North America, and NAD 1983 .prj. Dismiss the dialogs, and click OK on the warning message stating that the table does not have Object-ID field.

2. *snow.txt Events* is added to ArcMap. You can now project *snow.txt Events* and save the output to a shapefile. Click ArcToolbox window to open it in ArcMap. Double-click the Project tool in the Data Management Tools/Projections and Transformations/ Feature toolset. Select *snow.txt Events* for the input dataset, and specify *snowutm83.shp* for the output feature class. Click the button for the output coordinate system. Click Select in the Spatial Reference Properties dialog. Double-click Projected Coordinate Systems, UTM, NAD 1983, and NAD 1983 UTM Zone 11N.prj. Click OK to project the data set.

Q3. You did not have to ask for a geographic transformation in Step 2. Why?

Task 4 Convert from One Coordinate System to Another

What you need: *idtm.shp* from Task 1 and *snowutm83.shp* from Task 3.

Task 4 first shows you how on-the-fly projection works in ArcMap and then asks you to convert *idtm.shp* from the IDTM coordinate system to the UTM coordinate system.

1. Right-click Tasks 3&4, and select Properties. The Coordinate System tab shows GCS_ North_American_1983 to be the current

coordinate system. ArcMap assigns the coordinate system of the first layer (i.e., *snow.txt Events*) to be the data frame's coordinate system. You can change it by clicking Import in the Data Frame Properties dialog. In the next dialog, double-click *snowutm83.shp*. Dismiss the dialogs. Now Tasks 3&4 is based on the NAD 1983 UTM Zone 11N coordinate system.

2. Add *idtm.shp* to Tasks 3&4. Although *idtm* is based on the IDTM coordinate system, it registers spatially with *snowutm83* in ArcMap. (A couple of snow courses are supposed to be outside the Idaho border.) ArcGIS can reproject a data set on-the-fly (Section 2.5.3). It uses the spatial reference information available to project *idtm* to the coordinate system of the data frame.

3. The rest of Task 4 is to project *idtm.shp* to the UTM coordinate system and to create a new shapefile. Double-click the Project tool. Select *idtm* for the input feature class, specify *idutm83.shp* for the output feature class, and click the button for the output coordinate system. Click Select in the Spatial Reference Properties dialog. Double-click Projected Coordinate Systems, UTM, NAD 1983, and NAD 1983 UTM Zone 11N.prj. Click OK to dismiss the dialogs.

Q4. Can you use Import instead of Select in step 3? If yes, how?

4. Although *idutm83* looks exactly the same as *idtm* in ArcMap, it has been projected to the UTM grid system.

Challenge Task

What you need: *idroads.shp* and *mtroads.shp*.

The Chapter 2 database includes *idroads.shp* and *mtroads.shp,* the road shapefiles for Idaho and Montana respectively. *idroads.shp* is projected onto the IDTM, but it has the wrong false easting (500,000) and false northing (100,000) values. *mtroads.shp* is projected onto the NAD 1983 State

Plane Montana FIPS 2500 coordinate system in meters, but it does not have a projection file.

1. Use the Project tool and the IDTM information from Task 1 to reproject *idroads .shp* with the correct false easting (2,500,000) and false northing (1,200,000) values, while keeping the other parameters the same. Name the output *idroads2.shp*.

2. Use the Define Projection tool to first define the coordinate system of *mtroads.shp*. Then use the Project tool to reproject *mtroads.shp* to the IDTM and name the output *mtroads_ idtm.shp*.

3. Verify that *idroads2.shp* and *mtroads_idtm .shp* have the same spatial reference information.

REFERENCES

Bosowski, E. F., and T. G. Feeman. 1997. The User of Scale Factors in Map Analysis: An Elementary Approach. *Cartographica* 34: 35–44.

Burkard, R. K. 1984. *Geodesy for the Layman.* Washington, DC: Defense Mapping Agency. Available at **http://www .ngs.noaa.gov/PUBS_LIB/ Geodesy4Layman/TR80003A .HTM#ZZ0/.**

Dent, B. D. 1999. *Cartography: Thematic Map Design,* 5th ed. New York: McGraw-Hill.

Iliffe, J. 2000. *Datums and Map Projections for Remote Sensing, GIS, and Surveying.* Boca Raton, FL: CRC Press.

Kavanagh, B. F. 2003. *Geomatics.* Upper Saddle River, NJ: Prentice Hall.

Kumar, M. 1993. World Geodetic System 1984: A Reference Frame for Global Mapping, Charting

and Geodetic Applications. *Surveying and Land Information Systems* 53: 53–56.

Maling, D. H. 1992. *Coordinate Systems and Map Projections,* 2d ed. Oxford, England: Pergamon Press.

Moffitt, F. H., and J. D. Bossler. 1998. *Surveying,* 10th ed. Menlo Park, CA: Addison-Wesley.

Muehrcke, P. C., J. O. Muehrcke, and A. J. Kimerling. 2001. *Map Use: Reading, Analysis, and Interpretation.* Madison, WI: JP Publishers.

National Geospatial-Intelligence Agency. 2000. *Department of Defense World Geodetic System 1984: Its Definition and Relationships with Local Geodetic Systems,* 3rd ed. NIMA TR8350.2. Amendment 1, January 3, 2000. Available at **http://earth-info.nga.mil/ GandG/.**

Robinson, A. H., J. L. Morrison, P. C. Muehrcke, A. J. Kimerling, and S. C. Guptill. 1995. *Elements of Cartography,* 6th ed. New York: Wiley.

Slocum, T. A., R. B. McMaster, F. C. Kessler, and H. H. Howard. 2005. *Thematic Cartography and Geographic Visualization,* 2d ed. Upper Saddle River, NJ: Prentice Hall.

Snay, R. A., and T. Soler. 2000. Modern Terrestrial Reference Systems. Part 2: The Evolution of NAD83. *Professional Surveyor,* February 2000.

Snyder, J. P. 1987. *Map Projections—A Working Manual.* Washington, DC: U.S. Geological Survey Professional Paper 1395.

Snyder, J. P. 1993. *Flattening the Earth: Two Thousand Years of Map Projections.* Chicago: University of Chicago Press.

CHAPTER 3

VECTOR DATA MODEL

CHAPTER OUTLINE

3.1 Representation of Simple Features
3.2 Topology
3.3 Georelational Data Model
3.4 Object-Based Data Model
3.5 Representation of Composite Features

Looking at a paper map, we can tell what map features are like and how they are spatially related to one another. For example, we can see in Figure 3.1 that Idaho is bordered by Montana, Wyoming, Utah, Nevada, Oregon, Washington, and Canada, and contains several Native American reservations. How can the computer "see" the same features and their spatial relationships? Chapter 3 attempts to answer the question from the perspective of vector data.

The **vector data model** prepares data in two basic steps so that the computer can process the data. First, it uses points and their *x*-, *y*-coordinates to represent spatial features as points, lines, and polygons

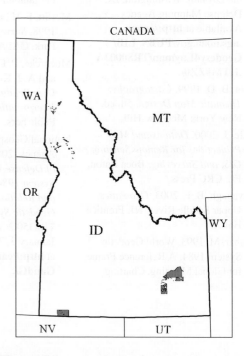

Figure 3.1
A reference map showing Idaho, lands held in trust by the United States for Native Americans in Idaho, and the surrounding states and country.

44

(areas). Second, it organizes geometric objects and their spatial relationships into digital data files that the computer can access, interpret, and process. Chapter 3 primarily deals with the second step.

The vector data model has undergone more changes over the past two decades than any other topics in GIS. As an example, Esri has introduced a new vector data model with each new software package: **coverage** with Arc/Info, **shapefile** with ArcView, and **geodatabase** with ArcGIS (Box 3.1). The coverage and shapefile are examples of the georelational data model, which uses a split system to store geometries and attributes. In contrast, the geodatabase is an example of the object-based data model, which stores geometries and attributes in a single system. The evolution of vector data models primarily reflects advances in computer technology and the competitive nature of the GIS market; but to GIS users, a new data model means accepting a whole new set of concepts, terms, and data file structures. Although some users are still using shapefiles, many have migrated to geodatabases to take advantage of object-oriented technology and new GIS functionalities developed by Esri.

Chapter 3 comprises the following five sections. Section 3.1 covers the representation of simple features as points, lines, and polygons. Section 3.2 explains the use of topology for expressing the spatial relationships in vector data and the importance of topology in GIS. Section 3.3 introduces the georelational data model, the coverage, and the shapefile. Section 3.4 introduces the object-based data model, the geodatabase, topology rules, and advantages of the geodatabase. Section 3.5 covers spatial features that are better represented as composites of points, lines, and polygons.

3.1 REPRESENTATION OF SIMPLE FEATURES

The vector data model uses the geometric objects of point, line, and polygon to represent simple spatial features (Figure 3.2). Dimensionality and property distinguish the three types of geometric objects as well as the features they represent.

A **point** has zero dimension and has only the property of location. A point feature is made of a point or a set of separate points. Wells, benchmarks, and gravel pits are examples of point features. A **line** is one-dimensional and has the property of length, in addition to location. A line has two end points and points in between to mark

---

Box 3.1	**Geodatabase, Shapefile, and Coverage for Esri Software**

The following table compares the terms used to describe geometries of the geodatabase, shapefile, and coverage.

Geodatabase	Shapefile	Coverage
Point	Point	Point
Multipoints	Multipoints	NA
Polyline	Polyline	Arc
Polygon	Polygon	Polygon
Polygon	Polygon	Region
Polyline with measure	Polyline with measure	Route
Feature class	Shapefile	Point, arc, polygon, tic
Feature dataset	NA	Coverage

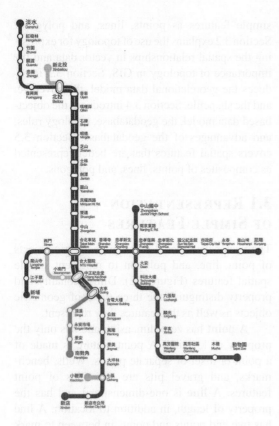

Figure 3.2
A subway map of Taipei, Taiwan.

the shape of the line. The shape of a line may be a smooth curve or a connection of straight-line segments. A polyline feature is made of lines. Roads, streams, and contour lines are examples of polyline features. A **polygon** is two-dimensional and has the properties of area (size) and perimeter, in addition to location. Made of connected, closed, nonintersecting line segments, a polygon may stand alone or share boundaries with other polygons. A polygon feature is made of a set of polygons. Examples of polygon features include timber stands, land parcels, and water bodies.

The representation of simple features using points, lines, and polygons is not always straightforward because it can depend on map scale. For example, a city on a 1:1,000,000 scale map may appear as a point, but the same city may appear

as a polygon on a 1:24,000 scale map. The representation of vector data can also depend on the criteria established by government mapping agencies (Robinson et al. 1995). The U.S. Geological Survey (USGS), for example, uses single lines to represent streams less than 40 feet wide on 1:24,000 scale topographic maps, and double lines (thus polygons) for larger streams.

3.2 TOPOLOGY

Vector data can be topological or nontopological, depending on whether topology is built into the data. The term **topology** refers to the study of those properties of geometric objects that remain invariant under certain transformations such as bending or stretching (Massey 1967). For example, a rubber band can be stretched and bent without losing its intrinsic property of being a closed circuit, as long as the transformation is within its elastic limits. An example of a topological map is a subway map (Figure 3.2).

To apply it to geospatial data, topology is often explained through **graph theory,** a subfield of mathematics that uses diagrams or graphs to study the arrangements of geometric objects and the relationships among objects (Wilson and Watkins 1990). Important to the vector data model are digraphs (directed graphs), which include points and directed lines. The directed lines are called **arcs,** and the points where arcs meet or intersect are called **nodes.** If an arc joins two nodes, the nodes are said to be *adjacent* and *incident* with the arc. Adjacency and incidence are two fundamental relationships that can be established between nodes and arcs in digraphs (Box 3.2).

3.2.1 TIGER

An early application of topology in geospatial technology is the Topologically Integrated Geographic Encoding and Referencing (TIGER) database from the U.S. Census Bureau (Broome and Meixler 1990). In the TIGER database, points are called 0-cells, lines 1-cells, and areas 2-cells (Figure 3.4). Each 1-cell in a TIGER file is a directed line, meaning that the line is directed

Box 3.2 | Adjacency and Incidence

If a line joins two points, the points are said to be *adjacent* and *incident* with the line, and the adjacency and incidence relationships can be expressed explicitly in matrices. Figure 3.3 shows an adjacency matrix and an incidence matrix for a digraph. The row and column numbers of the adjacency matrix correspond to the node numbers, and the numbers within the matrix refer to the number of arcs joining the corresponding nodes in the digraph. For example, 1 in (11,12) means one arc joint from node 11 to node 12, and 0 in (12,11) means no arc joint from node 12 to node 11. The direction of the arc determines whether 1 or 0 should be assigned.

The row numbers of the incidence matrix correspond to the node numbers in Figure 3.3, and the column numbers correspond to the arc numbers. The number 1 in the matrix means an arc is incident from a node, −1 means an arc is incident to a node, and 0 means an arc is not incident from or to a node. Take the example of arc 1. It is incident from node 13, incident to node 11, and not incident to all the other nodes. Thus, the matrices express the adjacency and incidence relationships mathematically.

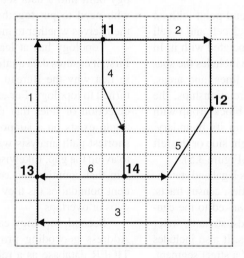

Adjacency matrix

	11	12	13	14
11	0	1	0	1
12	0	0	1	0
13	1	0	0	0
14	0	1	1	0

Incidence matrix

	1	2	3	4	5	6
11	−1	1	0	1	0	0
12	0	−1	1	0	−1	0
13	1	0	−1	0	0	−1
14	0	0	0	−1	1	1

Figure 3.3
The adjacency matrix and incidence matrix for a digraph.

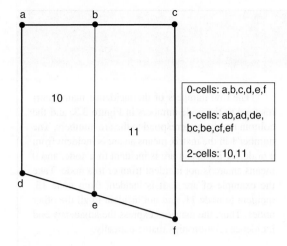

Figure 3.4
Topology in the TIGER database involves 0-cells or points, 1-cells or lines, and 2-cells or areas.

from a starting point toward an end point with an explicit left and right side. Each 2-cell and 0-cell has knowledge of the 1-cells associated with it. In other words, the TIGER database includes the spatial relationships among points, lines, and areas. Based on these built-in spatial relationships, the TIGER database can associate a block group with the streets or roads that make up its boundary. Thus, an address on either the right side or the left side of a street can be identified (Figure 3.5).

The TIGER database contains legal and statistical area boundaries such as counties, census tracts, and block groups, which can be linked to the census data, as well as roads, railroads, streams, water bodies, power lines, and pipelines. It also includes the address range on each side of a street segment. The Census Bureau updates the TIGER database regularly and makes current versions available for download at its website (**http://www.census.gov/).**

Besides the TIGER database, another well-known example of vector data with built-in topology is digital line graphs (DLGs) from the USGS. DLGs are digital representations of point, line, and area features from the USGS quadrangle maps, containing such data categories as contour lines, hydrography, boundaries, transportation, and the U.S. Public Land Survey System.

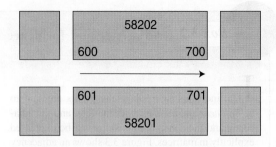

Figure 3.5
Address ranges and ZIP codes in the TIGER database have the right- or left-side designation based on the direction of the street.

3.2.2 Importance of Topology

Topology requires additional data files to store the spatial relationships. This naturally raises the question: What are the advantages of having topology built into a data set? Increasingly, GIS users who need to put together a database are asking themselves whether to include topology (Box 3.3).

Topology has at least two main advantages. First, it ensures data quality and integrity. This was in fact why the Bureau of the Census originally turned to topology. For example, topology enables detection of lines that do not meet correctly. If a gap exists on a supposedly continuous road, a shortest path analysis will take a circuitous route to avoid the gap. Likewise, topology can make certain that counties and census tracts share coincident boundaries (as they are supposed to) without gaps or overlaps.

Second, topology can enhance GIS analysis. Address geocoding providers typically use the TIGER database as a reference database because it not only has address ranges but also separates them according to the left or right side of the street (Chapter 16). The built-in topology in the TIGER database makes it possible to plot street addresses correctly. Other types of analyses can also benefit from using topological data. Analysis of traffic flow or stream flow is similar to address geocoding, because flow data are also directional (Regnauld and Mackaness 2006). Another example is wildlife habitat analysis involving edges between habitat types. If edges are coded with

Box 3.3 | **Topology or No Topology**

T he decision on topology depends on the GIS project. For some projects, topological functions are not necessary; for others they are a must. For example, a producer of GIS data will find it absolutely necessary to use topology for error checking and for ensuring that lines meet correctly and polygons are closed properly. Likewise, a GIS analyst working with transportation and utility networks will want to use topological data for data analysis.

Ordnance Survey (OS) is perhaps the first major GIS data producer to offer both topological and nontopological data to end users (Regnauld and Mackaness 2006). OS MasterMap is a framework for the referencing of geographic information in Great Britain (**http://www.ordnancesurvey.co.uk/ oswebsite/**). MasterMap has two types of polygon data: independent and topological polygon data. Independent polygon data duplicate the coordinate geometry shared between polygons. In contrast, topological polygon data include the coordinate geometry shared between polygons only once and reference each polygon by a set of line features. The referencing of a polygon by line features is similar to the polygon/arc list discussed in Section 3.3.2.

left and right polygons in a topology-based data set, specific habitat types (e.g., old growth and clear-cuts) along edges can easily be tabulated and analyzed (Chang, Verbyla, and Yeo 1995).

3.3 GEORELATIONAL DATA MODEL

The **georelational data model** stores geometries and attributes separately in a split system: geometries ("geo") in graphic files and attributes ("relational") in a relational database (Figure 3.6). Typically, a georelational data model uses the feature identification number (ID) to link the two components. The two components must be synchronized so that they can be queried, analyzed, and displayed in unison. The coverage and the shapefile, both Esri products, are examples of the georelational data model; however, the coverage is topological, and the shapefile is nontopological.

3.3.1 The Coverage

Esri introduced the coverage and its built-in topology in the 1980s to separate GIS from CAD (computer-aided design) at the time. AutoCAD by Autodesk was, and still is, the leading CAD package. A data format used by AutoCAD for transfer of data files

Graphic Files

| Polygon/arc list |
| Arc-coordinate list |
| Left/right list |
| : |

INFO File

Polygon-ID	Field 1	· · ·
1		
2		
3		

Figure 3.6
An example of the georelational data model, an ArcInfo coverage has two components: graphic files for spatial data and INFO files for attribute data. The label connects the two components.

is called DXF (drawing exchange format). DXF maintains data in separate layers and allows the user to draw each layer using different line symbols, colors, and text, but DXF files do not support topology.

The coverage supports three basic topological relationships (Environmental Systems Research Institute, Inc. 1998):

- **Connectivity:** Arcs connect to each other at nodes.

- **Area definition:** An area is defined by a series of connected arcs.
- **Contiguity:** Arcs have directions and left and right polygons.

Other than in the use of terms, these three topological relationships are similar to those in the TIGER database.

3.3.2 Coverage Data Structure

How are the topological relationships incorporated into the coverage's data structure? A point coverage is simple: It contains the feature IDs and pairs of *x*- and *y*-coordinates (Figure 3.7).

Figure 3.8 shows the data structure of a line coverage. The starting point of an arc is the from-node, and the end point is the to-node. The arc-node list sorts out the arc–node relationship. For example, arc 2 has 12 as the from-node and 13 as the to-node. The arc-coordinate list shows the *x*-, *y*-coordinates of the from-node, the to-node, and other points (called vertices) that make up each arc. For example, arc 3 consists of the from-node at (2, 9), the to-node at (4, 2), and two vertices

at (2, 6) and (4, 4). Arc 3 therefore has three line segments.

Figure 3.9 shows the data structure of a polygon coverage. The polygon/arc list shows the

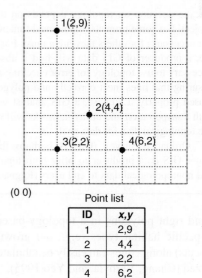

(0 0)

Point list

ID	x,y
1	2,9
2	4,4
3	2,2
4	6,2

Figure 3.7

The data structure of a point coverage.

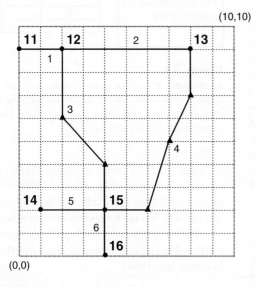

Figure 3.8

The data structure of a line coverage.

Arc-node list

Arc #	F-node	T-node
1	11	12
2	12	13
3	12	15
4	13	15
5	15	14
6	15	16

Arc-coordinate list

Arc #	x,y Coordinates
1	(0,9) (2,9)
2	(2,9) (8,9)
3	(2,9) (2,6) (4,4) (4,2)
4	(8,9) (8,7) (7,5) (6,2) (4,2)
5	(4,2) (1,2)
6	(4,2) (4,0)

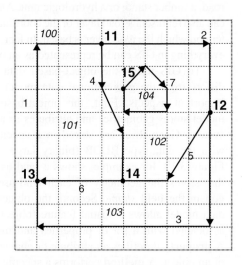

Left/right list

Arc #	L-poly	R-poly
1	100	101
2	100	102
3	100	103
4	102	101
5	103	102
6	103	101
7	102	104

Polygon-arc list

Polygon #	Arc #
101	1,4,6
102	4,2,5,0,7
103	6,5,3
104	7

Arc-coordinate list

Arc #	x,y Coordinates
1	(1,3) (1,9) (4,9)
2	(4,9) (9,9) (9,6)
3	(9,6) (9,1) (1,1) (1,3)
4	(4,9) (4,7) (5,5) (5,3)
5	(9,6) (7,3) (5,3)
6	(5,3) (1,3)
7	(5,7) (6,8) (7,7) (7,6) (5,6) (5,7)

Figure 3.9
The data structure of a polygon coverage.

relationship between polygons and arcs. For example, arcs 1, 4, and 6 connect to define polygon 101. Polygon 104 differs from the other polygons because it is surrounded by polygon 102. To show that polygon 104 is a hole within polygon 102, the arc list for polygon 102 contains a zero to separate the external and internal boundaries. Polygon 104 is also an isolated polygon consisting of only one arc (7). Therefore, a node (15) is placed along the arc to be the beginning and end node. Outside the mapped area, polygon 100 is the external or universe polygon. The left/right list in Figure 3.9 shows the relationship between arcs and their left and right polygons. For example, arc 1 is a directed line from node 13 to node 11 and has polygon 100 on the left and polygon 101 on the right. The arc-coordinate list in Figure 3.9 shows the nodes and vertices that make up each arc.

Lists such as the polygon/arc list are stored as graphic files in a coverage folder. Another folder, called INFO, which is shared by all coverages in the same workspace, stores attribute data files. Though more than two decades old, the coverage is still a standard topological vector data format for the Esri software. The graphic files such as the arc-coordinate list, the arc-node list, and the polygon-arc list are efficient in reducing data redundancy. A shared or common boundary between two polygons is stored in the arc-coordinate list once, not twice. This not only reduces the number of data entries but also makes it easier to update the polygons. For example, if arc 4 in Figure 3.9 is changed to a straight line between two nodes, only the coordinate list for arc 4 needs to be changed.

3.3.3 Nontopological Vector Data

In less than one decade after GIS companies introduced topology to separate GIS from CAD, the same companies adopted nontopological data format as a standard nonproprietary data format.

The shapefile is a standard nontopological data format used in Esri products. Although the shapefile treats a point as a pair of *x*-, *y*-coordinates, a line as a series of points, and a polygon as a series of line segments, no files describe the spatial relationships among these geometric objects. Shapefile polygons actually have duplicate arcs for the shared boundaries and can overlap one another. The geometry of a shapefile is stored in two basic files: The *.shp* file stores the feature geometry, and the *.shx* file maintains the spatial index of the feature geometry.

Nontopological data such as shapefiles have two main advantages. First, they can display more rapidly on the computer monitor than topology-based data (Theobald 2001). This advantage is particularly important for people who use, rather than produce, GIS data. Second, they are nonproprietary and interoperable, meaning that they can be used across different software packages (e.g., MapInfo can use shapefiles, and ArcGIS can use MapInfo Interchange Format files). GIS users have pushed for interoperability since the early 1990s. This push resulted in the establishment of Open GIS Consortium, Inc. (now Open Geospatial Consortium, Inc.), a nonprofit international voluntary consensus standards organization in 1994 (**http://www.opengeospatial .org/).** Interoperability was a primary mission of Open GIS Consortium, Inc. from the start. The introduction of nontopological data format in the early 1990s was perhaps a direct response to the call for interoperability.

3.4 OBJECT-BASED DATA MODEL

The latest entry in vector data models, the **object-based data model** treats geospatial data as objects. An **object** can represent a spatial feature such as a road, a timber stand, or a hydrologic unit. An object can also represent a road layer or the coordinate system on which the road layer is based. In fact, almost everything in GIS can be represented as an object.

To GIS users, the object-based data model differs from the georelational data model in two important aspects. First, the object-based data model stores geometries and attributes in a single system. Geometries are stored in a special field with the data type BLOB (binary large object). Figure 3.10, for example, shows a land-use layer that stores the geometry of each land-use polygon in the field *shape*. Second, the object-based data model allows a spatial feature (object) to be associated with a set of properties and methods. A **property** describes an attribute or characteristic of an object. A **method** performs a specific action. Therefore, as a feature layer object, a road layer can have the properties of shape and extent and can also have the methods of copy and delete. Properties and methods directly impact how GIS operations are performed. Work in an object-based GIS is in fact dictated by the properties and methods that have been defined for the objects in the GIS.

3.4.1 Classes and Class Relationships

A **class** is a set of objects with similar characteristics. A GIS package such as ArcGIS Desktop uses thousands of classes. To make it possible for software developers to systematically organize classes and their properties and methods, object-oriented

Objectid	Shape	Landuse_ID	Category	Shape_Length	Shape_Area
1	Polygon	1	5	14,607.7	5,959,800
2	Polygon	2	8	16,979.3	5,421,216
3	Polygon	3	5	42,654.2	21,021,728

Figure 3.10
The object-based data model stores each land-use polygon in a record. The Shape field stores the geometries of land-use polygons. Other fields store attribute data such as Landuse_ID and Category.

technology allows relationships such as association, aggregation, composition, type inheritance, and instantiation to be established between classes (Zeiler 2001; Larman 2001):

- *Association* defines how many instances of one class can be associated with another class through multiplicity expressions at both ends of the relationship. Common multiplicity expressions are 1 (default) and 1 or more (1..*). For example, an address is associated with one ZIP code, but the same address can be associated with one or more apartments.
- *Aggregation* describes the whole–part relationship between classes. Aggregation is a type of association except that the multiplicity at the composite ("whole") end is typically 1 and the multiplicity at the other ("part") end is 0 or any positive integer. For example, a census tract is an aggregate of a number of census blocks.
- *Composition* describes a type of association in which the parts cannot exist independently from the whole. For example, roadside rest areas along a highway cannot exist without the highway.
- *Type inheritance* defines the relationship between a superclass and a subclass. A subclass is a member of a superclass and inherits the properties and methods of the superclass, but a subclass can have additional properties and methods to separate itself from other members of the superclass. For example, residential area is a member of built-up area, but it can have properties such as lot size that separate residential area from commercial or industrial built-up area.
- *Instantiation* means that an object of a class can be created from an object of another class. For example, a high-density residential area object can be created from a residential area object.

3.4.2 Interface

An **interface** represents a set of externally visible operations of a class or object. Object-based technology uses a mechanism called **encapsulation** to

hide the properties and methods of an object so that the object can be accessed only through the predefined interfaces (Figure 3.11).

Figure 3.12 shows how two interfaces can be used to derive the area extent of a feature layer, which is a type of *Geodataset*. First, the *Extent* property is accessed via the *IGeodataset* interface that a *Geodataset* object, a feature layer in this case, supports. The *Extent* property returns an *Envelope* object,

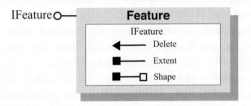

Figure 3.11

A *Feature* object implements the *IFeature* interface. *IFeature* has access to the properties of *Extent* and *Shape* and the method of *Delete*. Object-oriented technology uses symbols to represent interface, property, and method. The symbols for the two properties are different in this case because *Extent* is a read-only property, whereas *Shape* is a read-and-write (by reference) property.

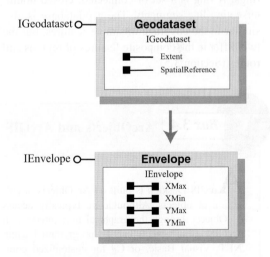

Figure 3.12

A *Geodataset* object supports *IGeodataset*, and an *Envelope* object supports *IEnvelope*. See the text for an explanation of how to use the interfaces to derive the area extent of a feature layer.

which implements the *IEnvelope* interface. The properties *XMin, XMax, YMin,* and *YMax* are then accessed on the interface to derive the area extent.

3.4.3 The Geodatabase

The geodatabase, an example of the object-based vector data model, is part of **ArcObjects** developed by Esri as the foundation for ArcGIS Desktop (Zeiler 2001; Ungerer and Goodchild 2002). ArcObjects consists of thousands of objects and classes. Most ArcGIS users do not have to deal with ArcObjects directly, because menus, icons, and dialogs have already been developed by Esri to access objects in ArcObjects and their properties and methods. Box 3.4 describes situations in which ArcObjects may be encountered while working with routine operations in ArcGIS.

Like the shapefile, the geodatabase uses points, polylines, and polygons to represent vector-based spatial features (Zeiler 1999). A point feature may be a simple feature with a point or a multipoint feature with a set of points. A polyline feature is a set of line segments that may or may not be connected. A polygon feature is made of one or many rings. A ring is a set of connected, closed nonintersecting line segments. The geodatabase is also similar to the coverage in simple features, but the two differ in the composite features of regions and routes (Section 3.5).

The geodatabase organizes vector data sets into feature classes and feature datasets (Figure 3.13). A **feature class** stores spatial features of the same geometry type. A **feature dataset** stores feature classes that share the same coordinate system and area extent. For example, a feature class may represent block groups, and a feature dataset may consist of block groups, census tracts, and counties for the same study area. Feature classes in a feature dataset often participate in topological relationships with one another, such as coincident boundaries between different levels of census units. If a feature class resides in a geodatabase but is not part of a feature dataset, it is called a

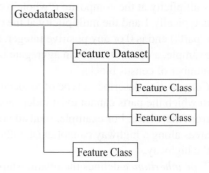

Figure 3.13
In a geodatabase, feature classes can be standalone feature classes or members of a feature dataset.

ArcGIS Desktop is built on ArcObjects, a collection of objects. Although we typically access ArcObjects through the graphical user interface in ArcGIS, these objects can be programmed using .NET, Visual Basic, or C# for customized commands, menus, and tools. Starting in ArcGIS 10.0, both ArcCatalog and ArcMap have the Python window that can run Python scripts. Python is a general-purpose high-level programming language and, in the case of ArcGIS, it is used as an extension language to provide a programmable interface for modules, or blocks of code, written with ArcObjects. Some dialogs in ArcMap have the advanced option that allows the user to enter Python scripts. For example, one can use the advanced option in the field calculator dialog to change the calculation expression. (Task 5 of the applications section of Chapter 8 uses this option.)

standalone feature class. Besides feature classes, a geodatabase can also store raster data, triangulated irregular networks (TINs, Section 3.5.1), location data, and attribute tables.

A geodatabase can be designed for single or multiple users. A single-user database can be a personal geodatabase or a file geodatabase. A personal geodatabase stores data as tables in a Microsoft Access database with the *mdb* extension. A file geodatabase, on the other hand, stores data in many small-sized files in a folder with the *gdb* extension. Unlike the personal geodatabase, the file geodatabase has no overall database size limit (as opposed to a 2-GB limit for the personal geodatabase) and can work across platforms (e.g., Windows as well as Linux). Esri also claims that, owing to its many small-sized files, the file geodatabase can provide better performance than the personal geodatabase in data access. A multiuser or ArcSDE geodatabase stores data in a database management system such as Oracle, Microsoft SQL Server, IBM DB2, or Informix. Chapter 3 deals mainly with single-user geodatabases.

3.4.4 Topology Rules

The geodatabase defines topology as relationship rules and lets the user choose the rules, if any, to be implemented in a feature dataset. In other words, the geodatabase offers on-the-fly topology. Table 3.1 shows over 25 topology rules by feature type. Some rules apply to features within a feature class, whereas others apply to two or more participating feature classes. Rules applied to the geometry of a feature class are functionally similar to the built-in topology for the coverage, but rules applied to two or more feature classes are new with the geodatabase.

The following are some real-world applications of topology rules:

- Counties must not overlap.
- County must not have gaps.
- County boundary must not have dangles (i.e., must be closed).
- Census tracts and counties must cover each other.
- Voting district must be covered by county.
- Contour lines must not intersect.
- Interstate route must be covered by feature class of reference line (i.e., road feature class).
- Milepost markers must be covered by reference line (i.e., road feature class).
- Label points must be properly inside polygons.

Some rules in the list such as no gap, no overlap, and no dangles are general in nature and can probably apply to many polygon feature classes. Some, such as the relationship between milepost markers and reference line, are specific to transportation applications. Examples of topology rules

TABLE 3.1	Topology Rules in the Geodatabase Data Model
Feature Type	**Rule**
Polygon	Must be larger than cluster tolerance, must not overlap, must not have gaps, must not overlap with, must be covered by feature class of, must cover each other, must be covered by, boundary must be covered by, area boundary must be covered by boundary of, and contains point
Line	Must be larger than cluster tolerance, must not overlap, must not intersect, must not have dangles, must not have pseudo-nodes, must not intersect or touch interior, must not overlap with, must be covered by feature class of, must be covered by boundary of, end point must be covered by, must not self-overlap, must not self-intersect, and must be single part
Point	Must be covered by boundary of, must be properly inside polygons, must be covered by end point of, and must be covered by line

that have been proposed for data models from different disciplines are available at the Esri website (**http://support.esri.com/datamodels**).

3.4.5 Advantages of the Geodatabase

ArcGIS can use coverages, shapefiles, and geodatabases. It can also export or import from one data format into another. A recent study shows that in a single-user environment coverages actually perform better than shapefiles and geodatabases for some spatial data handling (Batcheller, Gittings, and Dowers 2007). The question is then: What advantages are to be gained by migrating to the geodatabase? The following summarizes several advantages of the geodatabase:

First, the hierarchical structure of a geodatabase is useful for data organization and management (Gustavsson, Seijmonsbergen, and Kolstrup 2007). For example, if a project involves two study areas, two feature datasets can be used to store feature classes for each study area. This simplifies data management operations such as copy and delete (e.g., copy a feature dataset including its feature classes instead of copying individual feature classes). Moreover, any new data created through data query and analysis in the project will automatically be defined with the same coordinate system as the feature dataset, thus saving the time required to define the coordinate system of each new feature class. Recent trends suggest

that government agencies are taking advantage of this hierarchical structure of the geodatabase for data delivery. The National Hydrography Dataset (NHD) program, for example, distributes data in two feature datasets, one for hydrography and the other for hydrologic units (Box 3.5) (**http:// nhd.usgs.gov/data.html**). The program claims that the geodatabase is better than the coverage for Web-based data access and query and for data downloading. The New York State Department of Transportation also offers transportation data in geodatabase format (e.g., New York State Thruway Route System, **http://www.nysgis.state .ny.us/gisdata/**).

Second, the geodatabase, which is part of ArcObjects, can take advantage of object-oriented technology. For example, ArcGIS Desktop provides four general validation rules: attribute domains, relationship rules, connectivity rules, and custom rules (Zeiler 1999). Attribute domains group objects into subtypes by a valid range of values or a valid set of values for an attribute. (Task 1 in Chapter 8 uses an attribute domain for attribute data entry.) Relationship rules such as topology rules organize objects that are associated. Connectivity rules let users build geometric networks such as streams, roads, and water and electric utilities. (Chapter 17 deals with connectivity in the case of a street network.) Custom rules allow users to create custom features for advanced applications. (Task 2 in Chapter 9 uses

Box 3.5   NHDinGEO

The National Hydrography Data set (NHD) program uses the acronym NHDinGEO for their data in geodatabases. A sample NHD geodatabase includes two feature datasets and a number of attribute tables. The Hydrography feature dataset has feature classes such as NHDFlowline, NHDWaterbody, and NHDPoint for stream reach applications. The

Hydrologic Units feature dataset consists of basin, region, subbasin, subregion, subwatershed, and watershed classes. The NHD program has replaced the coverage (called NHDinARC) with the geodatabase. NHDinARC used to have regions and route subclasses to store some of the same feature classes in NHDinGEO.

cartographic representation rules for symbology.) Not available for shapefiles or coverages, these validation rules are useful for specific applications. Further developments based on object-based technology can be expected.

Third, the geodatabase offers on-the-fly topology, applicable to features within a feature class or between two or more participating feature classes. As discussed in Section 3.2.2, topology can ensure data integrity and can enhance certain types of data analyses. (As an example, Chapter 7 discusses how to use topological rules for correcting digitizing errors.) On-the-fly topology offers the choices to the users and lets them decide which topology rules, if any, are needed for their projects.

Fourth, thousands of objects, properties, and methods in ArcObjects are available for GIS users to develop customized applications (Burke 2003; Chang 2007). Customized applications can reduce the amount of repetitive work (e.g., define and project the coordinate system of each data set in a project), streamline the workflow (e.g., combine defining and projecting coordinate systems into one step), and even produce functionalities that are not easily available in ArcGIS.

Finally, ArcObjects provides a template for custom objects to be developed for different industries and applications. Real-world objects all have different properties and behaviors. It is therefore impossible to apply, for example, the properties and methods of transportation-related objects to forestry-related objects. As of November 2010, 33 industry-specific data models had been posted at the Esri website (**http://support.esri.com/datamodels**).

3.5 REPRESENTATION OF COMPOSITE FEATURES

Composite features refer to those spatial features that are better represented as composites of points, lines, and polygons for their applications. The following sections present TINs, regions, and routes as examples of composite features. They also discuss how regions and routes are handled differently in coverages, shapefiles, and geodatabases.

3.5.1 TINs

A triangulated irregular network (TIN) approximates the terrain with a set of nonoverlapping triangles (Figure 3.14). Each triangle in the TIN assumes a constant gradient. Flat areas of the land surface have fewer but larger triangles, whereas areas with higher variability in elevation have denser but smaller triangles. The TIN is commonly used for terrain mapping and analysis, especially for 3-D display (Chapter 13).

The inputs to a TIN include point, line, and polygon features. An initial TIN can be constructed from elevation points and contour lines. Its approximation of the surface can then be improved by incorporating line features such as streams, ridge lines, and roads and by polygon features such as lakes and reservoirs. A finished TIN comprises three types of geometric objects: triangles (faces), points (nodes), and lines (edges). Its data structure therefore includes the triangle number, the number of each adjacent triangle, and data files showing the lists of points, edges, as well as the x, y, and z values of each elevation point (Figure 3.15).

The coverage, the shapefile, and the geodatabase use the same method to create and modify TINs. ArcGIS 9.2 introduced a terrain data format, which can store elevation points along with line and polygon feature classes in a feature dataset.

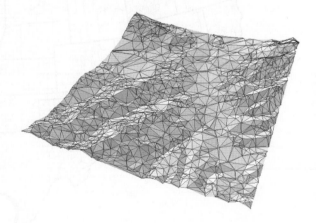

Figure 3.14
A TIN uses a series of nonoverlapping triangles to approximate the terrain.

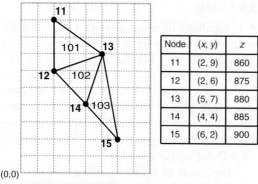

Node	(x, y)	z
11	(2, 9)	860
12	(2, 6)	875
13	(5, 7)	880
14	(4, 4)	885
15	(6, 2)	900

(0,0)

Triangle	Node list	Neighbors
101	11, 13, 12	--, 102, --
102	13, 14, 12	103, --, 101
103	13, 15, 14	--, --, 102

Figure 3.15
The data structure of a TIN.

Using the feature dataset and its contents, the user can construct a TIN on the fly. The terrain data format eases the task of putting together a TIN but does not change the basic data structure of the TIN.

3.5.2 Regions

A **region** is a geographic area with similar characteristics, and hierarchical regions can be formed by dividing the Earth's surface into progressively smaller uniform regions. The concept of regions is well established in geography, landscape ecology, and forestry (Berry 1968; Bailey 1983; Forman and Godron 1986; Cleland et al. 1997). Well-known examples of hierarchical regions include census units (Figure 3.16), hydrologic units, and ecological units.

A data model for regions must be able to handle two spatial characteristics: a region may have spatially joint or disjoint areas, and regions may overlap

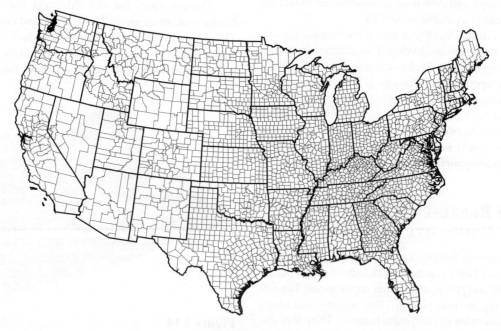

Figure 3.16
A hierarchy of counties and states in the conterminous United States.

or cover the same area (Figure 3.17). The simple polygon coverage cannot handle either characteristic; therefore, the coverage organizes regions as subclasses in a polygon coverage and, through additional data files, relates regions to the underlying polygons and arcs. Figure 3.18 shows the file structure for a regions subclass with two regions, four polygons, and five arcs. The region-polygon list relates the regions to the polygons. Region 101 consists of polygons 11 and 12. Region 102 has two components: one includes spatially joint polygons 12 and 13, and the other spatially disjoint polygon 14. Region 101 overlaps region 102 in polygon 12. The region-arc

list links the regions to the arcs. Region 101 has only one ring connecting arcs 1 and 2. Region 102 has two rings, one connecting arcs 3 and 4 and the other consisting of arc 5.

Because regions subclasses can be built on existing polygons and arcs, many government agencies have used them to create and store additional data layers for distribution. For example, the Clearwater National Forest uses regions subclasses to store historical fire perimeters, one subclass per year.

Neither the shapefile nor the geodatabase supports regions subclasses in its data structure, but both data formats support multipart polygons. Multipart polygons can have spatially joint or disjoint parts and can overlap each other. Therefore, multipart polygons can represent regions-like spatial features. For some GIS operations such as overlay, multipart polygons can actually simplify data records and management. (An example is included in Task 2 of the applications section of Chapter 11.)

3.5.3 Routes

A **route** is a linear feature such as a highway, a bike path, or a stream, but unlike other linear features, a route has a measurement system that allows linear measures to be used on a projected coordinate

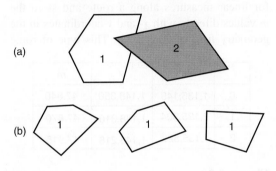

(a)

(b)

Figure 3.17
The regions subclass allows overlapped regions (*a*) and spatially disjoint polygons in regions (*b*).

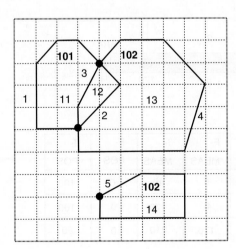

Figure 3.18
The data structure of a region subclass.

Region-polygon list

Region #	Polygon #
101	11
101	12
102	12
102	13
102	14

Region-arc list

Region #	Ring #	Arc #
101	1	1
101	1	2
102	1	3
102	1	4
102	2	5

system. Transportation agencies normally use linear measures from known points such as the beginning of a highway, a milepost, or a road intersection to locate accidents, bridges, and pavement conditions along roads. Natural resource agencies also use linear measures to record water quality data and fishery conditions along streams. These linear attributes, called **events,** must be associated with routes so that they can be displayed and analyzed with other spatial features.

Routes are stored as subclasses in a line coverage, similar to region subclasses in a polygon coverage. A route subclass is a collection of sections. A **section** refers directly to lines (i.e., arcs) in a line coverage and positions along lines. Because lines are a series of x-, y-coordinates based on a coordinate system, this means that a section is also measured in coordinates and its length can be derived from its reference lines. Figure 3.19 shows a route (Route-ID = 1) in a thick shaded line that is built on a line coverage. The route has three sections, and the section table relates them to the arcs in the line coverage. Section 1 (Section-ID = 1) covers the entire length of arc 7; therefore, it has a from-position (F-POS) of 0 percent and a to-position (T-POS) of 100 percent. Section 1 also has a from-measure (F-MEAS) of 0 (the beginning point of the route) and a to-measure (T-MEAS) of 40 units

measured from the line coverage. Section 2 covers the entire length of arc 8 over a distance of 130 units. Its from-measure and to-measure continue from section 1. Section 3 covers 80 percent of arc 9; thus, it has a to-position of 80 percent and a to-measure that is its from-measure plus 80 percent of the length of arc 9 (80% of 50, or 40, units). Combining the three sections, the route has a total length of 210 units (40 + 130 + 40).

Both the shapefile and the geodatabase use polylines with m (measure) values to replace route subclasses for GIS applications. Instead of working through sections and arcs, they use m values for linear measures along a route and store the m values directly with x- and y-coordinates in the geometry field (Figure 3.20). This type of route

	x	y	m
0	1,135,149	1,148,350	47.840
1	1,135,304	1,148,310	47.870
2	1,135,522	1,148,218	47.915

Figure 3.20
The linear measures (m) of a route are stored with x- and y-coordinates in a geodatabase. In this example, the m values are in miles, whereas the x- and y-coordinates are in feet.

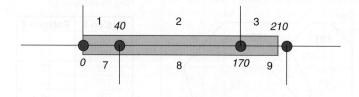

Route-ID	Section-ID	Arc-ID	F-MEAS	T-MEAS	F-POS	T-POS
1	1	7	0	40	0	100
1	2	8	40	170	0	100
1	3	9	170	210	0	80

Figure 3.19
The data structure of a route subclass.

object has been called *route dynamic location object* (Sutton and Wyman 2000). Figure 3.21 shows an example of a route in a geodatabase. The measure field directly records 0, 40, 170, and 210 along the route. These measures are based on a predetermined starting point, the end point on the left in this case.

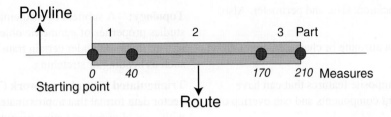

Figure 3.21
A route, shown here as a thicker, gray line, is built on a polyline with linear measures in a geodatabase.

KEY CONCEPTS AND TERMS

Arc: A line connected to two end points.

ArcObjects: A collection of objects used by ArcGIS Desktop.

Area definition: A topological relationship used in Esri's coverage data format that stipulates that an area is defined by a series of connected arcs.

Class: A set of objects with similar characteristics.

Connectivity: A topological relationship used in Esri's coverage data format that stipulates that arcs connect to each other at nodes.

Contiguity: A topological relationship used in Esri's coverage data format that stipulates that arcs have directions and left and right polygons.

Coverage: A topological vector data format used in Esri products.

Encapsulation: A principle used in object-oriented technology to hide the properties and methods of an object so that the object can be accessed only through the predefined interfaces.

Event: An attribute that can be associated and displayed with a route.

Feature class: A data set that stores features of the same geometry type in a geodatabase.

Feature dataset: A collection of feature classes in a geodatabase that share the same coordinate system and area extent.

Geodatabase: An object-based vector data model developed by Esri.

Georelational data model: A GIS data model that stores geometries and attributes in two separate but related file systems.

Graph theory: A subfield of mathematics that uses diagrams or graphs to study the arrangements of objects and the relationships among objects.

Interface: A set of externally visible operations of an object.

Line: A spatial feature that is represented by a series of points and has the geometric properties of location and length. Also called *arc* or *edge*.

Method: A specific action that an object can perform.

Node: The beginning or end point of a line.

Object: An entity such as a feature layer that has a set of properties and methods.

Object-based data model: A vector data model that uses objects to organize spatial data.

Point: A spatial feature that is represented by a pair of coordinates and has only the geometric property of location. Also called *node*.

Polygon: A spatial feature that is represented by a series of lines and has the geometric properties of location, size, and perimeter. Also called *area*.

Property: An attribute or characteristic of an object.

Regions: Composite features that can have spatially disjoint components and can overlap one another.

Route: A linear feature that allows linear measures to be used on a projected coordinate system.

Section: A part of a route that refers directly to the underlying arcs and positions along arcs in a coverage.

Shapefile: A nontopological vector data format used in Esri products.

Topology: A subfield of mathematics that studies properties of geometric objects that remain invariant under certain transformations such as bending or stretching.

Triangulated irregular network (TIN): A vector data format that approximates the terrain with a set of nonoverlapping triangles.

Vector data model: A data model that uses points and their x-, y-coordinates to construct spatial features.

REVIEW QUESTIONS

1. Find the GIS data clearinghouse for your state at the Geospatial One-stop website (**http://www.geodata.gov/**). Go to the clearinghouse website. What data format(s) does the website use for delivering vector data?

2. Name the three types of simple features used in GIS and their geometric properties.

3. Draw a stream coverage and show how the topological relationships of connectivity and contiguity can be applied to the coverage.

4. How many arcs connect at node 12 in Figure 3.8?

5. Suppose an arc (arc 8) is added to Figure 3.9 from node 13 to node 11. Write the polygon/ arc list for the new polygons and the left/ right list for arc 8.

6. Explain the importance of topology in GIS.

7. What are the main advantages of using shapefiles?

8. Explain the difference between the georelational data model and the object-based data model.

9. Describe the difference between the geodatabase and the coverage in terms of the geometric representation of spatial features.

10. Explain the relationship between the geodatabase, feature dataset, and feature class.

11. Feature dataset is useful for data management. Can you think of an example in which you would want to organize data by feature dataset?

12. Explain the difference between a personal geodatabase and a file geodatabase.

13. What is ArcObjects?

14. Provide an example of an object from your discipline and suggest the kinds of properties and methods that the object can have.

15. What is an interface?

16. Table 3.1 shows "must not overlap" as a topology rule for polygon features. Provide an example from your discipline that can benefit from enforcement of this topology rule.

17. "Must not intersect" is a topology rule for line features. Provide an example from your

discipline that can benefit from enforcement of this topology rule.

18. The text covers several advantages of adopting the geodatabase. Can you think of an example in which you would prefer the geodatabase to the coverage for a GIS project?

19. Compare Figure 3.19 with Figure 3.21, and explain the difference between the geodatabase and the coverage in handling the route data structure.

20. Draw a small TIN to illustrate that it is a composite of simple features.

APPLICATIONS: VECTOR DATA MODEL

This applications section consists of six tasks. In Task 1, you will convert a coverage into a shape-file and examine the data structure of the coverage and the shapefile. In Task 2, you will work with the basic elements of the file geodatabase. Task 3 shows how you can update the area and perimeter values of a polygon shapefile by converting it to a personal geodatabase feature class. In Task 4, you will view routes in the form of polylines with *m* values. In Task 5, you will view regions and route subclasses that reside in a hydrography coverage. Task 6 lets you view a TIN in ArcCatalog and ArcMap.

Task 1   Examine the Data File Structure of Coverage and Shapefile

What you need: *land*, a coverage.

In Task 1, you will view data layers (feature classes) associated with a coverage in ArcCatalog and examine its data structure using Windows Explorer. Then, you will convert the coverage into a shapefile and examine the shapefile's data structure.

1. Start ArcCatalog, and access the Chapter 3 database. Click the plus sign to expand the coverage *land* in the Catalog tree. The coverage contains four feature classes: *arc, label, polygon,* and *tic*. On the Preview tab, you can preview each feature class by first highlighting it in the Catalog tree. *arc* shows lines (arcs); *label*, the label points, one for each polygon; *polygon*, polygons; and *tic*, the tics or control points in *land*. Notice that the symbols for the four feature classes correspond to the feature type.

2. Right-click *land* in the Catalog tree and select Properties. The Coverage Properties dialog has two tabs: General, and Projection and Extent. The General tab shows the presence of topology for the polygon feature class. The Projection and Extent tab shows an unknown coordinate system and the area extent of the coverage.

3. Right-click *land polygon* and select Properties. The Coverage Feature Class Properties dialog has the General and Items tabs. The General tab shows 76 polygons. The Items tab describes the items or attributes in the attribute table.

4. Data files associated with *land* reside in two folders in the Chapter 3 database: land and info. You can use the Windows Explorer to view these files. The land folder contains arc data files (.adf). Some of these graphic files are recognizable by name, such as arc. adf for the arc-coordinate list and pal.adf for the polygon/arc list. The info folder, which is shared by other coverages in the same workspace, contains attribute data files of arcxxxx.dat and arcxxxx.nit. All files in both folders are binary files and cannot be read.

5. This step converts *land* to a polygon shapefile. Click ArcToolbox window to open it. There are at least two options for the conversion. First, you can use the Feature Class to Shapefile (multiple) tool in the Conversion Tools/To Shapefile toolset. The tool converts a coverage's feature classes to shapefiles. Second, you can use the Export function in a data set's context menu. Here you will use the second option. Right-click *land polygon*, point to Export, and select To Shapefile (single). In the next dialog, select the Chapter 3 database for the output location, and enter *land_polygon* for the output feature class name. Click OK. This conversion operation creates *land_polygon.shp* and adds the shapefile to the Catalog tree.

6. Right-click *land_polygon.shp* in the Catalog tree and select Properties. The Shapefile Properties dialog has the General, XY Coordinate System, Fields, and Indexes tabs. The XY Coordinate System tab shows an unknown coordinate system. The Fields tab describes the fields or attributes in the shapefile. The Indexes tab shows that the shapefile has a spatial index, which can increase the speed of drawing and data query.

7. The *land_polygon* shapefile is associated with a number of data files in the Chapter 3 database. Among these files, *land_polygon .shp* is the shape (geometry) file, *land_ polygon.dbf* is an attribute data file in dBASE format, and *land_polygon.shx* is the spatial index file.

Q1. Describe in your own words the difference between a coverage and a shapefile in terms of data structure.

Q2. The coverage data format uses a split system to store geometries and attributes. Use *land* as an example and name the two systems.

Task 2 Create File Geodatabase, Feature Dataset, and Feature Class

What you need: *elevzone.shp* and *stream.shp*, two shapefiles that have the same coordinate system and extent.

In Task 2, you will first create a file geodatabase and a feature dataset. You will then import the shapefiles into the feature dataset as feature classes and examine their data file structure. The name of a feature class in a geodatabase must be

unique. In other words, you cannot use the same name for both a standalone feature class and a feature class in a feature dataset.

1. Make sure that ArcCatalog is connected to the Chapter 3 database. This step creates a file geodatabase. Right-click the Chapter 3 database in the Catalog tree, point to New, and select File Geodatabase. Rename the new file geodatabase *Task2.gdb*.

2. Next, create a new feature dataset. Right-click *Task2.gdb*, point to New, and select Feature Dataset. In the next dialog, enter *Area_1* for the name (connect Area and 1 with an underscore; no space is allowed). Click Next. In the next dialog, select in sequence Projected Coordinate Systems, UTM, NAD 1927, and NAD 1927 UTM Zone 11N and click Next. Choose None in the next dialog and click Next. Accept the defaults on the tolerances and click Finish.

3. *Area_1* should now appear in *Task2.gdb*. Right-click *Area_1*, point to Import, and select Feature Class (multiple). Use the browse button or the drag-and-drop method to select *elevzone.shp* and *stream.shp* for the input features. Make sure that the output geodatabase points to *Area_1*. Click OK to run the import operation.

4. Right-click *Task2.gdb* in the Catalog tree and select Properties. The Database Properties dialog has the General and Domains tabs. A domain is a validation rule that can be used to establish valid values or a valid range of values for an attribute to minimize data entry errors.

5. Right-click *elevzone* and select Properties. The Feature Class Properties dialog has 10 tabs. Although some of these tabs such as Fields, Indexes, and XY Coordinate System are similar to those of a shapefile, others such as Subtypes, Domain, Representations, and Relationships are unique to a geodatabase feature class. These unique properties expand the functionalities of a geodatabase feature class.

6. You can find *Task2.gdb* in the Chapter 3 database. A file geodatabase, *Task2.gdb* has many small-sized files.

Task 3   Convert a Shapefile to a Personal Geodatabase Feature Class

What you need: *landsoil.shp*, a polygon shapefile that does not have the correct area and perimeter values.

When shapefiles are used as inputs in an overlay operation (Chapter 11), ArcGIS Desktop does not automatically update the area and perimeter values of the output shapefile. *landsoil.shp* represents such an output shapefile. In this task, you will update the area and perimeter values of *landsoil.shp* by converting it into a feature class in a personal geodatabase.

1. Click *landsoil.shp* in the Catalog tree. On the Preview tab, change the preview type to Table. The table shows two sets of area and perimeter values. Moreover, each field contains duplicate values. Obviously, *landsoil.shp* does not have the updated area and perimeter values.

2. Right-click the Chapter 3 database in the Catalog tree, point to New, and select Personal Geodatabase. Rename the new personal geodatabase *Task3.mdb*. Right-click *Task3.mdb*, point to Import, and select Feature Class (single). In the next dialog, select *landsoil.shp* for the input features. Make sure that *Task3.mdb* is the output location. Enter *landsoil* for the output feature class name. Click OK to create *landsoil* as a standalone feature class in *Task3.mdb*.

Q3. Besides shapefiles (feature classes), what other types of data can be imported to a geodatabase?

3. Now, preview the table of *landsoil* in *Task3. mdb*. On the far right of the table, the fields Shape_Length and Shape_Area show the correct perimeter and area values, respectively.

Task 4   Examine Polylines with Measures

What you need: *decrease24k.shp*, a shapefile showing Washington state highways.

decrease24k.shp is a shapefile downloaded from the Washington State Department of Transportation (WDOT) website. The shapefile contains polylines with measure (*m*) values. In other words, the shapefile contains highway routes. Originally in geographic coordinates, *decrease24k.shp* has been projected onto the Washington State Plane, South Zone, NAD83, and Units feet for Task 4.

1. Launch ArcMap. Rename the data frame Task 4, and add *decrease24k.shp* to Task 4. Open the attribute table *decrease24k*. The Shape field in the table suggests that *decrease24k* is a polyline shapefile with measures. The SR field stores the state route identifiers. Close the table. Right-click *decrease24k* and select Properties. On the Routes tab of the Layer Properties dialog, select SR for the Route Identifier. Click OK to dismiss the dialog.

2. This step is to add the Identify Route Locations tool. The tool does not appear on any toolbar by default. You need to add it. Select Customize Mode from the Customize menu. On the Commands tab, select the category Linear Referencing. The Commands frame shows five commands. Drag and drop the Identify Route Locations command to a toolbar in ArcMap. Close the Customize dialog.

3. Use the Select Features by Rectangle tool to select a highway from *decrease24k.shp*. Click the Identify Route Locations tool, and then click a point along the selected highway. This opens the Identify Route Location Results dialog and shows the measure value of the point you clicked as well as the minimum measure, maximum measure, and other information.

Q4. Can you tell the direction in which the route mileage is accumulated?

Task 5 View Regions and Routes

What you need: *nhd*, a hydrography data set for the 8-digit watershed (18070105) in Los Angeles, California.

nhd is a coverage with built-in regions and route subclasses. Task 5 lets you view these composite features as well as the simple features of arcs and polygons in the coverage.

1. Expand *nhd* in the Catalog tree. The *nhd* coverage contains 11 layers: *arc, label, node, polygon, region.lm, region.rch, region.wb, route.drain, route.lm, route.rch,* and *tic*. A *region* layer represents a regions subclass, and a *route* layer a route subclass.

2. Launch ArcMap, if necessary. Insert a data frame, rename it nhd1, and add *polygon, region.lm, region.rch*, and *region.wb* to nhd1. The *polygon* layer consists of all polygons on which the three regions subclasses are built. Right-click *nhd region.lm*, and select Open Attribute Table. The field FTYPE shows that *nhd region.lm* consists of inundation areas.

Q5. Regions from different regions subclasses may overlap. Do you see any overlaps among the three subclasses of the *nhd* coverage?

3. Insert a new data frame and rename it nhd2. Add *arc, route.drain, route.lm*, and *route.rch* to nhd2. The *arc* layer consists of all arcs on which the three route subclasses are built. Right-click *nhd route.rch*, and select Open Attribute Table. Each record in the table represents a reach, a segment of surface water that has a unique identifier.

Q6. Different route subclasses can be built on the arcs. Do you see any arcs used by different subclasses of the *nhd* coverage?

4. Each layer in *nhd* can be exported to a shapefile or a geodatabase feature class. For example, you can right-click *nhd route.rch*,

point to Export, and save the data set as either a shapefile or a geodatabase feature class.

Task 6 View TIN

What you need: *emidatin*, a TIN prepared from a digital elevation model.

1. Insert a new data frame in ArcMap. Rename the data frame Task 6, and add *emidatin* to Task 6. Right-click *emidatin*, and select Properties. On the Source tab, the Data Source frame shows the number of nodes and triangles as well as the Z (elevation) range in the TIN.

Q7. How many triangles does *emidatin* have?

3. On the Symbology tab, uncheck Elevation and click the Add button in the Show frame. In the next dialog, highlight Edges with the same symbol, click Add, and then click Dismiss. Click OK to dismiss the Layer Properties. The ArcMap window now shows the triangles (faces) that make up *emidatin*. You can follow the same procedure to view nodes that make up *emidatin*.

Challenge Task

NHD_Geo_July3 is a geodatabase downloaded from the National Hydrography Dataset program (**http://nhd.usgs.gov/data.html**).

Q1. Name the feature datasets included in the geodatabase.

Q2. Name the feature classes contained in each of the feature datasets.

Q3. *NHD_Geo_July3* contains the same types of hydrologic data as *nhd* in Task 5. *NHD_Geo_July3* is based on the geodatabase, whereas *nhd* is based on the coverage. Compare the two data sets and describe in your own words the difference between them.

REFERENCES

Bailey, R. G. 1983. Delineation of Ecosystem Regions. *Environmental Management* 7: 365–73.

Batcheller, J. K., B. M. Gittings, and S. Dowers. 2007. The Performance of Vector Oriented Data Storage Strategies in ESRI's ArcGIS. *Transactions in GIS* 11: 47–65.

Berry, B. J. L. 1968. Approaches to Regional Analysis: A Synthesis. In B. J. L. Berry and D. F. Marble, eds., *Spatial Analysis: A Reader in Statistical Geography*, pp. 24–34. Englewood Cliffs, NJ: Prentice-Hall.

Broome, F. R., and D. B. Meixler. 1990. The TIGER Data Base Structure. *Cartography and Geographic Information Systems* 17: 39–47.

Burke, R. 2003. *Getting to Know ArcObjects: Programming ArcGIS with VBA*. Redlands, CA: ESRI Press.

Chang, K. 2007. *Programming ArcObjects with VBA: A Task-Oriented Approach*, 2d ed. Boca Raton, FL: CRC Press.

Chang, K., D. L. Verbyla, and J. J. Yeo. 1995. Spatial Analysis of Habitat Selection by Sitka Black-Tailed Deer in Southeast Alaska. *Environmental Management* 19: 579–89.

Cleland, D. T., R. E. Avers, W. H. McNab, M. E. Jensen, R. G. Bailey, T. King, and

W. E. Russell. 1997. National Hierarchical Framework of Ecological Units. In M. S. Boyce and A. Haney, eds., *Ecosystem Management Applications for Sustainable Forest and Wildlife Resources*, pp. 181–200. New Haven, CT: Yale University Press.

Environmental Systems Research Institute, Inc. 1998. *Understanding GIS: The ARC/INFO Method*. Redlands, CA: ESRI Press.

Forman, R. T. T., and M. Godron. 1986. *Landscape Ecology*. New York: Wiley.

Gustavsson, M., A. C. Seijmonsbergen, and E. Kolstrup. 2007. Structure and Contents of a New Geomorphological GIS Database Linked to a Geomorphological Map—With an Example from Liden, central Sweden. *Geomorphology*, 95:335–49.

Larman, C. 2001. *Applying UML and Patterns: An Introduction to Object-Oriented Analysis and Design and the Unified Process*, 2d ed. Upper Saddle River, NJ: Prentice Hall PTR.

Massey, W. S. 1967. *Algebraic Topology: An Introduction*. New York: Harcourt, Brace & World.

Regnauld, N., and W. A. Mackaness. 2006. Creating a Hydrographic Network from Its Cartographic Representation: A Case Study Using Ordnance

Survey MasterMap Data. *International Journal of Geographical Information Science* 20: 611–731.

Robinson, A. H., J. L. Morrison, P. C. Muehrcke, A. J. Kimerling, and S. C. Guptill. 1995. *Elements of Cartography,* 6th ed. New York: Wiley.

Sutton, J. C., and M. M. Wyman. 2000. Dynamic Location: An Iconic Model to Synchronize Temporal and Spatial Transportation Data. *Transportation Research Part C* 8: 37–52.

Theobald, D. M. 2001. Topology Revisited: Representing Spatial Relations. *International Journal of Geographical Information Science* 15: 689–705.

Ungerer, M. J., and M. F. Goodchild. 2002. Integrating Spatial Data Analysis and GIS: A New Implementation Using the Component Object Model (COM). *International Journal of Geographical Information Science* 16: 41–53.

Wilson, R. J., and J. J. Watkins. 1990. *Graphs: An Introductory Approach.* New York: Wiley.

Zeiler, M. 1999. *Modeling Our World: The ESRI Guide to Geodatabase Design*. Redlands, CA: ESRI Press.

Zeiler, M., ed. 2001. *Exploring ArcObjects*. Redlands, CA: ESRI Press.

RASTER DATA MODEL

The vector data model uses the geometric objects of point, line, and polygon to represent spatial features. Although ideal for discrete features with well-defined locations and shapes, the vector data model does not work well with spatial phenomena that vary continuously over the space such as precipitation, elevation, and soil erosion (Figure 4.1). A better option for representing continuous phenomena is the raster data model. The **raster data model** uses a regular grid to cover the space. The value in each grid cell corresponds to the characteristic of a spatial phenomenon at the cell location. And the changes in the cell value reflect the spatial variation of the phenomenon.

Unlike the vector data model, the raster data model has not changed in terms of its concept or data format for the past three decades. Research on the raster data model has instead concentrated on data structure and data compression. A wide variety of data used in GIS are encoded in raster format. They include digital elevation data, satellite images, digital orthophotos, scanned maps, and graphic files. Raster data tend to require large amounts of the computer memory. Therefore, issues of data storage and retrieval are important to raster data users.

Commercial GIS packages can display raster and vector data simultaneously, and can easily convert between these two types of data. In many ways, raster and vector data complement each other. Integration of these two types of data has therefore become a common and desirable feature in a GIS project.

Chapter 4 is divided into the following five sections. Section 4.1 covers the basic elements of raster data including cell value, cell size, bands, and spatial reference. Section 4.2 presents different types of raster data. Section 4.3 describes

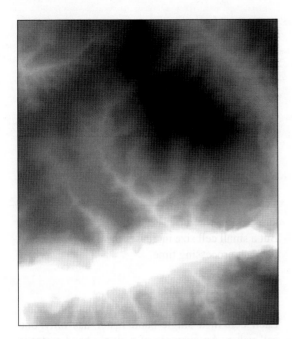

Figure 4.1
A continuous elevation raster with darker shades for higher elevations.

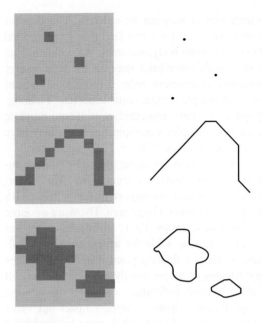

Figure 4.2
Representation of point, line, and polygon features: raster format on the left and vector format on the right.

three different raster data structures. Section 4.4 focuses on data compression methods. Section 4.5 discusses data conversion and integration of raster and vector data.

4.1 ELEMENTS OF THE RASTER DATA MODEL

A raster is also called a grid or an image in GIS. Raster is adopted in this chapter. A raster represents a continuous surface, but for data storage and analysis, a raster is divided into rows, columns, and cells. Cells are also called pixels with images. The origin of rows and columns is typically at the upper-left corner of the raster. Rows function as y-coordinates and columns as x-coordinates. Each cell in the raster is explicitly defined by its row and column position.

Raster data represent points with single cells, lines with sequences of neighboring cells, and polygons with collections of contiguous cells (Figure 4.2). Although the raster data model lacks

the vector model's precision in representing the location of spatial features, it has the distinct advantage of having fixed cell locations (Tomlin 1990). In computing algorithms, a raster can be treated as a matrix with rows and columns, and its cell values can be stored in a two-dimensional array. All commonly used programming languages can handle arrayed variables. Raster data are therefore much easier to manipulate, aggregate, and analyze than vector data.

4.1.1 Cell Value

Each cell in a raster carries a value, which represents the characteristic of a spatial phenomenon at the location denoted by its row and column. A cell value may be coded no data or null if the cell is outside the mapped spatial phenomenon. Depending on the coding of its cell values, a raster can be either an **integer** or a **floating-point** raster. An integer value has no decimal digits, whereas a floating-point value does. Integer cell values usually represent categorical data, which may or may not be ordered. A land cover raster may use 1 for urban land use, 2 for forested land, 3 for water body, and so on. A wildlife habitat raster, on the other hand, may use the same integer numbers to represent ordered categorical data of optimal, marginal, and unsuitable habitats. Floating-point cell values represent continuous, numeric data. For example, a precipitation raster may have precipitation values of 20.15, 12.23, and so forth.

A floating-point raster requires more computer memory than an integer raster. This difference can become an important factor for a GIS project that covers a large area. There are a couple of other differences. First, an integer raster has a value attribute table for access to its cell values, whereas a floating-point raster usually does not because of its potentially large number of cell values. Second, individual cell values can be used to query and display an integer raster but value ranges, such as 12.0 to 19.9, must be used on a floating-point raster. The chance of finding a specific value in a floating-point raster is very small.

Where does the cell value register within the cell? The answer depends on the type of raster data operation. Typically the cell value applies to the center of the cell in operations that involve distance measurements. Examples include resampling pixel values (Chapter 6) and calculating physical distances (Chapter 12). Many other raster data operations are cell-based, instead of point-based, and assume that the cell value applies to the entire cell.

4.1.2 Cell Size

The cell size determines the resolution of a raster. A cell size of 10 meters means that each cell measures 100 square meters (10×10 meters). A cell size of 30 meters, on the other hand, means that each cell measures 900 square meters (30×30 meters). Therefore a 10-meter raster has a finer (higher) resolution than a 30-meter raster.

A large cell size cannot represent the precise location of spatial features, thus increasing the chance of having mixed features such as forest, pasture, and water in a cell (Box 4.1). These problems lessen when a raster uses a smaller cell size. But a small cell size increases the data volume and the data processing time.

4.1.3 Raster Bands

A raster may have a single band or multiple bands. Each cell in a single-band raster has only one cell value. An example of a single-band raster is an elevation raster, with one elevation value at each cell location. Each cell in a multiband raster is associated with more than one cell value. An example of a multiband raster is a satellite image, which may have five, seven, or more bands at each cell location.

4.1.4 Spatial Reference

Raster data must have the spatial reference information so that they can align spatially with other

---

Box 4.1 **Rules in Determining a Categorical Cell Value**

If a large cell covers forest, pasture, and water on the ground, which category should be entered for the cell value? The most common method is to enter the category that occupies the largest percentage of the cell's area. But in some situations the majority rule may not be the best option. For example, studies of endangered species are more inclined to use the presence/absence rule than the majority rule (Chrisman 2001). As long as an endangered species has been recorded in a cell, no matter how much of the cell the species occupies, the cell will be coded with a value for presence. Similarly, the determination of cell values may be based on a ranking of the importance of spatial features to the study. If pasture is deemed to be more important than forest and water, a mixed cell of all three features will be coded as pasture.

data sets in a GIS. For example, to superimpose an elevation raster on a vector-based soil layer, we must first make sure that both data sets are based on the same coordinate system. A raster that has been processed to match a projected coordinate system is often called a **georeferenced raster.**

How does a raster match a projected coordinate system? First, the columns of the raster correspond to the *x*-coordinates, and the rows correspond to the *y*-coordinates. However, because the origin of the raster is at the upper-left corner, as opposed to the lower-left corner for the projected coordinate system, the row numbers increase in the direction opposite that of the *y*-coordinates. Second, the projected coordinates for each cell of the raster can be computed by using the *x*-, *y*-coordinates of the area extent of the raster. The following example is illustrative.

Suppose an elevation raster has the following information on the number of rows, number of columns, cell size, and area extent expressed in UTM (Universal Transverse Mercator) coordinates:

- Rows: 463, columns: 318, cell size: 30 meters
- *x*-, *y*-coordinates at the lower-left corner: 499995, 5177175
- *x*-, *y*-coordinates at the upper-right corner: 509535, 5191065

We can verify that the numbers of rows and columns are correct by using the bounding UTM coordinates and the cell size:

- Number of rows = (5191065 − 5177175)/30 = 463
- Number of columns = (509535 − 499995)/30 = 318

We can also derive the UTM coordinates that define each cell. For example, the cell of row 1, column 1 has the following UTM coordinates (Figure 4.3):

- 499995, 5191035 or (5191065 − 30) at the lower-left corner
- 500025 or (499995 + 30), 5191065 at the upper-right corner
- 500010 or (499995 + 15), 5191050 or (5191065 − 15) at the cell center

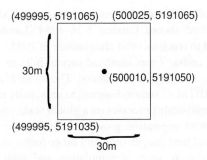

Figure 4.3
UTM coordinates for the extent and the center of a 30-meter cell.

4.2 TYPES OF RASTER DATA

A large variety of data that we use in GIS are encoded in raster format. These data all share the same basic elements of the raster data model.

4.2.1 Satellite Imagery

Remotely sensed satellite data are familiar to GIS users. The spatial resolution of a satellite image relates to the ground pixel size. For example, a spatial resolution of 30 meters means that each pixel in the satellite image corresponds to a ground pixel of 900 square meters. The pixel value, also called the brightness value, represents light energy reflected or emitted from the Earth's surface (Jensen 1996; Lillesand, Kiefer, and Chipman 2004). The measurement of light energy is based on spectral bands from a continuum of wavelengths known as the electromagnetic spectrum. Panchromatic images have a single spectral band, whereas multispectral images have multiple bands.

The U.S. **Landsat** program, started by the National Aeronautics and Space Administration (NASA) and the U.S. Geological Survey (USGS) in 1972, have produced the most widely used imagery worldwide. Landsat 1, 2, and 3 acquired images by the Multispectral Scanner (MSS) with a spatial resolution of about 79 meters. Aboard Landsat 4 in 1982, the Thematic Mapper (TM) scanner obtained images with seven spectral bands (blue, green, red, near infrared, midinfrared I, thermal infrared, and midinfrared II) and with a spatial resolution of 30 meters. A second TM was launched aboard Landsat 5 in 1984. Landsat 6 failed to reach its orbit after launch in 1993.

Landsat 7 was launched successfully in April 1999, carrying an Enhanced Thematic Mapper Plus (ETM+) sensor designed to seasonally monitor small-scale processes on a global scale, such as cycles of vegetation growth, deforestation, agricultural land use, erosion and other forms of land degradation, snow accumulation and melt, and urbanization (**http://landsat.usgs.gov/**). The spatial resolution of Landsat 7 imagery is 15 meters in the panchromatic band; 30 meters in the six visible, near infrared, and shortwave infrared bands; and 60 meters in the thermal infrared band. In June 2003, NASA discovered an instrument malfunction of the scan line corrector (SLC) onboard Landsat 7 and has since been collecting image data in the "SLC-off" mode. The Landsat program is now offering SLC-off gap-filled data, which use scenes from consecutive passes to fill the gaps of the target scene.

In December 1999, NASA's Earth Observing System launched the Terra spacecraft to study the interactions among the Earth's atmosphere, lands, oceans, life, and radiant energy (heat and light) (**http://terra.nasa.gov/About/**). Terra carries five sensors, of which ASTER (Advanced Spaceborne Thermal Emission and Reflection Radiometer) is the only high-spatial-resolution instrument. ASTER's spatial resolution is 15 meters in the visible and near infrared range, 30 meters in the shortwave infrared band, and 90 meters in the thermal infrared band. A major application of ASTER data products is land cover classification and change detection.

The U.S. National Oceanic and Atmospheric Administration (NOAA) uses weather satellites as an aid to weather prediction and monitoring. NOAA's Polar Orbiting Environmental Satellites (POES) carry the AVHRR (Advanced Very High Resolution Radiometer) scanner, which provides data useful for large-area land cover and vegetation mapping (**http://eros.usgs.gov/#/Find_Data/Products_and_Data_Available/AVHRR**). AVHRR data have a spatial resolution of 1.1 kilometers, which may be too coarse for some GIS projects. But the coarse

spatial resolution is offset by the daily coverage and reduced volume of AVHRR data.

A number of countries have programs similar to Landsat in the United States. The French **SPOT** satellite series began in 1986 (**http://www.spot.com/**). Each SPOT satellite carries two types of sensors. SPOT 1 to 4 acquire single-band imagery with a 10-meter spatial resolution and multiband imagery with a 20-meter resolution. SPOT 5, launched successfully in May 2002, sends back higher-resolution imagery: 5 and 2.5 meters in single-band, and 10 meters in multiband. Other important satellite programs have also been established since the late 1980s, for example, in India (**http://www.isro.org/**) and Japan (**http://www.jaxa.jp/index_e.html**).

The privatization of the Landsat program in the United States in 1985 opened the door for private companies to gather and market remotely sensed data using various platforms and sensors. GeoEye offers imagery collected by the IKONOS and OrbView-3 satellites: panchromatic at 1-meter resolution and multispectral at 4-meter resolution (**http://www.geoeye.com/**). DigitalGlobe's QuickBird collects panchromatic images with 61-centimeter resolution and multispectral images with 2.44-meter resolution (**http://www.digitalglobe.com/**). Higher- resolution color images can be produced by fusing the multispectral imagery with the panchromatic. Therefore, color images are offered at 1-meter resolution by GeoEye and 70-centimeter resolution by DigitalGlobe. On these high-resolution color images, ground objects such as cars, small houses, fires, and trees can be detected.

Landsat, IKONOS, and QuickBird are all passive systems that measure light energy reflected or emitted from the Earth's surface. In contrast, synthetic aperture radar (SAR) is an active remote sensing system that transmits its own signal and measures the energy reflected or scattered back from the Earth's surface. The chief advantage of SAR is that it works in the presence of clouds, rain, or darkness. An example of SAR is RADARSAT-2 launched by the Canadian Space Agency in December 2007, which gathers images at a 3-meter resolution useful for applications in agriculture, disaster management, forestry, geology, hydrology, ice, and marine surveillance (**http://www.radarsat2.info/**).

4.2.2 Digital Elevation Models (DEMs)

A **digital elevation model (DEM)** consists of an array of uniformly spaced elevation data. A DEM is point based, but it can easily be converted to raster data by placing each elevation point at the center of a cell. A traditional method for producing DEMs is to use a stereoplotter and aerial photographs with overlapped areas. The stereoplotter creates a 3-D model, allowing the operator to compile elevation data. Although this method can produce highly accurate DEM data, it is expensive for coverage of large areas. Another traditional method is to interpolate a DEM from the contour lines of a topographic map (Chapter 13).

More recently, GIS users have turned to remotely sensed data for making DEMs. One can extract elevation data and generate DEMs from high-resolution satellite imagery using software packages commercially available. Besides imagery data, the data extraction process requires ground control points, which can be measured in the field by GPS (global positioning system) with differential correction. The quality of such DEMs, however, depends on the software package and the quality of the inputs.

The National Center for Earth Resources Observation & Science (EROS) of the USGS offers SRTM (Shuttle Radar Topography Mission) DEMs, derived from SAR data collected in February 2000 (Rabus et al. 2003) (**http://edcsns17.cr.usgs.gov/ srtmdted2**). SRTM DEMs cover over 80 percent of the landmass of the Earth between 60° N and 56° S. For the United States and territorial islands, these DEMs have elevation data spaced 1 arc-second (about 30 meters in the midlatitudes) apart between 0° and 50° latitude and spaced 1 arc-second apart in latitude and 2 arc-seconds apart in longitude between 50° and 60° latitude. For other countries, SRTM DEMs are available at a 90-meter resolution. The absolute vertical accuracy of SRTM DEM values has been reported to be significantly better than the specification of 16 meters for the

mission (Bourgine and Baghdadi 2005). It should be noted that elevations of SRTM DEMs are with respect to the reflective surface, which may be vegetation, man-made features, or bare ground. SRTM DEMs are used in Earth browsers such as Google Earth and World Wind (Chapter 1).

Radar data can also be used to produce higher-resolution DEMs than SRTM DEMs. For example, Intermap Technologies derived elevation data from stereo radar data for the entire state of California with a reported vertical accuracy of 1 meter (**http:// intermap.com/**).

LIDAR (light detection and ranging) is a new technology for producing DEMs (Turner 2000; Brovelli, Longoni, and Cannata 2004). The basic components of a LIDAR system include a laser scanner mounted in an aircraft, GPS, and an Inertial Measurement Unit (IMU). A LIDAR sensor emits rapid laser pulses over an area and uses the time lapse of the pulse to measure distance. At the same time, the location and orientation of the laser source are determined by GPS and IMU, respectively. One application of LIDAR technology is the creation of high-resolution DEMs, with a spatial resolution of 0.5 to 2 meters and a vertical accuracy of ±15 centimeters (Flood 2001) (Figure 4.4). These DEMs are already georeferenced and can have different height levels such as ground elevation (from LIDAR last returns) and canopy elevation (from LIDAR first returns) (Suárez et al. 2005). With an initial release of LIDAR data for Puget Sound, Washington, the USGS now maintains a website for LIDAR information coordination and knowledge, where one can view and download DEMs compiled from LIDAR data (**http://lidar.cr.usgs.gov/**).

4.2.3 USGS DEMs

The USGS offers the National Elevation Dataset (NED), a nationwide coverage of DEM data in raster format (**http://ned.usgs.gov/Ned/about.asp**). To download DEM data, one can define the area of interest by entering geographic coordinates or by specifying state and/or county. (Task 1 of Chapter 5 uses geographic coordinates to download data from the NED website.)

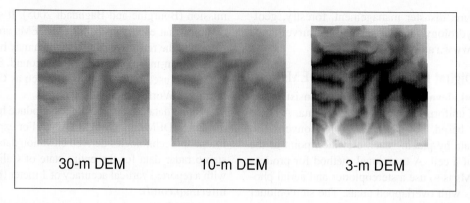

30-m DEM   10-m DEM   3-m DEM

Figure 4.4
DEMs at three resolutions: 30 meters, 10 meters, and 3 meters. The 30-meter and 10-meter DEMs are USGS DEMs. The 3-meter DEM, which contains more topographic details than the other two, is a derived product from LIDAR data.

NED has DEMs at a resolution of 1 arc-second (approximately 30 meters) for the conterminous United States, Hawaii, Puerto Rico, and the island territories and a resolution of 2 arc-seconds for Alaska. These DEMs are measured in geographic coordinates based on NAD83 (the North American Datum of 1983) for all areas except in Alaska, where NAD27 (the North American Datum of 1927) is used. The vertical accuracy of these DEMs is +/− 7 to 15 meters, depending on the data source, which includes 30-meter Level 2 USGS DEMs, 30-meter Level 1 USGS DEMs, 2-arc-second USGS DEMs, and 3-arc-second USGS DEMs. USGS DEMs are grouped into three levels by data accuracy and production, with level 1 having the least accurate data.

NED also has DEMs at higher resolutions of 1/3 arc-second (approximately 10 meters) and 1/9 arc-second (approximately 3 meters). However, these DEMs are not yet available for the entire United States. Both NED 1/3 arc-second and 1/9 arc-second are measured in geographic coordinates based on NAD83. NED 1/3 arc-second has a vertical accuracy of +/− 7 meters, which is better than NED 1 arc-second. The reason for this improvement in vertical accuracy, according to the USGS, is because data corrections are made in the production of NED 1/3 arc-second to minimize, but not eliminate, artifacts, perform edge matching, and fill sliver areas of missing data. NED 1/9 arc-second

has a vertical accuracy of +/− 1 meter because its data source is LIDAR data.

4.2.4 Global DEMs

DEMs at different resolutions are now available on the global scale. SRTM DEMs are available |for land areas outside the United States but at a coarser spatial resolution of 3 arc-seconds (90 meters) (**http://edcsns17.cr.usgs.gov/srtmdted2**). These global-scale DEMs are called SRTM DTED (digital terrain elevation data) Level 1 as opposed to DTED Level 2 for the United States and territorial islands. Because SRTM DTED Level 1 elevation values are derived from SRTM DTED Level 2 values, they have the same vertical accuracy of better than 16 meters at coincident points.

ETOPO5 (Earth Topography–5 Minute) data cover both the land surface and ocean floor of the Earth, with a grid spacing of 5 minutes of latitude by 5 minutes of longitude (**http://eros .usgs.gov//Find_Data/Products_and_Data_ Available/gtopo30_info**) and GLOBE (**http:// www.ngdc.noaa.gov/mgg/topo/globe.html/**) offer global DEMs with a horizontal grid spacing of 30 arc-seconds or approximately 1 kilometer. GTOPO30 and GLOBE were derived from raster data such as satellite imagery and vector data such as contour lines from the Digital Chart of the World. The vertical accuracy of GLOBE is estimated to be within 30 meters from raster sources and within 160 meters from vector sources.

4.2.5 Digital Orthophotos

A **digital orthophoto quad (DOQ)** is a digitized image of an aerial photograph or other remotely sensed data in which the displacement caused by camera tilt and terrain relief has been removed (Figure 4.5). The USGS began producing DOQs in 1991 from 1:40,000 scale aerial photographs of the National Aerial Photography Program. These USGS DOQs are georeferenced (NAD83 UTM coordinates) and can be registered with topographic and other maps.

The standard USGS DOQ format is either a 3.75-minute quarter quadrangle or a 7.5-minute quadrangle in black and white, color infrared, or natural color, with a 1-meter ground resolution. A black-and-white DOQ has 256 gray levels, similar to a single-band satellite image, whereas a color orthophoto is a multiband image, each band

representing red, green, or blue light. Easily integrated in a GIS, DOQs are the ideal background for data display and data editing.

4.2.6 Land Cover Data

Land cover data are typically classified and compiled from satellite imagery and are thus often presented as raster data. For example, the National Land Cover Dataset 1992 (NLCD 92), based primarily on the classification of Landsat TM 1992 imagery, has a cell size of 30 meters. The National Land Cover Database 2001 (NLCD 2001) that is being compiled across all 50 states and Puerto Rico also has a cell size of 30 meters.

4.2.7 Bi-Level Scanned Files

A **bi-level scanned file** is a scanned image containing values of 1 or 0 (Figure 4.6). In GIS, bi-level scanned files are usually made for the purpose of digitizing (Chapter 5). They are scanned from paper or Mylar maps that contain boundaries of soils, parcels, and other features. A GIS package usually has tools for converting bi-level scanned files into

Figure 4.5
USGS 1-meter black-and-white DOQ for Sun Valley, Idaho.

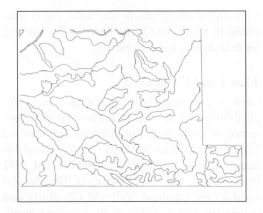

Figure 4.6
A bi-level scanned file showing soil lines.

vector-based features (Chapter 5). Maps to be digitized are typically scanned at 300 or 400 dots per inch (dpi).

4.2.8 Digital Raster Graphics (DRGs)

A **digital raster graphic (DRG)** is a scanned image of a USGS topographic map (Figure 4.7). The USGS scans the 7.5-minute topographic map at 250 dpi,

thus producing a DRG with a ground resolution of 2.4 meters. The USGS uses up to 13 colors on each 7.5-minute DRG. Because these 13 colors are based on an 8-bit (256) color palette, they may not look exactly the same as on the paper maps. USGS DRGs are georeferenced to the UTM coordinate system, based on either NAD27 or NAD83.

4.2.9 Graphic Files

Maps, photographs, and images can be stored as digital graphic files. Many popular graphic files are in raster format, such as TIFF (tagged image file format) GeoTIFF (a georeferenced version of TIFF), GIF (graphics interchange format), and JPEG (Joint Photographic Experts Group).

4.2.10 GIS Software-Specific Raster Data

GIS packages use raster data that are imported from DEMs, satellite images, scanned images, graphic files, and ASCII files or are converted from vector data. These raster data use different formats. For example, ArcGIS stores raster data in the **Esri Grid** format.

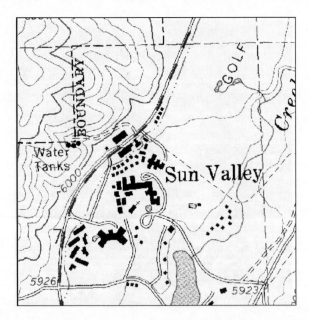

Figure 4.7
USGS DRG for Sun Valley, Idaho. This DRG is outdated compared to the DOQ in Figure 4.5.

4.3 RASTER DATA STRUCTURE

Raster data model describes how raster data is represented. Raster data structure, on the other hand, refers to the method or format for storing raster data in the computer. Three common methods are examined here: cell-by-cell encoding, run-length encoding, and quad tree.

4.3.1 Cell-by-Cell Encoding

The **cell-by-cell encoding** method provides the simplest raster data structure. A raster is stored as a matrix, and its cell values are written into a file by row and column (Figure 4.8). Functioning at the cell level, this method is an ideal choice if the cell values of a raster change continuously.

DEMs use the cell-by-cell data structure because the neighboring elevation values are rarely the same. Satellite images are also encoded cell by cell. With multiple spectral bands, however, a satellite image has more than one value for each pixel, thus requiring special handling. Multiband imagery is typically stored in the following three formats (Jensen 1996). The band sequential (.bsq) method stores the values of an image band as one file. Therefore, if an image has seven bands, the data set has seven consecutive

files, one file per band. The band interleaved by line (.bil) method stores, row by row, the values of all the bands in one file. Therefore the file consists of row 1, band 1; row 1, band 2 … row 2, band 1; row 2, band 2 … and so on. The band interleaved by pixel (.bip) method stores the values of all the bands by pixel in one file. The file is therefore composed of pixel (1, 1), band 1; pixel (1, 1), band 2 … pixel (2, 1), band 1; pixel (2, 1), band 2 … and so on.

4.3.2 Run-Length Encoding

Cell-by-cell encoding becomes inefficient if a raster contains many redundant cell values. For example, a bi-level scanned file from a soil map has many 0s representing non-inked areas and only occasional 1s representing the inked soil lines. Raster data with many repetitive cell values can be more efficiently stored using the **run-length encoding (RLE)** method, which records the cell values by row and by group. A group refers to a series of adjacent cells with the same cell value. Figure 4.9 shows the run-length encoding of the polygon in gray. For each row, the starting cell and

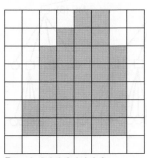

Row 1: 0 0 0 0 1 1 0 0
Row 2: 0 0 0 1 1 1 0 0
Row 3: 0 0 1 1 1 1 1 0
Row 4: 0 0 1 1 1 1 1 0
Row 5: 0 0 1 1 1 1 1 0
Row 6: 0 1 1 1 1 1 1 0
Row 7: 0 1 1 1 1 1 1 0
Row 8: 0 0 0 0 0 0 0 0

Figure 4.8
The cell-by-cell data structure records each cell value by row and column. The gray cells have the cell value of 1.

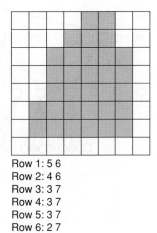

Row 1: 5 6
Row 2: 4 6
Row 3: 3 7
Row 4: 3 7
Row 5: 3 7
Row 6: 2 7
Row 7: 2 7

Figure 4.9
The run-length encoding method records the gray cells by row. Row 1 has two adjacent gray cells in columns 5 and 6. Row 1 is therefore encoded with one run, beginning in column 5 and ending in column 6. The same method is used to record other rows.

the end cell denote the length of the group ("run") that falls within the polygon.

A bi-level scanned file of a 7.5-minute soil quadrangle map, scanned at 300 dpi, can be over 8 megabytes (MB) if stored on a cell-by-cell basis. But using the RLE method, the file is reduced to about 0.8 MB at a 10:1 compression ratio. RLE is therefore a method for encoding as well as compressing raster data. Many GIS packages use RLE in addition to cell-by-cell encoding for storing raster data. They include GRASS, IDRISI, and ArcGIS.

4.3.3 Quad Tree

Instead of working along one row at a time, **quad tree** uses recursive decomposition to divide a raster into a hierarchy of quadrants (Samet 1990). Recursive decomposition refers to a process of continuous subdivision until every quadrant in a quad tree contains only one cell value.

Figure 4.10 shows a raster with a polygon in gray, and a quad tree that stores the feature. The quad tree contains nodes and branches (subdivisions). A node represents a quadrant. Depending

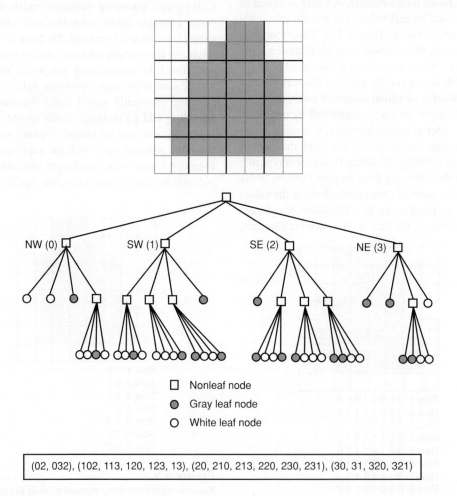

□ Nonleaf node
● Gray leaf node
○ White leaf node

(02, 032), (102, 113, 120, 123, 13), (20, 210, 213, 220, 230, 231), (30, 31, 320, 321)

Figure 4.10
The regional quad tree method divides a raster into a hierarchy of quadrants. The division stops when a quadrant is made of cells of the same value (gray or white). A quadrant that cannot be subdivided is called a leaf node. In the diagram, the quadrants are indexed spatially: 0 for NW, 1 for SW, 2 for SE, and 3 for NE. Using the spatial indexing method and the hierarchical quad tree structure, the gray cells can be coded as 02, 032, and so on. See Section 4.3.3 for more explanation.

on the cell value(s) in the quadrant, a node can be a nonleaf node or a leaf node. A nonleaf node represents a quadrant that has different cell values. A nonleaf node is therefore a branch point, meaning that the quadrant is subject to subdivision. A leaf node, on the other hand, represents a quadrant that has the same cell value. A leaf node is therefore an end point, which can be coded with the value of the homogeneous quadrant (gray or white). The depth of a quad tree, or the number of levels in the hierarchy, can vary depending on the complexity of the two-dimensional feature.

After the subdivision is complete, the next step is to code the two-dimensional feature using the quad tree and a spatial indexing method. For example, the level-1 NW quadrant (with the spatial index of 0) in Figure 4.10 has two gray leaf nodes. The first, 02, refers to the level-2 SE quadrant, and the second, 032, refers to the level-3 SE quadrant of the level-2 NE quadrant. The string of (02, 032) and others for the other three level-1 quadrants completes the coding of the two-dimensional feature.

Regional quad tree is an efficient method for storing area data, especially if the data contain few categories. This method is also efficient for data processing (Samet 1990). Quad tree has other uses in GIS as well. Researchers have proposed using a hierarchical quad tree structure for storing, indexing, and displaying global data (Tobler and Chen 1986; Dutton 1999; Ottoson and Hauska 2002; Platings and Day 2004). Quad tree is also useful for spatial index, which helps locate spatial data, both raster and vector, easily and quickly.

4.3.4 Header File

To import raster data from a DEM or a satellite image, a GIS package requires information about the raster, such as the data structure, area extent, cell size, number of bands, and value for no data. This information often is contained in a header file. Functionally, a header file is similar to a project file for a coordinate system (Box 4.2).

Other files besides the header file may accompany a raster data set. For example, a satellite image

---

Box 4.2   A Header File Example

The following is a header file example for a GTOPO30 DEM. The explanation of each entry in the file is given after /*.

BYTEORDER M /* byte order in which image pixel values are stored. M = Motorola byte order.

LAYOUT BIL /* organization of the bands in the file. BIL = band interleaved by line.

NROWS 6000 /* number of rows in the image.

NCOLS 4800 /* number of columns in the image.

NBANDS 1 /* number of spectral bands in the image. 1 = single band.

NBITS 16 /* number of bits per pixel.

BANDROWBYTES 9600 /* number of bytes per band per row.

TOTALROWBYTES 9600 /* total number of bytes of data per row.

BANDGAPBYTES 0 /* number of bytes between bands in a BSQ format image.

NODATA − 9999 /* value used for masking purpose.

ULXMAP − 99.9958333333334 /* longitude of the center of the upper-left pixel (decimal degrees).

ULYMAP 39.99583333333333 /* latitude of the center of the upper-left pixel (decimal degrees).

XDIM 0.00833333333333 /* x dimension of a pixel in geographic units (decimal degrees).

YDIM 0.00833333333333 /* y dimension of a pixel in geographic units (decimal degrees).

may have two optional files: the statistics file describes statistics such as minimum, maximum, mean, and standard deviation for each spectral band, and the color file associates colors with different pixel values. The Esri grid also has additional files to store information on the descriptive statistics of the cell values and, in the case of an integer grid, the number of cells that have the same cell value.

4.4 RASTER DATA COMPRESSION

Raster data typically require considerable memory space. Approximate file sizes are 1.1 megabytes (MB) for a 30-meter DEM, 9.9 MB for a 10-meter DEM, 5 to 15 MB for a 7.5-minute DRG, and 45 MB for a 3.75-minute quarter DOQ in black and white. An uncompressed 7-band TM scene requires almost 200 MB for storage (Sanchez and Canton 1999). The memory requirement becomes even higher for high-resolution DEMs and satellite images.

Data compression refers to the reduction of data volume, a topic particularly important for data delivery and Internet mapping. A variety of techniques are available for data compression. They can be lossless or lossy. A **lossless compression** preserves the cell or pixel values and allows the original raster or image to be precisely reconstructed. Therefore, lossless compression is desirable for raster data that are used for analysis or deriving new data. RLE is an example of lossless compression. Other methods include PackBits, a more efficient variation of RLE, and LZW (Lempel-Ziv-Welch) and its variations (e.g., LZ77). The TIFF data format offers both PackBits and LZW for image compression. The USGS uses GeoTIFF for delivering DRGs and DOQs.

A **lossy compression** cannot reconstruct fully the original image but can achieve higher compression ratios than a lossless compression. Lossy compression is therefore useful for raster data that are used as background images rather than for analysis. The traditional JPEG format uses a lossy compression method. The method divides an image into blocks of 64 (8×8) and processes each block independently. The colors in each block are

shifted and simplified to reduce the amount of data encoding. This block-based processing usually results in the "blocky" appearance. Image degradation through lossy compression can affect GIS-related tasks such as extracting ground control points from aerial photographs or satellite images for the purpose of georeferencing (Chapter 6).

Newer image compression techniques can be both lossless and lossy. An example is MrSID (Multiresolution Seamless Image Database) patented by LizardTech Inc. **(http://www.lizardtech.com/).** Multiresolution means that MrSID has the capability of recalling the image data at different resolutions or scales. Seamless means that MrSID can compress a large image such as a DOQ or a satellite image with subblocks and eliminates the artificial block boundaries during the compression process. Most commercial GIS packages can read MrSID files.

MrSID uses the wavelet transform for data compression. JPEG 2000, an updated version of the popular open format, also uses the wavelet transform (Acharya and Tsai 2005). The wavelet transform therefore appears to be the latest choice for image compression. The **wavelet transform** treats an image as a wave and progressively decomposes the wave into simpler wavelets (Addison 2002). Using a wavelet (mathematical) function, the transform repetitively averages groups of adjacent pixels (e.g., 2, 4, 6, 8, or more) and, at the same time, records the differences between the original pixel values and the average. The differences, also called wavelet coefficients, can be 0, greater than 0, or less than 0. In parts of an image that have few significant variations, most pixels will have coefficients of 0 or very close to 0. To save data storage, these parts of the image can be stored at lower resolutions by rounding off low coefficients to 0, but storage at higher resolutions is required for parts of the same image that have significant variations (i.e., more details). Box 4.3 shows a simple example of using the Haar function for the wavelet transform.

Both MrSID and JPEG 2000 can perform either lossless or lossy compression. A lossless compression saves the wavelet coefficients and uses them to reconstruct the original image. A lossy compression, on the other hand, stores only

Box **4.3** | A Simple Wavelet Example: The Haar Wavelet

A Haar wavelet consists of a short positive pulse followed by a short negative pulse (Figure 4.11a). Although the short pulses result in jagged lines rather than smooth curves, the Haar function is excellent for illustrating the wavelet transform because of its simplicity. Figure 4.11b shows an image with darker pixels near the center. The image is encoded as a series of numbers. Using the Haar function, we take the average of each pair of adjacent pixels. The averaging results in the string (2, 8, 8, 4) and retains the quality of the original image at a lower resolution. But if the process continues, the averaging results in the string (5, 6) and loses the darker center in the original image.

Suppose that the process stops at the string (2, 8, 8, 4). The wavelet coefficients will be -1 $(1 - 2)$, -1 $(7 - 8)$, 0 $(8 - 8)$, and 2 $(6 - 4)$. By rounding off these coefficients to 0, it would save the storage space by a factor of 2 and still retain the quality of the original image. If, however, a lossless

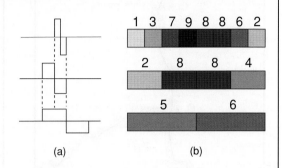

(a)   **(b)**

Figure 4.11
The Haar wavelet and the wavelet transform.
(a) Three Haar wavelets at three scales (resolutions).
(b) A simple example of the wavelet transform.

compression is needed, we can use the coefficients to reconstruct the original image. For example, $2 - 1 = 1$ (the first pixel), $2 - (-1) = 3$ (the second pixel), and so on.

the averages and those coefficients that did not get rounded off to 0. Trade reports have shown that JPEG 2000 can achieve a 20:1 compression ratio without a perceptible difference in the quality of an image (i.e., visually lossless). If JPEG 2000 compression is at or under a 10:1 ratio, it should be possible to extract ground control points from aerial photographs or satellite images for georeferencing (Li, Yuan, and Lam 2002).

4.5 DATA CONVERSION AND INTEGRATION

To take advantage of both vector and raster data for a GIS project, we must consider data conversion and integration.

4.5.1 Rasterization

Rasterization converts vector data to raster data (Piwowar, LeDraw, and Dudycha 1990; Congalton

1997) (Figure 4.12). Rasterization involves three basic steps (Clarke 1995). The first step sets up a raster with a specified cell size to cover the area extent of vector data and initially assigns all cell values as zeros. The second step changes the values of those cells that correspond to points, lines, or polygon boundaries. The cell value is set to 1 for a point, the line's value for a line, and the polygon's value for a polygon boundary. The third step fills the interior of the polygon outline with the polygon value. Errors from rasterization are usually related to the design of the computer algorithm, the size of the raster cell, and the boundary complexity (Bregt et al. 1991; Shortridge 2004).

4.5.2 Vectorization

Vectorization converts raster data to vector data (Figure 4.12). Vectorization involves three basic elements: line thinning, line extraction, and topological reconstruction (Clarke 1995). Lines in the

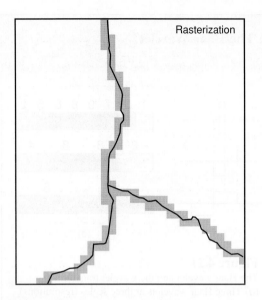

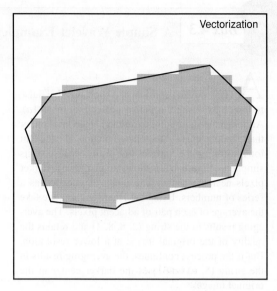

Figure 4.12
On the left is an example of conversion from vector to raster data, or rasterization. On the right is an example of conversion from raster to vector data, or vectorization.

vector data model have length but no width. Raster lines in a scanned file, however, usually occupy several pixels in width. Raster lines must be thinned, ideally to a 1-cell width, for vectorization. Line extraction is the process of determining where individual lines begin and end. Topological reconstruction connects extracted lines and shows where digitizing errors exist. Results of raster-to-vector conversion often exhibit steplike features along diagonal lines. A subsequent line smoothing operation can help reduce those artifacts from raster data.

4.5.3 Integration of Raster and Vector Data

Recent studies have approached integration of raster and vector data at the data model level, but they are still at the conceptual level (Kjenstad 2006; Goodchild, Yuan, and Cova 2007). For many GIS operations, especially data analysis, raster and vector data remain separate. How to use these two types of data together for projects is therefore of interest to GIS users.

DOQs, DRGs, and graphic files are useful as the background for data display and as the source for digitizing or editing vector data (Box 4.4). Bi-level scanned files are inputs for digitizing line or polygon features (Chapter 5). DEMs are the most important data source for deriving such topographic features as contour, slope, aspect, drainage network, and watershed (Chapters 13 and 14). These topographic features can be stored in either raster or vector format.

Perhaps the most promising area for integration is between GIS and image analysis. Georeferenced satellite images are like DOQs, useful for displaying with other spatial features such as business locations, roads, stands, and parcels (Box 4.5). Satellite images also contain quantitative spectral data that can be processed to create layers such as land cover, vegetation, urbanization, snow accumulation, and environmental degradation. For example, both the National Land Cover DataSet 1992 and the National Land Cover Database 2001 for the conterminous United States

Box 4.4 | Linking Vector Data with Images

Linking a vector-based feature with an image is probably the most common and useful type of data integration. An example of this hyperlink application is to link a map showing houses for sale with pictures of the houses so that, when a house (point) is clicked, it opens a picture of the house in a graphic file. The hyperlink tool in ArcMap can link a point, line, or polygon feature in vector format with an image. To open the image, the tool uses an attribute that specifies the path to the image.

Box 4.5 | Digital Earth

Formerly data for specialists, satellite images are now seen regularly on the Internet and the public media. This trend started with Digital Earth, a broad international initiative proposed by Al Gore in 1998 for an easy-to-use information system allowing users to view integrated images and vector data of the Earth. Many national agencies have since implemented Digital Earth online.

The concept of Digital Earth has been adopted by Google Maps, Yahoo! Maps, and Microsoft Virtual Earth (Chapter 1). In all three systems, georeferenced satellite images can be displayed with layers of boundaries, roads, shopping centers, schools, 3-D buildings (for major cities in the United States), and other types of vector data. These systems also provide such functions as "fly to," "local search," and "directions" for manipulating the display.

are based on Landsat TM imagery (Vogelmann et al. 2001; Homer et al. 2004).

Vector data are regularly used as ancillary information for processing satellite images (Ehlers, Edwards, and Bedard 1989; Hinton 1996; Rogan et al. 2003; Coppin et al. 2004). Image stratification is one example. It uses vector data to divide the landscape into major areas of different characteristics and then treats these areas separately in image processing and classification. Another example is the use of vector data in selecting control points for the georeferencing of remotely sensed data (Couloigner et al. 2002).

Recent developments suggest a closer integration of GIS and remote sensing. A GIS package can read files created in an image-processing package, and vice versa. For example, ArcGIS supports files created in ERDAS (IMAGINE, GIS, and LAN files) **(http://gis.leica-geosystems.com/)** and ER Mapper **(http://www.ermapper.com/).** ArcGIS 10 has introduced the Image Analysis window, which provides access to commonly used image processing techniques such as clipping, masking, orthorectification, convolution filters, and mosaicking. There are also extensions to a GIS package for processing satellite images. The Feature Analyst extension to ArcGIS, for example, can extract features such as buildings, roads, and water features directly from satellite images, especially those of high resolutions **(http://www .featureanalyst.com).** As high-resolution satellite images gain more acceptance among GIS users, an even stronger tie between GIS and remote sensing can be expected.

Bi-level scanned file: A scanned file containing values of 1 or 0.

Cell-by-cell encoding: A raster data structure that stores cell values in a matrix by row and column.

Data compression: Reduction of data volume, especially for raster data.

Digital elevation model (DEM): A digital model with an array of uniformly spaced elevation data in raster format.

Digital orthophoto quad (DOQ): A digitized image in which the displacement caused by camera tilt and terrain relief has been removed from an aerial photograph.

Digital raster graphic (DRG): A scanned image of a USGS topographic map.

Esri grid: A proprietary Esri format for raster data.

Floating-point raster: A raster that contains cells of continuous values.

Georeferenced raster: A raster that has been processed to align with a projected coordinate system.

Integer raster: A raster that contains cell values of integers.

Landsat: An orbiting satellite that provides repeat images of the Earth's surface. The latest Landsat 7 was launched in April 1999.

Lossless compression: One type of data compression that allows the original image to be precisely reconstructed.

Lossy compression: One type of data compression that can achieve high-compression ratios but cannot reconstruct fully the original image.

Quad tree: A raster data structure that divides a raster into a hierarchy of quadrants.

Raster data model: A data model that uses rows, columns, and cells to construct spatial features.

Rasterization: Conversion of vector data to raster data.

Run-length encoding (RLE): A raster data structure that records the cell values by row and by group. A run-length encoded file is also called a run-length compressed (RLC) file.

SPOT: A French satellite that provides repeat images of the Earth's surface. SPOT 5 was launched in May 2002.

Vectorization: Conversion of raster data to vector data.

Wavelet transform: An image compression technique that treats an image as a wave and progressively decomposes the wave into simpler wavelets.

REVIEW QUESTIONS

1. What are the basic elements of the raster data model?

2. Explain the advantages and disadvantages of the raster data model versus the vector data model.

3. Name two examples each for integer rasters and floating-point rasters.

4. Explain the relationship between cell size, raster data resolution, and raster representation of spatial features.

5. You are given the following information on a 30-meter DEM:
   - UTM coordinates in meters at the lower-left corner: 560635, 4816399

- UTM coordinates in meters at the upper-right corner: 570595, 4830380

How many rows does the DEM have? How many columns does the DEM have? What are the UTM coordinates at the center of the (row 1, column 1) cell?

6. Go to either the GeoEye website (**http://www.geoeye.com/**) or the DigitalGlobe website (**http://www.digitalglobe.com/**), and take a look at their high-resolution sample imagery.

7. Describe three common methods for producing DEMs.

8. You can find the status graphics for USGS DEMs, DRGs, and DOQs at the following website: **http://statgraph.cr.usgs.gov/viewer.htm.** How much of your state is covered for each of these USGS digital data products?

9. Find the GIS data clearinghouse for your state at the Geospatial One-Stop website (**http://www.geodata.gov/**). Go to the clearinghouse website. Does the website offer USGS DEMs, DRGs, and DOQs online? Does the website offer both 30-meter and 10-meter USGS DEMs?

10. How does NED 1/3 arc-second differ from NED 1 arc-second in terms of spatial resolution and vertical accuracy?

11. Use a diagram to explain how the run-length encoding method works.

12. Refer to the following figure, draw a quad tree, and code the spatial index of the shaded (spatial) feature.

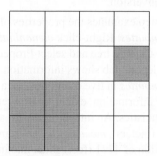

13. A number of government agencies have used MrSID to store digital images such as DOQs and DRGs. When these images are made available online, we can see them at different resolutions by zooming in or out. Go to the Massachusetts GIS website (**http://www.state.ma.us/mgis/mrsid.htm**) and view the DOQs and DRGs at different resolutions.

14. Explain the difference between lossless and lossy compression methods.

15. What is vectorization?

16. Use an example from your discipline and explain the usefulness of integrating vector and raster data.

APPLICATIONS: RASTER DATA MODEL

This applications section has three tasks. The first two tasks let you view two types of raster data: DEM and Landsat TM imagery. Task 3 covers the conversion of two shapefiles, one line and one polygon, to raster data.

Task 1 View and Import DEM Data

What you need: *menanbuttes.txt,* a text file containing elevation data.

1. This step lets you view the DEM data as stored in *menanbuttes.txt.* Use Notepad or WordPad to open *menanbuttes.txt.* The first six lines in *menanbuttes.txt* contain the header file information. They show that the DEM has 341 columns and 466 rows, that the lower-left corner of the DEM has the *x*-, *y*-coordinates of (419475, 4844265), that the cell size is 30 (meters), and that No Data cells are coded −9999. Elevation values are listed following the header file information.

2. Start ArcCatalog and connect to the Chapter 4 database. Click ArcToolbox window to open ArcToolbox. Double-click the ASCII to

Raster tool in the Conversion Tools/To Raster toolset. In the next dialog, select *menanbuttes.txt* for the input ASCII raster file, save the output raster as *menanbuttes* in the Chapter 4 database, and click OK to run the conversion.

3. This step examines the properties of *menanbuttes*. Right-click *menanbuttes* in the Catalog tree and select Properties. The General tab shows information about *menanbuttes* in five categories: data source, raster information, extent, spatial reference, and statistics. An integer grid with a cell size of 30 (meters), *menanbuttes* has a minimum elevation of 4771 (feet) and a maximum elevation of 5619. The spatial reference is listed as undefined.

Q1. What are the *x*- and *y*-coordinates at the upper-left corner of *menanbuttes*?

Q2. Can you verify your answer to Q1 is correct by referring to the *x*- and *y*-coordinates at the lower-left corner as listed in the header file information?

4. Launch ArcMap. Rename the data frame Task 1, and add *menanbuttes* to Task 1. Right-click *menanbuttes* and select Properties. On the Symbology tab, right-click on the Color Ramp box and uncheck Graphic View. Then select Elevation #1 from the Color Ramp's dropdown menu. Dismiss the Properties dialog. ArcMap now displays the dramatic landscape of the twin buttes.

Task 2 View a Satellite Image in ArcMap

What you need: *tmrect.bil,* a Landsat TM image comprised of the first five bands.

Task 2 lets you view a Landsat TM image with five bands. By changing the color assignments of the bands, you can alter the view of the image.

1. Right-click *tmrect.bil* in ArcCatalog and select Properties. The General tab shows that *tmrect .bil* has 366 rows, 651 columns, and 5 bands.

Q3. Can you verify that *tmrect.bil* is stored in the band interleaved by line format?

Q4. What is the pixel size (in meters) of *tmrect.bil*?

2. Launch ArcMap, if necessary. Insert a new data frame and rename it Task 2. Add *tmrect.bil* to Task 2. Ignore the missing spatial reference warning message. The table of contents shows *tmrect.bil* as an RGB Composite with Red for Band_1, Green for Band_2, and Blue for Band_3.

3. Select Properties from the context menu of *tmrect.bil*. On the Symbology tab, use the dropdown menus to change the RGB composite: Red for Band_3, Green for Band_2, and Blue for Band_1. Click OK. You should see the image as a color photograph.

4. Next, use the following RGB composite: Red for Band_4, Green for Band_3, and Blue for Band_2. You should see the image as a color infrared photograph.

Task 3 Convert Vector Data to Raster Data

What you need: *nwroads.shp* and *nwcounties.shp,* shapefiles showing major highways and counties in the Pacific Northwest, respectively.

In Task 3, you will convert a line shapefile (*nwroads.shp*) and a polygon shapefile (*nwcounties .shp*) to rasters. Covering Idaho, Washington, and Oregon, both shapefiles are projected onto a Lambert conformal conic projection and are measured in meters.

1. Insert a new data frame in ArcMap and rename it Task 3. Add *nwroads.shp* and *nwcounties.shp* to Task 3.

2. Click ArcToolbox window to open ArcToolbox. Double-click the Feature to Raster tool in the Conversion Tools/ To Raster toolset. Select *nwroads* for the input features, select RTE_NUM1 for the field, save the output raster as *nwroads_gd,* enter 5000 for the output cell size, and click

OK to run the conversion. *nwroads_gd* appears in the map in different colors. Each color represents a numbered highway. The highways look blocky because of the large cell size (5000 meters).

3. Double-click the Feature to Raster tool. Select *nwcounties* for the input features, select FIPS for the field, save the output raster as *nwcounties_gd,* enter 5000 for the output cell size, and click OK. *nwcounties_gd* appears in the map with symbols representing the classified values of 1 to 119 (119 is the total number of counties). Double-click *nwcounties_gd* in the table of contents. On the Symbology tab, select Unique Values in the Show frame and click OK. Now the map shows *nwcounties_gd* with a unique symbol for each county.

Q5. *nwcounties_gd* has 157 rows and 223 columns. If you had used 2500 for the Output cell size, how many rows would the output grid have?

Challenge Task

What you need: *emidalat,* an elevation raster; and *idtm.shp,* a polygon shapefile.

A USGS DEM, *emidalat* is projected onto the UTM coordinate system. *idtm.shp,* on the other hand, is based on the Idaho Transverse Mercator (IDTM) coordinate system. This challenge task asks you to project *emidalat* onto the IDTM coordinate system. It also asks you for the layer information about *emidalat.*

1. Launch ArcMap if necessary. Rename the new data frame Challenge, and add *idtm.shp* and *emidaltm* to Challenge. Read the spatial reference information about *emidalat* and *idtm.shp,* including the datum.

2. Use the Project Raster tool in the Data Management Tools/Projections and Transformations/Raster toolset to project *emidalat* onto the IDTM coordinate system. Use the default resampling technique and a cell size of 30 (meters). Rename the output raster *emidatm.*

3. After re-projection, *emidatm* should appear as a very small rectangle in northern Idaho.

Q1. What is the maximum elevation in *emidatm*?

Q2. Is *emidatm* a floating-point grid or an integer grid?

Q3. How many rows and columns does *emidatm* have?

REFERENCES

Acharya, T., and P. Tsai. 2005. *JPEG2000 Standard for Image Compression: Concepts, Algorithms and VLSI Architectures.* Hoboken, NJ: Wiley-Interscience.

Addison, P. S. 2002. *The Illustrated Wavelet Transform Handbook.* Bristol, UK: Institute of Physics Publishing.

Bourgine, B., and N. Baghdadi. 2005. Assessment of C-Band SRTM DEM in a Dense Equatorial Forest Zone. *Comptes Rendus Geoscience* 337: 1225–34.

Bregt, A. K., J. Denneboom, H. J. Gesink, and Y. Van Randen. 1991. Determination of Rasterizing Error: A Case Study with the Soil Map of the Netherlands. *International Journal of Geographical Information Systems* 5: 361–67.

Brovelli, M. A., U. M. Longoni, and M. Cannata. 2004. LIDAR Data Filtering and DTM Interpolation Within GRASS. *Transactions in GIS* 8: 155–74.

Chrisman, N. 2001. *Exploring Geographic Information Systems,* 2d ed. New York: Wiley.

Clarke, K. C. 1995. *Analytical and Computer Cartography,* 2d ed. Englewood Cliffs, NJ: Prentice Hall.

Congalton, R. G. 1997. Exploring and Evaluating the Consequences of Vector-to-Raster and

Raster-to-Vector Conversion. *Photogrammetric Engineering and Remote Sensing* 63: 425–34.

Coppin, P., I. Jonckheere, K. Nackaerts, B. Muys, and E. Lambin. 2004. Digital Change Detection Methods in Ecosystem Monitoring: A Review. *International Journal of Remote Sensing* 25: 1565–96.

Couloigner, I., K. P. B. Thomson, Y. Bedard, B. Moulin, E. LeBlanc, C. Djima, C. Latouche, and N. Spicher. 2002. Towards Automating the Selection of Ground Control Points in Radarsat Images Using a Topographic Database and Vector-Based Data Matching. *Photogrammetric Engineering and Remote Sensing* 68: 433–40.

Dutton, G. H. 1999. *A Hierarchical Coordinate System for Geoprocessing and Cartography.* Berlin: Springer-Verlag.

Ehlers, M., G. Edwards, and Y. Bedard. 1989. Integration of Remote Sensing with Geographical Information Systems: A Necessary Evolution. *Photogrammetric Engineering and Remote Sensing* 55: 1619–27.

Flood, M. 2001. Laser Altimetry: From Science to Commercial LIDAR Mapping. *Photogrammetric Engineering and Remote Sensing* 67: 1209–17.

Goodchild, M. F., M. Yuan, and T. Cova. 2007. Towards a General Theory of Geographic Representation in GIS. *International Journal of Geographical Information Science* 21: 239–60.

Hinton, J. E. 1996. GIS and Remote Sensing Integration for Environmental Applications. *International Journal of Geographical Information Systems* 10: 877–90.

Homer, C. C., L. Huang, B. Wylie, and M. Coan. 2004. Development of a 2001 National Landcover Database for the United States. *Photogrammetric Engineering and Remote Sensing* 70: 829–40.

Jensen, J. R. 1996. *Introductory Digital Image Processing: A Remote Sensing Perspective,* 2d ed. Upper Saddle River, NJ: Prentice Hall.

Kjenstad, K. 2006. On the Integration of Object-Based Models and Field-Based Models in GIS. *International Journal of Geographical Information Science* 20: 491–509.

Li, Z., X. Yuan, and K. W. K. Lam. 2002. Effects of JPEG Compression on the Accuracy of Photogrammetric Point Determination. *Photogrammetric Engineering and Remote Sensing* 68: 847–53.

Lillesand, T. M., R. W. Kiefer, and J. W. Chipman. 2004. *Remote Sensing and Image Interpretation,* 5th ed. New York: Wiley.

Ottoson, P., and H. Hauska. 2002. Ellipsoidal Quadtrees for Indexing of Global Geographical Data. *International Journal of Geographical Information Science* 16: 213–26.

Piwowar, J. M., E. F. LeDraw, and D. J. Dudycha. 1990.

Integration of Spatial Data in Vector and Raster Formats in a Geographic Information System Environment. *International Journal of Geographical Information Systems* 4: 429–44.

Platings, M., and A. M. Day. 2004. Comparison of Large-Scale Terrain Data for Real-Time Visualization Using a Tiled Quad Tree. *Computer Graphics Forum* 23: 741–59.

Rabus, B., M. Eineder, A. Roth, and R. Bamler. 2003. The Shuttle Radar Topography Mission—A New Class of Digital Elevation Models Acquired by Spaceborne Radar. *ISPRS Journal of Photogrammetry & Remote Sensing* 57: 241–62.

Rogan, J., J. Miller, D. Stow, J. Franklin, L. Levien, and C. Fischer. 2003. Land Cover Change Mapping in California using Classification Trees with Landsat TM and Ancillary Data. *Photogrammetric Engineering and Remote Sensing* 69: 793–804.

Samet, H. 1990. *The Design and Analysis of Spatial Data Structures.* Reading, MA: Addison-Wesley.

Sanchez, J., and M. P. Canton. 1999. *Space Image Processing.* Boca Raton, FL: CRC Press.

Shortridge, A. M. 2004. Geometric Variability of Raster Cell Class Assignment. *International Journal of Geographical Information Science* 18: 539–58.

Suárez, J. C., C. Ontiveros, S. Smith, and S. Snape. 2005. Use of Airborne LiDAR and Aerial Photography in the Estimation of Individual Tree Heights in Forestry. *Computers & Geosciences* 31: 253–62.

Tobler, W., and Z. Chen. 1986. A Quadtree for Global Information Storage. *Geographical Analysis* 18: 360–71.

Tomlin, C. D. 1990. *Geographic Information Systems and*

Cartographic Modeling. Englewood Cliffs, NJ: Prentice Hall.

Turner, A. K. 2000. LIDAR Provides Better DEM Data. *GEOWorld* 13(11): 30–31.

Vogelmann, J. E., S. M. Howard, L. Yang, C. R. Larson, B.

K. Wylie, and N. Van Driel. 2001. Completion of the 1990s National Land Cover Data Set for the Conterminous United States from Landsat Thematic Mapper Data and Ancillary Data Sources. *Photogrammetric Engineering and Remote Sensing* 67: 650–62.

CHAPTER 5

GIS DATA ACQUISITION

The most expensive part of a GIS project is database construction. Converting from paper maps to digital maps used to be the first step in constructing a database. But in recent years this situation has changed as digital data clearinghouses have become commonplace on the Internet. Now we can look at what data are available in the public domain before deciding to create new data. Many government agencies at the federal, state, regional, and local levels have set up clearinghouses for distributing GIS data. Private companies have also been active in marketing GIS data. Some of them produce new GIS data for their customers; others produce value-added GIS data from public data.

The proliferation of online GIS data, both vector and raster, has made it easier to organize a GIS project. But the topic of data acquisition is still important for two reasons. First, *public data*, as the term suggests, are intended for all GIS users rather than users of a particular software package. Therefore, we must be familiar with metadata and data conversion from one format to another. Second, the proliferation of online data suggests that government agencies and private companies are continually producing new data. New data must be produced to be put on the Internet or to be sold to customers. Even for GIS users, we may have to create our own data if we cannot find the digital data we need for a project.

New GIS data can be created from a variety of data sources. They include satellite images, field data, street addresses, text files with *x*- and *y*-coordinates, and paper maps. New data can also be created using manual digitizing, scanning, or on-screen digitizing. Our knowledge of data sources and production methods can better prepare us for putting together a GIS database.

Chapter 5 is presented in the following four sections. Section 5.1 discusses existing GIS data on the Internet, including examples from different levels of government and private companies. Sections 5.2 and 5.3 cover metadata and the data exchange methods, respectively. Section 5.4 provides an overview of creating new GIS data from different data sources and using different production methods.

5.1 EXISTING GIS DATA

To find existing GIS data for a project is often a matter of knowledge, experience, and luck. Since the early 1990s, government agencies at different levels in the United States as well as other countries have set up websites for sharing public data and for directing users to the source of the desired information (Masser 1999; Jacoby et al. 2002). The Internet is also a medium for finding existing data from nonprofit organizations and private companies. But searching for GIS data, especially data of different kinds for a GIS project, can be difficult (Falke 2002). A keyword search will probably result in thousands of matches, but most hits are irrelevant to the user. Internet addresses may have changed or are discontinued. Data on the Internet may be in a format incompatible with the GIS package used for a project, or to be usable for a project, the data may need extensive processing such as clipping the study area from a large data set or merging several data sets.

Common types of GIS data on the Internet are data that many organizations regularly use for GIS activities. These are called **framework data,** which typically include seven basic layers: geodetic control (accurate positional framework for surveying and mapping), orthoimagery (rectified imagery such as orthophotos), elevation, transportation, hydrography, governmental units, and cadastral information **(http://www.fgdc.gov/ framework/frameworkoverview).** In recent years some thematic data such as environmental data have also become available online.

Public data are downloadable from the Internet. Most data are free or available for fees that cover their cost of processing. All levels of government

let GIS users access their public data through clearinghouses in the United States. The following sections describe public data that are available at the federal, state, regional, metropolitan, and county levels as well as data from private companies.

5.1.1 Federal Geographic Data Committee

The **Federal Geographic Data Committee (FGDC)** is an interagency committee that leads the development of policies, metadata standards, and training to support the National Spatial Data Infrastructure (NSDI) and coordination efforts **(http://www.fgdc.gov/).** The NSDI is aimed at the sharing of geospatial data throughout all levels of government, the private and nonprofit sectors, and the academic community. The FGDC website provides a link to the NSDI Clearinghouse Network, a collection of hundreds of spatial data nodes in the United States and overseas.

5.1.2 Geospatial One-Stop

The **Geospatial One-Stop (GOS)** is a geospatial data portal established by the Federal Office of Management and Budget in 2003 as an e-government initiative **(http://gos2.geodata.gov/wps/portal/ gos)** (Box 5.1). The main objective of GOS is to expand collaborative partnerships at all levels of government to help leverage investments in geospatial data and to reduce the duplication of data. The initial GOS acted as a data clearinghouse for government agencies to post metadata describing their data resources. In the second phase of development launched in July 2005, GOS changed its function to that of an interactive portal, allowing users to access geospatial data from federal, state, local, and private sources and to use the data in their own environments (Goodchild, Fu, and Rich 2007).

5.1.3 U.S. Geological Survey

Through its National Map program, the U.S. Geological Survey (USGS) is the major provider of GIS data in the United States. Its website **(http://geography.usgs.gov/)** offers pathways to

Box 5.1   Clearinghouse and Portal

A clearinghouse refers to a website from which we can download GIS data. It usually does not provide additional services for data query. A clearinghouse is sometimes called a warehouse. In contrast, a portal provides multiple services, which typically include a directory of websites, links to other sites, news, references, a community forum, and, in the case of Geospatial One-Stop, interactive functionality (Maguire and Longley 2005; Goodchild, Fu, and Rich 2007). A portal is sometimes called a gateway.

USGS national mapping and remotely sensed data and to thematic data clearinghouses on biological, geologic, and water resources data. Public data available from the USGS include both vector and raster data.

Digital line graphs (DLGs) are digital representations of point, line, and area features from the USGS quadrangle maps at the scales of 1:24,000, 1:100,000, and 1:2,000,000. DLGs include such data categories as hypsography (i.e., contour lines and spot elevations), hydrography, boundaries, transportation, and the U.S. Public Land Survey System. DLGs contain attribute data and are topologically structured.

National Land Cover Data 1992 (NLCD 1992) include 21 thematic classes, compiled from the Thematic Mapper (TM) imagery of the early 1990s and other geospatial ancillary data sets. The 21 classes resemble the Anderson level II land use/land cover scheme used by the USGS in the 1970s and early 1980s (Anderson et al. 1976). A new project called **National Land Cover Database 2001 (NLCD 2001)** uses the Landsat 7 ETM+ imagery to compile land cover data for all 50 states and Puerto Rico. Information on both NLCD 1992 and 2001 is available at **http://landcover.usgs.gov/.**

USGS digital elevation models (DEMs) can be downloaded at three designated websites **(http://data.geocomm.com/, http://www.mapmart.com/, http://www.atdi-us.com/).** They include 7.5-minute, 15-minute, and 30-minute DEMs. The **National Elevation Dataset (NED)** compiled by the USGS provides 1:24,000 scale DEMs nationwide (1:63,360 scale DEMs for Alaska) **(http://ned.usgs.gov/).** The NED uses a seamless data distribution system based on user-defined areas. It also updates its data sets regularly to incorporate the highest-resolution, best-quality DEM data available. The horizontal datum for NED data is NAD83 (North American Datum of 1983) (Chapter 2). The vertical datum is NAVD88 (North American Vertical Datum of 1988), except for Alaska, which is NGVD29 (National Geodetic Vertical Datum of 1929). NGVD29 was based on observations at tidal stations on the Atlantic, Pacific, and Gulf of Mexico shorelines. NAVD88, referenced to a local mean sea-level-height value at Father Point, Quebec, Canada, is a refinement of the 1929 datum that includes gravimetric and other anomalies.

Other GIS-related data available from the USGS include Landsat 7 ETM+ data, TM data, digital orthophoto quads (DOQs), digital raster graphics (DRGs), and aerial photographs from the National Aerial Photography Program. In 2000, the USGS initiated AmericaView, a program designed to make satellite data from the U.S. government more accessible to the public through a network of state consortia **(http://americaview.usgs.gov/).** The pilot consortium, OhioView, offered Landsat 7 and *ASTER* data for the state of Ohio and elsewhere **(http://www.ohioview.org/).** The USGS expects to expand the program to all 50 states.

5.1.4 U.S. Census Bureau

The U.S. Census Bureau offers the TIGER/Line shapefiles, which are extracts of geographic/

cartographic information from its **MAF/TIGER (Master Address File/Topologically Integrated Geographic Encoding and Referencing)** database. Downloadable at the Census Bureau's website **(http://www.census.gov/),** the TIGER/Line shapefiles contain legal and statistical area boundaries such as counties, census tracts, and block groups, which can be linked to the census data, as well as roads, railroads, streams, water bodies, power lines, and pipelines (Sperling 1995). TIGER/Line attributes include the address range on each side of a street segment, useful for address matching (Chapter 16).

The U.S. Census Bureau also provides the cartographic boundary files, available in shapefile format, for small-scale (1:500,000 to 1:5,000,000), thematic mapping applications. These boundary files are files simplified and smoothed from the TIGER database.

5.1.5 Natural Resources Conservation Service

The Natural Resources Conservation Service (NRCS) of the U.S. Department of Agriculture distributes soils data nationwide through its website **(http://soils.usda.gov/).** Compiled at 1:250,000 scales, the **STATSGO (State Soil Geographic)** database is suitable for broad planning and management uses. Compiled from field mapping at scales ranging from 1:12,000 to 1:63,360, the **SSURGO (Soil Survey Geographic)** database is designed for uses at the farm, township, and county levels.

5.1.6 Statewide Public Data: An Example

The Geospatial One-Stop website provides a link to every state in the United States for statewide GIS data. An example is the Montana State Library **(http://www.nris.state.mt.us/).** This clearinghouse offers statewide and regional data downloadable in either ArcInfo export files or shapefiles. Statewide data include such categories as administrative and political boundary, biological and ecologic, environmental, inland water resources, and transportation networks.

5.1.7 Metropolitan Public Data: An Example

Sponsored by 18 local governments in the San Diego region, the San Diego Association of Governments (SANDAG) **(http://www.sandag .cog.ca.us/)** is an example of a metropolitan data clearinghouse. Data that can be downloaded from SANDAG's website include administrative boundaries, base map features, district boundaries, land cover and activity centers, transportation, and sensitive lands/natural resources.

5.1.8 County-Level Public Data: An Example

Many counties in the United States offer GIS data for sale. Clackamas County in Oregon, for example, distributes data in ArcInfo export files, shapefiles, and DXF files through its GIS division **(http://www.co.clackamas.us/gis/).** Examples of data sets include zoning boundaries, flood zones, tax lots, school districts, voting precincts, park districts, and fire districts.

5.1.9 GIS Data from Private Companies

Many GIS companies are engaged in software development, technical service, consulting, and data production. Some also provide free sample data or can direct GIS users to suitable sources. Esri, for example, offers ArcGIS Online with maps and satellite images to be used with ArcGIS.

Some companies provide specialized GIS data for their customers. An example is Tele Atlas **(http://www.teleatlas.com/),** which offers street, address, census, and postal databases for urban centers and rural areas. Another example is NAVTEQ **(http://www.navteq.com/),** which markets road networks, road attributes, and points of interest for location-based services and vehicle navigation systems. In contrast, online GIS data stores tend to carry a variety of geospatial data. Examples of GIS data stores include GIS Data Depot **(http://data.geocomm.com/),** Map-Mart **(http://www .mapmart.com/),** and LAND INFO International **(http://www.landinfo.com/).**

5.2 METADATA

Metadata provide information about geospatial data (Guptill 1999). They are therefore an integral part of GIS data and are usually prepared and entered during the data production process. Metadata are important to anyone who plans to use public data for a GIS project (Comber, Fisher, and Wadsworth 2005). First, metadata let us know if the data meet our specific needs for area coverage, data quality, and data currency. Second, metadata show us how to transfer, process, and interpret geospatial data. Third, metadata include the contact for additional information.

In 1998 the FGDC published the Content Standards for Digital Geospatial Metadata (CSDGM) **(http://www.fgdc.gov/metadata/ geospatial-metadata-standards).** These standards cover the following information: identification, data quality, spatial data organization, spatial reference, entity and attribute, distribution, metadata reference, citation, time period, and contact. In 2003 the International Organization of Standards (ISO) developed and approved ISO 19115, "Geographic Information—Metadata." The FGDC has since been revising the CSDGM in accord with the international standards. The ISO metadata standards include seven mandatory elements and 14 conditional elements (Box 5.2).

To assist in entering metadata, many metadata tools have been developed for different operating systems. Some tools are free, and some are designed for specific GIS packages. For example, ArcGIS has a metadata tool for creating and updating metadata, including CSDGM and ISO metadata.

5.3 CONVERSION OF EXISTING DATA

Public data are delivered in a variety of formats. Unless the data format is compatible with the GIS package in use, we must first convert the data. **Data conversion** is defined here as a mechanism for converting GIS data from one format to another. Data conversion can be easy or difficult, depending on the specificity of the data format. Proprietary data formats require special translators for data conversion, whereas neutral or public formats require a GIS package that has translators to work with the formats.

5.3.1 Direct Translation

Direct translation uses a translator in a GIS package to directly convert geospatial data from one format to another (Figure 5.1). Direct translation used to be the only method for data conversion before the development of data standards and open GIS. Many users still prefer direct translation because it is easier to use than other methods. ArcToolbox in ArcGIS, for example, can

Box 5.2   ISO Metadata Standards

ISO 19115 metadata standards include two elements: mandatory and conditional. The mandatory elements cover the following: dataset title, dataset reference date, dataset language, dataset topic category, abstract, metadata point of contact, and metadata date stamp. The conditional elements cover the following: dataset responsible party, geographic location by coordinates, dataset character set, spatial resolution, distribution format, spatial representation type, reference system, lineage statement, online resource, metadata file identifier, metadata standard name, metadata standard version, metadata language, and metadata character set.

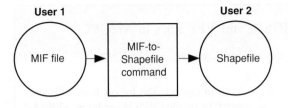

Figure 5.1
The MIF-to-Shapefile tool in ArcGIS converts a
MapInfo file to a shapefile.

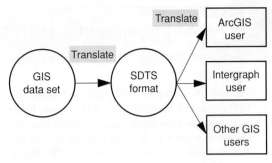

Figure 5.2
To accommodate users of different GIS packages,
a government agency can translate public data into
a neutral format such as SDTS format. Using the
translator in the GIS package, the user can convert the
public data into the format used in the GIS.

translate ArcInfo's interchange files, MGE and
Microstation's DGN files, AutoCAD's DXF and
DWG files, and MapInfo files into shapefiles or
geodatabases. Likewise, GeoMedia can access and
integrate data from ArcGIS, AutoCAD, MapInfo,
MGE, and Microstation.

5.3.2 Neutral Format

A **neutral format** is a public or de facto format
for data exchange. The **Spatial Data Transfer
Standard (SDTS)** is a neutral format approved
by the Federal Information Processing Standards
(FIPS) Program in 1992 **(http://mcmcweb.er
.usgs.gov/sdts/).** Several federal agencies have
converted some or all of their data to SDTS format. The USGS, for example, has converted many
DLG files into SDTS format. GIS vendors such as
Esri provide translators in their software packages
for importing SDTS data (Figure 5.2).

In practice, SDTS uses "profiles" to transfer spatial data. Each profile is targeted at a particular type of
spatial data. Currently there are five SDTS profiles:

- The Topological Vector Profile (TVP) covers
  DLG, TIGER, and other topology-based
  vector data.
- The Raster Profile and Extensions (RPE)
  accommodate DOQ, DEM, and other raster
  data.
- The Transportation Network Profile (TNP)
  covers vector data with network topology.
- The Point Profile supports geodetic control
  point data.
- The Computer Aided Design and Drafting
  Profile (CADD) supports vector-based
  CADD data, with or without topology.

Creating an elevation raster from an SDTS
raster profile is relatively straightforward. But
creating a topology-based vector data set from
an SDTS topological vector profile can be more
involved because the SDTS file may contain composite features such as routes and regions (Chapter 3)
in addition to topology.

The **vector product format (VPF),** used
by the U.S. Department of Defense, is a standard format, structure, and organization for large
geographic databases. The National Geospatial-
Intelligence Agency (NGA) uses VPF for digital
vector products developed at a variety of scales
(http://www.nga.mil/). NGA's vector products
for drainage systems, transportation, political
boundaries, and populated places are also part of
the global database that is being developed by
the International Steering Committee for Global
Mapping (ISCGM) **(http://www.iscgm.org/cgi-bin/
fswiki/wiki.cgi).** Similar to an SDTS topological
vector profile, a VPF file may contain composite
features of regions and routes. Box 5.3 summarizes tools in ArcGIS for importing DLG, SDTS,
TIGER, and VPF files.

Although a neutral format is typically used
for public data from government agencies, it can
also be found with "industry standards" in the private sector. A good example is the DXF (drawing

Box 5.3   Importing DLG, SDTS, TIGER, and VPF Files in ArcGIS

Tools for importing DLG, SDTS, TIGER, and VPF files are all available from the Coverage Tools/Conversion/To Coverage toolset. Coverage Tools are limited to users who have an ArcInfo license and ArcInfo Workstation on their computer. The Import From DLG tool can convert a DLG file, either standard or optional, into a coverage. The Import From SDTS tool can convert a Topological Vector Profile or a Point Profile into a coverage and a Raster Profile into an Esri grid. The Advanced Tiger Conversion tool can convert a set of TIGER/Line files into a set of coverages. The tool also has an option for the user to choose which version of TIGER/Line files

(e.g., 2000) to convert. The Import from VPF tool can convert an un-tiled VPF coverage or VPF tile into a coverage: a line coverage with route subclasses for multiple linear feature classes, and a polygon coverage with region subclasses for multiple area feature classes.

For ArcGIS users who do not have ArcInfo Workstation installed and licensed on their computer, there are options for obtaining data in formats other than SDTS or TIGER. As an example, Task 1 in Chapter 5 uses a USGS DEM in Esri grid format downloaded from the National Elevation Dataset (NED) website.

---

interchange file) format of AutoCAD. Another example is the ASCII format. Many GIS packages can import point data with *x*-, *y*-coordinates in ASCII format into digital data sets.

5.4 CREATING NEW DATA

Different data sources can be used for creating new geospatial data. Among these sources are street addresses from which point features can be created in address geocoding, a method to be covered in Chapter 16.

5.4.1 Remotely Sensed Data

Satellite images can be digitally processed to produce a wide variety of thematic data for a GIS project. Land use/land cover data such as USGS National Land Cover Database 2001 are typically derived from satellite images. Other types of data include vegetation types, crop health, eroded soils, geologic features, the composition and depth of water bodies, and even snowpack. Satellite images provide timely data and, if collected at regular intervals, they can also provide temporal data valuable for recording and monitoring changes in the terrestrial and aquatic environments.

Some GIS users felt in the past that satellite images did not have sufficient resolution, or were not accurate enough, for their projects. This is no longer the case with high-resolution satellite images (Chapter 4). IKONOS and QuickBird images can now be used to extract detailed features such as roads, trails, buildings, trees, riparian zones, and impervious surfaces.

DOQs are digitized aerial photographs that have been differentially rectified to remove image displacements by camera tilt and terrain relief. DOQs therefore combine the image characteristics of a photograph with the geometric qualities of a map. Black-and-white USGS DOQs have a 1-meter ground resolution (i.e., each pixel in the image measures 1-by-1 meter on the ground) and pixel values representing 256 gray levels (Figure 5.3). DOQs can be effectively used as a background for digitizing or updating of new roads, new subdivisions, and timber harvested areas.

5.4.2 Field Data

Two important types of field data are survey data and **global positioning system (GPS) data.** Survey data consist primarily of distances, directions, and elevations. Distances can be measured

Figure 5.3
A digital orthophoto quad (DOQ) can be used as the background for digitizing or updating geospatial data.

in feet or meters using a tape or an electronic distance measurement instrument. The direction of a line can be measured in azimuth or bearing using a transit, theodolite, or total station. An azimuth is an angle measured clockwise from the north end of a meridian to the line. Azimuths range in magnitude from 0° to 360°. A bearing is an acute angle between the line and a meridian. The bearing angle is always accompanied by letters that locate the quadrant (i.e., NE, SE, SW, or NW) in which the line falls. In the United States, most legal plans use bearing directions. An elevation difference between two points can be measured in feet or meters using levels and rods.

In GIS, field survey typically provides data for determining parcel boundaries. An angle and a distance can define a parcel boundary between two stations (points). For example, the description of N45°30'W 500 feet means that the course (line) connecting the two stations has a bearing angle of 45 degrees 30 minutes in the NW quadrant and a distance of 500 feet (Figure 5.4). A parcel represents a close traverse, that is, a series of established

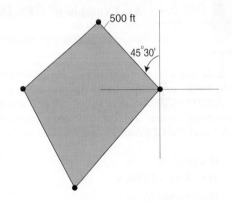

Figure 5.4
A bearing and a distance determine a course between two stations.

stations tied together by angle and distance (Kavanagh 2003). A close traverse also begins and ends at the same point. **Coordinate geometry (COGO),** a study of geometry and algebra, provides the methods for creating geospatial data of points, lines, and polygons from survey data.

Using GPS satellites in space as reference points, a GPS receiver can determine its precise position on the Earth's surface (Moffitt and Bossler 1998). GPS data include the horizontal location based on a geographic or projected coordinate system and, if chosen, the height of the point location (Box 5.4). A collection of GPS positions along a line can determine a line feature, and a series of lines measured by GPS can determine an area feature. This is why GPS has become a useful tool for collecting geospatial data (Kennedy 1996), for validating geospatial data such as road networks (Wu et al. 2005), and for tracking point objects such as vehicles and people (Laube, Imfeld, and Weibel 2005).

The GPS receiver measures its distance (range) from a satellite using the travel time and speed of signals it receives from the satellite. With three satellites simultaneously available, the receiver can determine its position in space (x, y, z) relative to the center of mass of the Earth. But to correct timing errors, a fourth satellite is required to get precise positioning (Figure 5.5). The receiver's position in space can then be converted to latitude, longitude, and height based on the World Geodetic System 1984 (WGS84).

The U.S. military maintains a constellation of 24 NAVSTAR (Navigation Satellite Timing and Ranging) satellites in space, and each satellite follows a precise orbit **(http://gps.losangeles.af.mil/jpo/).** This constellation gives GPS users between five and eight satellites visible from any point on the Earth's surface. Data transmitted by GPS satellites are modulated onto two carrier waves, referred to as L1 and L2. Two binary codes are in turn modulated onto the carrier waves. These two codes are the coarse acquisition (C/A) code, which

---

Box 5.4 **An Example of GPS Data**

T he following printout is an example of GPS data. The header information shows that the datum used is NAD27 (North American Datum of 1927) and the coordinate system is UTM (Universal Transverse Mercator). The GPS data include seven point locations.

The record for each point location includes the UTM zone number (i.e., 11), Easting (x-coordinate), and Northing (y-coordinate). The GPS data do not include the height or Alt value.

H R DATUM

M G NAD27 CONUS

H Coordinate System
U UTM UPS

H	IDNT	Zone	Easting	Northing	Alt	Description
W	001	11T	0498884	5174889	−9999	09-SEP-98
W	002	11T	0498093	5187334	−9999	09-SEP-98
W	003	11T	0509786	5209401	−9999	09-SEP-98
W	004	11T	0505955	5222740	−9999	09-SEP-98
W	005	11T	0504529	5228746	−9999	09-SEP-98
W	006	11T	0505287	5230364	−9999	09-SEP-98
W	007	11T	0501167	5252492	−9999	09-SEP-98

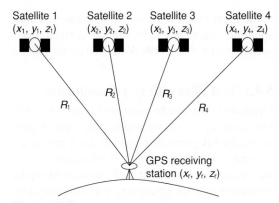

Satellite 1 (x_1, y_1, z_1) Satellite 2 (x_2, y_2, z_2) Satellite 3 (x_3, y_3, z_3) Satellite 4 (x_4, y_4, z_4)

R_1 R_2 R_3 R_4

GPS receiving station (x_r, y_r, z_r)

Figure 5.5
Use four GPS satellites to determine the coordinates of a receiving station. x_i, y_i, and z_i are coordinates relative to the center of mass of the Earth. R_i represents the distance (range) from a satellite to the receiving station.

is available to the public, and the precise (P) code, which is designed for military use exclusively. In addition to NAVSTAR satellites, there are also the Russian GLONASS system (**http://www.glonass-center.ru/**) and the Galileo system (**http://www.esa.int/esaNA/galileo.html**). (The first test satellite for the Galileo system was launched in December 2005.)

An important aspect of using GPS for spatial data entry is correcting errors in GPS data. One type of error is intentional. To make sure that no hostile force could get precise GPS readings, the U.S. military used to degrade their accuracy under the policy called "Selective Availability" or "SA" by introducing noise into the satellite's clock and orbital data. SA was switched off in May 2000, and GPS accuracies in basic point positioning improved from 100 meters to about 10 to 20 meters. Other types of errors may be described as noise errors, including ephemeris (positional) error, clock errors (orbital errors between monitoring times), atmospheric delay errors, and multipath errors (signals bouncing off obstructions before reaching the receiver).

With the aid of a reference or base station, **differential correction** can significantly reduce noise errors. Located at points that have been accurately surveyed, reference stations are operated by private companies and by public agencies such as

those participating in the National Geodetic Survey (NGS) Continuously Operating Reference System (CORS). Using its known position, the reference receiver can calculate what the travel time of the GPS signals should be. The difference between the predicted and actual travel times thus becomes an error correction factor. The reference receiver computes error correction factors for all visible satellites. These correction factors are then available to GPS receivers covered by the reference station. GIS applications usually do not need real-time transmission of error correction factors. Differential correction can be made later as long as records are kept of measured positions and the time each position is measured.

Equally as important as correcting errors in GPS data is the type of GPS receiver. Most GIS users use code-based receivers (Figure 5.6). With differential correction, code-based GPS readings can easily achieve an accuracy of 3 to 5 meters, and some newer receivers are even capable of submeter accuracy. Carrier phase receivers and dual-frequency receivers are mainly used in surveying and geodetic control. They are capable of subcentimeter differential accuracy (Lange and Gilbert 1999).

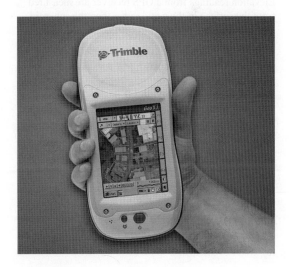

Figure 5.6
A portable GPS receiver. (Courtesy of Trimble.)

GPS data can include heights at point locations. Like x-, y-coordinates, heights (z) obtained from GPS are referenced to WGS84, a spheroid. Spheroidal heights can be transformed into elevations (also called *orthometric heights*) by using a geoid, a closer approximation of the Earth than a spheroid. The surface of a geoid is treated as the surface of mean sea level, and the separation between the geoid surface and the spheroid surface is called *geoid undulation*. As shown in Figure 5.7, a spheroidal height can be transformed into an elevation (h_1) by knowing the geoid undulation (h_2)

at the point location. In the United States, geoid–spheroid separation data can be derived from geoids such as GEOID99 developed by the NGS.

5.4.3 Text Files with x-, y-Coordinates

Geospatial data can be generated from a text file that contains x-, y-coordinates, either geographic (in decimal degrees) or projected. Each pair of x-, y-coordinates creates a point. Therefore, we can create spatial data from a file with recorded locations of weather stations, epicenters, or a hurricane track.

5.4.4 Digitizing Using a Digitizing Table

Digitizing is the process of converting data from analog to digital format. Manual digitizing uses a digitizing table (Figure 5.8). A **digitizing table** has a built-in electronic mesh, which can sense the position of the cursor. To transmit the x-, y-coordinates of a point to the connected computer, the operator simply clicks on a button on the cursor after lining up the cursor's cross hair with the point. Large-size digitizing tables typically have an absolute accuracy of 0.001 inch (0.003 centimeter).

Many GIS packages have a built-in digitizing module for manual digitizing. The module is likely

h_1 = elevation at point a

h_2 = geoid undulation at point a

$h_1 + h_2$ = spheroid height at point a

Figure 5.7
Elevation readings from a GPS receiver are measured from the surface of the geoid rather than the spheroid.

(a)

(b)

Figure 5.8
A large digitizing table (*a*) and a cursor with a 16-button keypad (*b*). (Courtesy of GTCO Calcomp, Inc.)

to have commands that can help move or snap a feature (i.e., a point or line) to a precise location. Figure 5.9 shows that a line can be snapped to an existing line within a user-specified tolerance or distance. Likewise, Figure 5.10 shows that a point can be snapped to another point, again within a specified distance.

Digitizing usually begins with a set of control points (also called tics), which are later used for converting the digitized map to real-world coordinates (Chapter 6). Digitizing point features is simple: each point is clicked once to record its location. Digitizing line features can follow either point mode or stream mode. The operator selects points to digitize in point mode. In stream mode, lines are digitized at a preset time or distance interval.

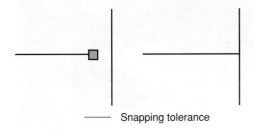

———— Snapping tolerance

Figure 5.9
The end of a new line can be automatically snapped to an existing line if the gap is smaller than the specified snapping tolerance.

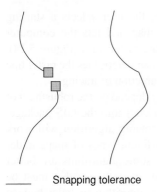

———— Snapping tolerance

Figure 5.10
A point can be automatically snapped to another point if the gap is smaller than the specified snapping tolerance.

For example, lines can be automatically digitized at a 0.01-inch interval. Point mode is preferred if features to be digitized have many straight-line segments. Because the vector data model treats a polygon as a series of lines, digitizing polygon features is the same as digitizing line features. Additionally, a polygon may have a label, which is treated as a point inside the polygon.

Although digitizing itself is mostly manual, the quality of digitizing can be improved with planning and checking. An integrated approach is useful in digitizing different layers of a GIS database that share common boundaries. For example, soils, vegetation types, and land-use types may share some common boundaries in a study area. Digitizing these boundaries only once and using them on each layer not only saves time in digitizing but also ensures coincident boundaries.

A rule of thumb in digitizing line or polygon features is to digitize each line once and only once to avoid duplicate lines. Duplicate lines are seldom on top of one another because of the high accuracy of a digitizing table. One way to reduce the number of duplicate lines is to put a transparent sheet on top of the source map and to mark off each line on the transparent sheet after the line is digitized. This method can also reduce the number of missing lines.

5.4.5 Scanning

A digitizing method, **scanning** uses a scanner (Figure 5.11) to convert an analog map into a scanned file in raster format, which is then converted back to vector format through tracing (Verbyla and Chang 1997). The simplest type of map to be scanned is a black-and-white map: black lines represent map features, and white areas represent the background. The map may be a paper or Mylar map, and it may be inked or penciled. The scanned image is binary: each pixel has a value of either 1 (map feature) or 0 (background). Map features are shown as raster lines, a series of connected pixels on the scanned file (Figure 5.12). The pixel size depends on the scanning resolution, which is often set at 300 dots per inch (dpi) or

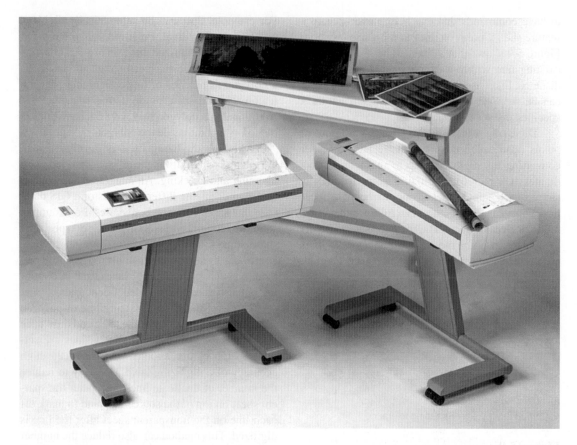

Figure 5.11
Large-format drum scanners. (Courtesy of GTCO Calcomp, Inc.)

400 dpi for digitizing. A raster line representing a thin inked line on the source map may have a width of 5 to 7 pixels (Figure 5.13).

Color maps, including historical maps (Leyk, Boesch, and Weibel 2005), can also be scanned by a scanner that can recognize colors. A DRG, for example, can have 13 different colors, each representing one type of map feature scanned from a USGS quadrangle map.

To complete the digitizing process, a scanned file must be vectorized. **Vectorization** turns raster lines into vector lines in a process called tracing. Tracing involves three basic elements: line thinning, line extraction, and topological reconstruction. Tracing can be semiautomatic or manual. In semiautomatic mode, the user selects a starting point on the image map and lets the computer trace all the connecting raster lines (Figure 5.14). In manual mode, the user determines the raster line to be traced and the direction of tracing.

Results of tracing depend on the robustness of the tracing algorithm built into the GIS package. Although no single tracing algorithm will work satisfactorily with different types of maps under different conditions, some algorithms are better than others. Examples of problems that must be solved by the tracing algorithm include: how to trace an intersection, where the width of a raster line may double or triple (Figure 5.15); how to continue when a raster line is broken or when two

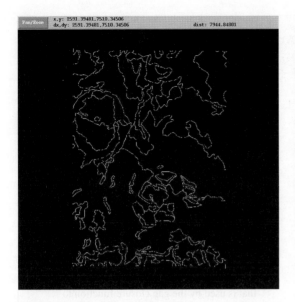

Figure 5.12
A binary scanned file: The lines are soil lines, and the black areas are the background.

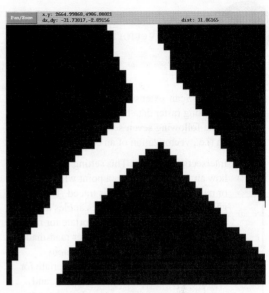

Figure 5.13
A raster line in a scanned file has a width of several pixels.

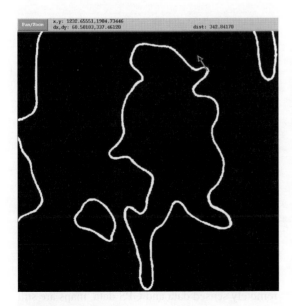

Figure 5.14
Semiautomatic tracing starts at a point (shown with an arrow) and traces all lines connected to the point.

Figure 5.15
The width of a raster line doubles or triples when lines meet or intersect.

Box **5.5** | **Vectorization Settings in ArcGIS**

The ArcScan extension to ArcGIS is designed for converting raster data to vector features. ArcScan offers the following seven settings for batch vectorization (i.e., vectorization of an entire raster).

- Intersection solution: This setting determines how an intersection (i.e., a point where three or more raster lines meet) is traced. The geometrical solution preserves angles and straight lines, which are appropriate for maps showing streets, canals, and other man-made features. The median solution has non-rectilinear angles and is more appropriate for natural features such as soils, streams, and vegetation covers. The none solution does not connect lines at an intersection.
- Maximum line width: This setting separates lines and polygons in vectorization. Centerline vectorization applies to raster lines that are less

than or equal to the maximum line width. All other lines are vectorized as polygon features.
- Compression tolerance: This setting determines the degree of generalization used in a vector postprocessing procedure. A Douglas-Peucker algorithm (Chapter 7) is used for generalization.
- Smoothing weight: This setting determines the degree of smoothing in vectorization. A larger weight results in smoother line features.
- Gap closure tolerance: This setting determines how many pixels can be jumped over in a broken raster line.
- Fan angle: This setting determines the angle that is used by the gap closure function to search for raster lines when jumping over gaps.
- Holes: This setting determines the size of holes in a raster line that can be ignored during tracing.

raster lines are close together; and how to separate a line from a polygon. A tracing algorithm normally uses the parameter values specified by the user in solving the problems. Box 5.5 lists the parameters that we can specify in ArcGIS.

Is scanning better than manual digitizing for data input? Large data producers apparently think so for the following reasons. First, scanning uses the machine and computer algorithm to do most of the work, thus avoiding human errors caused by fatigue or carelessness. Second, tracing has both the scanned image and vector lines on the same screen, making tracing more flexible than manual digitizing. With tracing, the operator can zoom in or out and can move around the raster image with ease. Manual digitizing requires the operator's attention on both the digitizing table and the computer monitor, and the operator can easily get tired. Third, scanning has been reported to be more cost effective than manual digitizing. In the United

States, the cost of scanning by a service company has dropped significantly in recent years.

5.4.6 On-Screen Digitizing

On-screen digitizing, also called *heads-up digitizing,* is manual digitizing on the computer monitor using a data source such as DOQ as the background. On-screen digitizing is an efficient method for editing or updating an existing layer such as adding new trails or roads from a DOQ. Likewise, we can use the method to update new clear-cuts or burned areas in a vegetation layer.

5.4.7 Importance of Source Maps

Despite the increased availability of high-resolution remotely sensed data and GPS data, maps are still a dominant source for creating new GIS data. Digitizing, either manual digitizing or scanning, converts an analog map to its digital format. The accuracy of the digitized map can be only as good or as accurate as its source map.

A number of factors can affect the accuracy of the source map. Maps such as USGS quadrangle maps are secondary data sources because these maps have gone through the cartographic processes of compilation, generalization, and symbolization. Each of these processes can in turn affect the accuracy of the mapped data. For example, if the compilation of the source map contains errors, these errors will be passed on to the digital map.

Paper maps generally are not good source maps for digitizing because they tend to shrink and expand with changes in temperature and humidity. In even worse scenarios, GIS users may use copies of paper maps or mosaics of paper map copies. Such source maps will not yield good results. Because of their plastic backing, Mylar maps are much more stable than paper maps for digitizing.

The quality of line work on the source map will determine not only the accuracy of the digital map but also the operator's time and effort in digitizing and editing. The line work should be thin, continuous, and uniform, as expected from inking or scribing—never use felt-tip markers to prepare the line work. Penciled source maps may be adequate for manual digitizing but are not recommended for scanning. Scanned files are binary data files, separating only the map feature from the background. Because the contrast between penciled lines and the background (i.e., surface of paper or Mylar) is not as sharp as inked lines, we may have to adjust the scanning parameters to increase the contrast. But the adjustment often results in the scanning of erased lines and smudges, which should not be in the scanned file. For supplemental information on the source map, words or symbols can be drawn in orange color, which will not be scanned.

KEY CONCEPTS AND TERMS

Coordinate geometry (COGO): A branch of geometry that provides the methods for creating geospatial data of points, lines, and polygons from survey data.

Data conversion: Conversion of geospatial data from one format to another.

Differential correction: A method that uses data from a base station to correct noise errors in GPS data.

Digital line graphs (DLGs): Digital representations of point, line, and area features from USGS quadrangle maps including contour lines, spot elevations, hydrography, boundaries, transportation, and the U.S. Public Land Survey System.

Digitizing: The process of converting data from analog to digital format.

Digitizing table: A table with a built-in electronic mesh that can sense the position of the cursor and can transmit its x-, y-coordinates to the connected computer.

Direct translation: The use of a translator or algorithm in a GIS package to directly convert geospatial data from one format to another.

Federal Geographic Data Committee (FGDC): A U.S. multiagency committee that coordinates the development of geospatial data standards.

Framework data: Data that many organizations regularly use for GIS activities.

Geospatial One-Stop (GOS): A portal established by the Federal Office of Management and Budget for accessing geospatial data.

Global positioning system (GPS) data: Longitude, latitude, and elevation data for point locations made available through a navigational satellite system and a receiver.

Metadata: Data that provide information about geospatial data.

National Elevation Dataset (NED): A USGS program that uses a seamless system for delivering

1:24,000 scale DEMs (1:63,360 scale DEMs for Alaska).

National Land Cover Data 1992 (NLCD 1992):   Land use/land cover data compiled from the Thematic Mapper imagery of the early 1990s for the conterminous United States.

National Land Cover Database 2001 (NLCD 2001):   Land use/land cover data compiled from the Landsat 7 ETM+ imagery for all 50 states of the United States and Puerto Rico.

Neutral format: A public format such as SDTS that can be used for data exchange.

On-screen digitizing: Manual digitizing on the computer monitor by using a data source such as DOQ as the background.

Scanning: A digitizing method that converts an analog map into a scanned file in raster format, which can then be converted back to vector format through tracing.

Spatial Data Transfer Standard (SDTS):   Public data format for transferring

geospatial data such as DLGs and DEMs from the USGS.

SSURGO (Soil Survey Geographic):   A soil database compiled from field mapping at scales from 1:12,000 to 1:63,360 by the Natural Resources Conservation Service of the U.S. Department of Agriculture.

STATSGO (State Soil Geographic):   A soil database compiled at 1:250,000 scale by the Natural Resources Conservation Service of the U.S. Department of Agriculture.

TIGER (Topologically Integrated Geographic Encoding and Referencing):   A database prepared by the U.S. Census Bureau that contains legal and statistical area boundaries, which can be linked to the census data.

Vectorization: The process of converting raster lines into vector lines through tracing.

Vector product format (VPF):   A standard format, structure, and organization for large geographic databases used by the U.S. military.

REVIEW QUESTIONS

1. What kinds of data are contained in the USGS DLG files?

2. What is SSURGO?

3. Suppose you want to make a map showing the rate of population change between 1990 and 2000 by county in your state. Describe (1) the kinds of digital data you will need for the mapping project, and (2) the website(s) from which you will download the data.

4. Find the GIS data clearinghouse for your state at the Geospatial One-Stop website (**http://gos2.geodata.gov/wps/portal/gos**). Go to the clearinghouse website. Select the metadata of a data set and go over the information in each category.

5. Describe the kinds of data that are contained in an SDTS topological vector profile, point profile, and raster profile.

6. What kinds of data are contained in the TIGER/Line files?

7. Go to ArcGIS Desktop Help > Data support in ArcGIS > Data formats supported in ArcGIS. The help page lists data formats developed by Esri, other GIS companies, or government agencies that can be used in ArcGIS. List three non-Esri data formats that can be used directly in ArcGIS, and three that can be imported into ArcGIS.

8. Describe two common types of field data that can be used in a GIS project.

9. Explain how differential correction works.

10. What types of GPS data errors can be corrected by differential correction?

11. What kinds of data must exist in a text file so that the text file can be converted to a shapefile?

12. What is COGO?

13. Suppose you are asked to convert a paper map to a digital data set. What methods can you use for the task? What are the advantages and disadvantages of each method?

14. Explain the difference between point mode and stream mode for digitizing.

15. The scanning method for digitizing involves both rasterization and vectorization. Why?

16. The source map can greatly influence the quality of a digitized map. Provide an example that supports the statement.

APPLICATIONS: GIS DATA ACQUISITION

This applications section covers three methods for data acquisition. In Task 1, you will download USGS DEM from the Internet. Task 2 covers on-screen digitizing. Task 3 uses a table with *x-, y*-coordinates. There are other methods for data acquisition: Chapter 6 covers the use of a scanned file for digitizing, and Chapter 16 covers address geocoding.

Task 1 Download USGS DEM from the Internet

What you need: access to the Internet and a unzip tool; *emidastrm.shp,* a stream shapefile.

1. Go to the National Elevation Dataset (NED), **http://seamless.usgs.gov/.** Click on "Continue to the Pre-packaged NED Download tool." On the next dialog for entering the area of interest, first click on Switch to Decimal Degrees. Then enter 47.124995 for top latitude, 47.067361 for bottom latitude, −116.62501 for left longitude, and −116.5516 for right longitude. Click Submit. The next dialog shows DEM data available for the area. Select 1 arc-second n48w117 in NED Pre-packaged ArcGrid format to download. Save the download bundle in the Chapter 5 database. Unzip the bundle.

2. The elevation data set *grdn48w117_1* is included in the *n48w117 folder.* Covering an area of 1 × 1 degree, *grdn48w117_1* is measured in longitude and latitude values based on NAD 1983.

3. Launch ArcMap, and rename the data frame as Task 1. Add *emidastrm.shp* to Task 1.

Then add *grdn48w117_1* to Task 1. Opt to build pyramid. The next dialog is a warning message because emidastrm.shp is based on NAD 1927, different from the datum for grdn48w117_1. Click on the Transformations button. In the next dialog, choose to convert from GCS_North_American_1983 into GCS_North_American_1927 by using NAD_1927_To_NAD_1983_NADCON and click OK. Close the warning message.

4. *emidastrm.shp* is now displayed on top of the elevation data sete.

Q1. What is the elevation range (in meters) in *grdn48w117_1*?

Q2. What is the cell size (in decimal degree) of *grdn48w117_1*?

Task 2 Digitize On-Screen in ArcMap

What you need: *land_dig.shp,* a background map for digitizing. *land_dig.shp* is based on UTM coordinates and is measured in meters.

On-screen digitizing is technically similar to manual digitizing. The differences are as follows: you use the mouse pointer rather than the digitizer's cursor for digitizing; you use a feature or image layer as the background for digitizing; and you must zoom in and out repeatedly while digitizing. Task 2 lets you digitize several polygons off *land_dig.shp* to make a new shapefile. You will digitize the new shapefile in a freehand style using the image as the background.

1. Make sure that ArcCatalog is connected to the Chapter 5 folder. First create a new shapefile for digitizing. Right-click the Chapter 5 folder, point to New, and select Shapefile. In the next dialog, enter *trial1* for the name, select Polygon for the feature type, and click the Edit button for the spatial reference. Import the coordinate system of *land_dig.shp* for *trial1*.

2. Insert a data frame in ArcMap and rename it Task 2. Add *trial1* and *land_dig.shp* to Task 2. Make sure that *trial1* is on top of *land_dig* in the table of contents. Before digitizing, you need to change the symbol of the two shapefiles, define the selection layer, and set up the digitizing environment. To facilitate digitizing, you want to draw *land_dig* in red with labels and *trial1* in black. Select Properties from the context menu of *land_dig*. On the Symbology tab, click Symbol and change it to a Hollow symbol with the Outline Color in red. On the Labels tab, check the box to label features in this layer and select LAND_DIG_I from the dropdown list for the label field. Click OK to dismiss the Layer Properties dialog. (You may need to right-click *land_dig* and select Zoom to Layer to see the layer.) Click the symbol of *trial1* in the table of contents. Select the Hollow symbol and the Outline Color of black.

3. Click the List by Selection tab in the table of contents. Click the icon to the right of *land_dig* to make it unselectable. This ensures that *trial1* is the only selectable layer during digitizing. Switch back to the List by Drawing Order tab.

4. Click the Customize menu, point to Toolbars, and check the Editor Toolbar. (The alternative is to click the Editor Toolbar button.) Select Start Editing from the Editor dropdown list. Click *Trial1* in the Create Features window, and Polygon is highlighted in the Construction Tools window. Click the Customize menu, point to Toolbars, and check Snapping. Select Options from the dropdown menu of Snapping, and enter 10 (pixels) in the Tolerance box of the Snapping Options dialog. Click the dropdown menu of Snapping again and make sure that Use Snapping is checked.

5. You are ready to digitize. Zoom to the area around polygon 72. Notice that polygon 72 in *land_dig* is made of a series of lines (edges), which are connected by points (vertices). Click Polygon in the Construction Tools window. (If closed, the Construction Tools window can be reopened through Editing Windows on the Editor pulldown menu.) Click the Straight Segment Tool on the Editor toolbar. Digitize a starting point of polygon 72 by left-clicking the mouse. Use *land_dig* as a guide to digitize the other vertices. When you come back to the starting point, right-click the mouse and select Finish Sketch. The completed polygon 72 appears in cyan with an x inside it. A feature appearing in cyan is an active feature. To unselect polygon 72, click the Edit tool and click a point outside the polygon. If you need to delete a polygon in *trial1* later, use the Edit tool to first select and activate the polygon and then press the Delete key.

6. Digitize polygon 73. You can zoom in and out, or use other tools, any time during digitizing. Click the Straight Segment tool whenever you are ready to resume digitizing.

7. Digitize polygons 74 and 75 next. The two polygons have a shared border. You will digitize one of the two polygons first, and then digitize the other polygon by using the Auto Complete Polygon option. Select the construction tool of Polygon, and then click the Straight Segment tool. Digitize polygon 75 first. After polygon 75 is digitized, select the construction tool of Auto Complete Polygon. Now digitize polygon 76, while polygon 75 is active. You start by left-clicking one end of the shared border with poygon 75, digitize the boundary that is not shared with polygon 75, and finish by double-clicking the other end of the shared border.

8. You are done with digitizing. Right-click *trial1* in the table of contents, and select Open Attribute Table. Click the first cell under Id and enter 72. Enter 73, 74, and 75 in the next three cells. (You can click the box to the left of a record and see the polygon that corresponds to the record.) Close the table.

9. Select Stop Editing from the Editor dropdown list. Save the edits.

Q3. Define the snapping tolerance. (Tip: Use the Index tab in ArcGIS Desktop Help.)

Q4. Will a smaller snapping tolerance give you a more accurate digitized map? Why?

Q5. What other construction tools are available, besides Polygon and Auto Complete Polygon?

Task 3 Add XY Data in ArcMap

What you need: *events.txt,* a text file containing GPS readings.

In Task 3, you will use ArcMap to create a new shapefile from *events.txt,* a text file that contains *x-, y*-coordinates of a series of points collected by GPS readings.

1. Insert a data frame in ArcMap and rename it Task 3. Add *events.txt* to Task 3. Right-click *events.txt* and select Display XY Data. Make sure that *events.txt* is the table to be added as a layer. Use the dropdown list to select EASTING for the X Field and NORTHING for the Y Field. Click the Edit button for the spatial reference of input coordinates. Select projected coordinate systems, UTM, NAD1927, and NAD 1927 UTM Zone 11N.prj. Click OK to dismiss the dialogs. Ignore the warning message that the table

does not have object-ID field. *events.txt Events* is added to the table of contents.

2. *events.txt Events* can be saved as a shapefile. Right-click *events.txt Events,* point to Data, and select Export Data. Opt to export all features and save the output as *events.shp* in the Chapter 5 database.

Challenge Task

What you need: *quake.txt.*

quake.txt in the Chapter 5 database contains earthquake data in northern California from January 2002 to August 2003. The quakes recorded in the file all had a magnitude of 4.0 or higher. The Northern California Earthquake Data Center maintains and catalogs the quake data (**http://quake.geo.berkeley.edu/**).

This challenge task asks you to perform two related tasks: (1) use the longitude (Lon) and latitude (Lat) readings in *quake.txt* to create a shapefile called *quake* and define its coordinate system as NAD27, and (2) download a shapefile of California counties. To download the shapefile of California counties, go to **http://www.atlas.ca.gov/.** On the Download tab, click Boundaries and then County Boundaries (1:24k). Download cnty24k09 1 shp.zip. Select geopolitical/counties.

Q1. How many records are in *quake*?

Q2. What is the maximum magnitude recorded in *quake*?

Q3. What coordinate system is *cnty24k09 1* based on?

Q4. What are the parameter values of the coordinate system in Q3

Q5. Were the quakes recorded in *quake* all on land?

REFERENCES

Anderson, J. R., E. E. Hardy, J. T. Roach, and R. E. Witmer. 1976. A Land Use and Land Cover Classification System for Use with Remote Sensor Data. *U.S. Geological Survey Professional Paper 964.* Washington, DC: U.S. Government Printing Office.

Comber, A., P. Fisher, and R. Wadsworth. 2005. Comparing Statistical and Semantic Approaches for Identifying Change from Land Cover Datasets. *Journal of Environmental Management* 77: 47–55.

Falke, S. R. 2002. Environmental Data: Finding It, Sharing It, and Using It. *Journal of Urban Technology* 9: 111–24.

Goodchild, M. F., P. Fu, and P. Rich. 2007. Sharing Geographic Information: An Assessment of the Geospatial One-Stop. *Annals of the Association of American Geographers* 97: 250–66.

Guptill, S. C. 1999. Metadata and Data Catalogues. In P. A. Longley, M. F. Goodchild, D. J. Maguire, and D. W. Rhind, eds., *Geographical Information Systems,* 2d ed., pp. 77–92. New York: Wiley.

Jacoby, S., J. Smith, L. Ting, and I. Williamson. 2002. Developing a Common Spatial Data Infrastructure Between State and Local Government—An Australian Case Study. *International Journal of Geographical Information Science* 16: 305–22.

Kavanagh, B. F. 2003. *Geomatics.* Upper Saddle River, NJ: Prentice Hall.

Kennedy, M. 1996. *The Global Positioning System and GIS.* Ann Arbor, MI: Ann Arbor Press.

Lange, A. F., and C. Gilbert. 1999. Using GPS for GIS Data Capture. In P. A. Longley, M. F. Goodchild, D. J. Maguire, and D. W. Rhind, eds., *Geographical Information Systems,* 2d ed., pp. 467–79. New York: Wiley.

Laube, P., S. Imfeld, and R. Weibel. 2005. Discovering Relative Motion Patterns in Groups of Moving Point Objects. *International Journal of Geographical Information Science* 19: 639–68.

Leyk, S., R. Boesch, and R. Weibel. 2005. A Conceptual Framework for Uncertainty Investigation in Map-Based Land Cover Change Modelling. *Transactions in GIS* 9: 291–322.

Maguire, D. J., and P. A. Longley. 2005. The Emergence of Geoportals and Their Role in Spatial Data Infrastructures. *Computers, Environment and Urban Systems* 29: 3–14.

Masser, I. 1999. All Shapes and Sizes: The First Generation of National Spatial Data Infrastructures. *International Journal of Geographical Information Science* 13: 67–84.

Moffitt, F. H., and J. D. Bossler. 1998. *Surveying,* 10th ed. Menlo Park, CA: Addison-Wesley.

Sperling, J. 1995. Development and Maintenance of the TIGER Database: Experiences in Spatial Data Sharing at the U.S. Bureau of the Census. In H. J. Onsrud and G. Rushton, eds., *Sharing Geographic Information,* pp. 377–96. New Brunswick, NJ: Center for Urban Policy Research.

Verbyla, D. L., and K. Chang. 1997. *Processing Digital Images in GIS.* Santa Fe, NM: OnWord Press.

Wu, J., T. H. Funk, F. W. Lurmann, and A. M. Winer. 2005. Improving Spatial Accuracy of Roadway Networks and Geocoded Addresses. *Transactions in GIS* 9: 585–601.

CHAPTER 6

GEOMETRIC TRANSFORMATION

CHAPTER OUTLINE

6.1 Geometric Transformation
6.2 Root Mean Square (RMS) Error
6.3 Interpretation of RMS Errors
on Digitized Maps
6.4 Resampling of Pixel Values

A newly digitized map has the same measurement unit as the source map used in digitizing or scanning. If manually digitized, the map is measured in inches, same as the digitizing table. If converted from a scanned image, the map is measured in dots per inch (dpi). Obviously this newly digitized map cannot be aligned spatially with GIS layers that are based on projected coordinate systems (Chapter 2). To make it usable, we must convert the newly digitized map into a projected coordinate system, whether it is the UTM (Universal Transverse Mercator) or State Plane Coordinate (SPC) system. This conversion is called geometric transformation, which, in this case, transforms map feature coordinates from digitizer units or dpi

into projected coordinates. Only through a geometric transformation can a newly digitized map align with other layers for data display or analysis.

Geometric transformation also applies to satellite images. Remotely sensed data are recorded in rows and columns. A geometric transformation can convert rows and columns into projected coordinates. Additionally, the transformation can correct geometric errors in the remotely sensed data, which are caused by the relative motions of a satellite (i.e., its scanners and the Earth) and uncontrolled variations in the position and altitude of the remote sensing platform. Although some of these errors (e.g., the Earth's rotation) can be removed systematically, they are typically removed through geometric transformation.

Chapter 6 shares the topic of projected coordinate systems with Chapter 2, but they are different in concept and process. Projection converts data sets from 3-D geographic coordinates to 2-D projected coordinates, whereas geometric transformation converts data sets from 2-D digitizer units or rows and columns to 2-D projected coordinates.

Chapter 6 has the following four sections. Section 6.1 reviews transformation methods, especially the affine transformation, which is commonly used in GIS and remote sensing. Section 6.2 examines the root mean square (RMS) error, a measure of the goodness of a geometric transformation, and how it is derived. Section 6.3 covers the interpretation of RMS errors on digitized maps. Section 6.4 deals with the resampling of pixel values for remotely sensed data after transformation.

6.1 GEOMETRIC TRANSFORMATION

Geometric transformation is the process of using a set of control points and transformation equations to register a digitized map, a satellite image, or an aerial photograph onto a projected coordinate system. As its definition suggests, geometric transformation is a common operation in GIS, remote sensing, and photogrammetry. But the mathematical aspects of geometric transformation are from coordinate geometry (Moffitt and Mikhail 1980).

6.1.1 Map-to-Map and Image-to-Map Transformation

A newly digitized map, either manually digitized or traced from a scanned file, is based on digitizer units. Digitizer units can be in inches or dots per inch. Geometric transformation converts the newly digitized map into projected coordinates in a process often called **map-to-map transformation.**

Image-to-map transformation applies to remotely sensed data (Jensen 1996; Richards and Jia 1999). The term suggests that the transformation changes the rows and columns (i.e., the image coordinates) of a satellite image into projected coordinates. Another term describing this kind of transformation is georeferencing (Verbyla and Chang 1997; Lillesand, Kiefer, and Chipman 2004). A georeferenced image can register spatially with other feature or raster layers in a GIS database, as long as the coordinate system is the same.

Whether map-to-map or image-to-map, a geometric transformation uses a set of control points to establish a mathematical model that relates the map coordinates of one system to another or image

coordinates to map coordinates. The use of control points makes the process somewhat uncertain. This is particularly true with image-to-map transformation because control points are selected directly from the original image. Misplacement of the control points can make the transformation result unacceptable.

The **root mean square (RMS) error** is a quantitative measure that can determine the quality of a geometric transformation. It measures the displacement between the actual and estimated locations of the control points. If the RMS error is acceptable, then a mathematical model derived from the control points can be used for transforming the entire map or image.

A map-to-map transformation automatically creates a new map that is ready to use. An image-to-map transformation, on the other hand, requires an additional step of resampling to complete the process. Resampling fills in each pixel of the transformed image with a value derived from the original image.

6.1.2 Transformation Methods

Different methods have been proposed for transformation from one coordinate system to another (Taylor 1977; Moffitt and Mikhail 1980). Each method is distinguished by the geometric properties it can preserve and by the changes it allows. The effect of transformation varies from changes of position and direction, to a uniform change of scale, to changes in shape and size (Figure 6.1). The following summarizes these transformation methods and their effect on a rectangular object:

- Equiarea transformation allows rotation of the rectangle and preserves its shape and size.
- Similarity transformation allows rotation of the rectangle and preserves its shape but not size.
- **Affine transformation** allows angular distortion of the rectangle but preserves the parallelism of lines (i.e., parallel lines remain as parallel lines).
- Projective transformation allows both angular and length distortions, thus allowing the rectangle to be transformed into an irregular quadrilateral.

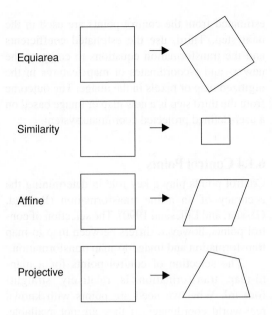

Figure 6.1
Different types of geometric transformations.

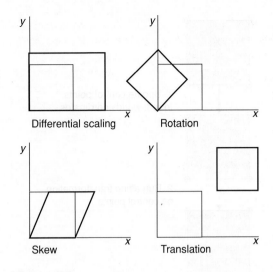

Figure 6.2
Differential scaling, rotation, skew, and translation in the affine transformation.

These transformation methods are available in GIS packages such as ArcGIS and MGE. The general rules suggest using the affine transformation for map-to-map or image-to-map transformations and the projective transformation for aerial photographs with relief displacement. Also available in GIS packages are general polynomial transformations that use surfaces generated from second- or higher-order polynomial equations to transform satellite images with high degrees of distortion and topographic relief displacement. The process of general polynomial transformations is commonly called *warping* (Jensen 1996).

6.1.3 Affine Transformation

The affine transformation allows rotation, translation, skew, and differential scaling on a rectangular object, while preserving line parallelism (Pettofrezzo 1978; Loudon, Wheeler, and Andrew 1980; Chen, Lo, and Rau 2003). Rotation rotates the object's *x*- and *y*-axes from the origin. Translation shifts its origin to a new location. Skew allows a nonperpendicularity (or affinity) between the axes, thus changing its shape

to a parallelogram with a slanted direction. And differential scaling changes the scale by expanding or reducing in the *x* and/or *y* direction. Figure 6.2 shows these four transformations graphically.

Mathematically, the affine transformation is expressed as a pair of first-order polynomial equations:

(6.1)

$$X = Ax + By + C$$

(6.1)

$$Y = Dx + Ey + F$$

where *x* and *y* are the input coordinates that are given; *X* and *Y* are the output coordinates to be determined; and *A, B, C, D, E,* and *F* are the transformation coefficients. The affine transformation is also called the *six-parameter transformation* because it involves six estimated coefficients.

The same equations apply to both digitized maps and satellite images. But there are two differences. First, *x* and *y* represent point coordinates in a digitized map, but they represent columns and rows in a satellite image. Second, the coefficient *E* is negative in the case of a satellite image. This is because the origin of a satellite image is located at

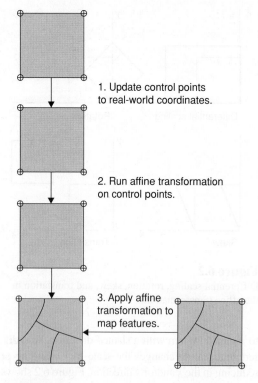

1. Update control points to real-world coordinates.

2. Run affine transformation on control points.

3. Apply affine transformation to map features.

Figure 6.3

A geometric transformation typically involves three steps. Step 1 updates the control points to real-world coordinates. Step 2 uses the control points to run an affine transformation. Step 3 creates the output by applying the transformation equations to the input features.

the upper-left corner, whereas the origin of a projected coordinate system is at the lower-left corner.

Operationally, an affine transformation of a digitized map or image involves three steps (Figure 6.3). First, update the x- and y-coordinates of selected control points to real-world (projected) coordinates. If real-world coordinates are not available, we can derive them by projecting the longitude and latitude values of the control points. Second, run an affine transformation on the control points and examine the RMS error. If the RMS error is higher than the expected value, select a different set of control points and rerun the affine transformation. If the RMS error is acceptable, then the six coefficients of the affine transformation

estimated from the control points are used in the next step. Third, use the estimated coefficients and the transformation equations to compute the new x- and y-coordinates of map features in the digitized map or pixels in the image. The outcome from the third step is a new map or image based on a user-defined projected coordinate system.

6.1.4 Control Points

Control points play a key role in determining the accuracy of an affine transformation (Bolstad, Gessler, and Lillesand 1990). The selection of control points, however, differs between map-to-map transformation and image-to-map transformation.

The selection of control points for a map-to-map transformation is relatively straightforward. What we need are points with known real-world coordinates. If they are not available, we can use points with known longitude and latitude values and project them into real-world coordinates. A USGS 1:24,000 scale quadrangle map has 16 points with known longitude and latitude values: 12 points along the border and 4 additional points within the quadrangle. (These 16 points divide the quadrangle into 2.5 minutes in longitude and latitude.) These 16 points are also called tics.

An affine transformation requires a minimum of three control points to estimate its six coefficients. But often four or more control points are used to reduce problems with measurement errors and to allow for a least-squares solution. After the control points are selected, they are digitized along with map features onto the digitized map. The coordinates of these control points on the digitized map are the x, y values in Eq. (6.1) and Eq. (6.2), and the real-world coordinates of these control points are the X, Y values. Box 6.1 shows an example of the use of a set of four control points to derive the six coefficients. Box 6.2 shows the output from the affine transformation and the interpretation of the transformation.

Control points for an image-to-map transformation are usually called ground control points. **Ground control points (GCPs)** are points where both image coordinates (in columns and rows)

Box 6.1 | **Estimation of Transformation Coefficients**

The example used here and later in Chapter 6 is a third-quadrangle soil map (one-third of a USGS 1:24,000 scale quadrangle map) scanned at 300 dpi. The map has four control points marked at the corners: Tic 1 at the NW corner, Tic 2 at the NE corner, Tic 3 at the SE corner, and Tic 4 at the SW corner. X and Y denote the control points' real-world (output) coordinates in meters based on the UTM coordinate system, and x and y denote the control points' digitized (input) locations. The measurement unit of the digitized locations is 1/300 of an inch, corresponding to the scanning resolution.

The following table shows the input and output coordinates of the control points:

Tic-Id	x	y	X	Y
1	465.403	2733.558	518843.844	5255910.5
2	5102.342	2744.195	528265.750	5255948.5
3	5108.498	465.302	528288.063	5251318.0
4	468.303	455.048	518858.719	5251280.0

We can solve for the transformation coefficients by using the following equation in matrix form:

$$\begin{bmatrix} C & F \\ A & D \\ B & E \end{bmatrix} = \begin{bmatrix} n & \Sigma x & \Sigma y \\ \Sigma x & \Sigma x^2 & \Sigma xy \\ \Sigma y & \Sigma xy & \Sigma y^2 \end{bmatrix}^{-1} \cdot \begin{bmatrix} \Sigma X & \Sigma Y \\ \Sigma xX & \Sigma xY \\ \Sigma yX & \Sigma yY \end{bmatrix}$$

where n is the number of control points and all other notations are the same as previously defined. The transformation coefficients derived from the equation show

$$A = 2.032,\ B = -0.004,\ C = 517909.198,$$
$$D = 0.004,\ E = 2.032,\ F = 5250353.802$$

Box 6.2 | **Output from an Affine Transformation**

Using the data from Box 6.1, we can interpret the geometric properties of the affine transformation. The coefficient C represents the translation in the x direction and F the translation in the y direction. Other properties such as rotation, skew, and scaling can be derived from the following equations:

$$A = Sx \cos (t)$$
$$B = Sy [k \cos (t) - \sin (t)]$$
$$D = Sx \sin (t)$$
$$E = Sy [k \sin (t) + \cos (t)]$$

where Sx is the change of scale in x, Sy is the change of scale in y, t is the rotation angle, and k is the shear factor. For example, we can use the equations for A and D to first derive the value of t and then use the

t value in either equation to derive Sx. The following lists the derived geometric properties of the affine transformation from Box 6.1:

Scale $(X, Y) = (2.032, 2.032)$
Skew (degrees) $= (-0.014)$
Rotation (degrees) $= (0.102)$
Translation $= (517909.198, 5250353.802)$

The positive rotation angle means a rotation counterclockwise from the x-axis, and the negative skew angle means a shift clockwise from the y-axis. Both angles are very small, meaning that the change from the original rectangle to a parallelogram through the affine transformation is very slight.

and real-world coordinates can be identified. The image coordinates are the x, y values, and their corresponding real-world coordinates are the X, Y values in Eq. (6.1) and Eq. (6.2).

GCPs are selected directly from a satellite image. Therefore the selection is not as straightforward as selecting four tics for a digitized map. Ideally, GCPs are those features that show up clearly as single distinct pixels. Examples include road intersections, rock outcrops, small ponds, or distinctive features along shorelines. Georeferencing a TM (Thematic Mapper) scene may require an initial set of 20 or more GCPs. Some of these points are eventually removed in the transformation process because they contribute to a large RMS. After GCPs are identified on a satellite image, their real-world coordinates can be obtained from digital maps or GPS (global positioning system) readings.

6.2 ROOT MEAN SQUARE (RMS) ERROR

The affine transformation uses the coefficients derived from a set of control points to transform a digitized map or a satellite image. The location of a control point on a digitized map or an image is an estimated location and can deviate from its actual location. A common measure of the goodness of the control points is the RMS error, which measures the deviation between the actual (true) and estimated (digitized) locations of the control points.

How is an RMS error derived from a digitized map? After the six coefficients have been estimated, we can use the digitized coordinates of the first control point as the inputs (i.e., the x and y values) to Eq. (6.1) and Eq. (6.2) and compute the X and Y values, respectively. If the digitized control point were perfectly located, the computed X and Y values would be the same as the control point's real-world coordinates. But this is rarely the case. The deviations between the computed (estimated) X and Y values and the actual coordinates then become errors associated with the first control point on the output. Likewise, to derive errors associated with a control point on the input, we can use the point's

real-world coordinates as the inputs and measure the deviations between the computed x and y values and the digitized coordinates.

The procedure for deriving RMS errors also applies to GCPs used in an image-to-map transformation. Again, the difference is that columns and rows of a satellite image replace digitized coordinates.

Mathematically, the input or ouput error for a control point is computed by:

(6.3)

$$\sqrt{\left(x_{act} - x_{est}\right)^2 + \left(y_{act} - y_{est}\right)^2}$$

where x_{act} and y_{act} are the x and y values of the actual location, and x_{est} and y_{est} are the x and y values of the estimated location.

The average RMS error can be computed by averaging errors from all control points:

(6.4)

$$\sqrt{\left[\sum_{i=1}^{n}\left(x_{act,\,i} - x_{est,\,i}\right)^2 + \sum_{i=1}^{n}\left(y_{act,\,i} - y_{est,\,i}\right)^2\right] / n}$$

where n is the number of control points, $x_{act,\,i}$ and $y_{act,\,i}$ are the x and y values of the actual location of control point i, and $x_{est,\,i}$ and $y_{est,\,i}$ are the x and y values of the estimated location of control point i. Box 6.3 shows an example of the average RMS errors and the output X and Y errors for each control point from an affine transformation.

To ensure the accuracy of geometric transformation, the RMS error should be within a tolerance value. The data producer defines the tolerance value, which can vary by the accuracy and the map scale or by the ground resolution of the input data. A RMS error (output) of <6 meters is probably acceptable if the input map is a 1:24,000 scale USGS quadrangle map. A RMS error (input) of <1 pixel is probably acceptable for a TM scene with a ground resolution of 30 meters.

If the RMS error based on the control points is within the acceptable range, then the assumption is that this same level of accuracy can also apply to the entire map or image. But this assumption may not be true under certain circumstances, as shown in Section 6.3.

Box 6.3 | RMS from an Affine Transformation

The following shows a RMS report using the data from Box 6.1.

RMS error (input, output) = (0.138, 0.281)

Tic-Id	Input x Output X	Input y Output Y	X Error	Y Error
1	465.403	2733.558		
	518843.844	5255910.5	−0.205	−0.192
2	5102.342	2744.195		
	528265.750	5255948.5	0.205	0.192
3	5108.498	465.302		
	528288.063	5251318.0	−0.205	−0.192
4	468.303	455.048		
	518858.719	5251280.0	0.205	0.192

The output shows that the average deviation between the input and output locations of the control points is 0.281 meter based on the UTM coordinate system, or 0.00046 inch (0.138 divided by 300) based on the digitizer unit. This RMS error is well within the acceptable range. The individual X and Y errors suggest that the error is slightly lower in the y direction than in the x direction and that the average RMS error is equally distributed among the four control points.

If the RMS error exceeds the established tolerance, then the control points need to be adjusted. For digitized maps, this means redigitizing the control points. For satellite images, the adjustment means removing the control points that contribute most to the RMS error and replacing them with new GCPs. Geometric transformation is therefore an iterative process of selecting control points, estimating the transformation coefficients, and computing the RMS error. This process continues until a satisfactory RMS is obtained.

6.3 INTERPRETATION OF RMS ERRORS ON DIGITIZED MAPS

If a RMS error is within the acceptable range, we usually assume that the transformation of the entire map is also acceptable. This assumption can be quite wrong, however, if gross errors are made in digitizing the control points or in inputting the longitude and latitude readings of the control points.

As an example, we can shift the locations of control points 2 and 3 (the two control points to the right) on a third quadrangle (similar to that in Box 6.1) by increasing their x values by a constant. The RMS error would remain about the same because the object formed by the four control points retains the shape of a parallelogram. But the soil lines would deviate from their locations on the source map. The same problem occurs if we increase the x values of control points 1 and 2 (the upper two control points) by a constant, and decrease the x values of control points 3 and 4 (the lower two control points) by a constant (Figure 6.4). In fact, the RMS error would be well within the tolerance value as long as the object formed by the shifted control points remains a parallelogram.

Longitude and latitude readings printed on paper maps are sometimes erroneous. This would lead to acceptable RMS errors but significant location errors for digitized map features. Suppose the latitude readings of control points 1 and 2 (the upper two control points) are off by 10″ (e.g., 47°27′20″ instead of 47°27′30″). The RMS error

Figure 6.4

Inaccurate location of soil lines can result from input tic location errors. The thin lines represent correct soil lines and the thick lines incorrect soil lines. In this case, the x values of the upper two tics were increased by 0.2″ while the x values of the lower two tics were decreased by 0.2″ on a third quadrangle (15.4″ × 7.6″).

Figure 6.5

Incorrect location of soil lines can result from output tic location errors. The thin lines represent correct soil lines and the thick lines incorrect soil lines. In this case, the latitude readings of the upper two tics were off by 10″ (e.g., 47°27′20″ instead of 47°27′30″) on a third quadrangle.

from transformation would be acceptable but the soil lines would deviate from their locations on the source map (Figure 6.5). The same problem occurs if the longitude readings of control points 2 and 3 (the two control points to the right) are off by 30″ (e.g., −116°37′00″ instead of −116°37′30″). Again, this happens because the affine transformation works with parallelograms. Although we tend to take for granted the accuracy of published maps, erroneous longitude and latitude readings are quite common, especially with inset maps (maps that are smaller than the regular size) and oversized maps (maps that are larger than the regular size).

We typically use the four corner points of the source map as control points. This practice makes sense because the exact readings of longitude and latitude are usually shown at those points. Moreover, using the corner points as control points helps in the process of joining the map with its adjacent maps. But the practice of using four corner control points does not preclude the use of more control points if additional points with known locations are available. The affine transformation uses a least-squares solution when more than three control points are present; therefore, the use of more control points means a better coverage of the entire map in transformation. In other situations, control points that are closer to the map features of interest should be used instead of the corner points. This ensures the location accuracy of these map features.

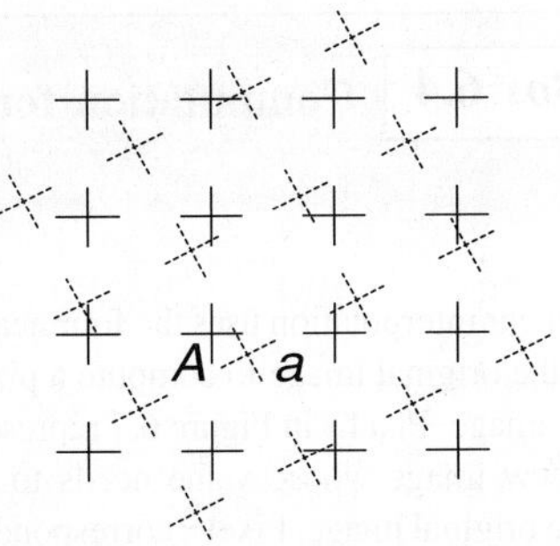

6.4 RESAMPLING OF PIXEL VALUES

The result of geometric transformation of a satellite image is a new image based on a projected coordinate system. But the new image has no pixel values. The pixel values must be filled through resampling. **Resampling** in this case means filling each pixel of the new image with a value or a derived value from the original image.

6.4.1 Resampling Methods

Three common resampling methods are nearest neighbor, bilinear interpolation, and cubic convolution. The **nearest neighbor** resampling method fills each pixel of the new image with the nearest pixel value from the original image. For example, Figure 6.6 shows that pixel A in the new image will take the value of pixel a in the original image because it is the closest neighbor. The nearest neighbor method is computationally efficient. The method has the additional property of preserving the original pixel values, which is important for categorical data such as land cover types and desirable for some image processing such as edge detection.

Both bilinear interpolation and cubic convolution fill the new image with distance-weighted

Figure 6.6

Because a in the original image is closest to pixel A in the new image, the pixel value at a is assigned to be the pixel value at A using the nearest neighbor technique.

averages of the pixel values from the original image. The **bilinear interpolation** method uses the average of the four nearest pixel values from three linear interpolations, whereas the **cubic convolution** method uses the average of the 16 nearest pixel values from five cubic polynomial interpolations (Richards and Jia 1999). Cubic convolution tends to produce a smoother output than bilinear interpolation but requires a longer processing time. Box 6.4 shows an example of bilinear interpolation.

6.4.2 Other Uses of Resampling

Geometric transformation of satellite images is not the only operation that requires resampling. Resampling is needed whenever there is a change of cell location or cell size between the input raster and the output raster. For example, projecting a raster from one coordinate system to another requires resampling to fill in the cell values of the output raster. Resampling is also involved when a raster changes from one cell size to another (e.g., from 10 to 15 meters). **Pyramiding** is a common technique for displaying large raster data sets (Box 6.5). Resampling is used with pyramiding to build different pyramid levels. Regardless of its application, resampling typically uses one of the three resampling methods to produce the output raster.

Box 6.4 | Computation for Bilinear Interpolation

Bilinear interpolation uses the four nearest neighbors in the original image to compute a pixel value in the new image. Pixel *x* in Figure 6.7 represents a pixel in the new image whose value needs to be derived from the original image. Pixel *x* corresponds to a location of (2.6, 2.5) in the original image. Its four nearest neighbors have the image coordinates of (2, 2), (3, 2), (2, 3), and (3, 3), and the pixel values of 10, 5, 15, and 10, respectively.

Using the bilinear interpolation method, we first perform two linear interpolations along the scan lines 2 and 3 to derive the interpolated values at *a* and *b*:

$$a = 0.6(5) + 0.4(10) = 7$$
$$b = 0.6(10) + 0.4(15) = 12$$

Then we perform the third linear interpolation between *a* and *b* to derive the interpolated value at *x*:

$$x = 0.5(7) + 0.5(12) = 9.5$$

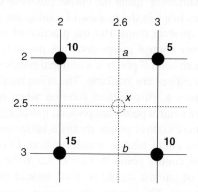

Figure 6.7
The bilinear interpolation method uses the value of the four closest pixels (black circles) in the original image to estimate the pixel value at *x* in the new image.

Box 6.5 | Pyramiding

GIS packages, including ArcGIS, have adopted pyramiding for displaying large raster data sets. Pyramiding builds different pyramid levels to represent reduced or lower resolutions of a large raster. Because a lower-resolution raster (i.e., a pyramid level of greater than 0) requires less memory space, it can display more quickly. Therefore, when viewing the entire raster, we view it at the highest pyramid level. And, as we zoom in, we view more detailed data at a finer resolution (i.e., a pyramid level of closer to 0). Resampling is involved in building different pyramid levels.

KEY CONCEPTS AND TERMS

Affine transformation: A geometric transformation method that allows rotation, translation, skew, and differential scaling on a rectangular object while preserving line parallelism.

Bilinear interpolation: A resampling method that uses the distance-weighted average of the four nearest pixel values to estimate a new pixel value.

Cubic convolution: A resampling method that uses the distance-weighted average of the 16 nearest pixel values to estimate a new pixel value.

Geometric transformation: The process of converting a map or an image from one coordinate system to another by using a set of control points and transformation equations.

Ground control points (GCPs):   Points used as control points for an image-to-map transformation.

Image-to-map transformation: One type of geometric transformation that converts the rows and columns of a satellite image into real-world coordinates.

Map-to-map transformation: One type of geometric transformation that converts a newly digitized map into real-world coordinates.

Nearest neighbor: A resampling method that uses the nearest pixel value to estimate a new pixel value.

Pyramiding: A technique that builds different pyramid levels for displaying large raster data sets at different resolutions.

Resampling: A process of filling each pixel of a newly transformed image with a value or a derived value from the original image.

Root mean square (RMS) error: A measure of the deviation between the actual location and the estimated location of the control points in geometric transformation.

REVIEW QUESTIONS

1. Explain map-to-map transformation.
2. Explain image-to-map transformation.
3. An image-to-map transformation is sometimes called an image-to-world transformation. Why?
4. The affine transformation allows rotation, translation, skew, and differential scaling. Describe each of these transformations.
5. Operationally, an affine transformation involves three sequential steps. What are these steps?
6. Explain the role of control points in an affine transformation.
7. How are control points selected for a map-to-map transformation?
8. How are ground control points chosen for an image-to-map transformation?

9. Define the root mean square (RMS) error in geometric transformation.
10. Explain the role of the RMS error in an affine transformation.
11. Describe a scenario in which the RMS error may not be a reliable indicator of the goodness of a map-to-map transformation.
12. Why do we have to perform the resampling of pixel values following an image-to-map transformation?
13. Describe three common resampling methods for raster data.
14. The nearest neighbor method is recommended for resampling categorical data. Why?
15. What is pyramiding?

APPLICATIONS: GEOMETRIC TRANSFORMATION

This applications section covers three tasks. Task 1 covers the affine transformation of a scanned file. In Task 2, you will use the transformed scanned file for vectorization. As covered in Chapter 5, scanning is a popular data acquisition method. Task 3 covers the affine transformation of a satellite image.

Task 1 Georeference and Rectify a Scanned Map

What you need: *hoytmtn.tif,* a TIFF file containing scanned soil lines.

The bi-level scanned file *hoytmtn.tif* is measured in inches. For this task, you will convert the scanned image into UTM coordinates. The conversion process involves two basic steps. First, you will georeference the image by using four control points, also called tics, which correspond to the four corner points on the original soil map. Second, you will transform the image by using the results from georeferencing. The four control points have the following longitude and latitude values in degrees-minutes-seconds (DMS):

Tic-Id	Longitude	Latitude
1	−116 00 00	47 15 00
2	−115 52 30	47 15 00
3	−115 52 30	47 07 30
4	−116 00 00	47 07 30

Projected onto the NAD 1927 UTM Zone 11N coordinate system, these four control points have the following *x*- and *y*-coordinates:

Tic-Id	x	y
1	575672.2771	5233212.6163
2	585131.2232	5233341.4371
3	585331.3327	5219450.4360
4	575850.1480	5219321.5730

Now you are ready to perform the georeferencing of *hoytmtn.tif.*

1. Launch ArcMap, and rename the data frame Task 1. Add *hoytmtn.tif* to Task 1. Ignore the missing spatial reference warning message. Click the Customize menu, point to Toolbars, and check Georeferencing. The Georeferencing toolbar should now appear in ArcMap, and the Layer dropdown list should list *hoytmtn.tif.*

2. Zoom in on *hoytmtn.tif* and locate the four control points. These control points are shown as brackets: two at the top and two at the bottom of the image. They are numbered 1 through 4 in a clockwise direction, with 1 at the upper-left corner.

3. Zoom to the first control point. Click (Activate) the Add Control Points tool on the Georeferencing toolbar. Click the intersection point where the centerlines of the bracket meet, and then click again. A plus-sign symbol at the control point turns from green to red. Use the same procedure to add the other three control points.

4. This step is to update the coordinate values of the four control points. Click the View Link Table tool on the Georeferencing toolbar. The link table lists the four control points at the top with their X Source, Y Source, X Map, Y Map, and Residual values. The X Source and Y Source values are the coordinates on the scanned image. The X Map and Y Map values are the UTM coordinates to be entered. The link table offers Auto Adjust, the Transformation method, and the Total RMS Error. Notice that the transformation method is 1st Order Polynomial (i.e., affine transformation). Click the first record, and enter 575672.2771 and 5233212.6163 for its X Map and Y Map values, respectively. (Ignore the warning message about the control points being collinear or not well distributed.) Enter the X Map and Y Map values for the other three records.

Q1. What is the total RMS error of your first trial?

Q2. What is the residual for the first record?

5. The total RMS error should be smaller than 4.0 (meters) if the digitized control points match their locations on the image. If the RMS error is high, highlight the record with a high residual value and delete it. Go back to the image and re-enter the control point. After you have come to an acceptable total RMS error, click OK on the Link Table dialog.

6. This step is to rectify (transform) *hoytmtn.tif.* Select Rectify from the Georeferencing dropdown menu. Take the defaults in the next dialog but save the rectified TIFF file as *rect_hoytmtn.tif* in the Chapter 6 database.

Task 2   Use ArcScan to Vectorize Raster Lines

What you need: *rect_hoytmtn.tif,* a rectified TIFF file from Task 1.

You need to use ArcScan, an extension to ArcGIS, for Task 2. First check the extension in ArcMap's Customize menu and then check the toolbar from the Toolbars pullright in the Customize pulldown menu. ArcScan can convert raster lines in a bi-level raster such as *rect_hoytmtn.tif* into line or polygon features. The output from vectorization can be saved into a shapefile or a geodatabase feature class. Task 2 is therefore an exercise for creating new spatial data from a scanned file. Scanning, vectorization, and vectorization parameters are topics already covered in Chapter 5.

Vectorization of raster lines can be challenging if a scanned image contains irregular raster lines, raster lines with gaps, or smudges. A poor-quality scanned image typically reflects the poor quality of the original map and, in some cases, the use of the wrong parameter values for scanning. The scanned image you will use for this task is of excellent quality. Therefore, the result from batch vectorization should be excellent as well.

1. This step creates a new shapefile that will store the vectorized features from *rect_hoytmtn.tif.* Right-click the Chapter 6 folder in the Catalog tree, point to New, and select Shapefile. In the Create New Shapefile dialog, enter *hoytmtn_trace.shp* for the name and polyline for the feature type. Click the Edit button in the Spatial Reference frame. Select NAD 1927 UTM Zone 11N for the new shapefile's coordinate system. Click OK to exit the dialogs.

2. Insert a new data frame in ArcMap and rename the data frame Task 2. Add *rect_hoytmtn.tif* and *hoytmtn_trace.shp* to Task 2. Change the line symbol of *hoytmtn_trace* to black. Select Properties from the context menu of *rect_hoytmtn.tif.* On the Symbology tab, choose Unique Values, opt to build the attribute table, and change the symbol for the value of 0 to red. Right-click *rect_hoytmtn.tif* and select Zoom to Layer. Because raster lines on *rect_hoytmtn.tif* are very thin, you do not see them at first on the monitor. Zoom in and you will see the red lines.

3. Click Editor Toolbar in ArcMap to open it. Select Start Editing from the Editor's dropdown menu. Select *hoytmtn_trace* to edit. The edit mode activates the ArcScan toolbar. The Raster dropdown list should show *rect_hoytmtn.tif.* (The edit mode also opens the Create Features window; you can close the window because you do not have to work with it in Task 2.)

4. This step is to set up the vectorization parameters, which are critical for batch vectorization. Select Vectorization Settings from the Vectorization menu. You can enter the parameter values in the settings dialog or choose a style with the predefined values. Click Styles. Choose Polygons and click OK in the next dialog. Click Apply and then Close to dismiss the Vectorization Settings dialog.

5. Select Generate Features from the Vectorization menu. Make sure that *hoytmtn_trace* is the layer to add the centerlines to. Notice that the tip in the dialog states that the command will generate features from the full extent of the raster. Click OK. The results of batch vectorization are now stored

in *hoytmtn_trace*. You can turn off *rect_ hoytmtn.tif* in the table of contents so that you can see the lines in *hoytmtn_trace*.

Q3. The Generate Features command adds the centerlines to *hoytmtn_trace*. Why are the lines called centerlines?

Q4. What are the other vectorization options besides batch vectorization?

6. The lower-left corner of *rect_hoytmtn.tif* has notes about the soil survey, which should be removed. Click the Select Features by Rectangle tool on the standard (Tools) toolbar, select the notes, and delete them. Use the same method to remove border lines on *hoytmtn_trace*.

7. Select Stop Editing from the Editor's menu and save the edits. Check the quality of the traced soil lines in *hoytmtn_trace*. Because the scanned image is of excellent quality, the soil lines should also be of excellent quality.

Task 3 Perform Image-to-Map Transformation

What you need: *spot-pan.bil*, a 10-meter SPOT panchromatic satellite image; *road.shp*, a road shapefile acquired with a GPS receiver and projected onto UTM coordinates.

You will perform an image-to-map transformation in Task 3. ArcMap provides the Georeferencing toolbar that has the basic tools for georeferencing and rectifying a satellite image.

1. Insert a new data frame in ArcMap and rename the data frame Task 3. Add *spot-pan. bil* and *road.shp* to Task 3. Click the symbol for *road,* and change it to orange. Make sure that the Georeferencing toolbar is available and that the Layer on the toolbar shows *spot-pan.bil*. Click View Link Table on the Georeferencing toolbar, select all existing links (from Task 1) in the table, and delete them. (If the toolbar does not show *spot-pan .bil*, right-click *spot-pan.bil*, point to Data, and select Export Data. In the next dialog, export

spot-pan.bil to a TIFF file called *spot-pan.tif* and use the TIFF file for the rest of the task.)

2. You can use the Zoom to Layer tool in the context menu to see *road,* or *spot-pan.bil,* but not both. This is because they are in different coordinates. To see both of them, you must have one or more links to initially georeference *spot-pan.bil.* Figure 6.8 marks the first four recommended links. They are all at road intersections. Examine these road intersections in both *spot-pan.bil* and *road* so that you know where they are.

3. Check Auto Adjust in the Georeferencing dropdown menu. Now you are ready to add links. If Task 3 shows *road,* right-click *spot-pan.bil* and select Zoom to Layer. Zoom to the first road intersection in the image, click Add Control Points on the Georeferencing toolbar, and click the intersection point. Right-click *road* and select Zoom to Layer. Zoom to the corresponding first road intersection in the layer, click the Add Control Points tool, and click the intersection point. The first link brings both the satellite image and the roads to view, but they

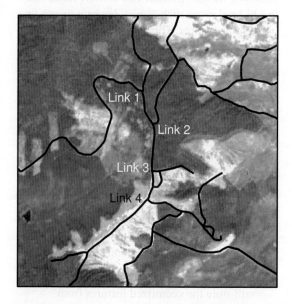

Figure 6.8
The four links to be created first.

are still far apart spatially. Repeat the same procedure to add the other three links. Each time you add a link, the Auto Adjust command uses the available links to develop a transformation.

4. Click View Link Table on the Georeferencing toolbar. The Link Table shows four records, one for each link you have added in Step 3. The X Source and Y Source values are based on the image coordinates of *spot-pan.bil.* The image has 1087 columns and 1760 rows. The X Source value corresponds to the column and the Y Source value corresponds to the row. Because the origin of the image coordinates is at the upper-left corner, the Y Source values are negative. The X Map and Y Map values are based on the UTM coordinates of *road.* The Residual value shows the RMS error of the control point. The Link Table dialog also shows the transformation method (i.e., affine transformation) and the total RMS error. You can save the link table as a text file at any time, and you can load the file next time to continue the georeferencing process.

Q5. What is the total RMS error from the four initial links?

5. An image-to-map transformation usually requires more than four control points. At the same time, the control points should cover the extent of the study area, rather than a limited portion. For this task, try to have a total of 10 links and keep the total RMS error to less than one pixel or 10 meters. If a link has a large residual value, delete it and add a new one. Each time you add or delete a link, you will see a change in the total RMS error.

6. This step is to rectify *spot-pan.bil* by using a link table you have created. Select Rectify from the Georeferencing menu. The next dialog lets you specify the cell size, choose a resampling method (nearest neighbor, bilinear interpolation, or cubic convolution), and specify the output name. For this task, you can specify 10 (meters) for the cell size, nearest neighbor for the resampling method,

TIFF for the format, and *rect_spot.tif* for the output. Click Save to dismiss the dialog.

7. Now you can add and view *rect_spot,* a georeferenced and rectified raster, with other georeferenced data sets for the study area. To delete the control points from *rect_spot,* select Delete Control Points from the Georeferencing menu.

8. If you have difficulty in getting enough links and an acceptable RMS error, first select Delete Control Points from the Georeferencing toolbar. Click View Link Table, and load *georef.txt* from the Chapter 6 database. *georef.txt* has 10 links and a total RMS error of 9.2 meters. Then use the link table to rectify *spot-pan.bil.*

9. *rect_spot* should have values ranging from 16 to 100. But if it has values ranging from 16 to 255 (and a black image), instead of the expected range of 16 to 100, it means the value of 255 has been assigned to the area outside the image. You can correct the problem by going through the following two steps. First, select Reclassify from the Spatial Analyst dropdown menu. In the Reclassify dialog, click Unique. Click the row with an old value of 255, change its new value to NoData, and click OK. The *Reclass of rect_spot* now has the correct value. Second, right-click *Reclass of rect_spot* and select Properties. Select Stretched in the Show frame. Change the Low Label value to 16, the High Label value to 100, and click OK. Now *Reclass of rect_spot* should look like *spot-pan.bil* except that it has been georeferenced. To save the corrected file, right-click *Reclass of rect_spot,* point to Data, and select Export Data. Then specify the output name and location.

Challenge Task

What you need: *cedarbt.tif.*

The Chapter 6 database contains *cedarbt.tif,* a bi-level scanned file of a soil map. This challenge question asks you to perform two operations.

First, convert the scanned file into UTM coordinates (NAD 1927 UTM Zone 12N) and save the result into *rec_cedarbt.tif*. Second, vectorize the raster lines in *rec_cedarbt.tif* and save the result into *cedarbt_trace.shp*. There are four tics on *cedarbt.tif*. Numbered clockwise from the upper left corner, these four tics have the following UTM coordinates:

Tic-Id	*x*	*y*
1	389988.78125	4886459.5
2	399989.875	4886299.5
3	399779.1875	4872416.0
4	389757.03125	4872575.5

Q1. What is the total RMS error for the affine transformation?

Q2. What problems, if any, did you encounter in vectorizing *rec_cedarbt.tif*?

REFERENCES

Bolstad, P. V., P. Gessler, and T. M. Lillesand. 1990. Positional Uncertainty in Manually Digitized Map Data. *International Journal of Geographical Information Systems* 4: 399–412.

Chen, L., C. Lo, and J. Rau. 2003. Generation of Digital Orthophotos from Ikonos Satellite Images. *Journal of Surveying Engineering* 129: 73–78.

Jensen, J. R. 1996. *Introductory Digital Image Processing: A Remote Sensing Perspective,*

2d ed. Upper Saddle River, NJ: Prentice Hall.

Lillesand, T. M., R. W. Kiefer, and J. W. Chipman. 2004. *Remote Sensing and Image Interpretation*, 5th ed. New York: Wiley.

Loudon, T. V., J. F. Wheeler, and K. P. Andrew. 1980. Affine Transformations for Digitized Spatial Data in Geology. *Computers & Geosciences* 6: 397–412.

Moffitt, F. H., and E. M. Mikhail. 1980. *Photogrammetry*, 3d ed. New York: Harper & Row.

Pettofrezzo, A. J. 1978. *Matrices and Transformations*. New York: Dover Publications.

Richards, J. A., and X. Jia. 1999. *Remote Sensing Digital Image Analysis: An Introduction,* 3d ed. Berlin: Springer-Verlag.

Taylor, P. J. 1977. *Quantitative Methods in Geography: An Introduction to Spatial Analysis.* Boston: Houghton Mifflin.

Verbyla, D. L., and Chang, K. 1997. *Processing Digital Images in GIS.* Santa Fe, NM: OnWord Press.

CHAPTER 7

SPATIAL DATA ACCURACY AND QUALITY

CHAPTER OUTLINE

7.1 Location Errors

7.2 Spatial Data Accuracy Standards

7.3 Topological Errors

7.4 Topological Editing

7.5 Nontopological Editing

7.6 Other Editing Operations

A basic requirement for GIS applications is accurate and good-quality spatial data. To meet the requirement, we rely on spatial data editing. Newly digitized layers, no matter how carefully prepared, always have some digitizing errors. They must be checked against the original sources to locate the errors. Existing layers may be outdated. They can be revised by using rectified aerial photographs or satellite images as references. As mobile GIS (Chapter 1) has become more common, data collected in the field can also be downloaded for updating the existing database. A new development in mobile GIS is web editing, which allows users to perform simple editing tasks such as adding, deleting, and modifying features online.

Because the raster data model uses a regular grid and fixed cells, spatial data editing does not apply to raster data. Vector data, on the other hand, can have location errors and topological errors. Location errors such as missing polygons or distorted lines relate to the geometric inaccuracies of spatial features, whereas topological errors such as dangling lines and unclosed polygons relate to the logical inconsistencies between spatial features. To correct location errors, we often have to reshape individual lines and digitize new lines. To correct topological errors, we must first learn about the topological relationships (Chapter 3) and then use a topology-based GIS package to help us make corrections.

Correcting digitizing errors may extend beyond individual layers. When a study area covers two or more source layers, we must match features across the layers. When two layers share some common boundaries, we must make sure that these boundaries are coincident. Spatial data editing can also take the form of simplification and smoothing of map features.

The use of both topological and nontopological data in GIS and the presence of the georelational and the object-based data models have significantly expanded the scope of spatial data editing. Thus, determining which method to use for a given data set has become a main challenge in spatial data editing.

Chapter 7 is organized into the following six sections. Section 7.1 describes location and topological errors. Section 7.2 discusses spatial data accuracy standards in the United States, which have gone through three development phases. Section 7.3 examines topological errors with simple features, and between layers. Section 7.4 introduces topological editing. Section 7.5 covers nontopological or basic editing. Section 7.6 includes edgematching, line simplification, and line smoothing.

7.1 LOCATION ERRORS

Location errors refer to the geometric inaccuracies of digitized features, which can vary by the data source used for digitizing.

7.1.1 Location Errors Using Secondary Data Sources

If the data source for digitizing is a secondary data source such as a paper map, the evaluation of location errors typically begins by comparing the digitized map with the source map. The obvious goal in digitizing is to duplicate the source map in digital format. To determine how well the goal has been achieved, we can make a **check plot** of the digitized map at the same scale as the source map, superimpose the plot on the source map, and see how well they match.

How well should the digitized map match the source map? There are no federal standards on the threshold value. A geospatial data producer can decide on the tolerance of location error. For example, an agency can stipulate that each digitized line shall be within 0.01-inch (0.254-millimeter) line width of the source map. At the scale of 1:24,000, this tolerance represents 20 feet (6 to 7 meters) on the ground.

Spatial features digitized from a source map can only be as accurate as the source map itself. A variety of factors can affect the accuracy of the source map. Perhaps the most important factor is the map scale. The accuracy of a map feature is less reliable on a 1:100,000 scale map than on a 1:24,000 scale map. Map scale also influences the level of detail on a published map. As the map scale becomes smaller, the number of map details decreases and the degree of line generalization increases (Monmonier 1996). As a result a meandering stream on a large-scale map becomes less sinuous on a small-scale map.

7.1.2 Causes of Digitizing Errors

Discrepancies between digitized lines and lines on the source map may result from three common scenarios. The first is human errors in manual digitizing. Human error is not difficult to understand: when a source map has hundreds of polygons and thousands of lines, one can easily miss some lines, connect the wrong points, or digitize the same lines twice or even more times. Because of the high resolution of a digitizing table, duplicate lines will not be on top of one another but will intersect to form a series of tiny polygons.

The second scenario consists of errors in scanning and tracing. A tracing algorithm usually has problems when raster lines meet or intersect, are too close together, are too wide, or are too thin and broken (Chapter 5). Digitizing errors from tracing include collapsed lines, misshapen lines, and extra lines (Figure 7.1). Duplicate lines can also occur in tracing because semiautomatic tracing follows continuous lines even if some of the lines have already been traced.

The third scenario consists of errors in converting the digitized map into real-world coordinates (Chapter 6). To make a check plot at the same scale as the source map, we must use a set of control points to convert the newly digitized map into real-world coordinates. With erroneous control points, this conversion can cause discrepancies between

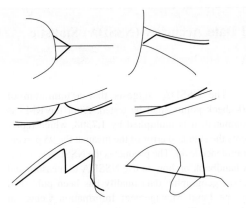

Figure 7.1
Common types of digitizing errors from tracing. The thin lines are lines on the source map, and the thick lines are lines from tracing.

digitized lines and source lines. Unlike seemingly random errors from the first two scenarios, discrepancies from geometric transformation often exhibit regular patterns. To correct these types of location errors, we must redigitize control points and rerun geometric transformation.

7.1.3 Location Errors Using Primary Data Sources

Although paper maps are still an important source for spatial data entry, new data entry methods using global positioning systems (GPS) and remote sensing imagery can bypass printed maps and map generalization practices. The resolution of the measuring instrument determines the accuracy of spatial data collected by GPS or satellite images; map scale has no meaning in this case. The spatial resolution of satellite images can range from less than 1 meter to 1 kilometer. Similarly, the spatial resolution of GPS point data can range from several millimeters to 10 meters or higher.

7.2 SPATIAL DATA ACCURACY STANDARDS

Discussions of location errors naturally lead to the topic of spatial data accuracy standards. As users of spatial data, we typically do not conduct the testing of location errors but rely on published standards in evaluating data accuracy. Spatial data accuracy standards have evolved as maps have changed from printed to digital format.

In the United States, the development of spatial data accuracy standards has gone through three phases. Revised and adopted in 1947, the U.S. National Map Accuracy Standard (NMAS) sets the accuracy standard for published maps such as topographic maps from the U.S. Geological Survey (USGS) (U.S. Bureau of the Budget 1947). The standards for horizontal accuracy require that no more than 10 percent of the well-defined map points tested shall be more than 1/30 inch at scales larger than 1:20,000, and 1/50 inch at scales of 1:20,000 or smaller. This means a threshold value of 40 feet on the ground for 1:24,000 scale maps and about 167 feet on the ground for 1:100,000 scale maps. But the direct linkage of the threshold values to map scales can be problematic in the digital age because digital spatial data can be easily manipulated and output to any scale.

In 1990 the American Society for Photogrammetry and Remote Sensing (ASPRS) published accuracy standards for large-scale maps (American Society for Photogrammetry and Remote Sensing 1990). The ASPRS defines the horizontal accuracy in terms of the root mean square (RMS) error, instead of fixed threshold values. The RMS error measures deviations between coordinate values on a map and coordinate values from an independent source of higher accuracy for identical points. Examples of higher-accuracy data sources may include digital or hard-copy map data, GPS, or survey data. The ASPRS standards stipulate the threshold RMS error of 16.7 feet for 1:20,000 scale maps and 2 feet for 1:2400 scale maps.

In 1998 the Federal Geographic Data Committee (FGDC), a committee representing 17 federal agencies at the time, established the National Standard for Spatial Data Accuracy (NSSDA) to replace the NMAS. The NSSDA follows the ASPRS accuracy standards but extends to map scales smaller than 1:20,000 (Federal Geographic Data Committee 1998) (Box 7.1). The NSSDA differs from the NMAS or the ASPRS accuracy

Box 7.1   **National Standard for Spatial Data Accuracy (NSSDA) Statistic**

To use the ASPRS standards or the NSSDA statistic, one must first compute the root mean square (RMS) error, which is defined by

$$\sqrt{\Sigma\left[\left(x_{\text{data},i} - x_{\text{check},i}\right)^2 + \left(y_{\text{data},i} - y_{\text{check},i}\right)^2\right]/n}$$

where $x_{\text{data},i}$ and $y_{\text{data},i}$ are the coordinates of the ith checkpoint in the data set; $x_{\text{check},i}$ and $y_{\text{check},i}$ are the coordinates of the ith check point in an independent source of higher accuracy; n is the number of check points tested; and i is an integer ranging from 1 to n.

The NSSDA suggests that a minimum of 20 check-points shall be tested. After the RMS is computed, it is multiplied by 1.7308, which represents the standard error of the mean at the 95 percent confidence level. The product is the NSSDA statistic. A handbook on how to use NSSDA to measure and report geographic data quality has been published by the Land Management Information Center at Minnesota Planning (**http://www.gda.state.mn.us/ pdf/1999/lmic/nssda_o.pdf**).

standards in that the NSSDA omits threshold accuracy values that spatial data, including paper maps and digital data, must achieve. Instead, agencies are encouraged to establish accuracy thresholds for their products and to report the NSSDA statistic, a statistic based on the RMS error. Given the new accuracy standards, we must ask for the product specifications and determine ourselves if the level of data accuracy is acceptable for a GIS project.

Data *accuracy* should not be confused with data *precision*. Spatial data accuracy measures how close the recorded location of a spatial feature is to its ground location, whereas **data precision** measures how exactly the location is recorded. Distances may be measured with decimal digits or rounded off to the nearest meter or foot. Likewise, numbers can be stored in the computer as integers or floating points. Moreover, floating-point numbers can be single precision with 7 significant digits or double precision with up to 15 significant digits. The number of significant digits used in data recording expresses the precision of a recorded location.

7.3 TOPOLOGICAL ERRORS

Topological errors violate the topological relationships that are either required by a GIS package or defined by the user. The coverage developed by Esri, incorporates the topological relationships of connectivity, area definition, and contiguity (Chapter 3). If digitized features did not follow these relationships, they would have topological errors. The geodatabase also from Esri, has 27 topology rules that govern the spatial relationships of point, line, and polygon features (Chapter 3). Some of these rules relate to features within a feature class, whereas others relate to two or more participating feature classes. Using a geodatabase, we can choose the topological relationships to implement in the data sets and define the kinds of topological errors that are important to a project.

7.3.1 Topological Errors with Geometric Features

Topological errors with polygon features include unclosed polygons, gaps between polygons, and overlapping polygons (Figure 7.2).

A common topological error with line features occurs when they do not meet perfectly at a point (node). This type of error becomes an **undershoot** if a gap exists between lines and an **overshoot** if a line is overextended (Figure 7.3). The result in both cases is a **dangling node** at the end of a **dangle.** Dangling nodes are, however, acceptable in special cases such as those attached to dead-end streets and small tributaries. A **pseudo node** appears along a

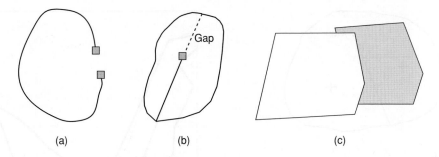

(a) (b) (c)

Figure 7.2
(*a*) An unclosed polygon, (*b*) a gap between two polygons, and (*c*) overlapped polygons.

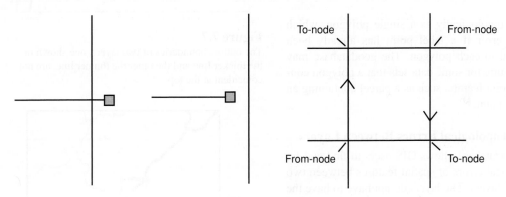

Figure 7.3
An overshoot (left) and an undershoot (right). Both types of errors result in dangling nodes.

Figure 7.5
The from-node and to-node of an arc determine the arc's direction.

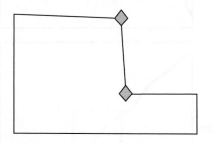

Figure 7.4
Pseudo nodes, shown by the diamond symbol, are nodes that are not located at line intersections.

continuous line and divides the line unnecessarily into separate lines (Figure 7.4). Some pseudo nodes are, however, acceptable. Examples include the insertion of pseudo nodes at points where the attribute values of a line feature change.

The direction of a line may become a topological error. For example, a hydrologic analysis project may stipulate that all streams must follow the downstream direction and that the starting point (the from-node) of a stream must be at a higher elevation than the end point (the to-node). Likewise, a traffic simulation project may require that all streets are clearly defined as two-way, one-way, or dead-end (Figure 7.5).

Point features have few topological errors. The georelational data model requires a label point to link a polygon to its attribute data. A polygon should therefore have one, and only one, label point. An error occurs if a polygon has zero or multiple label points. Unclosed polygons are often the cause of errors with multiple label points (Figure 7.6). When a gap exists between two polygons, they are

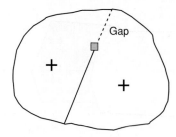

Figure 7.6
Multiple labels can be caused by unclosed polygons.

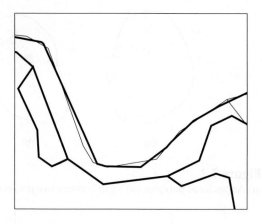

Figure 7.7
The outline boundaries of two layers, one shown in the thicker line and the other the thinner line, are not coincident at the top.

treated topologically as a single polygon, which has an error if a label point has already been assigned to each polygon. The geodatabase may also require for some data sets that a polygon contain a point feature, such as a parcel containing an address point.

7.3.2 Topological Errors Between Layers

The geodatabase allows GIS users to find and fix topological errors of spatial features between two or more layers. The layers do not have to have the same feature type, so that we can examine topological errors between two polygon feature classes or between a point and a polygon feature class.

A common error between two polygon layers is that their outline boundaries are not coincident (Figure 7.7). Suppose a GIS project uses a soil layer and a land-use layer for data analysis. Digitized separately, the two layers do not have coincident boundaries. If these two layers are later overlaid, the discrepancies between their boundaries become small polygons, which miss either soil or land-use attributes. Similar problems can occur with individual polygon boundaries. For example, census tracts are supposed to be nested within counties and subwatersheds within watersheds. Errors occur when the larger polygons (e.g., counties) do not share boundaries with the smaller polygons (e.g., census tracts).

One type of error with two line layers can occur when lines from one layer do not connect with those from another layer at end points (Figure 7.8). For example, when we merge two highway layers from

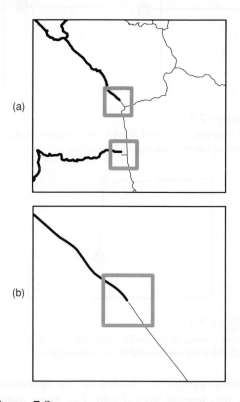

Figure 7.8
Line features from one layer do not connect perfectly with those from another layer at end points. (*b*) is an enlargement of the top error in (*a*).

two adjacent states, we expect highways to connect perfectly across the state border. Errors can happen if highways intersect, or overlap, or have gaps. Other errors with line layers include overlapping line features (e.g., rail lines on highways) and line features not covered by another set of line features (e.g., bus routes not covered by streets).

Errors with point features can occur if they do not fall along line features in another layer. For example, errors occur if gauge stations for measuring stream flow do not fall along streams. Likewise, errors occur if section corners do not fall on the polygon boundaries of the Public Land Survey System (PLSS).

7.4 TOPOLOGICAL EDITING

Topological editing ensures that digitized spatial features follow the topological relationships that are either built into a data model or specified by the user (e.g., Plümer and Gröger 1997). To perform topological editing, we must use a topology-based GIS package that can detect and display topological errors and has tools to remove them. Commercial GIS packages have similar capabilities in fixing topological errors with geometric features (Box 7.2).

Using ArcGIS as an example, the following groups topological editing tools into three general categories: topological editing on coverages, editing using a map topology, and editing using topology rules (Box 7.3).

7.4.1 Topological Editing on Coverages

Although coverage has been replaced by shapefile and geodatabase, many editing tools on coverages still remain in ArcGIS, albeit in different names. Topological editing on a coverage typically starts by constructing its topology. ArcGIS has the Clean command, which not only builds topology but also applies the dangle length and fuzzy tolerance to the entire coverage to remove some digitizing errors. The **dangle length** specifies the minimum length for dangling arcs on the output coverage. A dangling arc such as an overshoot is removed if it is shorter than the specified dangle length (Figure 7.9). The **fuzzy tolerance** specifies the minimum distance between vertices and arcs on the output coverage. It is useful for removing duplicate lines within the specified tolerance (Figure 7.10). Clean also automatically inserts a node at a line intersection to enforce the topological arc-node relationship.

Box 7.2 | **GIS Packages for Topological Editing**

ArcGIS, AutoCAD Map, and MGE are all capable of performing topological editing with simple features. Chapter 7 references only ArcGIS; the following list highlights the editing capabilities of the other two packages:

- AutoCAD Map: build topology, repair undershoots, snap clustered nodes, remove duplicate lines, simplify linear objects, and correct other errors.
- MGE (Base Mapper): build topology, remove redundant linear data segments, fix undershoots and overshoots, and create an intersection at a line crossing.

ArcGIS is not the only software package that uses topology rules for spatial data editing. Laser-Scan (now 1Spatial), a company in Cambridge, England, launched an extension to Oracle9i called Radius Topology in June 2002 (**http://www.1spatial.com/products/radius_topology/index.php**). Like the geodatabase, Radius Topology also implements topological relationships as rules between features and stores these rules as tables in the database. When these tables are activated, the topology rules automatically apply to spatial features in the enabled classes. An earlier version of Radius Topology was used to create the Ordnance Survey's MasterMap (**http://www.ordnancesurvey.co.uk/oswebsite/**).

Box 7.3 | **Topological Editing in ArcGIS**

Tools for topological editing are placed in different applications and on different toolbars in ArcGIS.

The Topology toolbar has the Map Topology button, which opens a dialog to enable the user to define a cluster tolerance and the feature classes to participate in a map topology. Tools for working with a map topology are the Topology Editing tool on the Topology toolbar and the Topology Tasks (Modify Edge and Reshape Edge) on the Editor toolbar.

Topology rules are defined through the properties dialog of a feature dataset in a geodatabase. The Topology toolbar has tools for validating, inspecting, and fixing topological errors. The options for actually fixing a topological error, such as subtract features and create features, are built into the topological error's context menu.

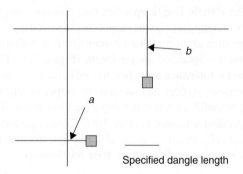

Figure 7.9
The dangle length can remove an overshoot if the overshoot, such as *a,* is smaller than the specified length. The overshoot *b* remains.

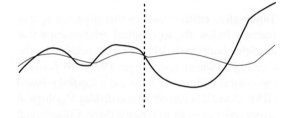

———— Specified fuzzy tolerance

Figure 7.10
The fuzzy tolerance can snap duplicate lines if the gap between the duplicate lines is smaller than the specified tolerance. In this diagram, the duplicate lines to the left of the dashed line will be snapped but not those to the right.

Because it applies to an entire coverage, Clean must be used cautiously. A large dangle length can remove undershoots as well as overshoots (Figure 7.11). Instead of being removed, undershoots should be extended to fill the gap. A large fuzzy tolerance will snap arcs that are not duplicate arcs, thus distorting the shape of map features (Figure 7.12).

For individual features rather than for the entire coverage, ArcGIS has commands that can snap points and lines within specified tolerances. Nodesnap can snap nodes, Arcsnap can snap the ends of arcs to existing arcs, and Extend can extend dangling arcs to intersect existing arcs. Again, these commands must be used cautiously. A large

nodesnap tolerance, for example, can snap the wrong nodes. Moreover, when a node is snapped to a new location within the nodesnap tolerance, the node moves the arc segment that is attached to it, thus altering the shape of the arc.

Combining the snapping commands with basic editing tools such as delete, move, add, split, unsplit, and flip, which are also available in ArcGIS, we can fix many types of digitizing errors. Some examples are listed in the following:

- Dangles: One can remove an undershoot by extending the dangling arc to meet with the target arc at a new node (Figure 7.13), and remove an overshoot by deleting the

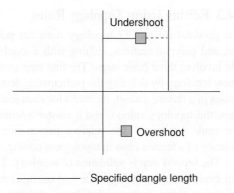

Figure 7.11
A large dangle length can remove the overshoot, which should be removed, and the undershoot, which should not be removed.

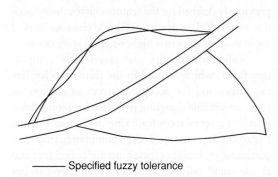

Figure 7.12
A large fuzzy tolerance can remove duplicate lines (top), which should be removed, as well as features such as a small stream channel (middle), which should not be removed.

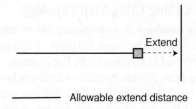

Figure 7.13
The allowable extend distance can remove the dangle by extending it to the line on the right.

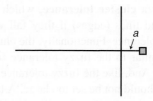

Figure 7.14
To remove the overshoot *a,* first select it and then delete it.

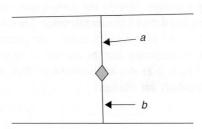

Figure 7.15
To remove the pseudo node, select *a* and *b,* assign the same ID value to both, and unsplit them.

extension (Figure 7.14). In general, it is easier to identify and remove an overshoot than an undershoot.

- Duplicate Arcs: One solution is to carefully select extra arcs and delete them, and the other is to delete all duplicate arcs within a box and redigitize.
- Wrong Arc Directions: One can alter the direction of an arc by flipping the arc, thus changing the relative position of the beginning and end nodes.
- Pseudo Nodes: One can remove a pseudo node by setting the two arcs on each side of the node to have the same ID value before unsplitting them (Figure 7.15).
- Label Errors: One can add new label points with proper IDs for missing labels in a polygon coverage. If a polygon has multiple labels, the extra labels can be selected and deleted. Multiple labels may also indicate an unclosed polygon.
- Reshaping Arcs: To reshape an arc, one can move, add, or delete vertices that make up the arc.

7.4.2 Editing Using Map Topology

A **map topology** is a temporary set of topological relationships between the parts of features that are supposed to be coincident. For example, a map topology can be built between a land-use layer and a soil layer to ensure that their outlines are coincident. Likewise, if a county borders a stream, a map topology can make the stream coincident with the county boundary. Layers participating in a map topology can be shapefiles or geodatabase feature classes, but not coverages.

Coincident features in a map topology are defined by a **cluster tolerance,** which can snap vertices and lines (edges) if they fall within the specified tolerance. Functionally, the cluster tolerance is similar to the fuzzy tolerance for editing coverages. And, like the fuzzy tolerance, a cluster tolerance should not be set too large. A large cluster tolerance can unintentionally alter the shapes of lines and polygons. A good strategy is to use a small cluster tolerance and to change it only to deal with more severe but localized errors.

To edit using map topology, we first create a map topology, specify the participating feature classes, and define a cluster tolerance. Then we use the editing tools in ArcGIS to force the geometries of the participating feature classes to be coincident. (Task 2 in the applications section uses a map topology for editing.)

7.4.3 Editing Using Topology Rules

The geodatabase has 27 topology rules for point, line, and polygon features. Editing with a topology rule involves three basic steps. The first step creates a new topology by defining the participating feature classes in a feature dataset, the ranks for each feature class, the topology rule(s), and a cluster tolerance. The rank determines the relative importance or accuracy of a feature class in topological editing.

The second step is validation of topology. This step evaluates the topology rule and creates errors indicating those features that have violated the topology rule. At the same time, the edges and vertices of features in the participating feature classes are snapped together if they fall within the specified cluster tolerance. The snapping uses the rankings previously defined for the feature classes: features of a lower-rank (less accurate) feature class are moved more than features of a higher-rank feature class.

Validation results are saved into a topology layer, which is used in the third step for fixing errors and for accepting errors as exceptions (e.g., acceptable dangling nodes). The geodatabase provides a set of tools for fixing topological errors. For example, if the study area boundaries from two participating feature classes are not coincident and create small polygons between them, we can opt to subtract (i.e., delete) these polygons, or create new polygons, or modify their boundaries until

---

Box 7.4 | Cluster Tolerance

The default cluster tolerance using ArcGIS is 0.001 meters. Unless specified otherwise, all vertices within the default tolerance will be moved to share the same location. Cluster processing applies to features in the same layer or features in different layers. If it is applied to different layers, features in the more important layer (specified by the user) are moved less.

Ordnance Survey's MasterMap uses a spatial tolerance of 2 millimeters on line and area features on their topography layer. Based on this tolerance, no geometry is permitted to be closer than 4 millimeters of other geometry (or itself). Ordnance Survey's website cautions users not to use a validation tolerance of greater than 4 millimeters to prevent points to be snapped together unintentionally.

they are coincident. (Tasks 3 and 4 in the applications section use some of these tools in fixing errors between layers.)

As with the editing of coverages, the steps of creating a topology, validating the topology rule, and fixing topological errors may have to be repeated before we are satisfied with the topological accuracy of a feature class or of the match between two or more feature classes.

7.5 NONTOPOLOGICAL EDITING

Nontopological editing refers to a variety of basic editing operations that can modify simple features and can create new features from existing features (Box 7.4). Like topological editing, many of these basic operations also use the snapping tolerances to snap points and lines and the line and polygon sketches to edit features. The difference is that the basic operations do not involve topology as defined in a coverage, a map topology, or a geodatabase.

7.5.1 Editing Existing Features

The following summarizes nontopological or basic editing operations on existing features.

- Extend/Trim Lines: One can extend or trim a line to meet a target line.
- Delete/Move Features: One can delete or move one or more selected features, which

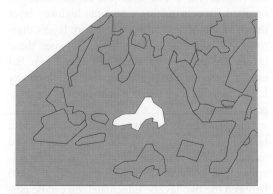

Figure 7.16
After a polygon of a shapefile is moved downward, a void area appears in its location.

may be points, lines, or polygons. Because each polygon in nontopological data is a unit, separate from other polygons, moving a polygon means placing the polygon on top of an existing polygon while creating a void area in its original location (Figure 7.16).

- Reshaping Features: One can alter the shape of a line by moving, deleting, or adding vertices on the line (Figure 7.17). The same method can be used to reshape a polygon. But, if the reshaping is intended

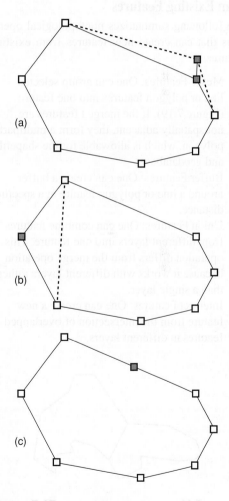

Figure 7.17
Reshape a line by moving a vertex (*a*), deleting a vertex (*b*), or adding a vertex (*c*).

for a polygon and its connected polygons, one must use a topological tool so that, when a boundary is moved, all polygons that share the same boundary are reshaped simultaneously.

- Split Lines and Polygons: One can sketch a new line that crosses an existing line to split the line, or sketch a split line through a polygon to split the polygon (Figure 7.18).

7.5.2 Creating Features from Existing Features

The following summarizes nontopological operations that can create new features from existing features.

- Merge Features: One can group selected line or polygon features into one feature (Figure 7.19). If the merged features are not spatially adjacent, they form a multipart polygon, which is allowable for the shapefile and geodatabase.
- Buffer Features: One can create a buffer around a line or polygon feature at a specified distance.
- Union Features: One can combine features from different layers into one feature. This operation differs from the merge operation because it works with different layers rather than a single layer.
- Intersect Features: One can create a new feature from the intersection of overlapped features in different layers.

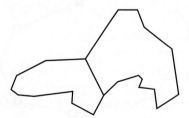

Figure 7.18
Sketch a line across the polygon boundary to split the polygon into two.

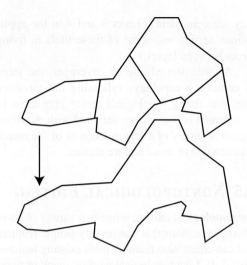

Figure 7.19
Merge four selected polygons into one.

7.6 OTHER EDITING OPERATIONS

Edgematching, line simplification, and line smoothing are examples of editing operations that cannot be classified as either topological or nontopological.

7.6.1 Edgematching

Edgematching matches lines along the edge of a layer to lines of an adjacent layer so that the lines are continuous across the border between the layers (Figure 7.20). For example, edgematching is required for constructing a regional highway layer that is made of several state highway layers digitized and edited separately. Errors between these layers are often very small (Figure 7.21), but unless they are removed, the regional highway layer cannot be used for such operations as shortest-path analysis.

Edgematching involves a source layer and a target layer. Features on the source layer are moved to match those on the target layer. A snapping tolerance can assist in snapping vertices (and lines) between the two layers. Edgematching can be performed on one pair of vertices, or multiple pairs, at a time. After edgematching is complete, the source layer and the target layer can be merged into a

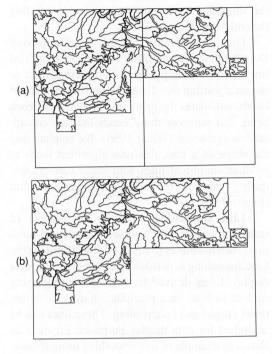

Figure 7.20
Edgematching matches the lines of two adjacent layers
(*a*) so that the lines are continuous across the border (*b*).

single layer and the artificial border separating
the two layers (e.g., the state boundary) can be
dissolved.

7.6.2 Line Simplification and Smoothing

Line simplification refers to the process of simplifying or generalizing a line by removing some of
its points. Line simplification is a common practice in map display (Robinson et al. 1995) and a
common topic under generalization in digital
cartography and GIS (McMaster and Shea 1992;
Weibel and Dutton 1999). When a map digitized
from the 1:100,000 scale source map is displayed
at the 1:1,000,000 scale, lines become jumbled and
fuzzy because of the reduced map space. Line simplification can also be important for GIS analysis
that uses every point that makes up a line. One
example is buffering, which measures a buffer distance from each point along a line (Chapter 11).

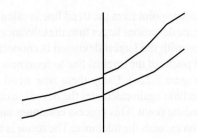

Figure 7.21
Mismatches of lines from two adjacent layers are only
visible after zooming in.

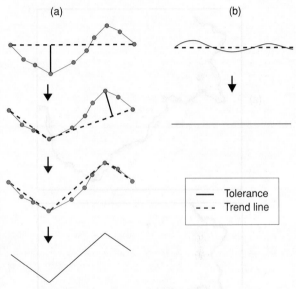

Figure 7.22
The Douglas-Peucker line simplification algorithm is
an iterative process that requires the use of a tolerance,
trend line, and the calculation of deviations of vertices
from the trend line. See Section 7.6.2 for explanation.

Lines with too many points do not necessarily
improve the result of analysis but will require
more processing time.

The **Douglas-Peucker algorithm** is a well-
known algorithm for line simplification (Douglas
and Peucker 1973). The algorithm works line by
line and with a specified tolerance. The algorithm
starts by connecting the end points of a line with
a trend line (Figure 7.22). The deviation of each

intermediate point from the trend line is calculated. If there are deviations larger than the tolerance, then the point with the largest deviation is connected to the end points of the original line to form new trend lines (Figure 7.22*a*). Using these new trend lines, the algorithm again calculates the deviation of each intermediate point. This process continues until no deviation exceeds the tolerance. The result is a simplified line that connects the trend lines. But if the initial deviations are all smaller than the tolerance,

the simplified line is the straight line connecting the end points (Figure 7.22*b*).

One shortcoming of the point-removing Douglas-Peucker algorithm is that the simplified line often has sharp angles. An alternative is to use an algorithm that dissects a line into a series of bends, calculates the geometric properties of each bend, and removes those bends that are considered insignificant (Wang 1996). By emphasizing the shape of a line, this new algorithm tends to produce simplified lines with better cartographic quality than does the Douglas-Peucker algorithm (Figure 7.23).

Line smoothing refers to the process of reshaping lines by using some mathematical functions such as splines (Saux 2003; Burghardt 2005). Line smoothing is perhaps most important for data display. Lines derived from computer processing such as isolines on a precipitation map are sometimes jagged and unappealing. These lines can be smoothed for data display purposes. Figure 7.24 shows an example of line smoothing using splines.

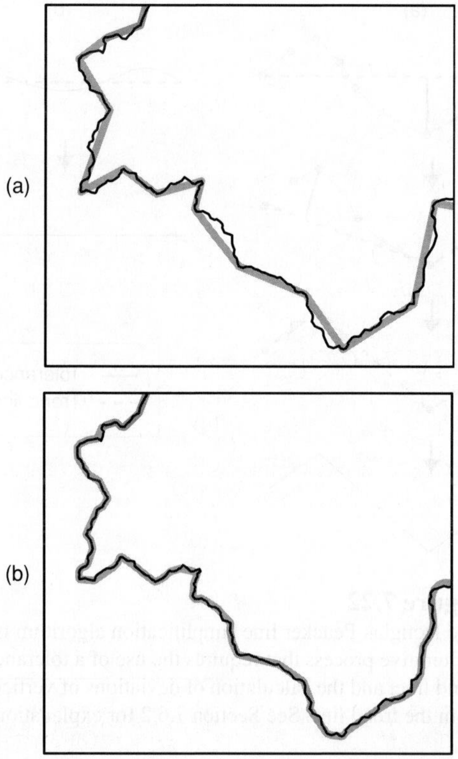

Figure 7.23
Result of line simplification can differ depending on the algorithm used: the Douglas-Peucker algorithm (*a*) and the bend-simplify algorithm (*b*).

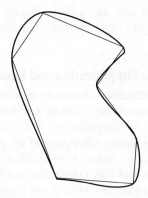

Figure 7.24
Line smoothing smoothes a line by generating new vertices mathematically and adding them to the line.

KEY CONCEPTS AND TERMS

Check plot: A plot of a digitized map for error checking.

Cluster tolerance: A tolerance for snapping points and lines. It is functionally similar to a fuzzy tolerance.

Dangle: An arc that has the same polygon on both its left and right sides and a dangling node at the end of the arc.

Dangle length: A tolerance that specifies the minimum length for dangling arcs on an output coverage.

Dangling node: A node at the end of an arc that is not connected to other arcs.

Data precision: A measure of how exactly data such as the location data of x- and y-coordinates are recorded.

Douglas-Peucker algorithm: A computer algorithm for line simplification.

Edgematching: An editing operation that matches lines along the edge of a layer to lines of an adjacent layer.

Fuzzy tolerance: A tolerance that specifies the minimum distance between vertices on a coverage.

Line simplification: The process of simplifying or generalizing a line by removing some of the line's points.

Line smoothing: The process of smoothing a line by adding new points, which are typically generated by a mathematical function such as spline, to the line.

Location errors: Errors related to the location of map features such as missing lines or missing polygons.

Map topology: A temporary set of topological relationships between coincident parts of simple features between layers.

Nontopological editing: Editing on nontopological data.

Overshoot: One type of digitizing error that results in an overextended arc.

Pseudo node: A node appearing along a continuous arc.

Topological editing: Editing on topological data to make sure that they follow the required topological relationships.

Topological errors: Errors related to the topology of map features such as dangling arcs and missing or multiple labels.

Undershoot: One type of digitizing error that results in a gap between arcs.

REVIEW QUESTIONS

1. Explain the difference between location errors and topological errors.
2. What are the primary data sources for digitizing?
3. Explain the importance of editing in GIS.
4. Although the U.S. National Map Accuracy Standard adopted in 1947 is still printed on USGS quadrangle maps, the standard is not really applicable to GIS data. Why?
5. According to the new National Standard for Spatial Data Accuracy, a geospatial data producer is encouraged to report a RMS statistic associated with a data set. In general terms, how does one interpret and use the RMS statistic?

6. Suppose a point location is recorded as (575729.0, 5228382) in data set 1 and (575729.64, 5228382.11) in data set 2. Which data set has a higher data precision? In practical terms, what does the difference in data precision in this case mean?

7. The ArcGIS Desktop Help has a poster illustrating topology rules (ArcGIS Desktop Help > Editing and data compilation > Editing topology > Topology rules). View the poster. Can you think of an example (other than those on the poster) that can use the polygon rule of "Must be covered by feature class of"?

8. Give an example (other than those on the poster) that can use the polygon rule of "Must not overlap with."

9. Give an example (other than those on the poster) that can use the line rule of "Must not intersect or touch interior."

10. Use a diagram to illustrate how a large nodesnap for editing can alter the shapes of line features.

11. Use a diagram to illustrate how a large cluster tolerance for editing can alter the shapes of line features.

12. Explain the difference between a dangling node and a pseudo node.

13. What is a map topology?

14. Describe the three basic steps in using a topology rule.

15. Some nontopological editing operations can create features from existing features. Give two examples of such operations.

16. Edgematching requires a source layer and a target layer. Explain the difference between these two types of layers.

17. The Douglas-Peucker algorithm typically produces simplified lines with sharp angles. Why?

APPLICATIONS: SPATIAL DATA ACCURACY AND QUALITY

This applications section covers four tasks. Task 1 lets you use basic editing tools on a shapefile. Task 2 asks you to use a map topology and a cluster tolerance to fix digitizing errors between two shapefiles. You will use topology rules in Tasks 3 and 4: fixing dangles in Task 3 and fixing outline boundaries in Task 4.

Task 1 Edit a Shapefile

What you need: *editmap2.shp* and *editmap3.shp*.

Task 1 covers three basic editing operations for shapefiles: merging polygons, splitting a polygon, and reshaping the polygon boundary. While working with *editmap2.shp* you will use *editmap3 .shp* as a reference, which shows how *editmap2.shp* will look after editing.

1. Start ArcCatalog and connect to the Chapter 7 database. Launch ArcMap. Change the name of the data frame to Task 1. Add *editmap3*.

shp and *editmap2.shp* to Task 1. To edit *editmap2* by using *editmap3* as a guide, you must show them with different outline symbols. Select Properties from the context menu of *editmap2*. On the Symbology tab, change the symbol to Hollow with the Outline Color in black. On the Labels tab, check the box to label features in this layer and select LANDED_ID to be the label field, and click OK to dismiss the dialog. Click the symbol of *editmap3* in the table of contents. Choose the Hollow symbol and the Outline Color of red. Right-click *editmap2*, point to Selection, and click on Make This The Only Selectable Layer.

2. Make sure that the Editor toolbar is checked. Click the Editor dropdown arrow and choose Start Editing. The Create Features window opens with *editmap2* highlighted. The first operation is to merge polygons 74 and 75.

Click the Edit tool on the Editor Toolbar. Click inside polygon 75, and then click inside polygon 74 while pressing the Shift key. The two polygons are highlighted in cyan. Click the Editor dropdown arrow and choose Merge. In the next dialog, choose the top feature and click OK to dismiss the dialog. Polygons 74 and 75 are merged into one with the label of 75.

Q1. List other editing operations besides Merge on the Editor menu.

3. The second operation is to cut polygon 71. Zoom to the area around polygon 71. Click the Edit Tool, and use it to select polygon 71 by clicking inside the polygon. Click the Cut Polygons tool on the Editor toolbar. Left-click the mouse where you want the cut line to start, click each vertex that makes up the cut line, and double-click the end vertex. Polygon 71 is cut into two, each labeled 71.

4. The third operation is to reshape polygon 73 by extending its southern border in the form of a rectangle. Because polygon 73 shares a border (i.e., edge) with polygon 59, you need to use a map topology to modify the border. Click the Editor's dropdown arrow, point to More Editing Tools, and check Topology. Click the Map Topology tool on the Topology toolbar. In the next dialog, check the *editmap2* box and click OK. Click the Topology Edit tool on the Topology toolbar, and then double-click on the southern edge of polygon 73. Now the outline of polygon 73 turns magenta with vertices in dark green and the end point in red. The Edit Vertices toolbar also appears on top of the Topology toolbar.

5. The strategy in reshaping the polygon is to add three new vertices and to drag the vertices to form the new shape. Click the Add Vertex tool on the Edit Vertices toolbar. Use the tool to click the midpoint of the southern border of polygon 73 and drag it to the midpoint of the new border (use *editmap3* as a guide). (The original border of polygon 73

remains in place as a reference. It will disappear when you click anywhere outside polygon 73.)

6. Next, use the Add Vertex tool to add another vertex (vertex 2) along the line that connects vertex 1 and the original SE corner of polygon 73. Drag vertex 2 to the SE corner of the new boundary. Add another vertex (vertex 3) and drag it the SW corner of the new boundary. The edge has been modified. Right-click the edge, and select Finish Sketch.

7. Select Stop Editing from the Editor dropdown list, and save the edits.

Task 2 Use Cluster Tolerance to Fix Digitizing Errors Between Two Shapefiles

What you need: *land_dig.shp,* a reference shapefile; and *trial_dig.shp,* a shapefile digitized off *land_dig.shp.*

There are discrepancies between *land_dig .shp* and *trial_dig.shp* due to digitizing errors. This task uses a cluster tolerance to force the boundaries of *trial_dig.shp* to be coincident with those of *land_dig.shp.* Both *land_dig.shp* and *trial_dig.shp* are measured in meters and in UTM (Universal Transverse Mercator) coordinates.

1. Insert a new data frame and rename it Task 2. Add *land_dig.shp* and *trial_dig.shp* to Task 2. Display *land_dig* in a black outline symbol and label it with the field of LAND_DIG_I. Display *trial_dig* in a red outline symbol. Right-click *trial_dig,* point to Selection, and click on Make This The Only Selectable Layer. Zoom in and use the Measure tool to check the discrepancies between the two shapefiles. Most discrepancies are smaller than 1 meter.

2. The first step is to create a map topology between the two shapefiles. Click the Customize menu and make sure that both the Editor and Topology toolbars are checked. Select Start Editing from the Editor's dropdown menu. Click the Map Topology

tool on the Topology toolbar. In the next dialog, select both *land_dig* and *trial_dig* to participate in the map topology and enter 1 (meter) for the Cluster Tolerance. Click OK to dismiss the dialog.

3. *trial_dig* has five polygons of which three are isolated polygons and two are spatially adjacent. Start editing with the polygon in the lower right that is supposed to be coincident with polygon 73 in *land_dig*. Zoom to the area around the polygon. Click the Topology Edit tool on the Topology toolbar. Then use the mouse pointer to double-click the boundary of the polygon. The boundary turns into an edit sketch with green squares representing vertices and a red square representing a node. Place the mouse pointer over a vertex until it has a new square symbol. Right-click on the symbol, and select Move from the context menu. Hit Enter to dismiss the next dialog. (You are using the specified cluster tolerance to snap the vertices and edges.) Click any point outside the polygon to unselect its boundary. The polygon now should coincide perfectly with polygon 73 in *land_dig*. Move to the other polygons in *trial_dig* and follow the same procedure to fix digitizing errors.

4. All discrepancies except one (polygon 76) are fixed between *trial_dig* and *land_dig*. The remaining discrepancy is larger than the specified cluster tolerance (1 meter). Rather than using a larger cluster tolerance, which may result in distorted features, you will use the basic editing operations to fix the discrepancy. Use the Edit tool to double-click the boundary with the discrepancy. Zoom in to the area of discrepancy. Use the Edit Tool to double-click the boundary of *trial_dig*. When the boundary turns into an edit sketch, you can drag a vertex to meet the target line.

5. After you have finished editing all five polygons, select Stop Editing from the Editor's dropdown menu and save the edits.

Q2. If you had entered 4 meters for the cluster tolerance in Step 2, what would have happened to *trial_dig.shp?*

Task 3 Use Topology Rule to Fix Dangles

What you need: *idroads.shp,* a shapefile of Idaho roads; *mtroads_idtm.shp,* a shapefile of Montana roads projected onto the same coordinate system as *idroads.shp*; and *Merge_result.shp,* a shapefile with merged roads from Idaho and Montana.

The two road shapefiles downloaded from the Internet are not perfectly connected across the state border. Therefore *Merge_result.shp* contains gaps. Unless the gaps are removed, *Merge_ result .shp* cannot be used for network applications such as finding the shortest path. This task asks you to use a topology rule to symbolize where the gaps are and then use the editing tools to fix the gaps.

1. The first step for this task is to prepare a personal geodatabase and a feature dataset, and to import *Merge_result.shp* as a feature class into the feature dataset. Right-click the Chapter 7 database in the Catalog tree, point to New, and select Personal Geodatabase. Rename the geodatabase *MergeRoads.mdb.* Right-click *MergeRoads.mdb,* point to New, and select Feature Dataset. Enter *Merge* for the Name of the feature dataset, and click Next. In the next dialog, click Projected Coordinate Systems and then click the Import button to import the coordinate system from *idroads.shp* for the feature dataset. Choose None for the vertical coordinate system. Change the XY tolerance to 1 meter, and click Finish. Right-click *Merge* in the Catalog tree, point to Import, and select Feature Class (single). In the next dialog, select *Merge_result.shp* for the input features and enter *Merge_result* for the output feature class name. Click OK to import the shapefile.

2. This step is to build a new topology. Right-click *Merge* in the Catalog tree in ArcMap, point to New, and select Topology. Click Next in the first two panels. Check the box

next to *Merge_result* in the third. Click Next in the fourth panel. Click the Add Rule button in the fifth panel. Select "Must Not Have Dangles" from the Rule dropdown list in the Add Rule dialog and click OK. Click Next and then Finish to complete the setup of the topology rule. After the new topology has been created, click Yes to validate it.

Q3. Each rule has a description in the Add Rule dialog. What is the rule description for "Must Not Have Dangles" in ArcGIS Desktop Help?

Q4. What is the rule description for "Must Not Have Pseudonodes"?

3. The validation results are saved in a topology layer called *Merge_Topology* in the *Merge* feature dataset. Select Properties from the context menu of *Merge_Topology*. The Topology Properties dialog has four tabs. The General, Feature Classes, and Rules tabs define the topology rule. Click the Errors tab and then Generate Summary. The summary report shows 96 errors, meaning that *Merge_result* has 96 dangling nodes.

4. Insert a new data frame in ArcMap, and rename the data frame Task 3. Add the *Merge* feature dataset to Task 3. Point errors in *Merge_Topology* are those 96 dangling nodes, most of which are the end points along the outside border of the two states and are thus acceptable dangling nodes. Only those nodes along the shared border of the two states need inspection and, if necessary, fixing. Add *idroads.shp* and *mtroads_idtm.shp* to Task 3. The two shapefiles can be used as references in inspecting and fixing errors. Use different colors to display *Merge_result, idroads,* and *mtroads_idtm* so that you can easily distinguish between them. Right-click *Merge_result,* point to Selection, and click on Make This The Only Selectable Layer.

5. Now you are ready to inspect and fix errors in *Merge_result.* Make sure that both the Editor toolbar and the Topology toolbar

are available. Select Start Editing from the Editor menu. Select *MergeRoads.mdb* as the database to edit data from. There are five places where the roads connect across the Montana-Idaho border. These places are shown with point errors. Zoom to the area around the first crossing near the top of the map until you see a pair of dangles, separated by a distance of about 5.5 meters. (Use the Measure tool on the standard toolbar to measure the distance.) Click the Fix Topology Error tool on the Topology toolbar, and then click a red square. The red square turns black after being selected. Click Error Inspector on the Topology toolbar. A report appears and shows the error type (Must Not Have Dangles). Use the Fix Topology Error tool and right-click on the black square. The context menu has the tools Snap, Extend, and Trim to fix errors. Select Snap, and a Snap Tolerance box appears. Enter 6 (meters) in the box. The two squares are snapped together into one square. Right-click the square again and select Snap. The square should now disappear. Remember that you have access to the Undo and Redo tools on the Edit menu as well as the standard toolbar.

6. The second point error, when zoomed in, shows a gap of 125 meters. There are at least two ways to fix the error. The first option is to use the Snap command of the Fix Topology Error tool by applying a snap tolerance of at least 125. Here you will use the second option, which uses the regular editing tools. First set up the editing environment. Point to Snapping on the Editor's menu and check Snapping toolbar to open the toolbar. Next select Options from the Snapping dropdown menu. In the General frame, enter 10 for the snapping tolerance. Make sure that Use Snapping is checked on the Snapping dropdown menu. Click *Merge_result* in the Create Features window, and select Line as a construction tool in the Create Features window. Right-click the

square on the right, point to Snap to Feature, and select Endpoint. Click the square on the left. Then right-click the it and select Finish Sketch. Now the gap is bridged with a new line segment. Click Validate Topology in Current Extent on the Topology toolbar. The square symbols disappear, meaning that the point error no longer exists.

7. You can use the preceding two options to fix the rest of the point errors.

8. After all point errors representing misconnections of roads across the state border have been fixed, select Stop Editing from the Editor's menu and save the edits.

Task 4 Use Topology Rule to Ensure That Two Polygon Layers Cover Each Other

What you need: *landuse.shp* and *soils.shp,* two polygon shapefiles based on UTM coordinates.

Digitized from different source maps, the outlines of the two shapefiles are not completely coincident. This task shows you how to use a topology rule to symbolize the discrepancies between the two shapefiles and use the editing tools to fix the discrepancies.

1. Similar to Task 3, the first step is to prepare a personal geodatabase and a feature dataset and to import *landuse.shp* and *soils.shp* as feature classes into the feature dataset. Right-click the Chapter 7 folder in the Catalog tree, point to New, and select Personal Geodatabase. Rename the geodatabase *Land.mdb*. Right-click *Land.mdb,* point to New, and select Feature Dataset. Enter *LandSoil* for the Name of the feature dataset, and click Next. In the next dialog, click Projected Coordinate Systems and then import the coordinate system from *landuse.shp* for the feature dataset. Choose None for the vertical coordinate system. Set the XY tolerance to be 0.001 meter, and click Finish. Right-click *LandSoil,* point to Import, and select Feature Class (multiple). In the next dialog, add

landuse.shp and *soils.shp* as the input features and click OK to import the feature classes.

2. Next build a new topology. Right-click *LandSoil,* point to New, and select Topology. Click Next in the first two panels. In the third panel, check both *landuse* and *soils* to participate in the topology. The fourth panel allows you to set ranks for the participating feature classes. Features in the feature class with a higher rank are less likely to move. Click Next because the editing operations for this task are not affected by the ranks. Click the Add Rule button in the fifth panel. Select *landuse* from the top dropdown list, select "Must Be Covered By Feature Class Of" from the Rule dropdown list, and select *soils* from the bottom dropdown list. Click OK to dismiss the Add Rule dialog. Click Next and then Finish to complete the setup of the topology rule. After the new topology has been created, click Yes to validate it.

Q5. What is the rule description for "Must Be Covered By Feature Class Of"?

3. Insert a new data frame in ArcMap, and rename the data frame Task 4. Add the *LandSoil* feature dataset to Task 4. Area errors are areas where the two shapefiles are not completely coincident. Use the outline symbols of different colors to display *landuse* and *soils*. Zoom to the area errors. Most deviations between the two feature classes are within 1 meter.

4. Select Start Editing from the Editor's menu. Click the Fix Topology Error tool on the Topology toolbar, and drag a box to select every area error. All area errors turn black. Right-click a black area and select Subtract. The Subtract command removes areas that are not common to both feature classes. In other words, Subtract makes sure that, after editing, every part of the area extent covered by *LandSoil* will have attribute data from both feature classes.

5. Select Stop Editing from the Editor dropdown menu. Save the edits.

Challenge Task

What you need: *idroads.shp, wyroads.shp,* and *idwyroads.shp.*

The Chapter 7 database contains *idroads.shp,* a major road shapefile for Idaho; *wyroads.shp,* a major road shapefile for Wyoming; and *idwyroads .shp,* a merged road shapefile of Idaho and Wyoming.

All three shapefiles are projected onto the Idaho Transverse Mercator (IDTM) coordinate system and measured in meters. This challenge task asks you to examine and fix gaps that are smaller than 200 meters on roads across the Idaho and Wyoming border. Use a topology rule in the geodatabase data model to fix dangles in *idwyroads.shp.*

REFERENCES

American Society for Photogrammetry and Remote Sensing. 1990. ASPRS Accuracy Standards for Large-Scale Maps. *Photogrammetric Engineering and Remote Sensing* 56: 1068–70.

Burghardt, D. 2005. Controlled Line Smoothing by Snakes. *GeoInformatica* 9: 237–52.

Douglas, D. H., and T. K. Peucker. 1973. Algorithms for the Reduction of the Number of Points Required to Represent a Digitized Line or Its Caricature. *The Canadian Cartographer* 10: 110–22.

Federal Geographic Data Committee. 1998. Part 3: *National Standard for Spatial Data Accuracy, Geospatial Positioning Accuracy Standards,*

FGDC-STD-007.3-1998. Washington, DC: Federal Geographic Data Committee.

McMaster, R. B., and K. S. Shea. 1992. *Generalization in Digital Cartography.* Washington, DC: Association of American Geographers.

Monmonier, M. 1996. *How to Lie with Maps,* 2d ed. Chicago: University of Chicago Press.

Plümer, L., and G. Gröger. 1997. Achieving Integrity in Geographic Information Systems—Maps and Nested Maps. *GeoInformatica* 1: 345–67.

Robinson, A. H., J. L. Morrison, P. C. Muehrcke, A. J. Kimerling, and S. C. Guptill. 1995. *Elements of Cartography,* 6th ed. New York: Wiley.

Saux, E. 2003. B-Spline Functions and Wavelets for Cartographic Line Generalization. *Cartography and Geographic Information Science* 30: 33–50.

U.S. Bureau of the Budget. 1947. *United States National Map Accuracy Standards.* Washington, DC: U.S. Bureau of the Budget.

Wang, Z. 1996. Manual versus Automated Line Generalization. *GIS/LIS '96 Proceedings:* 94–106.

Weibel, R., and G. Dutton. 1999. Generalizing Spatial Data and Dealing with Multiple Representations. In P. A. Longley, M. F. Goodchild, D. J. Maguire, and D. W. Rhind, eds., *Geographical Information Systems,* 2d ed., pp. 125–55. New York: Wiley.

CHAPTER 8

ATTRIBUTE DATA MANAGEMENT

CHAPTER OUTLINE

8.1 Attribute Data in GIS

8.2 The Relational Model

8.3 Joins, Relates, and Relationship Classes

8.4 Attribute Data Entry

8.5 Manipulation of Fields and Attribute Data

GIS involves both spatial and attribute data: spatial data relate to the geometries of spatial features, and attribute data describe the characteristics of the spatial features. Figure 8.1, for example, shows attribute data such as street name, address ranges, and ZIP codes associated with each street segment in the TIGER/Line files. Without the attributes, the TIGER/Line files will be of limited use.

The difference between spatial and attribute data is well defined with discrete features such as the TIGER/Line files. The georelational data model (e.g., a coverage) stores spatial data and

attribute data separately and links the two by the feature ID (Figure 8.2). The two data sets are synchronized so that they can be queried, analyzed, and displayed in unison. The object-based data model (e.g., a geodatabase) combines both geometries and attributes in a single system. Each spatial feature has a unique object ID and an attribute to store its geometry (Figure 8.3). Although the two data models handle the storage of spatial data differently, both operate in the same relational database environment. Therefore, materials covered in Chapter 8 can apply to both vector data models.

The raster data model presents a different scenario in terms of data management. The cell value corresponds to the value of a continuous feature at the cell location. But only an integer raster has a value attribute table, which summarizes cell values and their frequencies, rather than cell values by cell (Figure 8.4). If the cell value does represent some spatial unit such as the county FIPS (U.S. Federal Information Processing Standards) code, we can use the value attribute table to store

FEDIRP	FENAME	FETYPE	FRADDL	TOADDL	FRADDR	TOADDR	ZIPL	ZIPR
N	4th	St	6729	7199	6758	7198	83815	83815

Figure 8.1
Each street segment in the TIGER/Line shapefiles has a set of associated attributes. These attributes include street name, address ranges on the left side and the right side, and ZIP codes on both sides.

Record	Soil-ID	Area	Perimeter
1	1	106.39	495.86
2	2	8310.84	508,382.38
3	3	554.11	13,829.50
4	4	531.83	19,000.03
5	5	673.88	23,931.47

Figure 8.2
As an example of the georelational data model, the soils coverage uses Soil-ID to link spatial and attribute data.

ObjectID	Shape	Shape_Length	Shape_Area
1	Polygon	106.39	495.86
2	Polygon	8310.84	508,382.38
3	Polygon	554.11	13,829.50
4	Polygon	531.83	19,000.03
5	Polygon	673.88	23,931.47

Figure 8.3
The object-based data model uses the Shape field to store the geometries of soil polygons. The table therefore contains both spatial and attribute data.

county-level data and use the raster to display these county-level data. But the raster is always associated with the FIPS code for data query and analysis. This association between the raster and the spatial variable separates the raster data model from the vector data model and makes attribute data management a much less important topic for raster data than vector data.

ObjectID	Value	Count
0	160,101	142
1	160,102	1580
2	160,203	460
3	170,101	692
4	170,102	1417

Figure 8.4
A value attribute table lists the attributes of value and count. The Value field stores the cell values, and the Count field stores the number of cells in the raster.

With emphasis on vector data, Chapter 8 is divided into the following five sections. Section 8.1 provides an overview of attribute data in GIS. Section 8.2 discusses the relational model, data normalization, and types of data relationships. Section 8.3 explains joins, relates, and relationship classes. Sections 8.4 covers attribute data entry, including field definition, method, and verification. Section 8.5 discusses the manipulation of fields and the creation of new attribute data from existing attributes.

8.1 ATTRIBUTE DATA IN GIS

Attribute data are stored in tables. An attribute table is organized by row and column. Each row represents a spatial feature, each column describes a characteristic, and the intersection of a column and a row shows the value of a particular characteristic for a particular feature (Figure 8.5). A row is also called a **record,** and a column is also called a **field.**

8.1.1 Types of Attribute Tables

There are two types of attribute tables. The first type is called the **feature attribute table,** which has access to the feature geometry. Every vector data set must have a feature attribute table. In the case of the georelational data model, the feature attribute table uses the feature ID to link to the feature's geometry. In the case of the object-based data model, the feature attribute table has a field that stores the feature's geometry. Feature attribute tables also have the default fields that summarize the feature geometries such as length for line features and area and perimeter for polygon features.

A feature attribute table may be the only table needed if a data set has only several attributes. But this is often not the case. For example, a soil map unit can have over 100 soil interpretations, soil properties, and performance data. To store all these attributes in a feature attribute table will require many repetitive entries, a process that wastes both time and computer memory. Moreover, the table will be extremely difficult to use and update. This is why we need the second type of attribute table.

This second type of attribute table is nonspatial, meaning that the table does not have direct access to the feature geometry but has a field linking the table to the feature attribute table whenever necessary. Tables of nonspatial data may exist as delimited text files, dBASE files, Excel files, Access files, or files managed by database

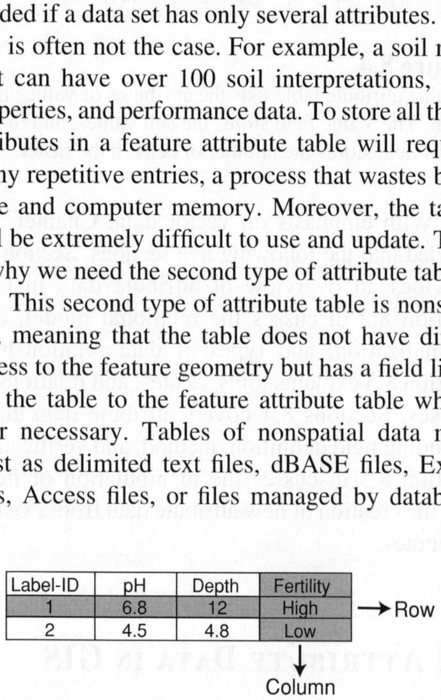

Figure 8.5

A feature attribute table consists of rows and columns. Each row represents a spatial feature, and each column represents a property or characteristic of the spatial feature.

software packages such as Oracle, Informix, SYBASE, SQL Server, and IBM DB2.

8.1.2 Database Management

The presence of feature attribute and nonspatial data tables means that a GIS requires a **database management system (DBMS)** to manage these tables. A DBMS is a software package that enables us to build and manipulate a database (Oz 2004). A DBMS provides tools for data input, search, retrieval, manipulation, and output. Most commercial GIS packages include database management tools for local databases. For example, ArcGIS Desktop uses Microsoft Access for managing personal geodatabases.

The use of a DBMS has other advantages beyond its GIS applications. Often a GIS is part of an enterprisewide information system, and attribute data needed for the GIS may reside in various departments of the same organization. Therefore, the GIS must function within the overall information system and interact with other information technologies.

The object-based data model such as the geodatabase has actually blurred the boundary between a GIS and a DBMS. Because a geodatabase stores both geometries and attributes as tables in a single database, it is basically the same as a database used in business or marketing. This has led some authors to refer to GIS as spatial database management systems (e.g., Shekhar and Chawla 2003).

Besides database management tools for managing local databases, many GIS packages also have database connection capabilities to access remote databases. This is important for GIS users who routinely access data from centralized databases. For example, GIS users at a ranger district office may regularly retrieve data maintained at the headquarters office of a national forest. This scenario represents a client-server distributed database system (Arvanitis et al. 2000). A client (e.g., a district office user) sends a request to the server, retrieves data from the server, and processes the data on the local computer.

Box 8.1 | **Categorical and Numeric Data**

Categorical data refer to data measured at nominal or ordinal scales, and numeric data refer to data measured at interval or ratio scales. These two types of data are clearly different. But GIS projects, especially those dealing with suitability analysis, commonly assign numeric scores to categorical data and use these scores in computation (Chrisman 2001). For example,

a suitability analysis for a housing development project may assign different scores to different soil types and then combine these scores with scores from other variables for computing the total suitability score. Assigning scores to categorical data requires information that is not in the base data. Scores in this case represent *interpreted* data, which are based on expert judgments.

8.1.3 Types of Attribute Data

One method for classifying attribute data is by data type. The data type determines how an attribute is stored in a GIS. Depending on the GIS package, the available data types can vary. Common data types are number, text (or character), date, and binary large object (BLOB). Numbers include integer (for numbers without decimal digits) and floats (for numbers with decimal digits). Moreover, depending on the designated computer memory, an integer can be short or long and a float can be single precision or double precision. BLOBs store images, multimedia, and feature geometrics as long sequences of binary numbers.

Another method is to define attribute data by measurement scale. The measurement scale concept groups attribute data into nominal, ordinal, interval, and ratio data, with increasing degrees of sophistication (Stevens 1946; Chang 1978). **Nominal data** describe different kinds or different categories of data such as land-use types or soil types. **Ordinal data** differentiate data by a ranking relationship. For example, soil erosion may be ordered from severe to moderate to light. **Interval data** have known intervals between values. For example, a temperature reading of 70°F is warmer than 60°F by 10°F. **Ratio data** are the same as interval data except that ratio data are based on a meaningful, or absolute, zero value. Population densities are an example of ratio data, because a density of 0 is an absolute zero. The distinction of measurement scales is important for statistical analysis as different types of tests (e.g., parametric vs. nonparametric tests) are designed

for data at different scales. It can also be important for data display because one of the determinants in selecting map symbols is the measurement scale of the data to be displayed (Chapter 9).

The measurement scales of nominal, ordinal, interval, and ratio are well defined in the literature. Occasionally, especially for image analysis, two general categories of categorical and numeric are also used: **categorical data** include nominal and ordinal scales, and **numeric data** include interval and ratio scales (Box 8.1).

Data type and measurement scale are obviously related. Character strings are appropriate for nominal and ordinal data. Integers and floats are appropriate for interval and ratio data, depending on whether decimal digits are included or not. But there are exceptions. For example, a study may classify the potential for groundwater contamination as high, medium, or low but may present the information as numeric data using a lookup table. The lookup table may show 1 for low, 2 for medium, and 3 for high. In this case, the numbers are simply the numeric codes for the ordered categorical data. We must pay attention to the nature of attribute data before using them in analysis.

8.2 THE RELATIONAL MODEL

A database is a collection of interrelated tables in digital format. At least four types of database designs have been proposed in the literature: flat file, hierarchical, network, and relational (Figure 8.6) (Jackson 1999).

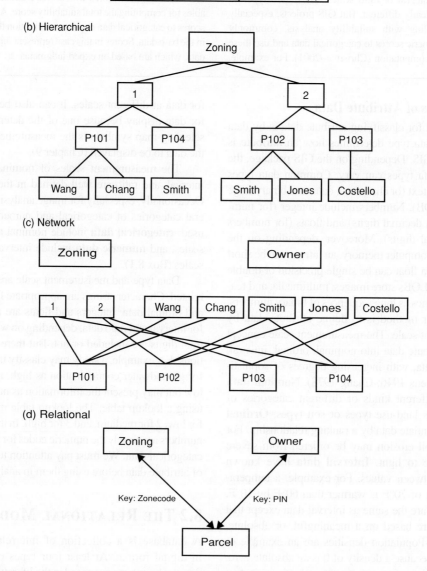

(a) Flat file

PIN	Owner	Zoning
P101	Wang	Residential (1)
P101	Chang	Residential (1)
P102	Smith	Commercial (2)
P102	Jones	Commercial (2)
P103	Costello	Commercial (2)
P104	Smith	Residential (1)

(b) Hierarchical

(c) Network

(d) Relational

Key: Zonecode   Key: PIN

Figure 8.6

Four types of database design: (*a*) flat file, (*b*) hierarchical, (*c*) network, and (*d*) relational.

A **flat file** contains all data in a large table. A feature attribute table is like a flat file. Another example is a spreadsheet with data only. A **hierarchical database** organizes its data at different levels and uses only the one-to-many association between levels. The simple example in Figure 8.6 shows the hierarchical levels of zoning, parcel, and owner. Based on the one-to-many association, each level is divided into different branches. A **network database** builds connections across tables, as shown by the linkages between the tables in Figure 8.6. A common problem with both the hierarchical and the network database designs is that the linkages (i.e., access paths) between tables must be known in advance and built into the database at design time (Jackson 1999). This requirement tends to make a complicated and inflexible database and limit the database applications.

GIS vendors typically use the relational model for database management (Codd 1970; Codd 1990; Date 1995). A **relational database** is a collection of tables or relations (the mathematical term for tables) that can be connected to each other by keys. A **primary key** represents one or more attributes whose values can uniquely identify a record in a table. Values of the primary key cannot be null and should never change. A **foreign key** is one or more attributes that refer to a primary key in another table. As long as they match in their function, the primary and foreign keys do not have to have the same name. But in GIS, they often have the same name, such as the feature ID. In that case, the feature ID is also called the *common field*. In Figure 8.6, Zonecode is the common field connecting zoning and parcel, and PIN (parcel ID number) is the common field connecting parcel and owner. When used together, the fields can relate zoning and owner.

Compared to other database designs, a relational database is simple and flexible (Carleton et al. 2005). It has two distinctive advantages. First, each table in the database can be prepared, maintained, and edited separately from other tables. This is important because, with the increased popularity of GIS technology, more data are being recorded and managed in spatial units. Second, the tables can remain separate until a query or an analysis requires that attribute data from different tables be linked together. Because the need for linking tables is often temporary, a relational database is efficient for both data management and data processing.

8.2.1 SSURGO: A Relational Database Example

The Natural Resources Conservation Service (NRCS) produces the **Soil Survey Geographic (SSURGO) database** nationwide **(http://soils.usda .gov/).** The NRCS collects SSURGO data from field mapping, archives the data in 7.5-minute quadrangle units, and organizes the database by soil survey area. A soil survey area may consist of a county, multiple counties, or parts of multiple counties. The SSURGO database represents the most detailed level of soil mapping by the NRCS in the United States.

The SSURGO database consists of spatial data and tabular data. For each soil survey area, the spatial data contain a detailed soil map. The soil map is made of *soil map units,* each of which may be composed of one or more noncontiguous polygons. As the smallest area unit for soil mapping, a soil map unit represents a set of geographic areas for which a common land-use management strategy is suitable. Interpretations and properties of soil map units are provided by links between soil maps and data that exist in more than 50 tables in the SSURGO database. The NRCS provides metadata that describe each table and the keys that link tables **(http://soildatamart.nrcs.usda.gov/ SSURGOMetada.aspx).**

The sheer size of the SSURGO database can be overwhelming at first. But the database is not difficult to use if we have a proper understanding of the relational model. In Section 8.2.3 we use the SSURGO database to illustrate types of relationships between tables. Chapter 10 uses the database as an example for data exploration.

8.2.2 Normalization

Preparing a relational database such as SSURGO involves following certain rules. An important rule is called normalization. **Normalization** is a process of decomposition, taking a table with all the attribute data and breaking it down into small tables while maintaining the necessary linkages between them (Vetter 1987). Normalization is designed to achieve the following objectives:

- To avoid redundant data in tables that waste space in the database and may cause data integrity problems;
- To ensure that attribute data in separate tables can be maintained and updated separately and can be linked whenever necessary; and
- To facilitate a distributed database.

An example of normalization is offered here. Table 8.1 shows attribute data for a parcel map. The table contains redundant data: owner addresses are repeated for Smith and residential and commercial zoning are entered twice. The table also contains uneven records: depending on the parcel, the fields of owner and owner address can have either one or two values. An unnormalized table such as Table 8.1

cannot be easily managed or edited. To begin with, it is difficult to define the fields of owner and owner address and to store their values. A change of the ownership requires that all attributes be updated in the table. The same difficulty applies to such operations as adding or deleting attribute values.

Table 8.2 represents the first step in normalization. Often called the first normal form, Table 8.2 no longer has multiple values in its cells, but the problem of data redundancy has increased. P101 and P102 are duplicated except for changes of the owner and the owner address. Smith's address is included twice. And the zoning descriptions of residential and commercial are listed three times each. Also, identifying the owner address is not possible with PIN alone but requires a compound key of PIN and owner.

Figure 8.7 represents the second step in normalization. In place of Table 8.2 are three small tables of parcel, owner, and address. PIN is the common field relating the parcel and owner tables. Owner name is the common field relating the address and owner tables. The relationship between

TABLE 8.1 | An Unnormalized Table

PIN	Owner	Owner Address	Sale Date	Acres	Zone Code	Zoning
P101	Wang Chang	101 Oak St 200 Maple St	1-10-98	1.0	1	Residential
P102	Smith Jones	300 Spruce Rd 105 Ash St	10-6-68	3.0	2	Commercial
P103	Costello	206 Elm St	3-7-97	2.5	2	Commercial
P104	Smith	300 Spruce Rd	7-30-78	1.0	1	Residential

TABLE 8.2 | First Step in Normalization

PIN	Owner	Owner Address	Sale Date	Acres	Zone Code	Zoning
P101	Wang	101 Oak St	1-10-98	1.0	1	Residential
P101	Chang	200 Maple St	1-10-98	1.0	1	Residential
P102	Smith	300 Spruce Rd	10-6-68	3.0	2	Commercial
P102	Jones	105 Ash St	10-6-68	3.0	2	Commercial
P103	Costello	206 Elm St	3-7-97	2.5	2	Commercial
P104	Smith	300 Spruce Rd	7-30-78	1.0	1	Residential

Parcel table

PIN	Sale date	Acres	Zone code	Zoning
P101	1-10-98	1.0	1	Residential
P102	10-6-68	3.0	2	Commercial
P103	3-7-97	2.5	2	Commercial
P104	7-30-78	1.0	1	Residential

Owner table

PIN	Owner name
P101	Wang
P101	Chang
P102	Smith
P102	Jones
P103	Costello
P104	Smith

Address table

Owner name	Owner address
Wang	101 Oak St
Chang	200 Maple St
Jones	105 Ash St
Smith	300 Spruce Rd
Costello	206 Elm St

Figure 8.7
Separate tables from the second step in normalization. The fields relating the tables are highlighted.

the parcel and address tables can be established through PIN and owner name. The only problem with the second normal form is data redundancy with the fields of zone code and zoning.

The final step in normalization is presented in Figure 8.8. A new table, zone, is created to take care of the remaining data redundancy problem with zoning. Zone code is the common field relating the parcel and zone tables. Unnormalized data in Table 8.1 are now fully normalized.

Higher normal forms than the third can achieve objectives consistent with the relational model, but they can slow down data access and create higher maintenance costs (Lee 1995). To find the addresses of parcel owners, for example, we must link three tables (parcel, owner, and address) and employ two fields (PIN and owner name). One way to increase the performance in data access is to reduce the level of normalization by, for example, removing the address table

and including the addresses in the owner table. Therefore, normalization should be maintained in the conceptual design of a database, but performance and other factors should be considered in its physical design (Moore 1997).

8.2.3 Types of Relationships

A relational database may contain four types of relationships or *cardinalities* between tables or, more precisely, between records in tables: one-to-one, one-to-many, many-to-one, and many-to-many (Figure 8.9). The **one-to-one relationship** means that one and only one record in a table is related to one and only one record in another table. The **one-to-many relationship** means that one record in a table may be related to many records in another table. For example, the street address of an apartment complex may include several households. In a reverse direction from the one-to-many

Parcel table

PIN	Sale date	Acres	Zone code
P101	1-10-98	1.0	1
P102	10-6-68	3.0	2
P103	3-7-97	2.5	2
P104	7-30-78	1.0	1

Address table

Owner name	Owner address
Wang	101 Oak St
Chang	200 Maple St
Jones	105 Ash St
Smith	300 Spruce Rd
Costello	206 Elm St

Owner table

PIN	Owner name
P101	Wang
P101	Chang
P102	Smith
P102	Jones
P103	Costello
P104	Smith

Zone table

Zone code	Zoning
1	Residential
2	Commercial

Figure 8.8

Separate tables after normalization. The fields relating the tables are highlighted.

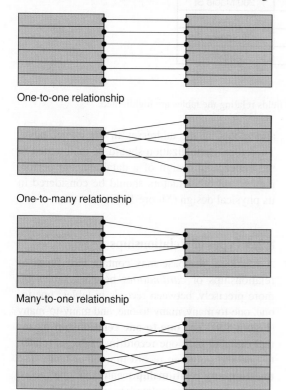

One-to-one relationship

One-to-many relationship

Many-to-one relationship

Many-to-many relationship

Figure 8.9

Four types of data relationships between tables: one-to-one, one-to-many, many-to-one, and many-to-many.

relationship, the **many-to-one relationship** means that many records in a table may be related to one record in another table. For example, several households may share the same street address. The **many-to-many relationship** means that many records in a table may be related to many records in another table. For example, a timber stand can grow more than one species and a species can grow in more than one stand.

To explain these relationships, especially one-to-many and many-to-one, the designation of the *origin* and *destination* can be helpful. For example, if the purpose is to join attribute data from a nonspatial table to a feature attribute table, then the feature attribute table is the origin and the other table is the destination (Figure 8.10). The feature attribute table has the primary key and the other table has the foreign key. Often, the designation of the origin and destination depends on the storage of data and the information sought. This is illustrated in the following two examples.

The first example refers to the four normalized tables of parcel, owner, address, and zone in Figure 8.8. Suppose the question is to find who owns a selected parcel. To answer this question, we can treat the parcel table as the origin and the owner table as the destination. The relationship between the tables is one-to-many: one record in

Primary key

Foreign key

Record	Soil-ID	Area	Perimeter
1	1	106.39	495.86
2	2	8310.84	508,382.38
3	3	554.11	13,829.50
4	4	531.83	19,000.03
5	5	673.88	23,931.47

Soil-ID	suit	soilcode
1	1	Id3
2	3	Sg
3	1	Id3
4	1	Id3
5	2	Ns1

Origin ◄————————————— Destination

Figure 8.10
The common field Soil-ID provides the linkage to join the table on the right to the feature attribute table on the left.

cotreestomng

cokey	plantcomna
79499:111020	western white pine
79499:111020	Douglas fir
79499:111020	grand fir

component

cokey	component
79499:111020	Helmer

Figure 8.11
This example of a many-to-one relationship in the SSURGO database relates three tree species in *cotreestomng* to the same soil component in *component*.

mapunit

musym	mukey
34	79523

component

mukey	component
79523	Helmer
79523	Thatuna

Figure 8.12
This example of a one-to-many relationship in the SSURGO database relates one soil map unit in *mapunit* to two soil components in *component*.

the parcel table may correspond to more than one record in the owner table.

Suppose the question is now changed to find land parcels owned by a selected owner. The owner table becomes the origin and the parcel table is the destination. The relationship is many-to-one: more than one record in the owner table may correspond to one record in the parcel table. The same is true between the parcel table and the zone table. If the question is to find the zoning code for a selected parcel, it is a many-to-one relationship; and if the question is to find land parcels that are zoned commercial, it is a one-to-many relationship.

The second example relates to the SSURGO database. One approach to better understanding the database is to sort out the relationships between tables. For example, a many-to-one relationship exists between the Component Trees To Manage table (*cotreastorang*) and the Component table (*component*) because different recommended tree species may be linked to the same soil component (Figure 8.11). On the other hand, the one-to-many relationship applies to the Mapunit table (*mapunit*) and the Component table because a map unit may be linked to multiple soil components (Figure 8.12).

Besides the obvious effect in linking tables, the type of relationship can also influence how data are displayed. Suppose the task is to display the parcel ownership. If the relationship between the parcel and ownership tables is one-to-one, each parcel can be displayed with a unique symbol. If the relationship is many-to-one, one or more parcels can be displayed with one symbol. But if the relationship is one-to-many, data display becomes a problem because it is not right to show a multiowner parcel with an owner who happens to be the first on the list. (One solution to the problem is to use a separate symbol design for the category of multiowner parcels.)

8.3 JOINS, RELATES, AND RELATIONSHIP CLASSES

To take advantage of a relational database, we can link tables in the database for the purpose of data query and management. Here we examine three ways of linking tables: join, relate, and relationship class.

8.3.1 Joins

A **join** operation brings together two tables by using a common field or a primary key and a foreign key. A typical example is to join attribute data from one or more nonspatial data tables to a feature attribute table for data query or analysis, such as the example in Figure 8.10, where the two tables can be joined by using the common field Soil-ID. A join operation is usually recommended for a one-to-one or many-to-one relationship. Given a one-to-one relationship, two tables are joined by record. Given a many-to-one relationship, many records in the origin have the same value from a record in the destination. A join operation is inappropriate with the one-to-many or many-to-many relationship because only the first matching record value from the destination is assigned to a record in the origin.

8.3.2 Relates

A **relate** operation temporarily connects two tables but keeps the tables physically separate. We can connect three or more tables simultaneously

by first establishing relates between tables in pairs. A Windows-based GIS package is particularly useful for working with relates because it allows the viewing of multiple tables. One advantage of relates is that they are appropriate for all four types of relationships. This is important for data query because a relational database is likely to include different types of relationships. But relates tend to slow down data access, especially if the data reside in a remote database.

8.3.3 Relationship Classes

The object-based data model such as the geodatabase can support relationships between objects. When used for attribute data management, a relationship is predefined and stored as a relationship class in a geodatabase. A relationship class can have the cardinalities of one-to-one, many-to-one, one-to-many, and many-to-many. For the first three relationships, records in the origin are directly linked to records in the destination, but for the many-to-many relationship, an intermediate table must first be set up to sort out the associations between records in the origin and the destination. When present in a geodatabase, a relationship class is automatically recognized and can be used in place of a relate operation. Task 7 in the applications section covers the creation and use of a relationship class.

8.4 ATTRIBUTE DATA ENTRY

Entering attribute data is like digitizing a paper map. The process requires the setup of attributes to be entered, the choice of a digitizing method, and the verification of attribute values.

8.4.1 Field Definition

The first step in attribute data entry is to define each field in the table. A field definition usually includes the field name, data width, data type, and number of decimal digits. The width refers to the number of spaces to be reserved for a field. The width should be large enough for the largest number, including the sign, or the longest string in the data. The data type must follow data types allowed in the GIS package. The number of decimal digits is part of the definition for the float data type. (In ArcGIS, the term *precision* defines the number of digits, and the term *scale* defines the number of decimal digits, for the float data type.)

The field definition becomes a property of the field. Therefore, it is important to consider how the field will be used before defining it. For example, the map unit key in the SSURGO database is defined as text, although it appears as numbers such as 79522 and 79523. Of course, these map unit key numbers cannot be used for computations.

8.4.2 Methods of Data Entry

Suppose a map has 4000 polygons, each with 50 fields of attribute data. This could require entering 200,000 values. How to reduce time and effort in attribute data entry is of interest to any GIS user.

Just as we look for existing spatial data, we should determine if an agency has already entered attribute data in digital format. If yes, we can simply import the digital data file into a GIS. For example, the tasks in the Applications section all use digital data files available from the U.S. Forest Service on vegetation and land type data and links them to a vegetation stand map. The data format is important for importing. Commercial GIS packages can import files in delimited text, dBASE, and Excel (introduced in ArcGIS 9.2). If attribute data files do not exist, then typing is the only option. But the amount of typing can vary depending on which method or command is used. For example, an editing command in a GIS package works with one record at a time. One way to save time is to follow the relational database design and to take advantage of common fields and lookup tables.

For map unit symbols or feature IDs, it is better to enter them directly in a GIS. This is because we can select a feature in the view window, see where the feature is located in the base map, and enter its unit symbol or ID in a dialog. But for nonspatial data, it is better to use a word processing (e.g., Notepad) or spreadsheet (e.g., Excel) package. These packages offer cut-and-paste, find-and-replace, and other

functions that are helpful to a typist, especially a poor typist. A GIS package may not have these options.

8.4.3 Attribute Data Verification

Attribute data verification has two steps. The first is to make sure that attribute data are properly linked to spatial data: the feature ID should be unique and should not contain null (empty) values. The second step is to verify the accuracy of attribute data. Data inaccuracies can be caused by a number of factors including observation errors, outdated data, and data entry errors.

An effective method for preventing data entry errors is to use attribute domains in the geodatabase (Zeiler 1999). An attribute domain allows the user to define a valid range of values or a valid set of values for an attribute. Suppose the field zoning has the value 1 for residential, 2 for commercial, and 3 for industrial for a land parcel feature class. This set of zoning values can be enforced whenever the field zoning is edited. Therefore, if a zoning value of 9 is entered, the value will be flagged or rejected because it is outside the valid set of values. A similar constraint using valid numeric ranges instead of a valid set of values can be applied to lot size or height of building. To ensure the accuracy of data entry, Task 1 of the applications section uses an attribute domain.

8.5 MANIPULATION OF FIELDS AND ATTRIBUTE DATA

Database management tasks involve tables as well as fields and field values in tables. Field management includes adding or deleting fields and creating new attributes through classification and computation of existing attribute data.

8.5.1 Adding and Deleting Fields

We regularly download data from the Internet for GIS projects. Often the downloaded data set contains far more attributes than we need. It is a good idea to delete those fields that are not needed. This not only reduces confusion in using the data set

Several options are available in ArcGIS for adding and deleting fields, two common operations in GIS. In Task 1, a new field is added via the property dialog of a feature class. In Task 4, ArcToolbox is used to add a new field. The third option to add a field is through the table options menu. ArcToolbox has tools for deleting and calculating a field. Another option to delete or calculate a field is to use the context menu of the field. Regardless of which method is used, the table on which the operation is performed should not be used by another application or program.

but also saves computer time for data processing. Deleting a field is straightforward. It requires specifying an attribute table and the field in the table to be deleted.

Adding a field is required for the classification or computation of attribute data. The new field is designed to receive the result of classification or computation. To add a field, we must define the field in the same way as for attribute data entry. Perhaps because adding or deleting a field is such a common operation in GIS, the operation can be performed in a number of places in ArcGIS Desktop (Box 8.2).

8.5.2 Classification of Attribute Data

New attribute data can be created from existing data through data classification. Based on an attribute or attributes, data classification reduces a data set to a small number of classes. Suppose you have a data set that describes the elevations of an area where you are conducting research. We can create new data by reclassifying these elevations into groups, such as elevations <500 meters, 500 to 100 meters, and so on.

Operationally, creating new attribute data by classification involves three steps: defining a new field for saving the classification result, selecting a data subset through query, and assigning a value to the selected data subset. The second and third steps are repeated until all records are classified and assigned new field values, unless a computer code

is written to automate the procedure (see Task 5 in the applications section). Data classification can simplify a data set so that the new data set can be more easily used in GIS analysis or modeling.

8.5.3 Computation of Attribute Data

New attribute data can also be created from existing data through computation. Operationally, it involves two steps: defining a new field, and computing the new field values from the values of an existing attribute or attributes. The computation is through a formula, which can be coded manually or by using a dialog with mathematical functions.

A simple example of computation is to create a new attribute called feet for a trail map that is measured in meters. This new attribute can be computed by "length" × 3.28, where length is an existing attribute. Another example is to create a new attribute that measures the quality of wildlife habitat by evaluating the existing attributes of slope, aspect, and elevation. The task first requires a scoring system for each variable. Once the scoring system is in place, we can compute the index value for measuring the quality of wildlife habitat by summing the scores of slope, aspect, and elevation. In some cases, different weights may be assigned to different variables. For example, if elevation is three times as important as slope and aspect, then we can compute the index value by using the following equation: slope score + aspect score + 3 × elevation score.

KEY CONCEPTS AND TERMS

Categorical data: Data measured at a nominal or an ordinal scale.

Database management system (DBMS): A software package for building and manipulating databases for such tasks as data input, search, retrieval, manipulation, and output.

Feature attribute table: An attribute table that has access to the geometries of features.

Field: A column in a table that describes an attribute of a spatial feature.

Flat file: A database that contains all data in a large table.

Foreign key: One or more attributes in a table that match the primary key in another table.

Hierarchical database: A database that is organized at different levels and uses the one-to-many association between levels.

Interval data: Data with known intervals between values, such as temperature readings.

Join: A relational database operation that brings together two tables by using keys or a field common to both tables.

Many-to-many relationship: One type of data relationship in which many records in a table are related to many records in another table.

Many-to-one relationship: One type of data relationship in which many records in a table are related to one record in another table.

Network database: A database based on the built-in connections across tables.

Nominal data: Data that show different kinds or different categories, such as land-use types or soil types.

Normalization: The process of taking a table with all the attribute data and breaking it down to small tables while maintaining the necessary linkages between them in a relational database.

Numeric data: Data measured at an interval or a ratio scale.

One-to-many relationship: One type of data relationship in which one record in a table is related to many records in another table.

One-to-one relationship: One type of data relationship in which one record in a table is related to one and only one record in another table.

Ordinal data: Data that are ranked, such as large, medium, and small cities.

Primary key: One or more attributes that can uniquely identify a record in a table.

Ratio data: Data that have known intervals between values and a meaningful zero value, such as population densities.

Record: A row in a table that represents a spatial feature.

Relate: A relational database operation that temporarily connects two tables by using keys or a field common to both tables.

Relational database: A database that consists of a collection of tables and uses keys to connect the tables.

Soil Survey Geographic (SSURGO) database: A database maintained by the Natural Resources Conservation Service, which archives soil survey data in 7.5-minute quadrangle units.

REVIEW QUESTIONS

1. What is a feature attribute table?
2. Provide an example of a nonspatial attribute table.
3. How does the geodatabase differ from the shapefile in terms of storage of attribute data?

4. Describe the four types of attribute data by measurement scale.

5. Can you convert ordinal data into interval data? Why, or why not?

6. Define a relational database.

7. Explain the advantages of a relational database.

8. Define a primary key.

9. What does the result of a join operation between the zone table and the parcel table in Figure 8.8 look like?

10. A fully normalized database may slow down data access. To increase the performance in data access, for example, one may remove the address table in Figure 8.8. How would the database look if the address table were removed?

11. Provide a real-world example (other than those of Chapter 8) of a one-to-many relationship.

12. Explain the similarity, as well as the difference, between a join operation and a relate operation.

13. Suppose you have downloaded a GIS data set. The data set has the length measure in meters instead of feet. Describe the steps you will follow to add a length measure in feet to the data set.

14. Suppose you have downloaded a GIS data set. The feature attribute table has a field that contains values such as 12, 13, and so on. How can you find out if these values represent numbers or text strings using ArcGIS?

15. Describe two ways of creating new attributes from the existing attributes in a data set.

APPLICATIONS: ATTRIBUTE DATA MANAGEMENT

This applications section has seven tasks. Task 1 covers attribute data entry using a geodatabase feature class. Tasks 2 and 3 cover joining tables and relating tables, respectively. Tasks 4 and 5 create new attributes by data classification. Task 4 uses the conventional method of repeatedly selecting a data subset and assigning a class value. Task 5, on the other hand, uses a Python script to automate the procedure. Task 6 shows how to create new attributes through data computation. Task 7 lets you create and use relationship classes in a file geodatabase.

Task 1 Enter Attribute Data of a Geodatabase Feature Class

What you need: *landat.shp,* a polygon shapefile with 19 records.

In Task 1, you will learn how to enter attribute data using a geodatabase feature class and a domain. The domain and its coded values can restrict values to be entered, thus preventing data entry errors.

1. Start ArcCatalog, and connect to the Chapter 8 database. First, create a personal geodatabase. Right-click the Chapter 8 database in the Catalog tree, point to New, and select Personal Geodatabase. Rename the new personal geodatabase *land.mdb.*

2. This step adds *landat.shp* as a feature class to *land.mdb.* Right-click *land.mdb,* point to Import, and select Feature Class (single). Use the browse button or the drag-and-drop method to add *landat.shp* as the input features. Name the output feature class *landat.* Click OK to dismiss the dialog.

3. Now create a domain for the geodatabase. Select Properties from the context menu of *land.mdb.* The Domains tab of the Database Properties dialog shows Domain Name, Domain Properties, and Coded Values. You will work with all three frames. Click the first cell under Domain Name, and enter lucodevalue. Click the cell next to Field Type, and select Short Integer. Click the cell next to

Domain Type, and select Coded Values. Click the first cell under Code and enter 100. Click the cell next to 100 under Description, and enter 100-urban. Enter 200, 300, 400, 500, 600, and 700 following 100 under Code and enter their respective descriptions of 200-agriculture, 300-brushland, 400-forestland, 500-water, 600-wetland, and 700-barren. Click Apply and OK to dismiss the Database Properties dialog.

4. This step is to add a new field to *landat* and to specify the field's domain. Right-click *landat* in *land.mdb* and select Properties. On the Fields tab, click the first empty cell under Field Name and enter lucode. Click the cell next to lucode and select Short Integer. Click the cell next to Domain in the Field Properties frame and select lucodevalue. Click Apply and OK to dismiss the Properties dialog.

Q1. List the data types available for a new field.

5. Launch ArcMap. Rename the data frame Task 1 and add *landat* to Task 1. Open the attribute table of *landat*. lucode appears with Null values in the last field of the table.

6. Click the Editor Toolbar button to open the toolbar. Click the Editor dropdown arrow and select Start Editing. Right-click the field of LANDAT-ID and select Sort Ascending. Now you are ready to enter the lucode values. Click the first cell under lucode and select forestland (400). Enter the rest of the lucode values according to the following table:

Landat-ID	Lucode	Landat-ID	Lucode
59	400	69	300
60	200	70	200
61	400	71	300
62	200	72	300
63	200	73	300
64	300	74	300
65	200	75	200
66	300	76	300
67	300	77	300
68	200		

Q2. Describe in your own words how the domain of coded values ensures the accuracy of the attribute data that you entered in Step 6.

7. When you finish entering the lucode values, select Stop Editing from the Editor dropdown list. Save the edits.

Task 2   Join Tables

What you need: *wp.shp,* a forest stand shapefile, and *wpdata.dbf,* an attribute data file that contains vegetation and land-type data.

Task 2 asks you to join a dBASE file to a feature attribute table. A join operation combines attribute data from different tables into a single table, making it possible to use all attribute data in query, classification, or computation.

1. Insert a new data frame in ArcMap and rename it Task 2. Add *wp.shp* and *wpdata .dbf* to Task 2.

2. Open the attribute table of *wp* and *wpdata.* The field ID is in both tables will be used as the field in joining the tables.

3. Now join *wpdata* to the attribute table of *wp.* Right-click *wp,* point to Joins and Relates, and select Join. At the top of the Join Data dialog, opt to join attributes from a table. Then, select ID in the first dropdown list, *wpdata* in the second list, and ID in the third list. Click OK to join the tables. Open the attribute table of *wp* to see the expanded table. Even though the two tables appear to be combined, they are actually linked via OLE (Object Linking and Embedding). To save the joined table permanently, you can export *wp* and save it with a different name.

Task 3   Relate Tables

What you need: *wp.shp, wpdata.dbf,* and *wpact.dbf.* The first two are the same as in Task 2. *wpact.dbf* contains additional activity records.

In Task 3, you will establish two relates among three tables.

1. Select Data Frame from the Insert menu in ArcMap. Rename the new data frame Tasks 3–6. Add *wp.shp, wpdata.dbf,* and *wpact.dbf* to Tasks 3–6.

2. Check the fields for relating tables. The field ID should appear in *wp*'s attribute table, *wpact,* and *wpdata.* Close the tables.

3. The first relate is between *wp* and *wpdata.* Right-click *wp,* point to Joins and Relates, and select Relate. In the Relate dialog, select ID in the first dropdown list, *wpdata* in the second list, and ID in the third list, and accept Relate1 as the relate name.

4. The second relate is between *wpdata* and *wpact.* Right-click *wpdata,* point to Joins and Relates, and select Relate. In the Relate dialog, select ID in the first dropdown list, *wpact* in the second list, and ID in the third list, and enter Relate2 as the relate name.

5. The three tables are now related. Right-click *wpdata* and select Open. Click the Select By Attributes button at the top of the table. In the next dialog, create a new selection by entering the following SQL statement in the expression box: "ORIGIN" > 0 AND "ORIGIN" <= 1900. Click Apply. Click Show selected records at the bottom of the table.

6. To see which records in the *wp* attribute table are related to the selected records in *wpdata,* go through the following steps. Click the dropdown arrow of Related Tables, and select Relate1: wp. The *wp* attribute table shows the related records. And the *wp* layer shows where those selected records are located.

7. You can follow the same procedure as in Step 6 to see which records in *wpact* are related to those selected polygons in *wp.*

Q3. How many records in *wpact* are selected in Step 7?

Task 4 Create New Attribute by Data Classification

What you need: *wpdata.dbf.*

Task 4 demonstrates how the existing attribute data can be used for data classification and the creation of a new attribute.

1. First click Clear Selected Features in the Selection menu in ArcMap to clear the selection. Click ArcToolbox window to open it. Double-click the Add Field tool in the Data Management Tools/ Fields toolset. Select *wpdata* for the input table, enter ELEVZONE for the field name, select SHORT for the type, and click OK.

2. Open *wpdata* in Tasks 3–6. ELEVZONE appears in *wpdata* with 0s. Click the Select By Attributes button. Make sure that the selection method is to create a new selection. Enter the following SQL statement in the expression box: "ELEV" > 0 AND "ELEV" <= 40. Click Apply. Click Show selected records. These selected records fall within the first class of ELEVZONE. Right-click the field ELEVZONE and select Field Calculator. Click Yes to proceed in the message box. Enter 1 in the expression box of the Field Calculator dialog, and click OK. The selected records in *wpdata* are now populated with the value of 1, meaning that they all fall within class 1.

3. Go back to the Select By Attributes dialog. Make sure that the method is to create a new selection. Enter the SQL statement: "ELEV" > 40 AND "ELEV" <= 45. Click Apply. Follow the same procedure to calculate the ELEVZONE value of 2 for the selected records.

4. Repeat the same procedure to select the remaining two classes of 46–50 and > 50, and to calculate their ELEVZONE values of 3 and 4, respectively.

Q4. How many records have the ELEVZONE value of 4?

Task 5 Use Advanced Method for Attribute Data Classification

What you need: *wpdata.dbf* and *Expression.cal*.

In Task 4 you have classified ELEVZONE in *wpdata.dbf* by repeating the procedure of selecting a data subset and calculating the class value. This task shows how to use a Python script and the advanced option to calculate the ELEVZONE values all at once.

1. Clear selected records in *wpdata*, if necessary. Double-click the Add Field tool in the Data Management Tools/ Fields toolset. Select *wpdata* for the input table, enter ELEVZONE2 for the field name, select SHORT for the type, and click OK.

2. Open *wpdata* in Tasks 3–6. ELEVZONE2 appears in *wpdata* with 0s. To use the advanced option, you will first copy the ELEV values to ELEVZONE2. Right-click ELEVZONE2 and select Field Calculator. Enter [ELEV] in the expression box.

3. Right-click ELEVZONE2 and select Field Calculator again. This time you will use the advanced option to classify the ELEV values and save the class values in ELEVZONE2. In the Field Calculator dialog, check Python as the Parser and check the box of Show Codeblock. Then click on the Load button and load *Expression.cal* as the calculation expression. After the file is loaded, you will see the following code in the Pre-Logic Script code box:

```
def Reclass (ELEVZONE2):
    if (ELEVZONE2 > 0 and ELEVZONE2
    <= 40):
        return 1
    elif (ELEVZONE2 > 40 and
    ELEVZONE2 <= 45):
        return 2
```

```
    elif (ELEVZONE2 > 45 and
    ELEVZONE2 <= 50):
        return 3
    elif (ELEVZONE2 > 50):
        return 4
```

and the expression, Reclass (!elevzone2!), in the box below "elevzone2=." Click OK to run the code.

4. ELEVZONE2 is now populated with values calculated by the Python code. They should be the same as those in ELEVZONE.

Task 6 Create New Attribute by Data Computation

What you need: *wp.shp* and *wpdata.dbf*.

You have created a new field from data classification in Tasks 4 and 5. Another common method for creating new fields is computation. Task 6 shows how a new field can be created and computed from existing attribute data.

1. Double-click the Add Field tool. Select *wp* for the input table, enter ACRES for the field name, select DOUBLE for the field type, enter 11 for the field precision, enter 4 for the field scale, and click OK.

2. Open the attribute table of *wp*. The new field ACRES appears in the table with 0s. Right-click ACRES to select Field Calculator. Click Yes in the message box. In the Field Calculator dialog, enter the following expression in the box below ACRES =: [AREA]/1000000 × 247.11. Click OK. The field ACRES now shows the polygons in acres.

Q5. How large is FID = 10 in acres?

Task 7 Create Relationship Class

What you need: *wp.shp*, *wpdata.dbf*, and *wpact. dbf*, same as in Task 3.

Instead of using on-the-fly relates as in Task 3, you will use the relationship classes in

Task 7 by first defining and saving them in a file geodatabase.

1. First, create a new file geodatabase in ArcCatalog. Right-click the Chapter 8 database in the Catalog tree, point to New, and select File Geodatabase. Rename the new geodatabase *relclass.gdb*.

2. This step adds *wp.shp* as a feature class to *relclass.gdb*. Right-click *relclass.gdb*, point to Import, and select Feature Class (single). Use the browse button to add *wp.shp* as the input features. Name the output feature class *wp*. Click OK to dismiss the dialog.

3. This step imports *wpdata.dbf* and *wpact.dbf* as tables to *relclass.gdb*. Right-click *relclass.gdb*, point to Import, and select Table (multiple). Use the browse button to add *wpdata.dbf* and *wpact.dbf* as input tables. Click OK to dismiss the dialog. Make sure that *relclass.gdb* now contains *wp*, *wpact*, and *wpdata*.

4. Right-click *relclass.gdb*, point to New, and select Relationship Class. You will first create a relationship class between *wp* and *wpdata* in several steps. Name the relationship class *wp2data*, select *wp* for the origin table, *wpdata* for the destination table, and click Next. Take the default of a simple relationship. Then, specify *wp* as a label for the relationship as it is traversed from the origin to the destination, specify *wpdata* as a label for the relationship as it is traversed from the destination to the origin, and opt for no messages propagated. In the next dialog, choose the one-to-one cardinality. Then, choose not to add attributes to the relationship class. In the next dialog, select ID for the primary key as well as for the foreign key. Review the summary of the relationship class before clicking Finish.

5. Follow the same procedure as in Step 4 to create the relationship class *data2act*

between *wpdata* and *wpact*. ID will again be the primary key as well as the foreign key.

6. This step shows how to use the relationship classes you have defined and stored in *relclass.gdb*. Insert a new data frame in ArcMap and rename it Task 7. Add *wp*, *wpact*, and *wpdata* from *relclass.gdb* to Task 7.

7. Right-click *wpdata* and select Open. Click Select By Attributes. In the next dialog, create a new selection by entering the following SQL statement in the expression box: "ORIGIN" > 0 AND "ORIGIN" <= 1900. Click Apply. Click Show selected records.

8. Select *wp2data* from the dropdown arrow of Related Tables. The *wp* attribute table shows the related records, and the *wp* layer shows where those selected records are located.

Q6. How many records in the *wp* attribute table are selected in Step 8?

9. You can use the relationship class *data2act* to find the related records in *wpact*.

Challenge Task

What you need: *bailecor_id.shp*, a shapefile showing Bailey's ecoregions in Idaho. The data set is projected onto the Idaho Transverse Mercator coordinate system and measured in meters.

This challenge task asks you to add a field to *bailecor_id* that shows the number of acres for each ecoregion in Idaho.

Q1. How many acres does the Owyhee Uplands Section cover?

Q2. How many acres does the Snake River Basalts Section cover?

REFERENCES

Arvanitis, L. G., B. Ramachandran, D. P. Brackett, H. Abd-El Rasol, and X. Du. 2000. Multiresource Inventories Incorporating GIS, GPS and Database Management Systems: A Conceptual Model. *Computers and Electronics in Agriculture* 28: 89–100.

Carleton, C. J., R. A. Dahlgren, and K. W. Tate. 2005. A Relational Database for the Monitoring and Analysis of Watershed Hydrologic Functions: I. Database Design and Pertinent Queries. *Computers & Geosciences* 31: 393–402.

Chang, K. 1978. Measurement Scales in Cartography. *The American Cartographer* 5: 57–64.

Chrisman, N. 2001. *Exploring Geographic Information*

Systems, 2d ed. New York: Wiley.

Codd, E. F. 1970. A Relational Model for Large Shared Data Banks. *Communications of the Association for Computing Machinery* 13: 377–87.

Codd, E. F. 1990. *The Relational Model for Database Management,* Version 2. Reading, MA: Addison-Wesley.

Date, C. J. 1995. *An Introduction to Database Systems.* Reading, MA: Addison-Wesley.

Jackson, M. 1999. Thirty Years (and More) of Databases. *Information and Software Technology* 41: 969–78.

Lee, H. 1995. Justifying Database Normalization: A Cost/Benefit Model. *Information Processing & Management* 31: 59–67.

Moore, R. V. 1997. The Logical and Physical Design of the Land Ocean Interaction Study Database. *The Science of the Total Environment* 194/195: 137–46.

Oz, E. 2004. *Management Information Systems,* 4th ed. Boston, MA: Course Technology.

Shekhar, S. and S. Chawla. 2003. *Spatial Databases: A Tour.* Upper Saddle River, NJ: Prentice Hall.

Stevens, S. S. 1946. On the Theory of Scales of Measurement. *Science* 103: 677–80.

Vetter, M. 1987. *Strategy for Data Modelling.* New York: Wiley.

Zeiler, M. 1999. *Modeling Our World: The ESRI Guide to Geodatabase Design.* Redlands, CA: Esri Press.

DATA DISPLAY AND CARTOGRAPHY

Maps are an interface to GIS (Kraak and Ormeling 1996). We view, query, and analyze maps. We plot maps to examine the results of query and analysis. And we produce maps for presentation and reports. As a visual tool, maps are most effective in communicating geospatial data, whether the emphasis is on the location or the distribution pattern of the data.

Common map elements are the title, body, legend, north arrow, scale bar, acknowledgment, and neatline/map border (Figure 9.1). Other elements include the graticule or grid, name of map projection, inset or location map, and data quality information. Together, these elements bring the map information to the map reader. As the most important part of a map, the map body contains the map information. The title suggests the subject matter. The legend relates map symbols to the map body. Other elements of the map support the communication process. In practical terms, mapmaking may be described as a process of assembling map elements.

Data display is one area in which commercial GIS packages have greatly improved in recent years. Desktop GIS packages with the graphical user interface are excellent for data display for two reasons. First, the mapmaker can simply point and click on the graphic icons to construct a map. Second, desktop GIS packages have incorporated some design options such as symbol choices and color schemes into menu selections.

For a novice mapmaker, these easy-to-use GIS packages and their "default options" can sometimes result in maps of questionable quality. Mapmaking should be guided by a clear idea of map design and map communication. A well-designed map is visually pleasing, is rich in data, and, at the same time, can help the mapmaker communicate the information to the map reader (Tufte 1983). In contrast, a poorly designed map

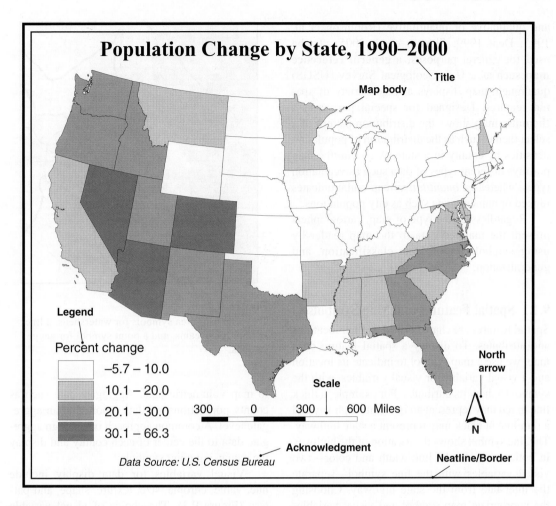

Figure 9.1
Common map elements.

can confuse the map reader and even distort the information.

Chapter 9 emphasizes maps for presentation and reports. Maps for data exploration and 3-D visualization are covered in Chapters 10 and 13, respectively. Chapter 9 is divided into the following five sections. Section 9.1 discusses cartographic representation including the symbology, use of color, data classification, and generalization. Section 9.2 considers different types of quantitative maps. Section 9.3 provides an overview of typography, the selection of type variations, and

the placement of text. Section 9.4 covers map design and the design elements of layout and visual hierarchy. Section 9.5 examines issues related to map production.

9.1 CARTOGRAPHIC REPRESENTATION

Cartography is the making and study of maps in all their aspects (Robinson et al. 1995). Cartographers classify maps into general reference or thematic, and qualitative or quantitative (Robinson et al. 1995; Dent 1999; Slocum et al. 2005). To be used for general purposes, a **general reference map** such as a U.S. Geological Survey (USGS) quadrangle map displays a large variety of spatial features. Designed for special purposes, a **thematic map** shows the distribution pattern of a select theme, such as the distribution of population densities by county in a state. A *qualitative* map portrays different types of data such as vegetation types, whereas a *quantitative* map communicates ranked or numeric data such as city populations.

Regardless of the type of map, cartographers present the mapped data to the map reader by using symbols, color, data classification, and generalization.

9.1.1 Spatial Features and Map Symbols

Spatial features are characterized by their locations and attributes. To display a spatial feature on a map, we use a map symbol to indicate its location and a visual variable, or visual variables, with the symbol to show its attributes. For example, a thick line in red may represent an interstate highway and a thin line in black may represent a state highway. The line symbol shows the location of the highway in both cases, but the line width and color—two visual variables with the line symbol—separate the interstate from the state highway. Choosing the appropriate map symbol and visual variables is therefore the main concern for data display and mapmaking (Robinson et al. 1995; Dent 1999; Slocum et al. 2005).

The choice of map symbol is simple for raster data: the map symbol applies to cells whether the spatial feature to be depicted is a point, line, or polygon. The choice of map symbol for vector data depends on the feature type (Figure 9.2). The general rule is to use point symbols for point features, line symbols for line features, and area symbols for polygon features. But this general rule does not apply to volumetric data or aggregate data. There are no volumetric symbols for data such as elevation, temperature, and precipitation. Instead, 3-D surfaces and isolines are often used

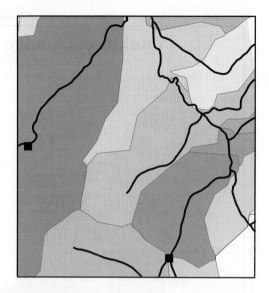

Figure 9.2
This map uses area symbols for watersheds, a line symbol for streams, and a point symbol for gauge stations.

to map volumetric data. Aggregate data such as county populations are data reported at an aggregate level. A common approach is to assign aggregate data to the center of each county and display the data using point symbols.

Visual variables for data display include hue, value, chroma, size, texture, shape, and pattern (Figure 9.3). The choice of visual variable depends on the type of data to be displayed. The measurement scale is commonly used for classifying attribute data (Chapter 8). Size (e.g., large vs. small circles) and texture (e.g., different spacing of symbol markings) are more appropriate for displaying ordinal, interval, and ratio data. For example, a map may use different-sized circles to represent different-sized cities. Shape (e.g., circle vs. square) and pattern (e.g., horizontal lines vs. crosshatching) are more appropriate for displaying nominal data. For example, a map may use different area patterns to show different land-use types. The use of hue, value, and chroma as visual variables is covered in Section 9.1.2. Most GIS packages organize choices of visual variables into palettes so that

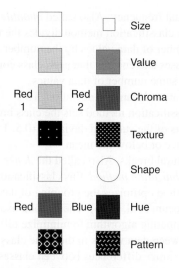

Figure 9.3
Visual variables for cartographic symbolization.

the user can easily select the variables to make up the map symbols. Some packages also allow custom pattern designs.

The choice of visual variables is limited in the case of raster data. The visual variables of shape and size do not apply to raster data because of the use of cells. Using texture or pattern is also difficult with small cells. Display options for raster data are therefore limited to different colors in most cases.

9.1.2 Use of Color

Because color adds a special appeal to a map, mapmakers will choose color maps over black and white maps whenever possible. But color is probably the most misused visual variable, according to published critiques of computer-generated maps (Monmonier 1996). The use of color in mapmaking must begin with an understanding of the visual dimensions of hue, value, and chroma.

Hue is the quality that distinguishes one color from another, such as red from blue. Hue can also be defined as the dominant wavelength of light making up a color. We tend to relate different hues with different kinds of data. **Value** is the lightness

or darkness of a color, with black at the lower end and white at the higher end. We generally perceive darker symbols on a map as being more important, or higher in magnitude (Robinson et al. 1995). Also called saturation or intensity, **chroma** refers to the richness, or brilliance, of a color. A fully saturated color is pure, whereas a low saturation approaches gray. We generally associate higher-intensity symbols with greater visual importance.

The first rule of thumb in the use of color is simple: hue is a visual variable better suited for qualitative (nominal) data, whereas value and chroma are better suited for quantitative (ordinal, interval, and ratio) data.

Mapping qualitative data is relatively easy. It is not difficult to find 12 or 15 distinctive hues for a map. If a map requires more symbols, we can add pattern or text to hue to make up more map symbols. Mapping quantitative data, on the other hand, has received much more attention in cartographic research. Over the years, cartographers have suggested general color schemes that combine value and chroma for displaying quantitative data (Cuff 1972; Mersey 1990; Brewer 1994; Robinson et al. 1995). A basic premise among these color schemes is that the map reader can easily perceive the progression from low to high values (Antes and Chang 1990). The following is a summary of these color schemes.

- The single hue scheme. This color scheme uses a single hue but varies the combination of value and chroma to produce a sequential color scheme such as from light red to dark red. It is a simple but effective option for displaying quantitative data (Cuff 1972).
- The hue and value scheme. This color scheme progresses from a light value of one hue to a darker value of a different hue. Examples are yellow to dark red and yellow to dark blue. Mersey (1990) finds that color sequences incorporating both regular hue and value variations outperform other color schemes on the recall or recognition of general map information.

- The diverging or double-ended scheme. This color scheme uses graduated colors between two dominant colors. For example, a diverging scheme may progress from dark blue to light blue and then from light red to dark red. The diverging color scheme is a natural choice for displaying data with positive and negative values, or increases and decreases. But Brewer et al. (1997) report that the diverging scheme is still better than other color schemes, even in cases where the map reader is asked to retrieve information from maps that do not include positive and negative values. Divergent color schemes are found in many maps of Census 2000 demographic data prepared by Brewer (2001) (**http://www.census.gov/population/www/ cen2000/atlas.html**).
- The part spectral scheme. This color scheme uses adjacent colors of the visible spectrum to show variations in magnitude. Examples of this color scheme include yellow to orange to red and yellow to green to blue.
- The full spectral scheme. This color scheme uses all colors in the visible spectrum. A conventional use of the full spectral scheme is found in elevation maps. Cartographers usually do not recommend this option for mapping other quantitative data because of the absence of a logical sequence between hues.

9.1.3 Data Classification

Data classification involves the use of a classification method and a number of classes for aggregating data and map features. A GIS package typically offers different data classification methods. The following summarizes six commonly used methods:

- Equal interval. This classification method divides the range of data values into equal intervals.
- Geometric interval. This method groups data values into classes of increasingly larger intervals.

- Equal frequency. Also called *quantile,* this classification method divides the total number of data values by the number of classes and ensures that each class contains the same number of data values.
- Mean and standard deviation. This classification method sets the class breaks at units of the standard deviation (0.5, 1.0, etc.) above or below the mean.
- Natural breaks. Also called the *Jenks optimization method,* this classification method optimizes the grouping of data values (Slocum et al. 2005). The method uses a computing algorithm to minimize differences between data values in the same class and to maximize differences between classes.
- User defined. This method lets the user choose the appropriate or meaningful class breaks. For example, in mapping rates of population change by state, the user may choose zero or the national average as a class break.

With changes in the classification method, the number of classes, or both, the same data can produce different-looking maps and different spatial patterns (Figure 9.4). This is why mapmakers usually experiment with data classification before deciding on a classification scheme for the final map. Although the decision is ultimately subjective, it should be guided by map purpose and map communication principles.

9.1.4 Generalization

Generalization is considered a necessary part of cartographic representation (Slocum et al. 2005). The use of map symbols to represent spatial features is a type of generalization; for example, the same point symbol may represent cities of different area extents. Data classification is another kind of generalization; for example, a class may consist of a group of cities of different population sizes.

Change of scale is often the reason for generalization. Many feature layers in the United States were digitized at a scale of 1:24000. When mapped at a scale of 1:250,000, the same features

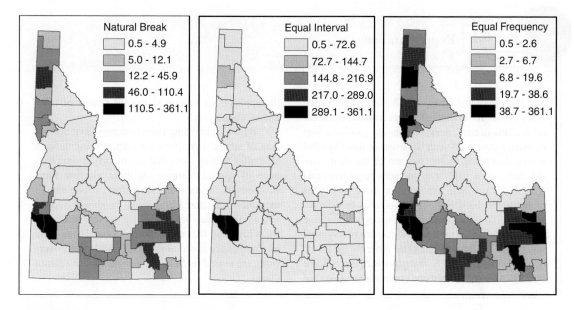

Figure 9.4
The three maps are based on the same data, but they look different because of the use of different classification methods.

require a greatly reduced amount of map space compared with the source map. As a result, map symbols become congested and may even overlap one another. The problem becomes more acute for maps that show a large variety of spatial features. How does a cartographer take care of the problem? Spatial features in close proximity may be grouped, merged, or collapsed into one single feature with a refined outline; a railroad and a highway that run parallel along a river may have to have their geometries altered to add more space between them; and the symbolization of an intersecting feature such as a bridge, an overpass, or an underpass may have to be interrupted or adjusted.

The vector data model emphasizes the geometric accuracy of geospatial data in GIS, but various generalization principles may have to be applied to their geometries to intelligibly represent spatial features on a map. Preserving the integrity of geometries while accommodating the mapping needs has been a constant challenge to GIS users. To help deal with the challenge, Esri

introduced *representations* as a new symbology option (Box 9.1). Representations offer rules and overrides for solving practical problems in symbology and generalization.

9.2 TYPES OF QUANTITATIVE MAPS

Figure 9.5 shows six common types of quantitative maps: the dot map, the choropleth map, the graduated symbol map, the pie chart map, the flow map, and the isarithmic map.

The **dot map** uses uniform point symbols to show geospatial data, with each symbol representing a unit value. One-to-one dot mapping uses the unit value of one, such as one dot representing one crime location. But in most cases, it is one-to-many dot mapping and the unit value is greater than one. The placement of dots becomes a major consideration in one-to-many dot mapping (Box 9.2).

Box 9.1   **Representations**

Symbology in ArcMap now has an option called *representations,* available to users with an ArcEditor or ArcInfo license. A representation can have one or more *rules.* A rule can have one or more symbol layers, depending on the makeup of the map symbol. Suppose a road layer includes interstates and U.S. highways. Map symbols for these two types of roads can be specified in two rules in a representation. To symbolize the interstates with red line symbols with black outlines will require two symbol layers: one for the outline and the other for the fill-in; but to symbolize the U.S. highways with thin red line symbols will require only one symbol layer. Task 2 of the applications section shows the details of rules and symbol layers.

Besides providing rules to control symbolization of spatial features on a map, representations also offer editing tools that can modify the appearance of spatial features—without altering their geometry in the database. Therefore, a river can be masked out so that a bridge can be shown over it, and a railroad can be shifted slightly to make room for a highway that runs parallel to it. These edits or changes are saved as *overrides.* Rules and overrides are stored as fields in a geodatabase feature class.

The **choropleth map** symbolizes, with shading, derived data based on administrative units (Box 9.3). An example is a map showing average household income by county. The derived data are usually classified prior to mapping and are symbolized using a color scheme for quantitative data. Therefore, the appearance of the choropleth map can be greatly affected by data classification. Cartographers often make several versions of the choropleth map from the same data and choose one—typically one with a good spatial organization of classes—for final map production.

The **dasymetric map** is a variation of the simple choropleth map. By using statistics and additional information, the dasymetric map delineates areas of homogeneous values rather than following administrative boundaries (Robinson et al. 1995) (Figure 9.6). Dasymetric mapping used to be a time-consuming task, but the analytical functions of a GIS have simplified the mapping procedure (Holloway, Schumacher and R. L. Redmond 1999; Eicher and Brewer 2001; Holt, Lo, and Hodler 2004).

A GIS package such as ArcGIS uses the term **graduated color map** to cover the choropleth and dasymetric maps because both map types use a graduated color scheme to show the variation in spatial data.

The **graduated symbol map** uses different-sized symbols such as circles, squares, or triangles to represent different ranges of values. For example, we can use graduated symbols to represent cities of different population ranges. Two important issues to this map type are the range of sizes and the discernible difference between sizes. Both issues are inextricably related to the number of graduated symbols on a map.

A **proportional symbol map** is a map that uses a specific symbol size for each numeric value rather than a range of values. Therefore, one circle size may represent a population of 10,000, another 15,000, and so on.

The **chart map** uses either pie charts or bar charts. A variation of the graduated circle, the pie chart can display two sets of quantitative data: the circle size can be made proportional to a value such as a county population, and the subdivisions can show the makeup of the value, such as the racial composition of the county population. Bar charts use vertical bars and their height to represent quantitative data. Bar charts are particularly useful for comparing data side by side.

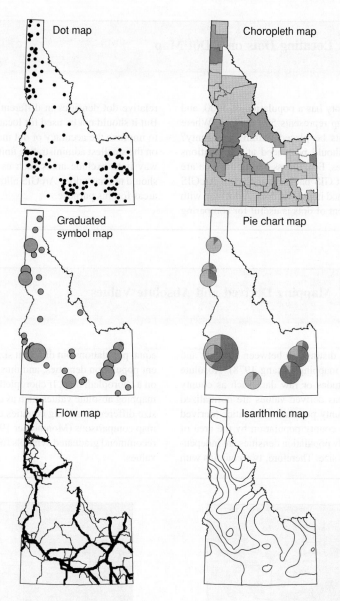

Figure 9.5
Six common types of quantitative maps.

The **flow map** displays different quantities of flow data such as traffic volume and stream flow by varying the line symbol width. Similar to the graduated symbols, the flow symbols usually represent ranges of values.

The **isarithmic map** uses a system of isolines to represent a surface. Each isoline connects points of equal value. GIS users often use the isarithmic map to display the terrain (Chapter 13) and the statistical surface created by spatial interpolation (Chapter 15).

Box 9.2   Locating Dots on a Dot Map

Suppose a county has a population of 5000, and a dot on a dot map represents 500 persons. Where should the 10 dots be placed within the county? Ideally the dots should be placed at the locations of populated places. But, unless additional data are incorporated, most GIS packages including ArcGIS use a random method in placing dots. A dot map with a random placement of dots is useful for comparing relative dot densities in different parts of the map. But it should not be used for locating data. One way to improve the accuracy of dot maps is to base them on the smallest administrative unit possible. Another way is to exclude areas such as water bodies that should not have dots. ArcGIS allows the use of mask areas for this purpose.

Box 9.3   Mapping Derived and Absolute Values

Cartographers distinguish between absolute and derived values in mapping (Chang 1978). Absolute values are magnitudes or raw data such as county population, whereas derived values are normalized values such as county population densities (derived from dividing the county population by the area of the county). County population densities are independent of the county size. Therefore, two counties with equal populations but different sizes will have different population densities, and thus different symbols, on a choropleth map. If choropleth maps are used for mapping absolute values such as county populations, size differences among counties can severely distort map comparisons (Monmonier 1996). Cartographers recommend graduated symbols for mapping absolute values.

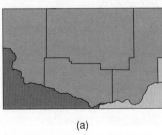
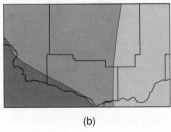

(a)   (b)

Figure 9.6
Map symbols follow the boundaries in the choropleth map (a) but not in the dasymetric map (b).

GIS has introduced a new classification of maps based on vector and raster data. Maps prepared from vector data are the same as traditional maps using point, line, and area symbols. Most of Chapter 9 applies to the display of vector data. Maps prepared from raster data, although they may look like traditional maps, are cell-based (Figure 9.7). But raster data can also be classified

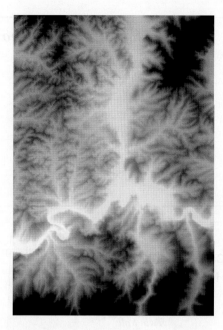

Figure 9.7
Map showing raster-based elevation data. Cells with higher elevations have darker shades.

as either quantitative or qualitative. Therefore, the guidelines for mapping qualitative and quantitative vector data can also be applied to raster data.

9.3 TYPOGRAPHY

A map cannot be understood without text on it. Text is needed for almost every map element. Mapmakers treat text as a map symbol because, like point, line, or area symbols, text can have many type variations. Using type variations to create a pleasing and coherent map is therefore part of the mapmaking process.

9.3.1 Type Variations

Type can vary in typeface and form. **Typeface** refers to the design character of the type. Two main groups of typefaces are **serif** (with serif) and **sans serif** (without serif) (Figure 9.8). Serifs are small, finishing touches at the ends of line strokes, which

Times New Roman

Tahoma

Figure 9.8
Times New Roman is a serif typeface, and Tahoma is a sans serif typeface.

Helvetica Normal

Helvetica Italic

Helvetica Bold

Helvetica Bold-Italic

Times Roman Normal

Times Roman Italic

Times Roman Bold

Times Roman Bold-Italic

Figure 9.9
Type variations in weight and roman versus italic.

tend to make running text in newspapers and books easier to read. Compared to serif types, sans serif types appear simpler and bolder. Although it is rarely used in books or other text-intensive materials, a sans serif type stands up well on maps with complex map symbols and remains legible even in small sizes. Sans serif types have an additional advantage in mapmaking because many of them come in a wide range of type variations.

Type form variations include **type weight** (bold, regular, or light), **type width** (condensed or extended), upright versus slanted (or roman versus italic), and uppercase versus lowercase (Figure 9.9). A **font** is a complete set of all variants of a given typeface. Fonts on a computer are those loaded from the printer manufacturer and software packages. They are usually enough for mapping purposes. If necessary, additional fonts can be imported into a GIS package.

Type can also vary in size and color. Type size measures the height of a letter in **points,** with 72 points to an inch. Printed letters look smaller than what their point sizes suggest. The point size is supposed to be measured from a metal type block, which must accommodate the lowest point of the descender (such as p or g) to the highest part of the ascender (such as d or b). But no letters extend to the very edge of the block. Text color is the color of letters. In addition to color, letters may also appear with drop shadow, halo, or fill pattern.

9.3.2 Selection of Type Variations

Type variations for text symbols can function in the same way as visual variables for map symbols. How does one choose type variations for a map? A practical guideline is to group text symbols into qualitative and quantitative classes. Text symbols representing qualitative classes such as names of streams, mountains, parks, and so on can vary in typeface, color, and upright versus italic. In contrast, text symbols representing quantitative classes such as names of different-sized cities can vary in type size, weight, and uppercase versus lowercase. Grouping text symbols into classes simplifies the process of selecting type variations.

Besides classification, cartographers also recommend legibility, harmony, and conventions for type selection (Dent 1999). Legibility is difficult to control on a map because it can be influenced not only by type variations but also by the placement of text and the contrast between text and the background symbol. As GIS users, we have the additional problem of having to design a map on a computer monitor and to print it on a larger plot. Experimentation may be the only way to ensure type legibility in all parts of the map.

Type legibility should be balanced with harmony. As a means to communicate the map content, text should be legible but should not draw too much attention. To achieve harmony, mapmakers should avoid using too many typefaces on a map (Figure 9.10) but instead use only one or two typefaces. For example, many mapmakers use a sans serif type in the body of a map and a serif type for

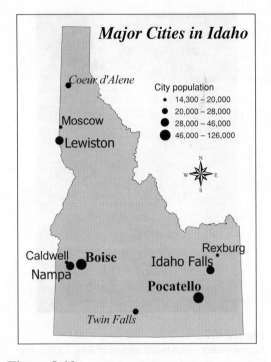

Figure 9.10
The look of the map is not harmonious because of too many typefaces.

the map's title and legend. The use of conventions can also lend support for harmony. Conventions in type selection include italic for names of water features, uppercase and letter spacing for names of administrative units, and variations in type size and form for names of different-sized cities.

9.3.3 Placement of Text in the Map Body

When considering placement of text, we must first recognize two types of text elements. Text elements in the map body, also called labels, are directly associated with the spatial features. In most cases, labels are names of the spatial features. But they can also be some attribute values such as contour readings or precipitation amounts. Other text elements on a map such as the title and the legend are not tied to any specific locations. Instead, the placement of these text elements (i.e., graphic elements) is related to the layout of the map (Section 9.4.1).

As a general rule, a label should be placed to show the location or the area extent of the named spatial feature. Cartographers recommend placing the name of a point feature to the upper right of its symbol, the name of a line feature in a block and parallel to the course of the feature, and the name of an area feature in a manner that indicates its area extent. Other general rules suggest aligning labels with either the map border or lines of latitude, and placing labels entirely on land or on water.

Implementing labeling algorithms in a GIS package is no easy task (Mower 1993; Chirié 2000). Automated name placement presents several difficult problems for the computer programmer: names must be legible, names cannot overlap other names, names must be clearly associated with their intended referent symbols, and name placement must follow cartographic conventions. These problems worsen at smaller map scales as competition for the map space intensifies between names. We should not expect labeling to be completely automated. Some interactive editing is usually needed to improve the final map's appearance. For this reason many GIS packages offer more than one method of labeling.

As an example, ArcGIS offers interactive and dynamic labeling. Interactive labeling works with one label at a time. If the placement does not work out well, the label can be moved immediately. Interactive labeling is ideal if the number of labels is small or if the location of labels must be exact. Dynamic labeling is probably the method of choice for most users because it can automatically label all or selected features. Using dynamic labeling, we can prioritize options for text placement and for solving potential conflicts (Box 9.4). For example, we can choose to place a line label in a block and parallel to the course of the line feature. We can also set rules to prioritize labels that compete for the same map space. By default, ArcGIS does not allow overlapped labels. This constraint, which is sensible, can impact the placement of labels and may require the adjustment of some labels (Figure 9.11).

Dynamic labels cannot be selected or adjusted individually. But they can be converted to text elements so that we can move and change them in the same way we change interactive labels (Figure 9.12). One way to take care of labels in a truly congested area is to use a leader line to link a label to its feature (Figure 9.13).

Perhaps the most difficult task in labeling is to label the names of streams. The general rule states that the name should follow the course of the

---

Box 9.4 | **Options for Dynamic Labeling**

When we choose dynamic labeling, we basically let the computer take over the labeling task. But the computer needs instructions for placement methods and for solving potential problems in labeling. ArcGIS uses the Placement Properties dialog to gather instructions from the user. The dialog has two tabs. The Placement tab deals with the placement method and duplicate labels. Different placement methods are offered for each feature type. For example, ArcGIS offers 36 options for placing a label relative to its point feature. The default placement method is usually what is recommended in cartography. Duplicate labels such as the same street name for a series of street segments can be either removed or reduced. The Conflict Detection tab handles overlapped labels and the overlapping of labels and feature symbols. ArcGIS uses a weighting scheme to determine the relative importance of a layer and its features. The Conflict Detection tab also has a buffer option that can provide a buffer around each label. We cannot always expect a "perfect" job from dynamic labeling. In most cases, we need to adjust some labels individually.

Figure 9.11

Dynamic labeling of major cities in the United States. The initial result is good but not totally satisfactory. Philadelphia is missing. Labels of San Antonio, Indianapolis, and Baltimore overlap slightly with point symbols. San Francisco is too close to San Jose.

Figure 9.12

A revised version of Figure 9.11. Philadelphia is added to the map, and several city names are moved individually to be closer to their point symbols.

stream, either above or below it. Both interactive labeling and dynamic labeling can curve the name if it appears along a curvy part of the stream. But the appearance of the curved name depends on the smoothness of the corresponding stream segment, the length of the stream segment, and the length of the name. Placing every name in its correct position the first time is nearly impossible

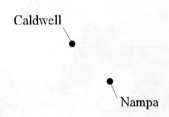

Figure 9.13
A leader line connects a point symbol to its label.

(Figure 9.14a). Problem names must be removed and relabeled using the **spline text** tool, which can align a text string along a curved line that is digitized on-screen (Figure 9.14b).

9.4 MAP DESIGN

Like graphic design, **map design** is a visual plan to achieve a goal. The purpose of map design is to enhance map communication, which is particularly important for thematic maps. A well-designed map is balanced, coherent, ordered, and interesting to look at, whereas a poorly designed map is confusing and disoriented (Antes, Chang, and Mullis 1985). Map design is both an art and a science. There may not be clear-cut right or wrong designs for maps, but there are better or more effective maps and worse or less effective maps.

Map design overlaps with the field of graphic arts, and many map design principles have their

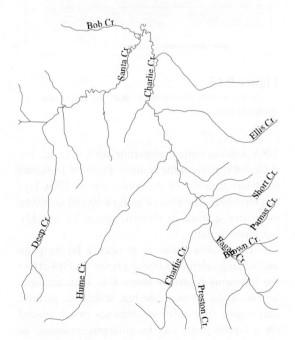

Figure 9.14a
Dynamic labeling of streams may not work for every label. Brown Cr. overlaps with Fagan Cr., and Pamas Cr. and Short Cr. do not follow the course of the creek.

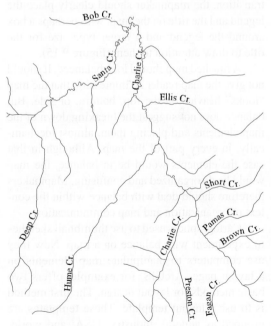

Figure 9.14b
Problem labels in Figure 9.14a are redrawn with the spline text tool.

origin in visual perception. Cartographers usually study map design from the perspectives of layout and visual hierarchy.

9.4.1 Layout

Layout, or planar organization, deals with the arrangement and composition of various map elements. Major concerns with layout are focus, order, and balance. A thematic map should have a clear focus, which is usually the map body or a part of the map body. To draw the map reader's attention, the focal element should be placed near the optical center, just above the map's geometric center. The focal element should be differentiated from other map elements by contrasts in line width, texture, value, detail, and color.

After viewing the focal element, the reader should be directed to the rest of the map in an ordered fashion. For example, the legend and the title are probably the next elements that the viewer needs to look at after the map body. To smooth the transition, the mapmaker should clearly place the legend and the title on the map, with perhaps a box around the legend and a larger type size for the title to draw attention to them (Figure 9.15).

A finished map should look balanced. It should not give the map reader an impression that the map "looks" heavier on the top, bottom, or side. But balance does not suggest the breaking down of the map elements and placing them, almost mechanically, in every part of the map. Although in that case the elements would be in balance, the map would be disorganized and confusing. Mapmakers therefore should deal with balance within the context of organization and map communication.

Cartographers used to use thumbnail sketches to experiment with balance on a map. Now they use computers to manipulate map elements on a layout page. ArcGIS, for example, offers two basic methods for layout design. The first method is to use a layout template. These templates are grouped as general, industry, USA, and world. Each group has a list of choices. For example, the layout templates for the United States include

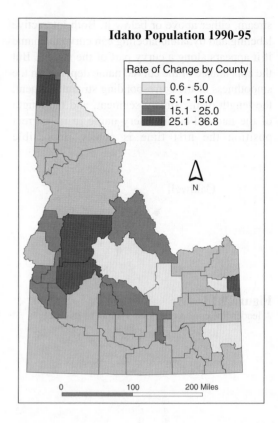

Figure 9.15
Use a box around the legend to draw the map reader's attention to it.

USA, USA counties, conterminous USA, and five different regions of the country. Figure 9.16 shows the basic structure of the conterminous USA layout template. The idea of using a layout template is to use a built-in design option to quickly compose a map.

The second option is to open a layout page and add map elements one at a time. ArcGIS offers the following frame elements: title, text, neatline, legend, north arrow, scale bar, scale text, picture, and object. These frame elements, when placed on a layout page, can be enlarged, reduced, or moved around. Other map elements that ArcGIS offers are inset box (extent rectangle), map border (frame), and graticule (grid) (Box 9.5). A layout

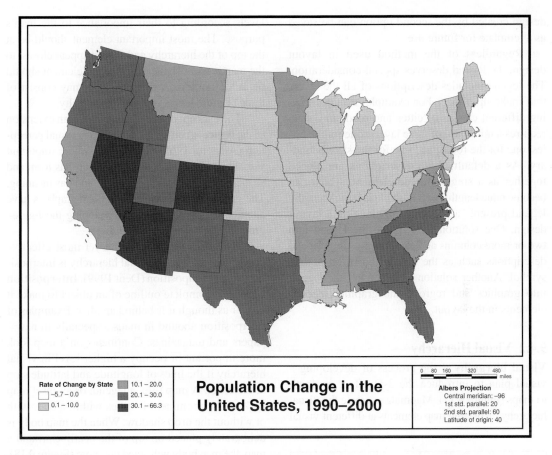

Figure 9.16
The basic structure of the conterminous USA layout template in ArcMap.

Box 9.5 | **Wizards for Adding Map Elements**

ArcMap provides wizards and selectors for adding map elements to a map. The Grids and Graticules wizard is one of them. The graticule refers to lines or tics of longitude and latitude. The grid can be a measured grid or a reference grid. A measured grid shows lines or tics of projected coordinates. A reference grid divides a map into a grid for spatial indexing. The Grids and Graticules wizard guides the user through a series of dialogs to choose the intervals, symbol design, and placement option. Wizards are useful for the beginner but can be tedious and limiting for experienced users. We can choose to turn off the wizard and to work directly with the main property dialog of a map element.

design created by the second option can be saved as a template for future use.

Regardless of the method used in layout design, the legend deserves special consideration. The legend includes descriptions of all the layers that make up a map. For example, a map showing different classes of cities and roads in a state requires a minimum of three layers: one for the cities, one for the roads, and one for the state boundary. As a default, these descriptions are placed together as a single graphic element, which can become quite lengthy with multiple layers. A lengthy legend presents a problem in balancing a layout design. One solution is to divide the legend into two or more columns and to remove useless legend descriptions such as the description of an outline symbol. Another solution is to convert the legend into graphics and regroup the graphic (legend) elements in the layout.

9.4.2 Visual Hierarchy

Visual hierarchy is the process of developing a visual plan to introduce the 3-D effect or depth to maps (Figure 9.17). Mapmakers create the visual hierarchy by placing map elements at different visual levels according to their importance to the map's purpose. The most important element should be at the top of the hierarchy and should appear closest to the map reader. The least important element should be at the bottom. A thematic map may consist of three or more levels in a visual hierarchy.

The concept of visual hierarchy is an extension of the **figure–ground relationship** in visual perception (Arnheim 1965). The figure is more important visually, appears closer to the viewer, has form and shape, has more impressive color, and has meaning. The ground is the background. Cartographers have adopted the depth cues for developing the figure–ground relationship in map design.

Probably the simplest and yet most effective principle in creating a visual hierarchy is interposition or superimposition (Dent 1999). **Interposition** uses the incomplete outline of an object to make it appear as though it is behind another. Examples of interposition abound in maps, especially in newspapers and magazines. Continents on a map look more important or occupy a higher level in visual hierarchy if the lines of longitude and latitude stop at the coast. A map title, a legend, or an inset map looks more prominent if it lies within a box, with or without the drop shadow. When the map body is deliberately placed on top of the neatline around a map, the map body will stand out more (Figure 9.18). Because interposition is so easy to use, it can be

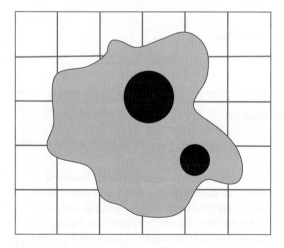

Figure 9.17
A visual hierarchy example. The two black circles are on top (closest to the map reader), followed by the gray polygon. The grid, the least important, is on the bottom.

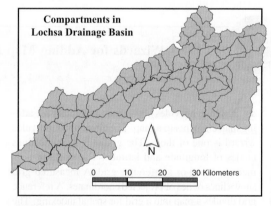

Figure 9.18
The interposition effect in map design.

overused or misused. A map looks confusing if several of its elements compete for the map reader's attention simultaneously (Figure 9.19).

Subdivisional organization is a map design principle that groups map symbols at the primary and secondary levels according to the intended visual hierarchy (Robinson et al. 1995). Each primary symbol is given a distinctive hue, and the differentiation among the secondary symbols is based on color variation, pattern, or texture. For example, all tropical climates on a climate map are shown in red, and different tropical climates (e.g., wet equatorial climate, monsoon climate, and wet–dry tropical climate) are distinguished by different shades of red. Subdivisional organization is most useful for maps with many map symbols, such as climate, soil, geology, and vegetation maps.

Contrast is a basic element in map design, important to layout as well as to visual hierarchy. Contrast in size or width can make a state outline look more important than county boundaries and larger cities look more important than smaller ones (Figure 9.20). Contrast in color can separate the figure from the ground. Cartographers often use a warm color (e.g., orange to red) for the figure and a cool color (e.g., blue) for the ground. Contrast in texture can also differentiate between the figure and the ground because the area containing more details or a greater amount of texture tends to stand out on a map. Like the use of interposition, too much contrast can create a confusing map appearance. For instance, if bright red and green are used side by side as area symbols on a map, they appear to vibrate.

A tool that many GIS packages offer for data display is called **transparency,** which controls the percentage of a layer that is transparent. Transparency can be useful in creating a visual hierarchy by "toning down" the symbols of a background layer. Suppose we want to superimpose a layer showing major cities on top of a layer showing rates of population change by county. We can apply transparency to the county layer so that the city layer will stand out more.

9.5 MAP PRODUCTION

GIS users design and make maps on the computer screen. These soft-copy maps can be used in a variety of ways. They can be printed, exported for use on the Internet, used in overhead computer projection systems, exported to other software packages, or further processed for publishing (Box 9.6).

Map production is a complex topic. We are often surprised that color symbols from the color printers do not exactly match those on the computer screen. This discrepancy results from the use of different media and color models.

Data display on the computer screen uses either **CRT (cathode ray tube)** or **LCD (liquid crystal display).** It used to be that desktop

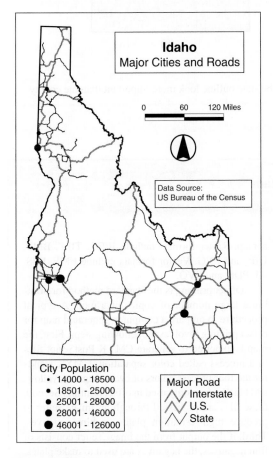

Figure 9.19
A map looks confusing if it uses too many boxes to highlight individual elements.

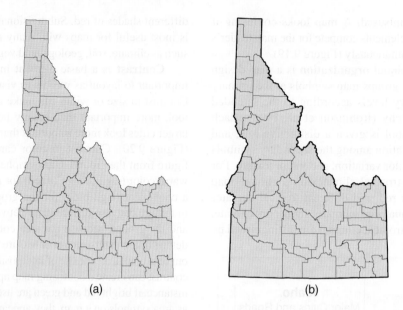

(a)   (b)

Figure 9.20
Contrast is missing in (a), whereas the line contrast makes the state outline look more important than the county boundaries in (b).

Box 9.6   Working with Soft-Copy Maps

When completed, a soft-copy map can be either printed or exported. Printing a map from the computer screen requires the software interface (often called the driver) between the operating system and the print device. ArcMap uses Enhanced Metafiles (EMF) as the default and offers two additional choices of PostScript (PS) and ArcPress. EMF files are native to the Windows operating system for printing graphics. PS, developed by Adobe Systems Inc., in the 1980s, is an industry standard for high-quality printing. Developed by Esri, ArcPress is PS-based and useful for printing maps containing raster data sets, images, or complex map symbols.

One must specify an export format to export a computer-generated map for other uses. ArcMap offers both raster and vector formats for exporting

a map. Raster formats include JPEG, TIFF, BMP, GIF, and PNG. Vector formats include EMF, EPS, AI, PDF, and SVG.

Offset printing is the standard method for printing a large number of copies of a map. A typical procedure to prepare a computer-generated map for offset printing involves the following steps. First, the map is exported to separate CMYK PostScript files in a process called color separation. CMYK stands for the four process colors of cyan, magenta, yellow, and black, commonly used in offset printing. Second, these files are processed through an image setter to produce high-resolution plates or film negatives. Third, if the output from the image setter consists of film negatives, the negatives are used to make plates. Finally, offset printing runs the plates on the press to print color maps.

computers used CRTs and laptop or portable computers used LCDs. But now more desktop computers are also using LCDs to take advantage of the thin, flat screen. A CRT screen has a built-in fine mesh of pixels, and each pixel has colored dots called phosphors. When struck by electrons from an electron gun, a dot lights up and slowly grows dim. An LCD screen uses two sheets of polarizing materials with a liquid crystal solution between them. Each pixel on an LCD screen can be turned on or off independently. Besides being thinner and lighter, LCD screens have two other advantages over CRT screens: they consume less power, and they produce sharper, flicker-free images.

With either a CRT or an LCD, a color symbol we see on a screen is made of pixels, and the color of each pixel is a mixture of **RGB** (red, green, and blue). The intensity of each primary color in a mixture determines its color. The number of intensity levels each primary color can have depends on the number of bit-planes assigned to the electron gun in a CRT screen or the variation of the voltage applied in an LCD screen. Typically, the intensity of each primary color can range over 256 shades. Combining the three primary colors produces a possible palette of 16.8 million colors ($256 \times 256 \times 256$).

Many GIS packages offer the RGB color model for color specification. But color mixtures of RGB are not intuitive and do not correspond to the human dimensions of color perception (Figure 9.21) (MacDonald 1999). For example, it is difficult to perceive that a mixture of red and green at full intensity is a yellow color. This is why other color models have been developed to specify colors based on the visual perception of hue, value, and chroma. ArcGIS, for example, has the **HSV** (hue/saturation/value) color model in addition to the RGB color model for specifying custom colors.

Printed color maps differ from color displays on a computer screen in two ways: color maps reflect rather than emit light; and the creation of colors on color maps is a subtractive rather than an additive process. The three primary subtractive colors are cyan, magenta, and yellow. In printing, these three primary colors plus black form the four process colors of **CMYK.**

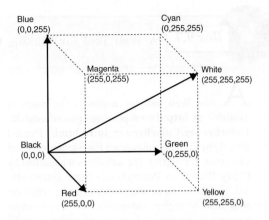

Figure 9.21
The RGB (red, green, and blue) color model.

Color symbols are produced on a printed map in much the same way as on a computer screen. In place of pixels are color dots, and in place of varying light intensities or voltages are percentages of area covered by color dots. A deep orange color on a printed map may represent a combination of 60% magenta and 80% yellow, whereas a light orange color may represent a combination of 30% magenta and 90% yellow. To match a color symbol on the computer screen with a color symbol on the printed map requires a translation from the RGB color model to the CMYK color model. As yet there is no exact translation between the two, and therefore the color map looks different when printed (Fairchild 2005; Slocum et al. 2005). The International Color Consortium, a consortium of over 70 companies and organizations worldwide, has been working since 1993 on a color management system that can be used across different platforms and media (**http://www.color .org/**). Until such a color management system is developed and accepted, we must experiment with colors on different media.

Map production, especially the production of color maps, can be a challenging task to GIS users. Box 9.7 describes ColorBrewer, a free Web tool that can help GIS users choose color symbols that are appropriate for a particular mode of map production (Brewer, Hatchard, and Harrower 2003).

Box 9.7   A Web Tool for Making Color Maps

A free Web tool for making color maps is available at **http://www.personal.psu.edu/cab38/ColorBrewer/ColorBrewer_intro.html.** Funded initially by the National Science Foundation's Digital Government Program, the website is maintained by Cindy Brewer of Pennsylvania State University. The tool offers three main types of color schemes: sequential, diverging, and qualitative. One can select a color scheme and see how the color scheme looks on a sample choropleth map. One can also add point and line symbols to the sample map and change the colors for the map border and background. Then, for each color selected, the tool shows its color values in terms of CMYK, RGB, and other models. And, for each color scheme selected, the tool rates its potential uses including photocopy in black and white, CRT display, color printing, LCD projector, and laptop LCD display.

KEY CONCEPTS AND TERMS

Cartography: The making and study of maps in all their aspects.

Chart map: A map that uses charts such as pie charts or bar charts as map symbols.

Choropleth map: A map that applies shading symbols to data or statistics collected for enumeration units such as counties or states.

Chroma: The richness or brilliance of a color. Also called *saturation* or *intensity*.

CMYK: A color model in which colors are specified by the four process colors: cyan (C), magenta (M), yellow (Y), and black (K).

Contrast: A basic element in map design that enhances the look of a map or the figure–ground relationship by varying the size, width, color, and texture of map symbols.

CRT (cathode-ray tube) screen: A display device for a personal computer that uses electron guns and color dots.

Dasymetric map: A map that uses statistics and additional information to delineate areas of homogeneous values, rather than following administrative boundaries.

Dot map: A map that uses uniform point symbols to show spatial data, with each symbol representing a unit value.

Figure–ground relationship: A tendency in visual perception to separate more important objects (figures) in a visual field from the background (ground).

Flow map: A map that displays different quantities of flow data by varying the width of the line symbol.

Font: A complete set of all variants of a given typeface.

General reference map: One type of map used for general purposes such as the USGS topographic map.

Graduated color map: A map that uses a progressive color scheme such as light red to dark red to show the variation in geospatial data.

Graduated symbol map: A map that uses different-sized symbols such as circles, squares, or triangles to represent different magnitudes.

HSV: A color model in which colors are specified by their hue (H), saturation (S), and value (V).

Hue: The quality that distinguishes one color from another, such as red from blue. Hue is the dominant wavelength of light.

Interposition: A tendency for an object to appear as though it is behind another because of its incomplete outline.

Isarithmic map: A map that uses a system of isolines to represent a surface.

Layout: The arrangement and composition of map elements on a map.

LCD (liquid crystal display) screen: A display device for a personal computer that uses electric charge through a liquid crystal solution between two sheets of polarizing materials.

Map design: The process of developing a visual plan to achieve the map's purpose.

Point: Measurement unit of type, with 72 points to an inch.

Proportional symbol map: A map that uses a specific-sized symbol for each numeric value.

RGB: A color model in which colors are specified by their red (R), green (G), and blue (B) components.

Sans serif: Without serif.

Serif: Small, finishing touches added to the ends of line strokes in a typeface.

Spline text: A text string aligned along a curved line.

Subdivisional organization: A map design principle that groups map symbols at the primary and secondary levels according to the intended visual hierarchy.

Thematic map: One type of map that emphasizes the spatial distribution of a theme, such as a map that shows the distribution of population densities by county.

Transparency: A display tool that controls the percentage of a layer to appear transparent.

Typeface: A particular style or design of type.

Type weight: Relative blackness of a type such as bold, regular, or light.

Type width: Relative width of a type such as condensed or extended.

Value: The lightness or darkness of a color.

Visual hierarchy: The process of developing a visual plan to introduce the 3-D effect or depth to maps.

REVIEW QUESTIONS

1. What are the common elements on a map for presentation?

2. Why is it important to pay attention to map design?

3. Mapmakers apply visual variables to map symbols. What are visual variables?

4. Name common visual variables for data display.

5. Describe the three visual dimensions of color.

6. Use an example to describe a "hue and value" color scheme.

7. Use an example to describe a "diverging" color scheme.

8. Define the choropleth map.

9. Why is data classification important in mapping, especially in choropleth mapping?

10. ArcMap offers the display options of graduated colors and graduated symbols. How do these two options differ?

11. Suppose you are asked to redo Figure 9.10. Provide a list of type designs, including typeface, form, and size, that you will use for the four classes of cities.

12. What are the general rules for achieving harmony with text on a map?

13. ArcGIS offers interactive labeling and dynamic labeling for the placement of text. What are the advantages and disadvantages of each labeling method?

14. Figure 9.15 shows a layout template available in ArcMap for the conterminous USA. Will you consider using the layout template for future projects? Why, or why not?

15. What is visual hierarchy in map design? How is the hierarchy related to the map purpose?

16. Figure 9.17 shows an example of using interposition in map design. Does the map achieve the intended 3-D effect?

17. What is subdivisional organization in map design? Can you think of an example, other than the climate map example in Chapter 9, to which you can apply the principle?

18. Explain why color symbols from a color printer do not exactly match those on the computer screen.

19. Define the RGB and CMYK color models.

APPLICATIONS: DATA DISPLAY AND CARTOGRAPHY

This applications section consists of three tasks. Task 1 guides you through the process of making a choropleth map. Task 2 introduces cartographic representations and lets you experiment with text labeling and highway shield symbols. Task 3 focuses on the placement of text. Because a layout in ArcMap will include all data frames, you must exit ArcMap at the end of each task to preserve the layout design. Making maps for presentation can be tedious. You must be patient and willing to experiment.

Task 1 Make a Choropleth Map

What you need: *us.shp,* a shapefile showing population change by state in the United States between 1990 and 2000. The shapefile is projected onto the Albers equal-area conic projection and is measured in meters.

Choropleth maps display statistics by administrative unit. For Task 1 you will map the rate of population change between 1990 and 2000 by state. The map consists of the following elements: a map of the conterminous United States and a scale bar, a map of Alaska and a scale bar, a map of Hawaii and a scale bar, a title, a legend, a north arrow, a data source statement, a map projection statement, and a neatline around all elements. The basic layout of the map is as follows: The map page is 11″ (width) × 8.5″ (height), or letter size, with a landscape orientation. One-third of the map on the left, from top to

bottom, has the title, map of Alaska, and map of Hawaii. Two-thirds of the map on the right, from top to bottom, has the map of the conterminous United States and all the other elements.

1. Start ArcCatalog, and connect to the Chapter 9 database. Launch ArcMap. Maximize the view window of ArcMap. Add *us.shp* to the new data frame, and rename the data frame Conterminous. Zoom in on the lower 48 states.

2. This step symbolizes the rate of population change by state. Right-click *us* and select Properties. On the Symbology tab, click Quantities and select Graduated colors. Click the Value dropdown arrow and choose ZCHANGE (rate of population change from 1990 to 2000). Cartographers recommend the use of round numbers and logical breaks such as 0 in data classification. Click the first cell under Range and enter 0. The new range should read −5.7–0.0. Enter 10, 20, and 30 for the next three cells, and click the empty space below the cells to unselect. Next, change the color scheme for ZCHANGE. Right-click the Color Ramp box and uncheck Graphic View. Click the dropdown arrow and choose Yellow to Green to Dark Blue. The first class from −5.7 to 0.0 is shown in yellow, and the other classes are shown in green to dark blue. Click OK to dismiss the Layer Properties dialog.

Q1. How many records in *us.shp* have *ZCHANGE* < 0?

3. The Alaska map is next. Insert a new data frame and rename it Alaska. Add *us.shp* to Alaska. Zoom in on Alaska. Follow the same procedure as in Step 2, or click the Import button on the Symbology tab in the Layer Properties dialog, to display ZCHANGE. Another option is to copy and paste *us.shp* from the Conterminous map to the Alaska map and zoom in on Alaska.

4. The Hawaii map is next. Select Data Frame from the Insert menu. Rename the new data frame Hawaii. Add *us.shp* to Hawaii. Zoom in on Hawaii. Display the map with ZCHANGE.

5. The table of contents in ArcMap now has three data frames: Conterminous, Alaska, and Hawaii. Select Layout View from the View menu. Click the Zoom Whole Page button. Select Page and Print Setup from the File menu. Check Landscape for the page orientation, uncheck the box for Use Printer Paper Settings, set the page size to be 11.0″ × 8.5″, and click OK.

6. The three data frames are stacked up on the layout. You will rearrange the data frames according to the basic layout plan. Click the Select Elements button in the ArcMap toolbar. Click the Conterminous data frame. Use the handles around the data frame to move and adjust the data frame so that it occupies about two-thirds of the layout in both width and height on the upper-right side of the layout. Click the Alaska data frame, and move it to the center left of the layout. Click the Hawaii data frame, and place it below the Alaska data frame.

7. Now you will add a scale bar to each data frame. Begin with the Conterminous data frame by clicking on it. Select Scale Bar from the Insert menu. Click the selection of Alternating Scale Bar 1, and then Properties. The Scale Bar dialog has three tabs: Scale and Units, Numbers and Marks, and Format. On the Scale and Units tab, start with the middle part of the dialog: select Adjust width when

resizing, choose Kilometers for division units, and enter km for the label. Now work with the upper half of the dialog: enter 1000 (km) for the division value, select 2 for the number of divisions, select 0 for the number of subdivisions, and opt not to show one division before zero. On the Numbers and Marks tab, select divisions from the frequency dropdown list and Above bar from the position dropdown list. On the Format tab, select Times New Roman from the font dropdown list. Click OK to dismiss the dialogs. The scale bar appears in the map with the handles. Move the scale bar to the lower-left corner of the Conterminous data frame. The scale bar should have the divisions of 1000 and 2000 kilometers. (You can set Zoom Control on the Layout toolbar at 100% and then use the Pan tool to check the scale bar.) Separate scale bars are needed for the other two data frames. Click the Alaska data frame. Use the same procedure to add its scale bar with 500 kilometers for the division value. Click the Hawaii data frame. Add its scale bar by using 100 kilometers for the division value.

Q2. Explain in your own words the number of divisions and the number of subdivisions on a scale bar.

Q3. In Step 7, you have chosen the option of "Adjust width when resizing." What does this option mean?

8. So far, you have completed the data frames of the map. The map must also have the title, legend, and other elements. Select Title from the Insert menu. Enter "Title" in the Insert Title box. "Title" appears on the layout with the outline of the box shown in cyan. Double-click the title box. A Properties dialog appears with two tabs. On the Text tab, enter two lines in the text box: "Population Change" in the first line, and "by State, 1990–2000" in the second line. Click Change Symbol. The Symbol Selector dialog lets you choose color, font, size, and style. Select black, Bookman Old

Style (or another serif type), 20, and B (bold) respectively. Click OK to dismiss the dialogs. Move the title to the upper left of the layout above the Alaska data frame.

9. The legend is next. Because all three data frames in the table of contents use the same legend, it does not matter which data frame is used. Click ZCHANGE of the active data frame, and click it one more time so that ZCHANGE is highlighted in a box. Delete ZCHANGE. (Unless ZCHANGE is removed in the table of contents, it will show up in the layout as a confusing legend descriptor.) Select Legend from the Insert menu. The Legend Wizard uses five panels. In the first panel, make sure *us* is the layer to be included in the legend. The second panel lets you enter the legend title and its type design. Delete Legend in the Legend Title box, and enter "Rate of Population Change (%)." Then choose 14 for the size, choose Times New Roman for the font, and uncheck B (Bold). Skip the third and fourth panels, and click Finish in the fifth panel. Move the legend to the right of the Hawaii data frame.

10. A north arrow is next. Select North Arrow from the Insert menu. Choose ESRI North 6, a simple north arrow from the selector, and click OK. Move the north arrow to the upper right of the legend.

11. Next is the data source. Select Text from the Insert menu. An Enter Text box appears on the layout. Click outside the box. When the outline of the box is shown in cyan, double-click it. A Properties dialog appears with two tabs. On the Text tab, enter "Data Source: US Census 2000" in the Text box. Click on Change Symbol. Select Times New Roman for the font and 14 for the size. Click OK in both dialogs. Move the data source statement below the north arrow in the lower right of the layout.

12. Follow the same procedure as for the data source to add a text statement about the map projection. Enter "Albers Equal-Area Conic

Projection" in the text box, and change the symbol to Times New Roman with a size of 10. Move the projection statement below the data source.

13. Finally, add a neatline to the layout. Select Neatline from the Insert menu. Check to place inside margins, select Double, Graded from the Border dropdown list, and select Sand from the Background dropdown list. Click OK.

14. The layout is now complete. If you want to rearrange a map element, select the element and then move it to a new location. You can also enlarge or reduce a map element by using its handles or properties.

15. If your PC is connected to a color printer, you can print the map directly by selecting Print in the File menu. There are two other options on the File menu: save the map as an ArcMap document, or export the map as a graphic file (e.g., EPS, JPEG, TIFF, PDF, etc.). Exit ArcMap.

Q4. In Task 1, why did you have to prepare three data frames (i.e., Conterminous, Alaska, and Hawaii)?

Task 2 Use Graduated Symbols, Line Symbols, Highway Shield Symbols, and Text Symbols

What you need: *Task2.mdb* with three feature classes: *idlcity,* showing the 10 largest cities in Idaho; *idhwy,* showing interstate and U.S. highways in Idaho; and *idoutl,* containing the outline of Idaho.

Task 2 introduces cartographic representations for symbolizing *idoutl* and *idhwy.* Because cartographic representations require the use of an ArcEditor or ArcInfo license, separate instructions are provided for ArcView users. Task 2 also lets you experiment with text labeling and highway shield symbols.

1. Make sure that ArcCatalog is still connected to the Chapter 9 database. Launch ArcMap. Rename the data frame Task 2, and add *idlcity, idhwy, and idoutl* from *Task 2.mdb*

to Task 2. Select Page and Print Setup from the File menu. Make sure that the page has a width of 8.5 (inches), a height of 11 (inches), and a portrait orientation.

2. Select the properties of *idoutl*. The Symbology tab has Representations in the Show list. The cartographic representation idoutl_Rep has only one rule, which consists of one stroke (squiggle) symbol layer and one fill symbol layer. Click the stroke symbol layer. This is an outline symbol in black with a width of 0.4 (point). Click the fill symbol layer. This is a fill symbol in gray. The cartographic representation therefore displays *idoutl* in gray with a thin black outline. Click OK. (For ArcView users, click the symbol for *idoutl* in the table of contents. Select gray for the fill color and black with a width of 0.4 for the outline.)

3. Select Properties from the context menu of *idhwy*. The cartographic representation idhwy_Rep has two rules, one for Interstate and the other for U.S. Click Rule 1; rule 1 consists of two stroke symbol layers. Click the first stroke symbol layer, which shows a red line symbol with a width of 2.6. Click the second stroke symbol layer, which shows black line symbol with a width of 3.4. The superimposition of the two line symbols result in a red line symbol with a black casing for the interstate highways. Click Rule 2; rule 2 consists of one stroke symbol layer, which shows a red line symbol with a width of 2. Click OK. (For ArcView users, symbolize *idhwy* as follows. On the Symbology tab, select Categories and Unique values for the show option and select ROUTE_DESC from the Value Field dropdown list. Click Add All Values at the bottom. Interstate and U.S. appear as the values. Uncheck all other values. Double-click the Symbol next to Interstate and select the Freeway symbol in the Symbol Selector box. Double-click the Symbol next to U.S. Select the Major Road symbol but change its color to Mars Red. Click OK in both dialogs.)

4. Select Properties from the context menu of *idlcity*. On the Symbology tab, select the show option of Quantities and Graduated Symbols and select POPULATION for the Value field. Next change the number of classes from 5 to 3. Change the Range values by entering 28000 in the first class and 46000 in the second class. Change the labels to read 14302–28000, 28001–46000, and 46001–125659. Click Template, and choose Solar Yellow for the color. Click OK to dismiss the dialog.

5. Labeling the cities is next. Click the Customize menu, point to Toolbars, and check Labeling to open the Labeling toolbar. Click the Label Manager button on the Labeling toolbar. In the Label Manager dialog, click *idlcity* in the Label Classes frame and click the Add button in the Add label classes from symbology categories frame. Click Yes to overwrite the existing labeling classes. Expand *idlcity* in the Label Classes frame. You should see the three label classes by population.

6. Click the first label class of *idlcity* (14302–28000). Make sure that the label field is CITY_NAME. Select Century Gothic (or another sans serif type) and 10 (size) for the text symbol. Click the SQL Query button. Change the first part of the query expression from "POPULATION" > 14302 to "POPULATION" >= 14302. Unless the change is made, the label for the city with the population of 14302 (Rexburg) will not appear. Click the second label class (28001–46000). Select Century Gothic and 12 for the text symbol. Click the third label class (46001–125659). Select Century Gothic, 12, and B (bold) for the text symbol. Make sure that *idlcity* is checked in the Label Classes frame. Click OK to dismiss the dialog.

7. All city names should now appear in the map, but it is difficult to judge the quality of labeling in Data View. You must switch to Layout View to see how the labels will appear on a plot. Select Layout View from the View menu. Click Full Extent on the ArcMap toolbar.

Select 100% from the Zoom Control list on the Layout toolbar. Use the Pan tool to see how the labels will appear on an 8.5-by-11-inch plot.

8. All labels except Nampa are well placed. But to alter the label position of Nampa, you have to convert labels to annotation. Right-click *idlcity* in the table of contents and click Convert Labels to Annotation. In the next dialog, select to store annotation in the map, rather than in the database. Click Convert. (The conversion operation may take a while.) To move the label for Nampa, click the Select Elements tool on the standard toolbar, click Nampa to select it, and then move the label to below its point symbol. (Nampa is between Boise and Caldwell. You can also use the Identify tool to check which city is Nampa.) You may also want to move other city labels to better positions.

9. The last part of this task is to label the interstates and U.S. highways with the highway shield symbols. Switch to Data View. Right-click *idhwy,* and select Properties. On the Labels tab, make sure that the Label Field is MINOR1, which lists the highway number. Then click Symbol, select the U.S. Interstate HWY shield from the Category list, and dismiss the Symbol Selector dialog. Click Placement Properties in the Layer Properties dialog. On the Placement tab, check Horizontal for the orientation. Click OK to dismiss the dialogs. You are ready to plot the interstate shield symbols. Click the Customize menu, point to Toolbars, and check the Draw toolbar. Click the New Text (A) dropdown arrow on the Draw toolbar and choose the Label tool. Opt to place label at position clicked, and close the Label Tool Options dialog. Move the Label tool over an interstate in the map, and click a point to add the label at its location. (The highway number may vary along the same interstate because the interstate has multiple numbers such as 90 and 10, or 80 and 30.) While the label is still active,

you can move it for placement at a better position. Add one label to each interstate.

10. Follow the same procedure as in Step 9 but use the U.S. Route HWY shield to label U.S. highways. Add one label to each U.S. highway. Switch to Layout View and make sure that the highway shield symbols are labeled appropriately. Because you have placed the highway shield symbols interactively, these symbols can be individually adjusted.

11. To complete the layout, you must add the title, legend, and other map elements. Start with the title. Select Title from the Insert menu. Enter "Title" in the Insert Title box. "Title" appears on the layout with the outline of the box shown in cyan. Double-click the box. On the Text tab of the Properties dialog, enter "Idaho Cities and Highways" in the text box. Click Change Symbol. Select Bookman Old Style (or another serif type), 24, and B for the text symbol. Move the title to the upper right of the layout.

12. Next is the legend. But before plotting the legend, you want to remove the layer names of *idlcity* and *idhwy*. Click *idlcity* in the table of contents, click it again, and delete it. Follow the same procedure to delete *idhwy*. Select Legend from the Insert menu. By default, the legend includes all layers from the map. Because you have removed the layer names of *idlcity* and *idhwy,* they appear as blank lines. *idoutl* shows the outline of Idaho and does not have to be included in the legend. You can remove *idoutl* from the legend by clicking *idoutl* in the Legend Items box and then the left arrow button. Click Next. In the second panel, highlight Legend in the Legend Title box and delete it. (If you want to keep the word Legend on the map, do not delete it.) Skip the next two panels, and click Finish in the fifth panel. Move the legend to the upper right of the layout below the title.

13. The labels in the legend are Population and Representation: idhwy_Rep (or Route_Desc for ArcView Users). To change them to more

descriptive labels, you can first convert the legend to graphics. Right-click the legend, and select Convert To Graphics. Right-click the legend again, and select Ungroup. Select the label Population, and then double-click it to open the Properties dialog. Type City Population in the Text box, and click OK. Use the same procedure to change Representation: idhwy_Rep to Highway Type. To regroup the legend graphics, you can use the Select Elements tool to drag a box around the graphics and then select Group from the context menu.

14. A scale bar is next. Select Scale Bar from the Insert menu. Click Alternating Scale Bar 1, and then click Properties. On the Scale and Units tab, first select Adjust width when resizing and select Miles for division units. Then enter the division value of 50 (miles), select 2 for the number of divisions, and select 0 for the number of subdivisions. On the Numbers and Marks tab, select divisions from the Frequency dropdown list. On the Format tab, select Times New Roman from the Font dropdown list. Click OK to dismiss the dialogs. The scale bar appears in the map. Use the handles to place the scale bar below the legend.

15. A north arrow is next. Select North Arrow from the Insert menu. Choose ESRI North 6, a simple north arrow from the selector, and click OK. Place the north arrow below the scale bar.

16. Finally, change the design of the data frame. Right-click Task 2 in the table of contents, and select Properties. Click the Frame tab. Select Double, Graded from the Border dropdown list. Click OK.

17. You can print the map directly, save the map as an ArcMap document, or export the map as a graphic file. Exit ArcMap.

Task 3 Label Streams

What you need: *charlie.shp,* a shapefile showing Santa Creek and its tributaries in north Idaho.

Task 3 lets you try the dynamic labeling method in ArcMap. Although the method can label all features on a map and remove duplicate names, it requires adjustments on some individual labels and overlapped labels. Therefore, Task 3 also requires you to use the Spline Text tool.

1. Launch ArcMap. Rename the data frame Task 3, and add *charlie.shp* to Task 3. Select Page and Print Setup from the File menu. Uncheck the box for Use Printer Page Settings. Enter 5 (inches) for Width and 5 (inches) for Height. Click OK to dismiss the dialog.

2. Click the Customize menu, point to Toolbars, and check Labeling to open the Labeling toolbar. Click the Label Manager button on the Labeling toolbar. In the Label Manager dialog, click Default under *charlie* in the Label Classes frame. Make sure that the label field is NAME. Select Times New Roman, 10, and *I* for the text symbol. Notice that the default placement properties include a parallel orien tation and an above position. Click Properties. The Placement tab repeats more or less the same information as in the Label Manager dialog. The Conflict Detection tab lists label weight, feature weight, and buffer for resolving the potential problem of overlapped labels. Check *charlie* in the Label Classes frame. Click OK to dismiss the dialogs. Stream names are now placed on the map.

Q5. List the position options available for line features.

3. Switch to the layout view. Click the Zoom Whole Page button. Use the control handles to fit the data frame within the specified page size. Select 100% from the Zoom Control dropdown list, and use the Pan tool to check the labeling of stream names. The result is generally satisfactory. But you may want to change the position of some labels such as Fagan Cr., Pamas Cr., and Short Cr. Consult Figure 9.14 for possible label changes.

4. Dynamic labeling, which is what you have done up to this point, does not allow individual labels to be selected and modified. To fix the placement of individual labels, you must convert labels to annotation. Right-click *charlie* in the table of contents, and select Convert Labels to Annotation. Select to save annotation in the map. Click Convert.

5. To make sure that the annotation you add to the map has the same look as other labels, you must specify the drawing symbol options. Make sure that the Draw toolbar is open. Click the Drawing dropdown arrow, point to Active Annotation Target, and check charlie anno. Click the Drawing arrow again, and select Default Symbol Properties. Click Text Symbol. In the Symbol Selector dialog, select Times New Roman, 10, and *I*. Click OK to dismiss the dialogs.

6. Switch to Data View. The following instructions use Fagan Cr. as an example. Zoom to the lower right of the map. Use the Select Elements tool to select Fagan Cr., and delete it. Click the New Text (A) dropdown arrow on the Draw toolbar and choose the Spline Text tool. Move the mouse pointer to

below the junction of Brown Cr. and Fagan Cr. Click along the course of the stream, and double-click to end the spline. Enter Fagan Cr. in the Text box. Fagan Cr. appears along the clicked positions. You can follow the same procedure to change other labels.

Challenge Task

What you need: *country.shp,* a world shapefile that has attributes on population and area of over 200 countries.

This challenge task asks you to map the population density distribution of the world.

1. Use POP_CNTRY (population) and SQMI_ CNTRY (area in square miles) in *country.shp* to create a population density field. Name the field POP_DEN and calculate the field value by [POP_CNTRY]/[SQMI_CNTRY].

2. Classify POP_DEN into five classes by using class breaks of your choice.

3. Prepare a layout of the map, complete with a title ("Population Density Map of the World"), a legend (with a legend description of "Persons per Square Mile"), and a neatline around the map.

REFERENCES

Antes, J. R., and K. Chang. 1990. An Empirical Analysis of the Design Principles for Quantitative and Qualitative Symbols. *Cartography and Geographic Information Systems* 17: 271–77.

Antes, J. R., K. Chang, and C. Mullis. 1985. The Visual Effects of Map Design: An Eye Movement Analysis. *The American Cartographer* 12: 143–55.

Arnheim, R. 1965. *Art and Visual Perception.* Berkeley, CA: University of California Press.

Brewer, C. A. 1994. Color Use Guidelines for Mapping and Visualization. In A. M. MacEachren and D. R. F. Taylor, eds., *Visualization in Modern Cartography,* pp. 123–47. Oxford: Pergamon Press.

Brewer, C. A. 2001. Reflections on Mapping Census 2000. *Cartography and Geographic Information Science* 28: 213–36.

Brewer, C. A., G. W. Hatchard, and M. A. Harrower. 2003. ColorBrewer in Print: A Catalog

of Color Schemes for Maps. *Cartography and Geographic Information Science* 30: 5–32.

Brewer, C. A., A. M. MacEachren, L. W. Pickle, and D. Herrmann. 1997. Mapping Mortality: Evaluating Color Schemes for Choropleth Maps. *Annals of the Association of American Geographers* 87: 411–38.

Chang, K. 1978. Measurement Scales in Cartography. *The American Cartographer* 5: 57–64.

Chirié, F. 2000. Automated Name Placement with High

Cartographic Quality: City Street Maps. *Cartography and Geographic Information Science* 27: 101–10.

Cuff, D. J. 1972. Value versus Chroma in Color Schemes on Quantitative Maps. *Canadian Cartographer* 9: 134–40.

Dent, B. D. 1999. *Cartography: Thematic Map Design,* 5th ed. New York: McGraw-Hill.

Eicher, C. L., and C. A. Brewer. 2001. Dasymetric Mapping and Areal Interpolation: Implementation and Evaluation. *Cartography and Geographic Information Science* 28: 125–38.

Fairchild, M. D. 2005. *Color Appearance Models,* 2d ed. New York: Wiley.

Holloway, S., J. Schumacher, and R. L. Redmond. 1999. People and Place: Dasymetric Mapping Using ARC/INFO. In S. Morain, ed., *GIS Solutions in Natural Resource Management: Balancing the Technical–Political Equation,* pp. 283–91, Santa Fe, NM: OnWord Press.

Holt, J. B., C. P. Lo, and T. W. Hodler. 2004. Dasymetric Estimation of Population Density and Areal Interpolation of Census Data. *Cartography and Geographic Information Science* 31: 103–21.

Kraak, M. J., and F. J. Ormeling. 1996. *Cartography: Visualization of Spatial Data.* Harlow, England: Longman.

MacDonald, L. W. 1999. Using Color Effectively in Computer Graphics. *IEEE Computer Graphics and Applications* 19: 20–35.

Mersey, J. E. 1990. Colour and Thematic Map Design: The Role of Colour Scheme and Map Complexity in Choropleth Map Communication. *Cartographica* 27(3): 1–157.

Monmonier, M. 1996. *How to Lie with Maps,* 2d ed. Chicago: University of Chicago Press.

Mower, J. E. 1993. Automated Feature and Name Placement on Parallel Computers. *Cartography and Geographic Information Systems* 20: 69–82.

Robinson, A. H., J. L. Morrison, P. C. Muehrcke, A. J. Kimerling, and S. C. Guptill. 1995. *Elements of Cartography,* 6th ed. New York: Wiley.

Slocum, T. A., R. B. McMaster, F. C. Kessler, and H. H. Howard. 2005. *Thematic Cartography and Geographic Visualization,* 2d ed. Upper Saddle River, NJ: Prentice Hall.

Tufte, E. R. 1983. *The Visual Display of Quantitative Information.* Cheshire, CT: Graphics Press.

CHAPTER 10

DATA EXPLORATION

CHAPTER OUTLINE

10.1 Data Exploration
10.2 Map-Based Data Manipulation
10.3 Attribute Data Query
10.4 Spatial Data Query
10.5 Raster Data Query

Starting data analysis in a GIS project can be overwhelming. The GIS database may have dozens of layers and hundreds of attributes. Where do you begin? What attributes do you look for? What data relationships are there? One way to ease into the analysis phase is data exploration. Centered on the original data, data exploration allows you to examine the general trends in the data, to take a close look at data subsets, and to focus on possible relationships between data sets. The purpose of data exploration is to better understand the data and to provide a starting point in formulating research questions and hypotheses.

Perhaps the most celebrated example of data exploration is Dr. John Snow's study of the cholera outbreak of 1854 in London (Vinten-Johansen et al.

2003). There were 13 pumps supplying water from wells in the Soho area of London. When cholera outbreak happened, Snow mapped the locations of the homes of those who had died from cholera. Primarily based on the map, Snow was able to determine that the culprit was the pump in Broad Street. After the pump handle was removed, the number of infections and deaths dropped rapidly. Interestingly enough, a 2009 study still followed Dr. Snow's approach, but with modern geospatial technology, to assess the role of drinking water in sporadic enteric disease in British Columbia, Canada (Uhlmann et al. 2009).

An important part of modern-day data exploration is the use of interactive and dynamically linked visual tools. Maps (both vector- and raster-based), graphs, and tables are displayed in multiple windows and dynamically linked so that selecting records from a table will automatically highlight the corresponding features in a graph and a map. Data exploration allows data to be viewed from different perspectives, making it easier for information processing and synthesis. Windows-based GIS packages, which can work with maps, graphs, charts, and tables in different windows simultaneously, are well suited for data exploration.

Chapter 10 is organized into the following five sections. Section 10.1 discusses elements of data exploration. Section 10.2 deals with map-based data manipulation, using maps as a tool for data exploration. Sections 10.3 and 10.4 cover methods for exploring vector data. Section 10.3 focuses on attribute data query, and Section 10.4 covers spatial data query and the combination of attribute and spatial data queries. Section 10.5 turns to raster data query.

10.1 DATA EXPLORATION

Statisticians have traditionally used a variety of graphic techniques and descriptive statistics to examine data prior to more formal and structured data analysis (Tukey 1977; Tufte 1983). The Windows operating system, with multiple and dynamically linked windows, has further assisted exploratory data analysis by allowing the user to directly manipulate data points in charts and diagrams (Cleveland and McGill 1988; Cleveland 1993). More recently, **data visualization** has emerged as a discipline that uses a variety of exploratory techniques and graphics to understand and gain insight into the data (Buja, Cook, and Swayne 1996) (Box 10.1).

Similar to statistics, data exploration in GIS lets the user view the general patterns in a data set, query data subsets, and hypothesize about possible relationships between data sets (Anselin 1999; Andrienko et al. 2001; Haining 2003). But there are two important differences. First, data exploration in a GIS involves both spatial and attribute data. Second, the media for data exploration in GIS include maps and map features. For example, in studying soil conditions, we want to know not only how much of the study area is rated poor but also where those poor soils are distributed on a map. Therefore, besides descriptive statistics and graphics, data exploration in GIS must also cover map-based data manipulation, attribute data query, and spatial data query.

10.1.1 Descriptive Statistics

Descriptive statistics summarize the values of a data set. They include the following:

- The *range* is the difference between the minimum and maximum values.
- The *median* is the midpoint value, or the 50th percentile.
- The *first quartile* is the 25th percentile.
- The *third quartile* is the 75th percentile.

Box 10.1 | **Data Visualization**

Buja, Cook, and Swayne (1996) divide data visualization activities into two areas: rendering and manipulation. Rendering deals with the decision about what to show in a graphic plot and what type of plot to make. Manipulation refers to how to operate on individual plots and how to organize multiple plots. Buja, Cook, and Swayne (1996) further identify three fundamental tasks in data visualization: finding Gestalt, posing queries, and making comparisons. Finding Gestalt means finding patterns and properties in the data set. Posing queries means exploring data characteristics in more detail by examining data subsets. And making comparisons refers to comparisons between variables or between data subsets. Many software packages have been specially designed for data visualization. An example is Aviation Weather Data Visualization Environment (AWE), which focuses on the graphic displays of weather data important to pilots (Spirkovska and Lodha 2002).

- The *mean* is the average of data values. The mean can be calculated by

$$\sum_{i=1}^{n} x_i / n,$$ where x_i is the ith value and n is the number of values.

- The *variance* is the average of the squared deviations of each data value about the mean. The variance can be calculated by

$$\sum_{i=1}^{n} (x_i - \text{mean})^2 / n$$

- The *standard deviation* is the square root of the variance.

- A *Z score* is a standardized score that can be computed by $(x - \text{mean})/s$, where s is the standard deviation.

GIS packages typically offer descriptive statistics in a menu selection that can apply to a numeric field. Box 10.2 includes descriptive statistics of the rate of population change by state in the United States between 1990 and 2000. This data set is frequently used as an example in Chapter 10.

10.1.2 Graphs

Different types of graphs are used for data exploration. A graph may involve a single variable or

---

Box 10.2 Descriptive Statistics

The following table shows, in ascending order, the rate of population change by state in the United States from 1990 to 2000. The data set exhibits a skewness toward the higher end.

DC	−5.7	VT	8.2	AL	10.1	NM	20.1
ND	0.5	NE	8.4	MS	10.5	OR	20.4
WV	0.8	KS	8.5	MD	10.8	WA	21.1
PA	3.4	SD	8.5	NH	11.4	NC	21.4
CT	3.6	IL	8.6	MN	12.4	TX	22.8
ME	3.8	NJ	8.6	MT	12.9	FL	23.5
RI	4.5	WY	8.9	AR	13.7	GA	26.4
OH	4.7	HI	9.3	CA	13.8	ID	28.5
IA	5.4	MO	9.3	AK	14.0	UT	29.6
MA	5.5	KY	9.6	VA	14.4	CO	30.6
NY	5.5	WI	9.6	SC	15.1	AZ	40.0
LA	5.9	IN	9.7	TN	16.7	NV	66.3
MI	6.9	OK	9.7	DE	17.6		

The descriptive statistics of the data set are as follows:

- Mean: 13.45
- Median: 9.7
- Range: 72.0
- 1st quartile: 7.55, between MI (6.9) and VT (8.2)
- 3rd quartile: 17.15, between TN (16.7) and DE (17.6)
- Standard deviation: 11.38
- Z score for Nevada (66.3): 4.64

multiple variables, and it may display individual values or classes of values. An important guideline in choosing a graph is to let the data tell their story through the graph (Tufte 1983; Wiener 1997).

A line graph displays data as a line. The line graph example in Figure 10.1 shows the rate of population change in the United States along the *y*-axis and the state along the *x*-axis. Notice a couple of "peaks" in the line graph.

A bar chart, also called a histogram, groups data into equal intervals and uses bars to show the number or frequency of values falling within each class. A bar chart may have vertical bars or horizontal bars. Figure 10.2 uses a vertical bar chart to group rates of population change in the United States into six classes. Notice one bar at the high end of the histogram.

A cumulative distribution graph is one type of line graph that plots the ordered data values against the cumulative distribution values. The cumulative distribution value for the *i*th ordered value is typically calculated as $(i - 0.5)/n$, where n is the number of values. This computational formula converts the values of a data set to within the range of 0.0 to 1.0. Figure 10.3 shows a cumulative distribution graph.

A scatterplot uses markings to plot the values of two variables along the *x*- and *y*-axes. Figure 10.4 plots percent population change 1990–2000 against

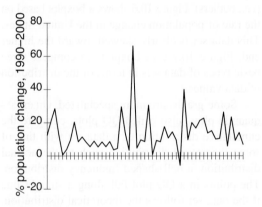

Figure 10.1
A line graph.

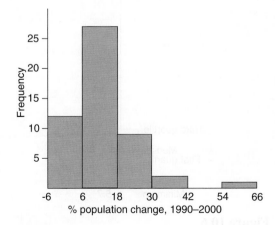

Figure 10.2
A histogram (bar chart).

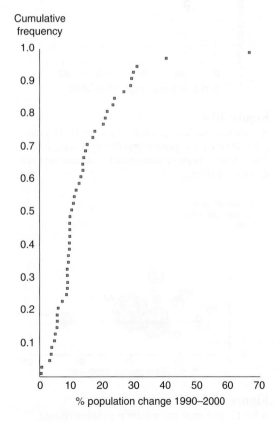

Figure 10.3
A cumulative distribution graph.

percent persons under 18 years old in 2000 by state in the United States. The scatterplot suggests a weak positive relationship between the two variables. Given a set variables, a scatterplot matrix can show the pair-wise scatter plots of the variables in a matrix format.

Bubble plots are a variation of scatterplots. Instead of using constant symbols as in a scatterplot, a bubble plot has varying-sized bubbles that are made proportional to the value of a third variable.

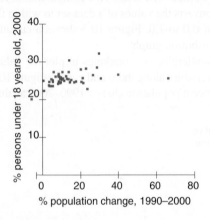

Figure 10.4
A scatterplot plotting percent persons under 18 years old in 2000 against percent population change, 1990–2000. A weak positive relationship is present between the two variables.

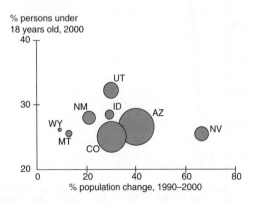

Figure 10.5
A bubble plot showing percent population change 1990–2000, percent persons under 18 years old in 2000, and state population in 2000.

Figure 10.5 is a variation of Figure 10.4: the additional variable shown by the bubble size is the state population in 2000. As an illustration, Figure 10.5 only shows states in the Mountain region, one of the nine regions defined by the U.S. Census Bureau.

Boxplots, also called the "box and whisker" plots, summarize the distribution of five statistics from a data set: the minimum, first quartile, median, third quartile, and maximum. By examining the position of the statistics in a boxplot, we can tell if the distribution of data values is symmetric or skewed and if there are unusual data points (i.e., outliers). Figure 10.6 shows a boxplot based on the rate of population change in the United States. This data set is clearly skewed toward the higher end. Figure 10.7 uses boxplots to compare three basic types of data sets in terms of the distribution of data values.

Some graphs are more specialized. Quantile–quantile plots, also called QQ plots, compare the cumulative distribution of a data set with that of some theoretical distribution such as the normal distribution, a bell-shaped frequency distribution. The points in a QQ plot fall along a straight line if the data set follows the theoretical distribution.

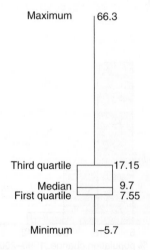

Figure 10.6
A boxplot based on the percent population change, 1990–2000, data set.

Figure 10.8 plots the rate of population change against the standardized value from a normal distribution. It shows that the data set is not normally distributed. The main departure occurs at the two highest values, which are also highlighted in previous graphs.

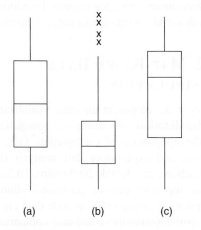

Figure 10.7
Boxplot (*a*) suggests that the data values follow a normal distribution. Boxplot (*b*) shows a positively skewed distribution with a higher concentration of data values near the high end. The x's in (*b*) may represent outliers, which are more than 1.5 box lengths from the end of the box. Boxplot (*c*) shows a negatively skewed distribution with a higher concentration of data values near the low end.

Some graphs are designed for spatial data. Figure 10.9, for example, shows a plot of spatial data values by raising a bar at each point location so that the height of the bar is proportionate to its value. This kind of plot allows the user to see the general trends among the data values in both the *x*-dimension (east–west) and *y*-dimension (north–south). Given spatial data, there are also descriptive spatial statistics such as centroid (Chapter 12) and root mean square (Chapter 15).

Most GIS packages provide tools for making graphs and charts. ArcGIS, for example, has a chart engine that offers bar graphs, line graphs, scatterplots, scatterplot matrix, boxplots, and pie charts for mapping as well as for exporting to other software packages (Task 2 of the applications section uses the chart engine). Commercial statistical analysis packages such as SAS, SPSS, SYSTAT, S-PLUS, and Excel also offer tools for making graphs and charts. There are also open-source software packages such as gnuplot (**http://www .gnuplot.info/**) for scientific data visualization.

10.1.3 Dynamic Graphics

When graphs are displayed in multiple and dynamically linked windows, they become **dynamic graphs.** We can directly manipulate data points in dynamic graphs. For example, we can pose a query in one

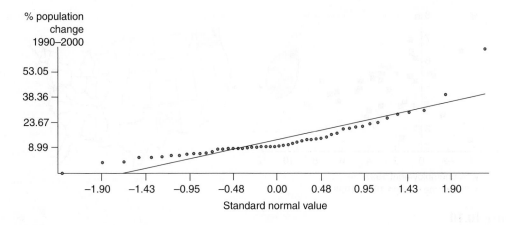

Figure 10.8
A QQ plot plotting percent population change, 1990–2000 against the standardized value from a normal distribution.

window and get the response in other windows, all in the same visual field. By viewing selected data points highlighted in multiple windows, we can hypothesize any patterns or relationships that may exist in the data. This is why multiple linked views have been described as the optimal framework for posing queries about data (Buja, A., D. Cook, and D. F. Swayne 1996).

A common method for manipulating dynamic graphs is **brushing,** which allows the user to graphically select a subset of points from a chart and view related data points in other graphics (Becker and Cleveland 1987). Brushing can be extended to linking a chart and a map (Monmonier 1989). Figure 10.10 illustrates a brushing example that

links a scatterplot and a map. Brushing is included in Task 3 of the applications section.

Other methods that can be used for manipulating dynamic graphics include rotation, deletion, and transformation of data points. Rotation of a 3-D plot lets the viewer see the plot from different perspectives. Deletions of data points (e.g., outliers) and data transformations (e.g., logarithmic transformation) are both useful for uncovering data relationships.

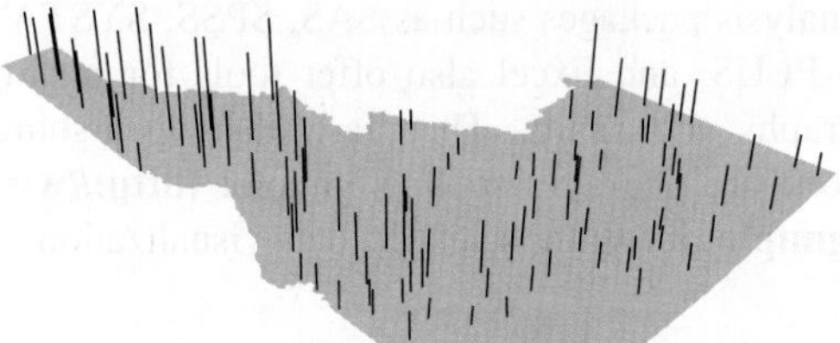

10.2 MAP-BASED DATA MANIPULATION

A recent development in data visualization is **geovisualization**—focusing on geospatial data and the integration of cartography, GIS, image analysis, and exploratory data analysis (Dykes, MacEachren, and Kraak 2005) (Box 10.3). Maps are an important part of geovisualization. Data manipulations using maps include data classification, spatial aggregation, and map comparison.

10.2.1 Data Classification

Data classification can be a tool for data exploration, especially if the classification is based on descriptive statistics. Suppose we want to explore rates of unemployment by state in the United States. To get a preliminary look at the data, we may place rates

Figure 10.9
A 3-D plot showing annual precipitation at 105 weather stations in Idaho. A north-to-south decreasing trend is apparent in the plot.

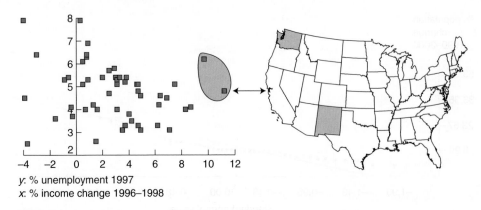

y: % unemployment 1997
x: % income change 1996–1998

Figure 10.10
The scatterplot on the left is dynamically linked to the map on the right. The "brushing" of two data points in the scatterplot highlights the corresponding states (Washington and New Mexico) on the map.

Geographic visualization used to focus on the use of maps for setting up a context for processing visual information and for formulating research questions or hypotheses (MacEachren et al. 1998; Crampton 2002). A number of software packages have been developed specially for geographic visualization such as EXPLOREMAP (Egbert and Slocum 1992) and CommonGIS (Andrienko et al. 2002). Recent developments have broadened the scope of

geographic visualization and introduced the new term geovisualization. Geovisualization emphasizes the integration of cartography, GIS, image analysis, and exploratory data analysis for the visual exploration, analysis, synthesis, and presentation of geospatial data (Dykes, MacEachren, and Kraak 2005) and for spatial decision support (Andrienko et al. 2007). One might consider geovisualization as a special type of data visualization.

of unemployment into classes of above and below the national average (Figure 10.11a). Although generalized, the map divides the country into contiguous regions, which may suggest some regional factors for explaining unemployment.

To isolate those states that are way above or below the national average, we can use the mean and standard deviation method to classify rates of unemployment and focus our attention on states that are more than one standard deviation above the mean (Figure 10.11b).

Classified maps can be linked with tables, graphs, and statistics for more data exploration activities (Egbert and Slocum 1992). For example, we can link the maps in Figure 10.11 with a table showing percent change in median household income and determine whether states with lower unemployment rates also tend to have higher rates of income growth, and vice versa.

10.2.2 Spatial Aggregation

Spatial aggregation is functionally similar to data classification except that it groups data spatially. Figure 10.12 shows percent population change in the United States by state and by region. Used by the U.S. Census Bureau for data collection, regions are spatial aggregates of states. Compared with a map by state, a map by region gives a more general view of population growth in the country

(Figure 10.12). Other geographic levels used by the U.S. Census Bureau are county, census tract, block group, and block. Because these levels of geographic units form a hierarchical order, we can explore the effect of spatial scaling by examining data at different spatial scales.

If distance is the primary factor in a study, spatial data can be aggregated by distance measures from points, lines, or areas. An urban area, for example, may be aggregated into distance zones away from the city center or from its streets (Batty and Xie 1994). Unlike the geography of census, these distance zones require additional data processing such as buffering and areal interpolation (Chapter 11).

Spatial aggregation for raster data means aggregating cells of the input raster to produce a coarser-resolution raster. For example, a raster can be aggregated by a factor of 3. Each cell in the output raster corresponds to a 3-by-3 matrix in the input raster, and the cell value is a computed statistic such as the mean, median, minimum, maximum, or sum from the nine input cell values (Chapter 12).

10.2.3 Map Comparison

Map comparison can help a GIS user sort out the relationship between different data sets. For example, to examine the association between wildlife

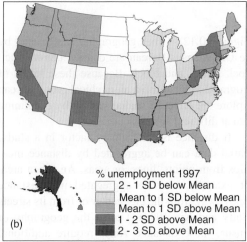

Figure 10.11

Two classification schemes: above or below the national average (*a*), and mean and standard deviation (SD) (*b*).

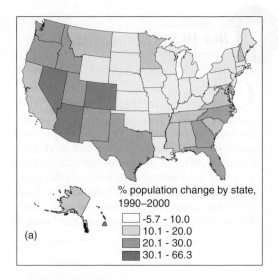

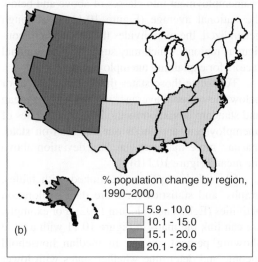

Figure 10.12

Two levels of spatial aggregation: by state (*a*), and by region (*b*).

locations and streams, you can plot wildlife locations (point features) directly on a stream layer (line features). The two layers can also be grouped together (called "group layer" in ArcGIS) so that they both can be plotted on a vegetation layer (polygon features) for comparison. Comparing layers that are polygon or raster layers is difficult. One option is to turn these layers on and off. ArcGIS has a Swipe tool that can show what is underneath a particular layer without having to turn it off. Another option is to use transparency (Chapter 9). Thus, if

two raster layers are to be compared, one layer can be displayed in a color scheme and the other in semitransparent shades of gray. The gray shades simply darken the color symbols and do not produce confusing color mixtures.

If the layers to be compared represent temporal changes of a theme (e.g., land cover maps from 1970 to 2000), they can be viewed in animation

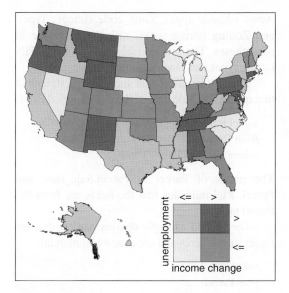

Figure 10.13

A bivariate map: rate of unemployment in 1997, either above or below the national average, and rate of income change, 1996–1998, either above or below the national average.

(Peterson 1995). ArcGIS can export animations as sequential images in bitmap or JPEG format. These sequential images can then be used as input frames to create videos in AVI or QuickTime format.

Finally, there are map symbols that can show multiple data sets. A bar chart placed in each polygon can have two or more bars representing, for example, amounts of timber harvest from two or more years for comparison. Another example is the bivariate choropleth map (Meyer, Broome, and Schweitzer 1975). Figure 10.13, for example, classifies rates of unemployment and rates of income change into either greater than (>) or less than or equal to (< =) the national average and then shows the combinations of the two variables on a single map. To be readable, a bivariate map cannot have too many classes for each variable. Otherwise, the lack of logical progression between the symbols, even color symbols, can become a problem (Olson 1981; Wiener 1997).

10.3 ATTRIBUTE DATA QUERY

Attribute data query retrieves a data subset by working with attribute data. The selected data subset can be simultaneously examined in the table, displayed in charts, and linked to the highlighted features in the map. The selected data subset can also be printed or saved for further processing.

Attribute data query requires the use of expressions, which must be interpretable by a GIS or a database management system. The structure of these expressions varies from one system to another, although the general concept is the same. ArcGIS, for example, uses SQL (Structured Query Language) for query expressions (Box 10.4).

Box 10.4 | **Query Methods in ArcGIS**

In ArcMap, query operations use SQL if the data set is a geodatabase feature class but a limited version of SQL if the data set is a coverage or a shapefile. Therefore, a geodatabase feature class allows greater flexibility and more query operations than a coverage or a shapefile. Additionally, the notation can be different. For example, the fields in a query expression have square brackets around them if the fields belong to a geodatabase feature class but have double quotes if the fields belong to a coverage or a shapefile.

10.3.1 SQL (Structured Query Language)

SQL is a data query language designed for manipulating relational databases (Chapter 8). For GIS applications, SQL is a command language for a GIS (e.g., ArcGIS) to communicate with a database (e.g., Microsoft Access). IBM developed SQL in the 1970s, and many commercial database management systems such as Oracle, Informix, DB2, Access, and Microsoft SQL Server have since adopted the query language. A new development is to extend SQL to object-oriented database management systems and to geospatial data (Shekhar and Chawla 2003).

To use SQL to access a database, we must follow the structure (i.e., syntax) of the query language. The basic syntax of SQL, with the keywords in italic, is

select <attribute list>
from <relation>
where <condition>

The *select* keyword selects field(s) from the database, the *from* keyword selects table(s) from the database, and the *where* keyword specifies the condition or criterion for data query. The following shows three examples of using SQL to query the tables in Figure 10.14. The Parcel table has the fields PIN (string type), Sale_date (date type),

Acres (double type), Zone_code (integer type), and Zoning (string type), with the data type in parentheses. The Owner table has the fields PIN (string type) and Owner_name (string type).

The first example is a simple SQL statement that queries the sale date of the parcel coded P101:

select Parcel.Sale_date
from Parcel
where Parcel.PIN = 'P101'

The prefix of Parcel in Parcel.Sale_date and Parcel. PIN indicates that the fields are from the Parcel table.

The second example queries parcels that are larger than 2 acres and are zoned commercial:

select Parcel.PIN
from Parcel
where Parcel.Acres > 2 AND
Parcel.Zone_code = 2

The fields used in the expression are all present in the Parcel table.

The third example queries the sale date of the parcel owned by Costello:

select Parcel.Sale_date
from Parcel, Owner
where Parcel.PIN = Owner.PIN AND
Owner_name = 'Costello'

PIN	Owner_name
P101	Wang
P101	Chang
P102	Smith
P102	Jones
P103	Costello
P104	Smith

Relation 1: Owner

PIN	Sale_date	Acres	Zone_code	Zoning
P101	1-10-98	1.0	1	Residential
P102	10-6-68	3.0	2	Commercial
P103	3-7-97	2.5	2	Commercial
P104	7-30-78	1.0	1	Residential

Relation 2: Parcel

Figure 10.14
PIN (parcel ID number) relates the Owner and Parcel tables and allows the use of SQL with both tables.

This query involves two tables, which are joined first before the query. The *where* clause consists of two parts: the first part states that Parcel.PIN and Owner.PIN are the keys for the join operation (Chapter 8), and the second part is the actual query expression.

SQL can be used to query a local database or an external database. But there are a couple of procedural differences. First, a GIS package may have already prepared the keywords of *select, from,* and *where* in the dialog for querying a local database. Therefore, we only have to enter the *where* clause (commonly called the query expression) in the dialog box. Second, an attribute query dialog in a GIS package is designed for a single table. If a query involves two tables, as in the third example, they must be joined first. Therefore, the dialog does not show the prefix to a field (e.g., Parcel. PIN) if data query involves only one table. The prefixes only appear in the dialog if two or more tables have been joined for data query.

10.3.2 Query Expressions

Query expressions, or the *where* conditions, consist of Boolean expressions and connectors. A simple **Boolean expression** contains two operands and a logical operator. For example, Parcel.PIN = 'P101' is an expression in which PIN and P101 are operands and = is a logical operator. In this example, PIN is the name of a field, P101 is the field value used in the query, and the expression selects the record that has the PIN value of P101. Operands may be a field, a number, or a string. Logical operators may be equal to (=), greater than (>), less than (<), greater than or equal to (>=), less than or equal to (<=), or not equal to (<>).

Boolean expressions may contain calculations that involve operands and the arithmetic operators +, −, ×, and /. Suppose length is a field measured in feet. We can use the expression, "length" × 0.3048 > 100, to find those records that have the length value of greater than 100 meters. Longer calculations, such as "length" × 0.3048 − 50 > 100, evaluate the × and / operators first

from left to right and then the + and − operators. We can use parentheses to change the order of evaluation. For example, to subtract 50 from length before multiplication by 0.3048, we can use the following expression: ("length" − 50) × 0.3048 > 100.

Boolean connectors are AND, OR, XOR, and NOT, which are used to connect two or more expressions in a query statement. For example, AND connects two expressions in the following statement: Parcel.Acres > 2 AND Parcel. Zone_ code = 2. Records selected from the statement must satisfy both Parcel.Acres > 2 and Parcel. Zone_code = 2. If the connector is changed to OR in the example, then records that satisfy either one or both of the expressions are selected. If the connector is changed to XOR, then records that satisfy one and only one of the expressions are selected. (XOR is functionally opposite to AND.) The connector NOT negates an expression so that a true expression is changed to false and vice versa. The statement, NOT Parcel.Acres > 2 AND Parcel. Zone_code = 2, for example, selects those parcels that are not larger than 2 acres and are zoned commercial.

Boolean connectors of NOT, AND, and OR are actually keywords used in the operations of Complement, Intersect, and Union on sets in probability. The operations are illustrated in Figure 10.15, with A and B representing two subsets of a universal set.

- The Complement of A contains elements of the universal set that do NOT belong to A.
- The Union of A and B is the set of elements that belong to A OR B.
- The Intersect of A and B is the set of elements that belong to both A AND B.

10.3.3 Type of Operation

Attribute data query begins with a complete data set. A basic query operation selects a subset and divides the data set into two groups: one containing selected records and the other unselected records. Given a selected data subset, three types

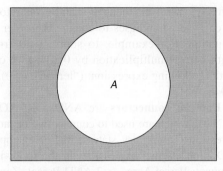

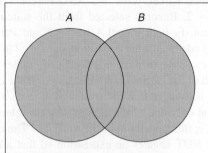

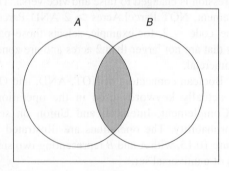

Figure 10.15

The shaded portion represents the complement of data subset *A* (top), the union of data subsets *A* and *B* (middle), and the intersection of *A* and *B* (bottom).

of operations can act on it: add more records to the subset, remove records from the subset, and select a smaller subset (Figure 10.16). Operations can also be performed between the selected and unselected subsets. We can switch between the selected and the unselected subsets, or we can clear the selection by bringing back all records.

These different types of operations allow greater flexibility in data query. For example,

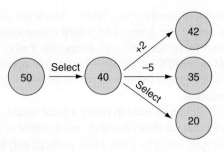

Figure 10.16

Three types of operations may be performed on the selected subset of 40 records: add more records to the subset ($+2$), remove records from the subset (-5), or select a smaller subset (20).

instead of using an expression of Parcel.Acres > 2 AND Parcel.Zone_code $= 2$, we can first use Parcel.Acres > 2 to select a subset and then use Parcel.Zone_code $= 2$ to select a subset from the previously selected subset. Although this example may be trivial, the combination of query expressions and operations can be quite useful for examining various data subsets in data exploration.

10.3.4 Examples of Query Operations

The following examples show different query operations using data from Table 10.1, which has 10 records and 3 fields.

Example 1: Select a Data Subset and Then Add More Records to It

[Create a new selection] "cost" $>= 5$ AND "soiltype" = 'Ns1'
0 of 10 records selected
[Add to current selection] "soiltype" = 'N3'
3 of 10 records selected

Example 2: Select a Data Subset and Then Switch Selection

[Create a new selection] "cost" $>= 5$ AND "soiltype" = 'Tn4' AND "area" $>= 300$
2 of 10 records selected
[Switch Selection]
8 of 10 records selected

TABLE 10.1	A Data Set for Query Operation Examples				
Cost	Soiltype	Area	Cost	Soiltype	Area
1	Ns1	500	6	Tn4	300
2	Ns1	500	7	Tn4	200
3	Ns1	400	8	N3	200
4	Tn4	400	9	N3	100
5	Tn4	300	10	N3	100

Example 3: Select a Data Subset and Then Select a Smaller Subset from It

[Create a new selection] "cost"> 8 OR "area" > 400

4 of 10 records selected

[Select from current selection] "soiltype" = 'Ns1'

2 of 10 records selected

10.3.5 Relational Database Query

Relational database query works with a relational database, which may consist of many separate but interrelated tables. A query of a table in a relational database not only selects a data subset in the table but also selects records related to the subset in other tables. This feature is desirable in data exploration because it allows the user to examine related data characteristics from multiple linked tables.

To use a relational database, we must be familiar with the overall structure of the database, the designation of keys in relating tables, and a data dictionary listing and describing the fields in each table. For data query involving two or more tables, we can choose to either join or relate the tables. A join operation combines attribute data from two or more tables into a single table. A relate operation dynamically links the tables but keeps the tables separate. When a record in one table is selected, the link will automatically select and highlight the corresponding record or records in the related tables. An important consideration in choosing a join or relate operation is the type of data relationship between tables. A join operation is appropriate for the one-to-one or many-to-one relationship but inappropriate for the one-to-many

or many-to-many relationship. A relate operation, on the other hand, can be used with all four types of relationships.

The Soil Survey Geographic (SSURGO) database is a relational database developed by the Natural Resources Conservation Service (NRCS). The database contains soil maps and soil properties and interpretations in more than 50 tables. Sorting out where each soil attribute resides and how tables are linked can therefore be a challenge. Suppose we ask the following question: What types of plants, in their common names, are found where annual flooding frequency is rated as either frequent or occasional? To answer this question, we need the following four SSURGO tables: a soil map and its attribute table; the Component Month table or *comonth,* which contains data on the annual probability of a flood event in the field flodfreqcl; the Component Existing Plants table or *coeplants,* which has common plant names in the field plantcomna; and the Component table or *component,* which includes the keys to link to the other tables (Figure 10.17).

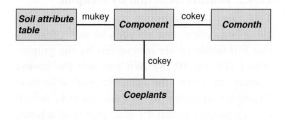

Figure 10.17

The keys relating three dBASE files in the SSURGO database and the soil map attribute table.

After the tables are related, we can first issue the following query statement to *comonth*: "flodfreqcl" = 'frequent' OR "flodfreqcl" = 'occasional'. Evaluation of the query expression selects and highlights records in *comonth* that meet the criteria. And through relates, the corresponding records in *coeplants, component,* and the soil attribute table are highlighted. So are the corresponding soil polygons in the map. This dynamic selection is possible because the tables are interrelated in the SSURGO database and are dynamically linked to the map. A detailed description of relational database query is included in Task 4 of the applications section.

10.4 SPATIAL DATA QUERY

Spatial data query refers to the process of retrieving a data subset from a layer by working directly with feature geometries. Results of a spatial data query like those of an attribute data query, can be simultaneously inspected in the map, linked to the highlighted records in the table, and displayed in charts. They can also be saved as a new data set for further processing. To select features spatially, a cursor, a graphic, or the spatial relationship between features can be used.

10.4.1 Feature Selection by Cursor

The simplest spatial data query is to select a feature by pointing at it or to select features by dragging a box around them.

10.4.2 Feature Selection by Graphic

This query method uses a graphic such as a circle, a box, a line, or a polygon to select features that fall inside or are intersected by the graphic object (Figure 10.18). We can use the mouse pointer to draw the graphic for selection. Examples of query by graphic include selecting restaurants within a 1-mile radius of a hotel, selecting land parcels that intersect a proposed highway, and finding owners of land parcels within a proposed nature reserve.

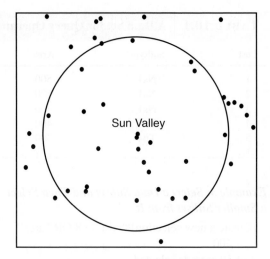

Figure 10.18
Select features by a circle centered at Sun Valley.

10.4.3 Feature Selection by Spatial Relationship

This query method selects features based on their spatial or topological relationships (Chapter 3) to other features. *Features to be selected* may be in the same layer as *features for selection*. Or, more commonly, they are in different layers. An example of the first type of query is to find roadside rest areas within a radius of 50 miles of a selected rest area; in this case, features to be selected and for selection are in the same layer. An example of the second type of query is to find rest areas within each county. Two layers are required for this query: one layer showing county boundaries and the other roadside rest areas.

Spatial relationships used for query include the following:

- **Containment**—selects features that fall completely within features for selection. Examples include finding schools within a selected county, and finding state parks within a selected state.
- **Intersect**—selects features that intersect features for selection. Examples include selecting land parcels that intersect a

proposed road, and finding urban areas that intersect an active fault line.

- **Proximity**—selects features that are within a specified distance of features for selection. Examples include finding state parks within 10 miles of an interstate highway, and finding pet shops within 1 mile of selected streets. If features to be selected and features for selection share common boundaries and if the specified distance is 0, then proximity becomes **adjacency.** Examples of spatial adjacency include selecting land parcels that are adjacent to a flood zone, and finding vacant lots that are adjacent to a new theme park.

Box 10.5 shows the types of spatial relationship that can be used for spatial data query in ArcMap.

10.4.4 Combining Attribute and Spatial Data Queries

So far we have approached data exploration through attribute data query or spatial data query. In many cases data exploration requires both types of queries. For example, both are needed to find gas stations that are within 1 mile of a freeway exit in southern California and have an annual revenue exceeding $2 million each. Assuming that the layers of gas stations and freeway exits are available, there are at least two ways to answer the question.

1. Locate all freeway exits in the study area, and draw a circle around each exit with a 1-mile radius. Select gas stations within the circles through spatial data query. Then use attribute data query to find gas stations that have annual revenues exceeding $2 million.

2. Locate all gas stations in the study area, and select those stations with annual revenues exceeding $2 million through attribute data query. Next, use spatial data query to narrow the selection of gas stations to those within 1 mile of a freeway exit.

The first option queries spatial data and then attribute data. The process is reversed with the second option. Assuming that there are many more gas stations than freeway exits, the first option may be a better option, especially if the gas station layer must be linked to other attribute tables for getting the revenue data.

Box 10.5   Expressions of Spatial Relationships in ArcMap

ArcMap handles feature selection by spatial relationship through the Select By Location dialog. The dialog requires the user to specify one or more layers whose features will be selected, and a layer whose features will be used for selection. Thirteen expressions of spatial relationships connect features to be selected and used for selection. These expressions may be subdivided by the relationships of containment, intersect, and proximity/adjacency:

- Containment: "are completely within," "are within (Clementini)," "completely contain,"

"have their centroid in," "contain," "contain (Clementini)," and "are contained by."
- Intersect: "intersect" and "are crossed by the outline of."
- Proximity/adjacency: "are within a distance of," "share a line segment with," "touch the boundary of," and "are identical to."

A complete query expression in the Select By Location dialog may read as follows: "I want to select features from *city* that are completely within the features in *quake*," where *city* is a city layer and *quake* is an earthquake-prone layer.

The combination of spatial and attribute data queries opens wide the possibilities of data exploration. Some GIS users might even consider this kind of data exploration to be data analysis because that is what they need to do to solve most of their routine tasks.

10.5 RASTER DATA QUERY

Although the concept and even some methods for data query are basically the same for both raster data and vector data, there are enough practical differences to warrant a separate section on raster data query.

10.5.1 Query by Cell Value

The cell value in a raster represents the value of a continuous feature (e.g., elevation) at the cell location. Therefore, to query the feature, we use the raster itself, rather than a field, in the operand.

One type of raster data query uses a Boolean statement to separate cells that satisfy the query statement from cells that do not. The expression, [road] = 1, queries a road raster that has the cell value of 1. The operand [road] refers to the raster and the operand 1 refers to a cell value, which may represent the interstate category. This next expression, [elevation] > 1243.26, queries a floating-point elevation raster that has the cell value greater than 1243.26. Because a floating-point elevation raster contains continuous values, querying a specific value is not likely to find any cell in the raster.

Raster data query can also use the Boolean connectors of AND, OR, and NOT to string together separate expressions. A compound statement with separate expressions usually applies to multiple rasters, which may be integer, or floating point, or a mix of both types. For example, the statement, ([slope] = 2) AND ([aspect] = 1), selects cells that have the value of 2 (e.g., 10–20% slope) in the slope raster, and 1 (e.g., north aspect) in the aspect raster (Figure 10.19). Those cells that satisfy the statement have the cell value

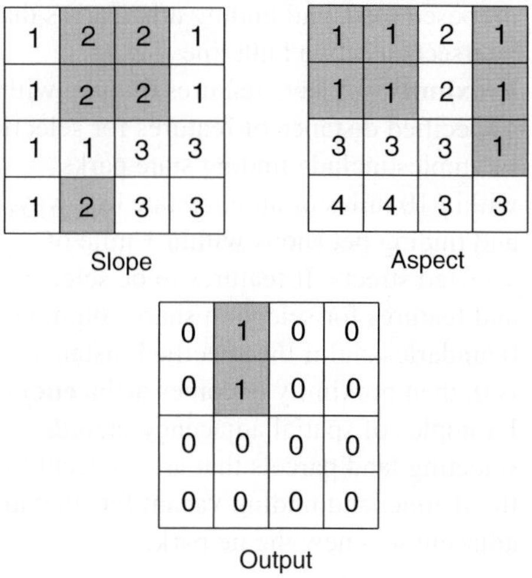

Figure 10.19
Raster data query involving two rasters: slope = 2 and aspect = 1. Selected cells are coded 1 and others 0 in the output raster.

of 1 on the output, while other cells have the cell value of 0.

Querying multiple rasters directly is unique to raster data. For vector data, all attributes to be used in a compound expression must be from the same table or tables that have been joined. Another difference is that a GIS package such as ArcGIS has dialogs specifically designed for vector data query but does not have them for raster data query. The user interface for raster data query is often mixed with commands for raster data analysis.

10.5.2 Query by Select Features

Features such as points, circles, boxes, or polygons can be used directly to query a raster. The query returns an output raster with values for cells that correspond to the point locations or fall within the features for selection. Other cells on the output raster carry no data. Again, this type of raster data query shares the same user interface as commands for data analysis. Chapter 12 provides more information on this topic.

Adjacency: A spatial relationship that can be used to select features that share common boundaries.

Attribute data query: The process of retrieving data by working with attributes.

Boolean connector: A keyword such as AND, OR, XOR, or NOT that is used to construct compound expressions.

Boolean expression: A combination of a field, a value, and a logical operator, such as "class" = 2, from which an evaluation of True or False is derived.

Brushing: A data exploration technique for selecting and highlighting a data subset in multiple views.

Containment: A spatial relationship that can be used in data query to select features that fall within specified features.

Data visualization: The process of using a variety of exploratory techniques and graphics to understand and gain insight into the data.

Dynamic graphics: A data exploration method that lets the user manipulate data points in charts

and diagrams that are displayed in multiple and dynamically linked windows.

Geovisualization: Visualization of geospatial data by integrating approaches from cartography, GIS, image analysis, and exploratory data analysis.

Intersect: A spatial relationship that can be used in data query to select features that intersect specified features.

Proximity: A spatial relationship that can be used in data query to select features within a distance of specified features.

Relational database query: Query in a relational database, which not only selects a data subset in a table but also selects records related to the subset in other tables.

Spatial data query: The process of retrieving data by working with spatial features.

SQL (Structured Query Language): A data query and manipulation language designed for relational databases.

1. Give an example of data exploration from your own experience.

2. Download % population change by county for your state between 2000 and 2009 (or the latest). (You can get the data from the American FactFinder link at the Census Bureau's website, **http://www.census.gov/,** or from the GIS data clearinghouse website in your state.) Use the data to compute the median, first quartile, third quartile, mean, and standard deviation.

3. Use the county data and descriptive statistics from Question 2 to draw a boxplot. What kind of data distribution does the boxplot show?

4. Among the graphics presented in Section 10.1.2, which are designed for multivariate (i.e., two or more variables) visualization?

5. Figure 10.4 exhibits a weak positive relationship between % population change, 1990–2000 and % persons under 18 years old, 2000. What does a positive relationship mean in this case?

6. Describe brushing as a technique for data exploration.

7. Describe an example of using spatial aggregation for data exploration.

8. Explain how an animation for showing temporal changes of a theme is made.

9. Refer to Figure 10.14, and write an SQL statement to query the owner name of parcel P104.

10. Refer to Figure 10.14, and write an SQL statement to query the owner name of parcel P103 OR parcel P104.

11. Refer to Table 10.1, and fill in the blank for each of the following query operations:

    [Create a new selection] "cost" > 8

    _____ of 10 records selected

    [Add to current selection] "soiltype" = 'N3' OR "soiltype" = 'Ns1'

    _____ of 10 records selected

    [Select from current selection] "area" > 400

    _____ of 10 records selected

    [Switch Selection]

    _____ of 10 records selected

12. Refer to Box 10.5, and describe an example of using "intersect" for spatial data query.

13. Refer to Box 10.5, and describe an example of using "are contained by" for spatial data query.

14. You are given two digital maps of New York City: one shows landmarks, and the other shows restaurants. One of the attributes of the restaurant layer lists the type of food (e.g., Japanese, Italian, etc.). Suppose you want to find a Japanese restaurant within 2 miles of Times Square. Describe the steps you will follow to complete the task.

15. Can you think of another solution for Question 14?

16. Refer to Figure 10.19. If the query statement is ([slope] = 1) AND ([aspect] = 3), how many cells in the output will have the cell value of 1?

APPLICATIONS: DATA EXPLORATION

This applications section covers six tasks. Task 1 uses the select feature by location tool. Task 2 lets you use the chart engine to create a scatterplot and link the plot to a table and a map. In Task 3, you will query a joint attribute table, examine the query results using a magnifier window, and bookmark the area with the query results. Task 4 covers relational database query. Task 5 combines spatial and attribute data queries. Task 6 deals with raster data query.

Task 1 Select Feature by Location

What you need: *idcities.shp,* with 654 places in Idaho; and *snowsite.shp,* with 206 snow courses in Idaho and the surrounding states.

Task 1 lets you use the select feature by location method to select snow courses within 40 miles of Sun Valley, Idaho, and plot the snow station data in charts.

1. Start ArcCatalog, and connect to the Chapter 10 database. Launch ArcMap. Add *idcities.shp* and *snowsite.shp* to Layers. Right-click Layers and select Properties. On the General tab, rename the data frame Tasks 1&2 and select Miles from the Display dropdown list.

2. Step 2 selects Sun Valley from *idcities.* Choose Select By Attributes from the Selection menu. Select *idcities* from the layer dropdown list and "Create a new selection" from the method list. Then enter in the expression box the following SQL statement: "CITY_NAME" = 'Sun Valley'. (You can click Get Unique Values to get Sun Valley from the list.) Click Apply and close the dialog. Sun Valley is now highlighted in the map.

3. Choose Select By Location from the Selection menu. In the Select By Location dialog, choose "select features from" for the selection method, check *snowsite* as the target layer, choose *idcities* as the source layer, make sure that the box is checked for using selected features, choose "Target layer(s) are within a distance of the Source

layer feature" for the spatial selection method, check the box for applying a search distance, enter 40 miles for the distance to buffer, and click OK. Snow courses that are within 40 miles of Sun Valley are highlighted in the map.

4. Right-click *snowsite* and select Open Attribute Table. Click Show selected records to show only the selected snow courses.

Q1. How many snow courses are within 40 miles of Sun Valley?

5. You can do a couple of things with the selected records. First, the Table Options menu has options for you to print or export the selected records. Second, you can highlight records (and features) from the currently selected records. For example, Vienna Mine Pillow has the highest SWE_MAX among the selected records. To see where Vienna Mine Pillow is located on the map, you can click on the far left column of its record. Both the record and the point feature are highlighted in yellow.

6. Leave the table and the selected records open for Task 2.

Task 2 Make Dynamic Chart

What you need: *idcities.shp* and *snowsite.shp,* as in Task 1.

Task 2 lets you create a scatterplot from the selected records in Task 1 and take advantage of a live connection among the plot, the attribute table, and the map.

1. Make sure that the *snowsite* attribute table shows the selected records from Task 1. This step exports the selected snow courses to a new shapefile. Right-click *snowsite* in Tasks 1&2, point to Data, and select Export Data. Save the output shapefile as *svstations* in the Chapter 10 workspace. Add *svstations* to Tasks 1&2. Select Clear Selected Features from the Selection menu for *snowsite,* and close the *snowsite* attribute table.

2. Next, create a chart from *svstations.* Open the attribute table of *svstations.* Select Create Graph from the Table Options menu. In the Create Graph Wizard, select ScatterPlot for the graph type, *svstations* for the layer/table, ELEV for the Y field, and SWE_MAX for the X field. Click Next. In the second panel, enter Elev-SweMax for the title. Click Finish. A scatterplot of ELEV against SWE_MAX appears.

Q2. Describe the relationship between ELEV and SWE_MAX.

3. The scatterplot is dynamically linked to the *svstations* attribute table and the map. Click a point in the scatterplot. The point, as well as its corresponding record and feature, is highlighted. You can also use the mouse pointer to select two or more points within a rectangle in the scatterplot. This kind of interaction can also be initiated from either the attribute table or the map.

4. Right-click the scatterplot. The contact menu offers various options for the plot, such as print, save, export, and add to layout.

Task 3 Query Attribute Data from a Joint Attribute Table

What you need: *wp.shp,* a timber-stand shapefile; and *wpdata.dbf,* a dBASE file containing stand data.

Data query can be approached from either attribute data or spatial data. Task 3 focuses on attribute data query.

1. Insert a new data frame in ArcMap and rename it Task 3. Add *wp.shp* and *wpdata.dbf* to Task 3. Next join *wpdata* to *wp* by using ID as the common field. Right-click *wp,* point to Joins and Relates, and select Join. In the Join Data dialog, opt to join attributes from a table, select ID for the field in the layer, select *wpdata* for the table, select ID for the field in the table, and click OK.

2. *wpdata* is now joined to the *wp* attribute table. Open the attribute table of *wp.* The

table now has two sets of attributes. Click
the Table Options dropdown arrow and
choose Select By Attributes. In the Select By
Attributes dialog, make sure that the method
is to create a new selection. Then enter the
following SQL statement in the expression
box: "wpdata.ORIGIN" > 0 AND
"wpdata.ORIGIN" <= 1900. Click Apply.

Q3. How many records are selected?

3. Click Show selected records at the bottom of
the table so that only the selected records are
shown. Polygons of the selected records are
also highlighted in the *wp* layer. To narrow
the selected records, again choose Select By
Attributes from the Table Options dropdown
menu. In the Select By Attributes dialog,
make sure that the method is to select from
current selection. Then prepare an SQL
statement in the expression box that reads:
"wpdata.ELEV" <= 30. Click Apply.

Q4. How many records are in the subset?

4. To take a closer look at the selected
polygons in the map, click the Window
menu in ArcMap and select Magnifier. When
the magnifier window appears, click the
window's title bar, drag the window over
the map, and release the title bar to see a
magnified view.

5. Before moving to the next part of the task,
select Clear Selection from the Table Options
menu in the *wp* attribute table. Then choose
Select By Attributes from the same menu.
Enter the following SQL statement in the
expression box: ("wpdata.ORIGIN" > 0
AND "wpdata.ORIGIN" <= 1900) AND
"wpdata.ELEV" > 40. (The pair of
parentheses is for clarity; it is not necessary
to have them.) Click Apply. Four records
are selected. The selected polygons are
all near the top of the map. Zoom to the
selected polygons. You can bookmark the
zoom-in area for future reference. Click
the Bookmarks menu, and select Create.
Enter *protect* for the Bookmark Name.

To view the zoom-in area next time, click the
Bookmarks menu, and select *protect.*

Task 4 Query Attribute Data from a Relational Database

What you need: *mosoils.shp,* a soil map shapefile; *component.dbf, coeplants.dbf,* and *comonth*
.dbf, three dBASE files derived from the SSURGO
database developed by the Natural Resources
Conservation Service (NRCS).

Task 4 lets you work with the SSURGO
database. By linking the tables in the database
properly, you can explore many soil attributes
in the database from any table. And, because the
tables are linked to the soil map, you can also see
where selected records are located.

1. Insert a new data frame in ArcMap and
rename it Task 4. Add *mosoils.shp,*
component.dbf, coeplants.dbf, and
comonth.dbf to Task 4.

2. First, relate *mosoils* to *component.* Right-click
mosoils in the table of contents, point to Joins
and Relates, and click Relate. In the Relate
dialog, select mukey from the first dropdown
list, select *component* from the second list,
select mukey from the third list, enter soil_
comp for the relate name, and click OK.

3. Next prepare two other relates: comp_plant,
relating *component* to *coeplants* by using
cokey as the common field; and comp_
month, relating *component* to *comonth* by
using cokey as the common field.

4. The four tables (the *mosoils* attribute table,
component, coeplants, and *comonth*) are now
related in pairs by three relates. Right-click
comonth and select Open. Click the Table
Options dropdown arrow and choose Select
By Attributes. In the next dialog, create a
new selection by entering the following SQL
statement in the expression box: "flodfreqcl" =
'Frequent' OR "flodfreqcl" = 'Occasional'.
Click Apply. Click Show selected records so
that only the selected records are shown.

Q5. How many records are selected in *comonth?*

5. To see which records in *component* are related to the selected records in *comonth,* go through the following steps: Click the Related Tables dropdown arrow at the top of the *comonth* table, and click comp_month: component. The *component* table appears with the related records. You can find which records in *coeplants* are related to those records that have frequent or occasional annual flooding by using comp_plant: coeplants with the *component* table.

6. To see which polygons in *mosoils* are subject to frequent or occasional flooding, do the following: Click the Related Tables dropdown menu, and click soil_comp: mosoils. The attribute table of *mosoils* appears with the related records. And the *mosoils* map shows where those selected records are located.

Q6. How many polygons in *mosoils.shp* have a plant species with the common plant name of "Idaho fescue"?

Task 5 Combine Spatial and Attribute Data Queries

What you need: *thermal.shp,* a shapefile with 899 thermal wells and springs; *idroads.shp,* showing major roads in Idaho.

Task 5 assumes that you are asked by a company to locate potential sites for a hot-spring resort in Idaho. You are given two criteria for selecting potential sites:

- The site must be within 2 miles of a major road.
- The temperature of the water must be greater than 60°C.

The field TYPE in *thermal.shp* uses *s* to denote springs and *w* to denote wells. The field TEMP shows the water temperature in °C.

1. Insert a new data frame in ArcMap. Add *thermal.shp* and *idroads.shp* to the new data frame. Right-click the new data frame and select Properties. On the General tab, rename the data frame Task 5 and choose Miles from the Display dropdown list.

2. First select thermal springs and wells that are within 2 miles of major roads. Choose Select By Location from the Selection menu. Do the following in the Select By Location dialog: choose "select features from" for the selection method, check *thermal* as the target layer, select *idroads* for the source layer, select "Target layer(s) are within a distance of the Source layer feature" for the spatial selection method, check the box for applying a search distance, and enter 2 (miles) for the distance to buffer. Click OK. Thermal springs and wells that are within 2 miles of roads are highlighted in the map.

Q7. How many thermal springs and wells are selected?

3. Next, narrow the selection of map features by using the second criterion. Choose Select By Attributes from the Selection menu. Select *thermal* from the Layer dropdown list and "Select from current selection" from the Method list. Then enter the following SQL statement in the expression box: "TYPE" = 's' AND "TEMP" > 60. Click OK.

4. Open the attribute table of *thermal.* Click Show selected records so that only the selected records are shown. The selected records all have TYPE of *s* and TEMP above 60.

5. Map tips are useful for examining the water temperature of the selected hot springs. Right-click *thermal* in the table of contents and select Properties. On the Display tab, select TEMP from the Field dropdown menu and check the box to Show Map Tips using the display expression. Click OK to dismiss the Properties dialog. Click Select Elements on the standard toolbar. Move the mouse pointer to a highlighted hot-spring location, and a map tip will display the water temperature of the spring.

Q8. How many hot wells and springs are within 5 kilometers of *idroads* and have temperatures above 70?

Task 6 Query Raster Data

What you need: *slope_gd,* a slope raster; and *aspect_gd,* an aspect raster.

Task 6 shows you different methods for querying a single raster or multiple rasters.

1. Select Data Frame from the Insert menu in ArcMap. Rename the new data frame Task 6, and add *slope_gd* and *aspect_gd* to Task 6.

2. Select Extension from the Customize menu and make sure that Spatial Analyst is checked. Click ArcToolbox window to open it. Double-click Raster Calculator in the Spatial Analyst Tools/Map Algebra toolset. In the Raster Calculator dialog, prepare the following map algebra expression in the expression box: "slope_gd"=2.(== is the same as=.) Save the output raster as *slope2* and click OK to run the operation. *slope2* is added to the table of contents. Cells with the value of 1 are areas with slopes between 10 and 20 degrees.

Q9. How many cells in *slope2* have the cell value of 1?

3. Go back to the Raster Calculator tool, and prepare the following map algebra expression in the expression box: ("slope_gd"==2) & ("aspect_gd"==4). (& is the same as AND.)

Save the output raster as *asp_slp,* and click OK. Cells with the value of 1 in *asp_slp* are areas with slopes between 10 and 20 degrees and the south aspect.

Q10. What percentage of area covered by the above two rasters has slope = 3 AND aspect = 3?

Challenge Task

What you need: *cities.shp,* a shapefile with 194 cities in Idaho; *counties.shp,* a county shapefile of Idaho; and *idroads.shp,* same as Task 5.

cities.shp has an attribute called CityChange, which shows the rate of population change between 1990 and 2000. *counties.shp* has attributes on 1990 county population (pop1990) and 2000 county population (pop2000). Add a new field to *counties.shp* and name the new field CoChange. Calculate the field values of CoChange by using the following expression: ([pop2000] − [pop1990]) × 100/[pop1990]. CoChange therefore shows the rate of population change between 1990 and 2000 at the county level.

Q1. What is the average rate of population change for cities that are located within 50 miles of Boise?

Q2. How many counties that intersect an interstate highway have CoChange >= 30?

Q3. How many cities with CityChange >= 50 are located within counties with CoChange >= 30?

REFERENCES

Andrienko, G., N. Andrienko, P. Jankowski, D. Keim, M.-J. Kraak, A. MacEachren, and S. Wrobel. 2007. Geovisual Analytics for Spatial Decision Support: Setting the Research Agenda. *International Journal of Geographical Information Science* 21: 839–57.

Andrienko, N., G. Andrienko, A. Savinov, H. Voss, and D. Wettschereck. 2001. Exploratory Analysis of Spatial Data Using Interactive Maps and Data Mining. *Cartography and Geographic Information Science* 28: 151–65.

Andrienko, N., G. Andrienko, H. Voss, F. Bernardo, J. Hipolito, and U. Kretchmer. 2002. Testing the Usability of Interactive Maps in CommonGIS. *Cartography and Geographic Information Science* 29: 325–42.

Anselin, L. 1999. Interactive Techniques and Exploratory

Spatial Data Analysis. In P. A. Longley, M. F. Goodchild, D. J. Maguire, and D. W. Rhind, eds., *Geographical Information Systems,* 2d ed., pp. 253–66. New York: Wiley.

Batty, M., and Y. Xie. 1994. Modelling Inside GIS: Part I. Model Structures, Exploratory Spatial Data Analysis and Aggregation. *International Journal of Geographical Information Systems* 8: 291–307.

Becker, R. A., and W. S. Cleveland. 1987. Brushing Scatterplots. *Technometrics* 29: 127–42.

Buja, A., D. Cook, and D. F. Swayne. 1996. Interactive High-Dimensional Data Visualization. *Journal of Computational and Graphical Statistics* 5: 78–99.

Cleveland, W. S. 1993. *Visualizing Data.* Summit, NJ: Hobart Press.

Cleveland, W. S., and M. E. McGill, eds. 1988. *Dynamic Graphics for Statistics.* Belmont, CA: Wadsworth.

Crampton, J. W. 2002. Interactivity Types in Geographic Visualization. *Cartography and Geographic Information Science* 29: 85–98.

Dykes, J., A. M. MacEachren, and M.–J. Kraak (eds.). 2005. *Exploring Geovisualization.* Amsterdam: Elsevier.

Egbert, S. L., and T. A. Slocum. 1992. EXPLOREMAP: An Exploration System for Choropleth Maps. *Annals of the Association of American Geographers* 82: 275–88.

Haining, R. 2003. *Spatial Data Analysis: Theory and Practice.* Cambridge, England: Cambridge University Press.

MacEachren, A. M., F. P. Boscoe, D. Haug, and L. W. Pickle. 1998. Geographic Visualization: Designing Manipulable Maps for Exploring Temporally Varying Georeferenced Statistics. *Proceedings of the IEEE Symposium on Information Visualization,* pp. 87–94.

Meyer, M. A., F. R. Broome, and R. H. J. Schweitzer. 1975. Color Statistical Mapping by the U.S. Bureau of the Census. *American Cartographer* 2: 100–17.

Monmonier, M. 1989. Geographic Brushing: Enhancing Exploratory Analysis of the Scatterplot Matrix. *Geographical Analysis* 21: 81–84.

Olson, J. 1981. Spectrally Encoded Two-Variable Maps. *Annals of the Association of American Geographers* 71: 259–76.

Peterson, M. P. 1995. *Interactive and Animated Cartography.*

Englewood Cliffs, NJ: Prentice Hall.

Shekhar, S., and S. Chawla. 2003. *Spatial Databases: A Tour.* Upper Saddle River, NJ: Prentice Hall.

Spirkovska, L., and S. K. Lodha. 2002. AWE: Aviation Weather Data Visualization Environment. *Computers & Graphics* 26: 169–91.

Tufte, E. R. 1983. *The Visual Display of Quantitative Information.* Cheshire, CT: Graphics Press.

Tukey, J. W. 1977. *Exploratory Data Analysis.* Reading, MA: Addison-Wesley.

Uhlmann, S., E. Galanis, T. Takaro, S. Mak, L. Gustafson, G. Embree, N. Bellack, K. Corbett, and J. Isaac-Renton. 2009. Where's the Pump? Associating Sporadic Enteric Disease with Drinking Water Using a Geographic Information System, in British Columbia, Canada, 1996–2005. *Journal of Water and Health* 7: 692–98.

Vinten-Johansen, P., H. Brody, N. Paneth, S. Rachman, and M. Rip. 2003. Cholera, Chloroform, and the Science of Medicine: A Life of John Snow. New York: Oxford University Press.

Wiener, H. 1997. *Visual Revelations.* New York: Copernicus.

CHAPTER 11

VECTOR DATA ANALYSIS

CHAPTER OUTLINE

11.1 Buffering

11.2 Overlay

11.3 Distance Measurement

11.4 Pattern Analysis

11.5 Feature Manipulation

The scope of GIS analysis varies among disciplines that use GIS. GIS users in hydrology will likely emphasize the importance of terrain analysis and hydrologic modeling, whereas GIS users in wildlife management will be more interested in analytical functions dealing with wildlife point locations and their relationship to the environment. This is why GIS companies have taken two general approaches in packaging their products. One prepares a set of basic tools used by most GIS users, and the other prepares extensions designed for specific applications such as hydrologic modeling. Chapter 11 covers basic analytical tools for vector data analysis.

The vector data model uses points and their x-, y-coordinates to construct spatial features of points, lines, and polygons. Vector data analysis uses the geometric objects of point, line, and polygon as inputs. Therefore, the accuracy of data analysis depends on the accuracy of these objects in terms of their location and shape and whether they are topological or not. Additionally, it is important to note that an analysis may apply to all, or selected, features in a layer.

As more tools are introduced in a GIS package, we must avoid confusion with the use of these tools. A number of analytical tools such as Union and Intersect also appear as editing tools. Although the terms are the same, they perform different functions. As overlay tools, Union and Intersect work with both geometries and attributes. But as editing tools, they work only with geometries. Whenever appropriate, comments are made throughout Chapter 11 to note the differences.

Chapter 11 is grouped into the following five sections. Section 11.1 covers buffering and its applications. Section 11.2 discusses overlay, types of overlay, problems with overlay, and applications of overlay. Section 11.3 covers tools for measuring distances between points and between points and lines. Section 11.4 examines spatial statistics for pattern analysis. Section 11.5 includes tools for feature manipulation.

11.1 BUFFERING

Based on the concept of proximity, **buffering** creates two areas: one area that is within a specified distance of select features and the other area that is beyond. The area within the specified distance is the buffer zone. A GIS typically varies the value of an attribute to separate the buffer zone (e.g., 1) from the area beyond the buffer zone (e.g., 0).

Features for buffering may be points, lines, or polygons (Figure 11.1). Buffering around points creates circular buffer zones. Buffering around lines creates a series of elongated buffer zones around each line segment. And buffering around polygons creates buffer zones that extend outward from the polygon boundaries.

11.1.1 Variations in Buffering

There are several variations in buffering from those of Figure 11.1. The buffer distance or buffer size does not have to be constant; it can vary according to the values of a given field (Figure 11.2). For example, the width of the riparian buffer can vary depending on its expected function and the intensity of adjacent land use (Box 11.1). A feature may have more than one buffer zone. As an example, a nuclear power plant may be buffered with distances of 5, 10, 15, and 20 miles, thus forming multiple rings around the plant (Figure 11.3). Although the interval of each ring is the same at 5 miles, the rings are not equal in area. The second ring from the plant in fact covers an area about three times larger than the first ring. One must consider this area difference if the buffer zones are part of an evacuation plan.

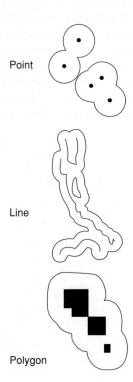

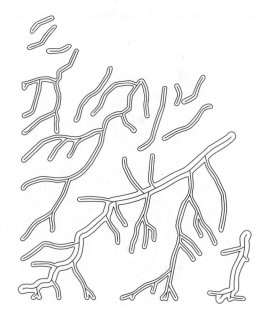

Figure 11.1
Buffering around points, lines, and polygons.

Figure 11.2
Buffering with different buffer distances.

Box 11.1 Riparian Buffer Width

Riparian buffers are strips of land along the banks of rivers and streams that can filter polluted runoff and provide a transition zone between water and human land use. Riparian buffers are also complex ecosystems that can protect wildlife habitat and fisheries. Depending on what the buffer is supposed to protect or provide, the buffer width can vary. According to various reports, a width of 100 feet is necessary for filtering dissolved nutrients and pesticides from runoff. A minimum of 100 feet is usually recommended for protecting fisheries, especially cold-water fisheries. And at least 300 feet is required for protecting wildlife habitat. Many states in the United States have adopted policies of grouping riparian buffers into different classes by width.

Buffering around line features does not have to be on both sides of the lines, it can be on either the left side or the right side of the line feature. (The left or right side is determined by the direction from the starting point to the end point of a line.) Likewise, buffer zones around polygons can be extended either outward or inward from the polygon boundaries. Boundaries of buffer zones may remain intact so that each buffer zone is a separate polygon. Or these boundaries may be dissolved, leaving no overlapped areas between buffer zones (Figure 11.4). Even the ends of buffer zones can be either rounded or flat.

Regardless of its variations, buffering uses distance measurements from select features to create the buffer zones. We must therefore know the measurement unit of select features (e.g., meters or feet) and, if necessary, input that information prior to buffering. ArcGIS, for example, can use the map unit (i.e., the measurement unit defined for the data set) as the default distance unit or allow the user to select a different distance unit such as miles instead of feet. Because buffering uses distance measurements from spatial features,

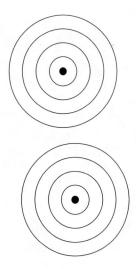

Figure 11.3
Buffering with four rings.

Figure 11.4
Buffer zones not dissolved (top) or dissolved (bottom).

the positional accuracy of spatial features in a data set also determines the accuracy of buffer zones.

Most GIS packages offer buffering as an analysis tool. ArcGIS, for example, offers a buffer tool and a multiple ring buffer tool. Because buffering requires only distance measurements, some GIS packages including ArcGIS also offer buffering as an editing tool (Chapter 7). Such a tool can create a buffer zone around each select feature but does not have the buffering options such as dissolving overlapping buffer zone boundaries.

11.1.2 Applications of Buffering

Buffering creates a buffer zone data set, which sets the buffering operation apart from the use of proximity measures for spatial data query (Chapter 10). Spatial data query using the proximity relationship can select spatial features that are located within a certain distance of other features but cannot create a buffer zone data set.

A buffer zone is often treated as a protection zone and is used for planning or regulatory purposes:

- A city ordinance may stipulate that no liquor stores or pornographic shops shall be within 1000 feet of a school or a church.
- Government regulations may set 2-mile buffer zones along streams to minimize sedimentation from logging operations.
- A national forest may restrict oil and gas well drilling within 500 feet of roads or highways.
- A planning agency may set aside land along the edges of streams to reduce the effects of nutrient, sediment, and pesticide runoff; to maintain shade to prevent the rise of stream temperature; and to provide shelter for wildlife and aquatic life (Thibault 1997; Dosskey 2002; Qiu 2003).
- A resource agency may establish stream buffers or vegetated filter strips to protect aquatic resources from adjacent agricultural land-use practices (Castelle et al. 1994; Daniels and Gilliam 1996; Zimmerman et al. 2003).

A buffer zone may be treated as a neutral zone and as a tool for conflict resolution. In controlling protesting groups, police may require protesters to be at least 300 feet from a building. Perhaps the best-known neutral zone is the demilitarized zone separating North Korea from South Korea along the 38°N parallel.

Sometimes buffer zones may represent the inclusion zones in GIS applications. For example, the siting criteria for an industrial park may stipulate that a potential site must be within 1 mile of a heavy-duty road. In this case, the 1-mile buffer zones of all heavy-duty roads become the inclusion zones. A city may also create buffer zones around available open access points (i.e., hot spots) to see the area coverage of wireless connection.

Rather than serving as a screening device, buffer zones themselves may become the object for analysis. A forest management plan may define areas that are within 300 meters of streams as riparian zones and evaluate riparian habitat for wildlife (Iverson et al. 2001). Another example comes from urban planning in developing countries, where urban expansion typically occurs near existing urban areas and major roads. Management of future urban growth therefore tends to concentrate on the buffer zones of existing urban areas and major roads.

Finally, buffering with multiple rings can be useful as a sampling method. A stream network can be buffered at a regular interval so that the composition and pattern of woody vegetation can be analyzed as a function of distance from the stream network (Schutt et al. 1999). One can also apply incremental banding to studies of land-use changes around urban areas.

11.2 OVERLAY

An **overlay** operation combines the geometries and attributes of two feature layers to create the output. The geometry of the output represents the geometric intersection of features from the input layers. Figure 11.5 illustrates an overlay operation with two polygon layers. Each feature on the output contains a combination of attributes from the input layers, and this combination differs from its neighbors.

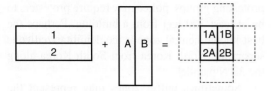

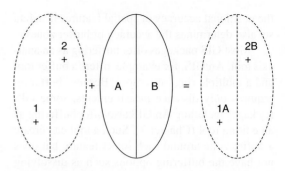

Figure 11.5

Overlay combines geometries and attributes from two layers into a single layer. The dashed lines are for illustration only and are not included in the output.

Feature layers to be overlaid must be spatially registered and based on the same coordinate system. In the case of the UTM (Universal Transverse Mercator) coordinate system or the SPC (State Plane Coordinate) system, the layers must also be in the same zone and have the same datum (e.g., NAD27 or NAD83).

11.2.1 Feature Type and Overlay

In practice, the first consideration for overlay is feature type. There are two groups of overlay operations. The first group uses two polygon layers as inputs. The second group uses one polygon layer and another layer, which may contain points or lines. Overlay operations can therefore be classified by feature type into point-in-polygon, line-in-polygon, and polygon-on-polygon. To distinguish the layers in the following discussion, the layer that may be a point, line, or polygon layer is called the input layer, and the layer that is a polygon layer is called the overlay layer.

In a **point-in-polygon overlay** operation, the same point features in the input layer are included in the output but each point is assigned with attributes of the polygon within which it falls (Figure 11.6). For example, a point-in-polygon overlay can find the association between wildlife locations and vegetation types.

In a **line-in-polygon overlay** operation, the output contains the same line features as in the input layer but each line feature is dissected by the polygon boundaries on the overlay layer (Figure 11.7). Thus the output has more line segments than does the input layer. Each line segment on the output

Figure 11.6

Point-in-polygon overlay. The input is a point layer. The output is also a point layer but has attribute data from the polygon layer.

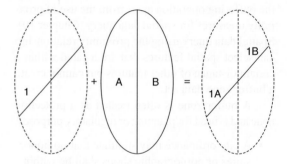

Figure 11.7

Line-in-polygon overlay. The input is a line layer. The output is also a line layer. But the output differs from the input in two aspects: the line is broken into two segments, and the line segments have attribute data from the polygon layer.

combines attributes from the input layer and the underlying polygon. For example, a line-in-polygon overlay can find soil data for a proposed road. The input layer includes the proposed road. The overlay layer contains soil polygons. And the output shows a dissected proposed road, each road segment having a different set of soil data from its adjacent segments.

The most common overlay operation is **polygon-on-polygon,** involving two polygon layers. The output combines the polygon boundaries from the input and overlay layers to create a new set of polygons (Figure 11.8). Each new polygon carries attributes from both layers, and these attributes differ from those of adjacent polygons. For

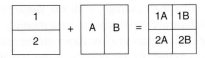

Figure 11.8

Polygon-on-polygon overlay. In the illustration, the two layers for overlay have the same area extent. The output combines the geometries and attributes from the two layers into a single polygon layer.

example, a polygon-on-polygon overlay can analyze the association between elevation zones and vegetation types.

11.2.2 Overlay Methods

Although they may appear in different names in different GIS packages, all overlay methods are based on the Boolean connectors AND, OR, and XOR. Intersect uses the AND connector. Union uses the OR connector. Symmetrical Difference or Difference uses the XOR connector. And Identity or Minus uses the following expression: [(input layer) AND (identity layer)] OR (input layer). The following explains in more detail these four common overlay methods.

Union preserves all features from the inputs (Figure 11.9). The area extent of the output combines the area extents of both input layers. Union requires that both input layers be polygon layers.

Intersect preserves only those features that fall within the area extent common to the inputs (Figure 11.10). The input layers may contain different feature types, although in most cases, one of them (the input layer) is a point, line, or polygon layer and the other (the overlay layer) is a polygon layer. Intersect is often a preferred method of overlay because any feature on its output has attribute data from both of its inputs. For example, a forest management plan may call for an inventory of vegetation types within riparian zones. Intersect will be a more efficient overlay method than Union in this case because the output contains only riparian zones with vegetation types.

Symmetrical Difference preserves features that fall within the area extent that is common to only one of the inputs (Figure 11.11). In

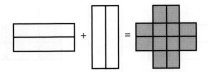

Figure 11.9

The Union method keeps all areas of the two input layers in the output.

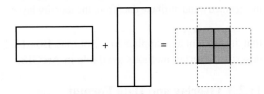

Figure 11.10

The Intersect method preserves only the area common to the two input layers in the output.

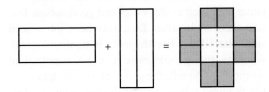

Figure 11.11

The Symmetrical Difference method preserves areas common to only one of the input layers in the output.

other words, Symmetrical Difference is opposite to Intersect in terms of the output's area extent. Symmetrical Difference requires that both input layers be polygon layers.

Identity preserves only features that fall within the area extent of the layer defined as the input layer (Figure 11.12). The other layer is called the identity layer. The input layer may contain points, lines, or polygons, and the identity layer is a polygon layer.

The choice of an overlay method becomes relevant only if the inputs have different area extents. If the input layers have the same area extent, then

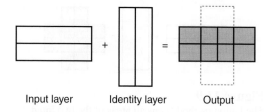

Input layer   Identity layer   Output

Figure 11.12
The Identity method produces an output that has the same extent as the input layer. But the output includes the geometry and attribute data from the identity layer.

that area extent also applies to the output. Box 11.2 describes overlay methods available in ArcGIS.

11.2.3 Overlay and Data Format

As the most recognizable, if not the most important, tool in GIS, overlay has a long history in GIS. Many concepts and methods developed for overlay are based on traditional, topological vector data such as coverages from Esri (Chapter 3). Other data such as shapefiles and geodatabase feature classes, also from Esri, have been introduced since the 1990s. The shapefile is nontopological, whereas the geodatabase can be topological on-the-fly. Overall, these newer vector data have not required a new way of dealing with overlay. But they have introduced some changes.

Unlike the coverage, both the shapefile and geodatabase allow polygons to have multiple components, which may also overlap with one another. This means that overlay operations can actually be applied to a single feature layer: Union creates a new feature by combining different polygons, and Intersect creates a new feature from the area where polygons overlap. But when used on a single layer, Union and Intersect are basically editing tools for creating new features (Chapter 7). They are not overlay tools because they do not perform geometric intersections or combine attribute data from different layers.

Many shapefile users are aware of a problem with the overlay output: the area and perimeter values are not automatically updated. In fact, the output contains two sets of area and perimeter values, one from each input layer. Task 1 of the applications section shows how to use a simple tool to update these values. A geodatabase feature class, on the other hand, does not have the same problem because its default fields of area (shape_area) and perimeter (shape_length) are automatically updated.

11.2.4 Slivers

A common error from overlaying polygon layers is **slivers,** very small polygons along correlated or shared boundary lines (e.g., the study area boundary)

Box 11.2 | **Overlay Methods in ArcGIS**

The offering of overlay methods differs in ArcGIS depending on whether the software in use is ArcView, the simplest version, or ArcInfo, the complete version. The Analysis Tools/Overlay toolset of ArcToolbox in ArcInfo includes Identity, Intersect, Symmetrical Difference, and Union. The same toolset in ArcView only has Intersect and Union. Another difference applies to the Intersect tool. ArcInfo accepts two or more input layers for an Intersect operation but ArcView can only have two input layers.

Because the inputs to an Intersect operation can be of different feature types, ArcInfo stipulates that the feature type of the output corresponds to the lowest dimension geometry of the inputs. For example, if the inputs consist of a point layer and two polygon layers, then the output is a point layer. But through overlay, the output point layer has attributes from the two polygon layers.

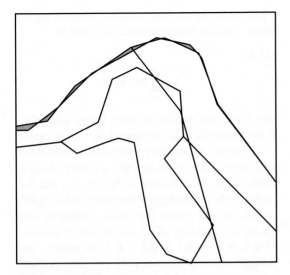

Figure 11.13
The top boundary has a series of slivers (shaded areas).
These slivers are formed between the coastlines from the
input layers in overlay.

of the input layers (Figure 11.13). The existence of
slivers often results from digitizing errors. Because
of the high precision of manual digitizing or scanning, the shared boundaries on the input layers are
rarely on top of one another. When the layers are
overlaid, the digitized boundaries intersect to form
slivers. Other causes of slivers include errors in
the source map or errors in interpretation. Polygon
boundaries on soil and vegetation maps are usually
interpreted from field survey data, aerial photographs, and satellite images. Wrong interpretations
can create erroneous polygon boundaries.

Most GIS packages incorporate some kind
of tolerance in overlay operations to remove slivers. ArcGIS, for example, uses the **cluster tolerance,** which forces points and lines to be snapped
together if they fall within the specified distance
(Figure 11.14). The cluster tolerance is either
defined by the user or based on a default value.
Slivers that remain on the output of an overlay
operation are those beyond the cluster tolerance.
Therefore, one option to reduce the sliver problem
is to increase the cluster tolerance. But because the
cluster tolerance applies to the entire layer, large

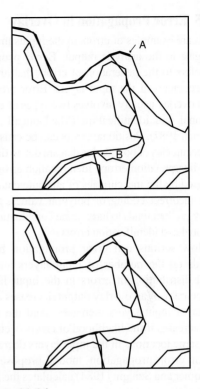

Figure 11.14
A cluster tolerance can remove many slivers along the
top boundary (A) but can also snap lines that are not
slivers (B).

tolerances will likely snap shared boundaries as
well as lines that are not shared on the input layers and eliminate legitimate small polygons on the
overlay output (Wang and Donaghy 1995).

Other options for dealing with the sliver problem include data preprocessing and postprocessing.
We can apply topology rules available to the geodatabase to the input layers, for example, to make sure
that their shared boundaries are coincident before the
overlay operation (Chapter 7). We can also apply the
concept of minimum mapping unit after the overlay
operation to remove slivers. The **minimum mapping unit** represents the smallest area unit that will
be managed by a government agency or an organization. For example, if a national forest adopts 5 acres
as its minimum mapping unit, then we can eliminate
any slivers smaller than 5 acres by combining them
with adjacent polygons (Section 11.5).

11.2.5 Error Propagation in Overlay

Slivers are examples of errors in the inputs that can propagate to the analysis output. **Error propagation** refers to the generation of errors that are due to inaccuracies of the input layers. Error propagation in overlay usually involves two types of errors: positional and identification (MacDougall 1975; Chrisman 1987). Positional errors can be caused by the inaccuracies of boundaries that are due to digitizing or interpretation errors. Identification errors can be caused by the inaccuracies of attribute data such as the incorrect coding of polygon values. Every overlay product tends to have some combinations of positional and identification errors.

How serious can error propagation be? It depends on the number of input layers and the spatial distribution of errors in the input layers. The accuracy of an overlay output decreases as the number of input layers increases. And the accuracy decreases if the likelihood of errors occurring at the same locations in the input layers decreases.

An error propagation model proposed by Newcomer and Szajgin (1984) calculates the probability of the event that the inputs are correct on an overlay output. The model suggests that the highest accuracy that can be expected of the output is equal to that of the least accurate layer among the inputs, and the lowest accuracy is equal to:

(11.1)

$$1 - \sum_{i=1}^{n} \Pr\left(E_i'\right)$$

where n is the number of input layers and $\Pr(E_i')$ is the probability that the input layer i is incorrect.

Suppose an overlay operation is conducted with three input layers and the accuracy levels of these layers are estimated to be 0.9, 0.8, and 0.7, respectively. According to Newcomer and Szajgin's model, we can expect the overlay output to have the highest accuracy of 0.7 and the lowest accuracy of 0.4, or $1 - (0.1 + 0.2 + 0.3)$. Newcomer and Szajgin's model illustrates the potential problem with error propagation in overlay. But it is a simple model, which, as shown in Box 11.3, can deviate significantly from the real-world operations.

11.2.6 Applications of Overlay

An overlay operation combines features and attributes from the input layers. The overlay output is useful for query and modeling purposes. Suppose an investment company is looking for a land parcel

***

Box 11.3   **Error Propagation Models**

N ewcomer and Szajgin's model on error propagation in overlay is simple and easy to follow. But it remains largely a conceptual model. First, the model is based on square polygons, a much simpler geometry than real data sets used in overlay. Second, the model deals only with the Boolean operation of AND, that is, input layer 1 is correct AND input layer 2 is correct. The Boolean operation of OR in overlay is different because it requires that only one of the input layers be correct (i.e., input layer 1 is correct OR input layer 2 is correct). Therefore, the probability of the event that the overlay output is correct actually increases as more input layers are overlaid (Veregin 1995). Third, Newcomer and Szajgin's model applies only to binary data, meaning that an input layer is either correct or incorrect. The model does not work with interval or ratio data and cannot measure the magnitude of errors. Modeling error propagation with numeric data is more difficult than with binary data (Arbia et al. 1998; Heuvelink 1998).

that is zoned commercial, not subject to flooding, and not more than 1 mile from a heavy-duty road. The company can first create the 1-mile road buffer and overlay the buffer zone layer with the zoning and floodplain layers. A subsequent query of the overlay output can select land parcels that satisfy the company's selection criteria. Many other examples of this type of application are included in the applications section of Chapter 11 and in Chapter 18.

A more specific application of overlay is to help solve the areal interpolation problem (Goodchild and Lam 1980). **Areal interpolation** involves transferring known data from one set of polygons (source polygons) to another (target polygons). For example, census tracts may represent source polygons with known populations in each tract from the U.S. Census Bureau, and school districts may represent target polygons with unknown population data. A common method for estimating the populations in each school district is called area-weighting, which includes the following steps (Figure 11.15):

- Overlay the layers of census tracts and school districts.
- Query the areal proportion of each census tract that is within each school district.

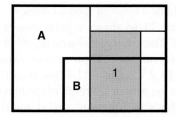

Figure 11.15

An example of areal interpolation. Thick lines represent census tracts and thin lines school districts. Census tract A has a known population of 4000 and B has 2000. The overlay result shows that the areal proportion of census tract A in school district 1 is 1/8 and the areal proportion of census tract B, 1/2. Therefore, the population in school district 1 can be estimated to be 1500, or [(4000 × 1/8) + (2000 × 1/2)].

- Apportion the population for each census tract to school districts according to the areal proportion.
- Sum the apportioned population from each census tract for each school district.

This method for areal interpolation, however, assumes a uniform population distribution within each census tract, which is usually unrealistic.

Recent studies have proposed statistical methods and the use of ancillary information such as street segments, satellite images, and land cover for improving areal interpolation (Flowerdew and Green 1989; Goodchild, Anselin, and Deichmann 1993; Flowerdew and Green 1995; Xie 1995; Fisher and Langford 1996; Holt, Lo, and Hodler 2004; Cai et al. 2006; Reibel and Agrawal 2007). In two recent studies, the street-weighted method has been shown to be more accurate than other areal interpolation methods (Hawley and Moellering 2005; Reibel and Bufalino 2005).

11.3 DISTANCE MEASUREMENT

Distance measurement refers to measuring straight-line (euclidean) distances between features. Measurements can be made from points in a layer to points in another layer, or from each point in a layer to its nearest point or line in another layer. In both cases, distance measures are stored in a field. A GIS package such as ArcGIS may include distance measurement as an analysis tool or as a join operation between two tables (Box 11.4).

Distance measures can be used directly for data analysis. Chang, Verbyla, and Yeo (1995), for example, use distance measures to test whether deer relocation points are closer to old-growth/clear-cut edges than random points located within the deer's relocation area. Fortney et al. (2000) use distance measures between home locations and medical providers to evaluate geographic access to health services.

Distance measures can also be used as inputs to data analysis. The gravity model, a spatial interaction model commonly used in migration studies and business applications, uses distance measures between points as the input. Pattern analysis covered in Section 11.4 also uses distance measures as inputs.

Box 11.4   **Distance Measurement Using ArcGIS**

One option for distance measurement in ArcGIS is called spatial join, or join data by location. The layer to assign data to must be a point layer. The layer to assign data from may be a point or line layer. And the spatial relationship between the two layers is nearest. In contrast to the join operation covered in Chapter 8, which joins attributes between tables, the spatial join method joins attributes of the closest point or line to the appropriate record of the point attribute table. In addition,

a new field called distance is added to the table after spatial join, showing the distance measurement.

ArcGIS also has the Near and Point Distance tools for distance measurement. Near calculates the distance between each feature in a feature layer and its nearest features in another layer. Point Distance measures the distance between each point in a point layer and all points in another layer. Point Distance does not use the nearest relationship.

11.4 PATTERN ANALYSIS

Pattern analysis refers to the use of quantitative methods for describing and analyzing the distribution pattern of spatial features. At the general level, a pattern analysis can reveal if a distribution pattern is random, dispersed, or clustered. A random pattern is a pattern in which the presence of a point at a location does not encourage or inhibit the occurrence of neighboring points. This spatial randomness separates a random pattern from a dispersed or clustered pattern. At the local level, a pattern analysis can detect if a distribution pattern contains local clusters of high or low values. Because pattern analysis can be a precursor to more formal and structured data analysis, some researchers have included pattern analysis as a data exploration activity (Murray et al. 2001; Haining 2003).

11.4.1 Point Pattern Analysis

A classic technique for point pattern analysis, **nearest neighbor analysis** uses the distance between each point and its closest neighboring point in a layer to determine if the point pattern is random, regular, or clustered (Clark and Evans 1954). The nearest neighbor statistic is the ratio (R) of the observed average distance between nearest

neighbors (d_{obs}) to the expected average for a hypothetical random distribution (d_{exp}):

(11.2)

$$R = \frac{d_{obs}}{d_{exp}}$$

The R ratio is less than 1 if the point pattern is more clustered than random, and greater than 1 if the point pattern is more dispersed than random. Nearest neighbor analysis can also produce a Z score, which indicates the likelihood that the pattern could be a result of random chance.

Figure 11.16 shows a point pattern of deer locations. The result of a nearest neighbor analysis shows an R ratio of 0.58 (more clustered than random) and a Z score of -11.4 (less than 1% likelihood that the pattern is a result of random chance).

Ripley's K-function, called multidistance spatial cluster analysis in ArcGIS, is another popular method for analyzing point patterns (Ripley 1981; Boots and Getis 1988; Bailey and Gatrell 1995). It can identify clustering or dispersion over a range of distances, thus setting it apart from nearest-neighbor analysis. In practice and for simplification of interpretation, a standardized version of Ripley's K-function, called L function, is commonly used. The observed L function at distance d,

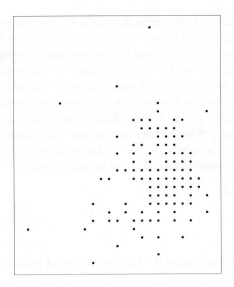

Figure 11.16
A point pattern showing deer locations.

without edge correction, can be computed by (Ripley 1981):

(11.3)

$$L(d) = \sqrt{ \left[A \sum_{i=1}^{N} \sum_{i=1, j \neq i}^{N} k(i, j) \right] \Big/ \left[\pi N(N-1) \right] }$$

where A is the size of the study area, N is the number of points, and π is a mathematical constant. In Eq. (11.3), the summation of $k(i, j)$ measures the number of j points within distance d of all i points: $k(i, j)$ is 1 when the distance between i and j is less than or equal to d and 0 when the distance is greater than d. The expected $L(d)$ for a random point pattern is d (Boots and Getis 1988). A point pattern is more clustered than random at distance d if the computed $L(d)$ is higher than expected, and a point pattern is more dispersed than random if the computed $L(d)$ is less than expected.

The computed $L(d)$ can be affected by points near the edges of the study area. Different algorithms for the edge correction are available (Li and Zhang 2007). ArcGIS, for example, offers the following three methods: simulated outer boundary values, reduced analysis area, and Ripley's edge

correction formula. Because of the edge correction, it is difficult to formally assess the statistical significance of the computed $L(d)$. Instead, the lower and upper envelopes of $L(d)$ can be derived from executing a number of simulations, starting each simulation by randomly placing N points within the study area. If the computed $L(d)$ is above the upper simulation envelope, then it is highly likely that the point pattern is clustered at distance d; if the computed $L(d)$ is below the lower simulation envelope, then it is highly likely that the point pattern is dispersed at distance d.

Table 11.1 shows the expected and computed $L(d)$ and the difference between them for the deer location data in Figure 11.16. Distance d ranges from 100 to 750 meters, with an increment of 50 meters. Clustering is observed at all distances from 100 to 700 meters, but its peak occurs at 250 meters. Figure 11.17 plots the computed $L(d)$ and the lower and upper simulation envelopes. The computed $L(d)$ lies above the upper envelope from 100 to 650 meters, thus confirming empirically the clustered point pattern.

TABLE 11.1	**Expected $L(d)$, Observed $L(d)$, and Their Difference for Deer Location Data**	
Expected $L(d)$ (1)	**Observed $L(d)$** (2)	(2)−(1)
100	239.3	139.3
150	323.4	173.4
200	386.8	186.8
250	454.5	204.5
300	502.7	202.7
350	543.9	193.9
400	585.1	185.1
450	621.5	171.5
500	649.5	149.5
550	668.3	118.3
600	682.9	82.9
650	697.1	47.1
700	704.9	4.9
750	713.7	−36.3

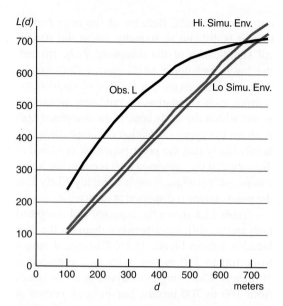

Figure 11.17
The computed $L(d)$ and the lower and upper simulation envelopes.

11.4.2 Moran's I for Measuring Spatial Autocorrelation

Point pattern analysis uses only distances between points as inputs. Analysis of spatial autocorrelation, on the other hand, considers both the point locations and the variation of an attribute at the locations. **Spatial autocorrelation** therefore measures the relationship among values of a variable according to the spatial arrangement of the values (Cliff and Ord 1973). The relationship may be described as highly correlated if like values are spatially close to each other, and independent or random if no pattern can be discerned from the arrangement of values. Spatial autocorrelation is also called *spatial association* or *spatial dependence*.

A popular measure of spatial autocorrelation is **Moran's I,** which can be computed by:

(11.4)

$$I = \frac{\sum_{i=1}^{n}\sum_{j=1}^{m} w_{ij}\left(x_i - \bar{x}\right)\left(x_j - \bar{x}\right)}{s^2 \sum_{i=1}^{n}\sum_{j=1}^{m} w_{ij}}$$

where x_i is the value at point i, x_j is the value at point i's neighbor j, w_{ij} is a coefficient, n is the number of points, and s^2 is the variance of x values with a mean of $\bar{x}$. The coefficient w_{ij} is the weight for measuring spatial autocorrelation. Typically, w_{ij} is defined as the inverse of the distance (d) between points i and j, or $1/d_{ij}$. Other weights such as the inverse of the distance squared can also be used.

The values Moran's I takes on are anchored at the expected value $E(I)$ for a random pattern:

(11.5)

$$E(I) = \frac{-1}{n - 1}$$

$E(I)$ approaches 0 when the number of points n is large.

Moran's I is close to $E(I)$ if the pattern is random. It is greater than $E(I)$ if adjacent points tend to have similar values (i.e., are spatially correlated) and less than $E(I)$ if adjacent points tend to have different values (i.e., are not spatially correlated). Similar to nearest neighbor analysis, we can compute the Z score associated with a Moran's I. The Z score indicates the likelihood that the point pattern could be a result of random chance.

Figure 11.18 shows the same deer locations as Figure 11.16 but with the number of sightings as the attribute. This point distribution produces a Moran's I of 0.1, which is much higher than $E(I)$ of 0.00962, and a Z score of 11.7, which suggests that the likelihood of the pattern being a result of random chance is less than 1 percent.

Moran's I can also be applied to polygons. Eq. (11.4) remains the same for computing the index value, but the w_{ij} coefficient is based on the spatial relationship between polygons. One option is to assign 1 to w_{ij} if polygon i is adjacent to polygon j and 0 if i and j are not adjacent to each other. Another option, which is used in ArcGIS, converts polygons to polygon centroids (the geometric centers of polygons) and then assigns the distance measures between centroids to w_{ij}.

Figure 11.19 shows the percentage of Latino population in Ada County, Idaho, by block group. This data set produces a Moran's I of 0.05, which

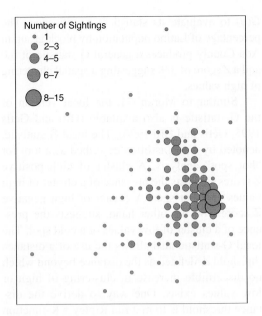

Figure 11.18
A point pattern showing deer locations and the number of sightings at each location.

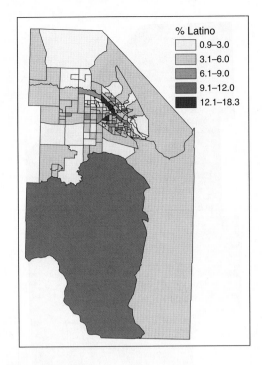

Figure 11.19
Percent Latino population by block group in Ada County, Idaho. Boise is located in the upper center of the map with small-sized block groups.

is higher than $E(I)$ of 0.00685, and a Z score of 6.7, which suggests that the likelihood of the pattern being a result of random chance is less than 1 percent. We can therefore conclude that adjacent block groups tend to have similar percentages of Latino population, either high or low. (With the exception of the two large but sparsely populated block groups to the south, the high percentages of Latino population are clustered near Boise.)

Recent developments in spatial statistics have included **Local Indicators of Spatial Association (LISA)** (Anselin 1995). A local version of Moran's I, LISA calculates for each feature (point or polygon) an index value and a Z score. A high positive Z score suggests that the feature is adjacent to features of similar values, either above the mean or below the mean. A high negative Z score indicates that the feature is adjacent to features of dissimilar values. Figure 11.20 shows a cluster of highly similar values (i.e., high percentages of Latino population) near Boise. Around the cluster are small pockets of highly dissimilar values.

11.4.3 G-Statistic for Measuring High/Low Clustering

Moran's I, either general or local, can only detect the presence of the clustering of similar values. It cannot tell whether the clustering is made of high values or low values. This has led to the use of the **G-statistic,** which can separate clusters of high values from clusters of low values (Getis and Ord 1992). The general G-statistic based on a specified distance, d, is defined as:

(11.6)

$$G(d) = \frac{\sum \sum w_{ij}(d)x_i x_j}{\sum \sum x_i x_j}, \ i \neq j$$

where x_i is the value at location i, x_j is the value at location j if j is within d of i, and $w_{ij}(d)$ is the spatial weight. The weight can be based on some

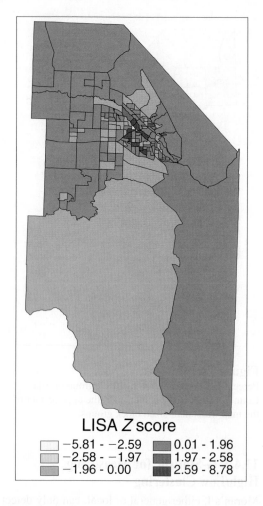

LISA Z score

☐ −5.81 - −2.59	■ 0.01 - 1.96
▨ −2.58 - −1.97	■ 1.97 - 2.58
▨ −1.96 - 0.00	■ 2.59 - 8.78

Figure 11.20
Z scores for the Local Indicators of Spatial Association (LISA) by block group in Ada County, Idaho.

weighted distance such as inverse distance or 1 and 0 (for adjacent and nonadjacent polygons).

The expected value of $G(d)$ is:

(11.7)

$$E\left(G\right) = \frac{\sum\sum w_{ij}(d)}{n(n-1)}$$

$E(G)$ is typically a very small value when n is large.

A high $G(d)$ value suggests a clustering of high values, and a low $G(d)$ value suggests a clustering of low values. A Z score can be computed for a $G(d)$ to evaluate its statistical significance. The percentage of Latino population by block group in Ada County produces a general G-statistic of 0.0 and a Z score of 3.9, suggesting a spatial clustering of high values.

Similar to Moran's I, the local version of the G-statistic is also available (Ord and Getis 1995; Getis and Ord 1996). The local G-statistic, denoted by $G_i(d)$, is often described as a tool for "hot spot" analysis. A cluster of high positive Z scores suggests the presence of a cluster of high values or a hot spot. A cluster of high negative Z scores, on the other hand, suggests the presence of a cluster of low values or a cold spot. The local G-statistic also allows the use of a distance threshold d, defined as the distance beyond which no discernible increase in clustering of high or low values exists. One way to derive the distance threshold is to first run Ripley's K-function (Section 11.4.1) and to determine the distance where the clustering ceases.

Figure 11.21 shows the Z scores of local G-statistics for the distribution of Latino population in Ada County. The map shows a clear hot spot in Boise but no cold spots in the county.

11.4.4 Applications of Pattern Analysis

Pattern analysis has many applications. Nearest neighbor analysis and Ripley's K-function are standard methods for analyzing the spatial distribution and structure of plant species (Wiegand and Moloney 2004; Li and Zhang 2007). K-function has also been applied to other types of spatial data, including industrial firms (Marcon and Puech 2003). Hot spot analysis is a standard tool for mapping and analyzing crime locations (Murray et al. 2001; Ceccato, Haining, and Signoretta 2002; Chainey, Thompson, and Uhlig 2008). Standard software packages for crime mapping and analysis, such as CrimeStat **(http:// www.icpsr.umich.edu/NACJD/crimestat .htm)**, include the local G-statistic for hot spot analysis. Pattern analysis is also important for analyzing public health data (Lam, Fan, and Liu 1996; Jacquez and Greiling 2003; Greiling et al. 2005), postwar health symptoms (Proctor et al. 2005), pest

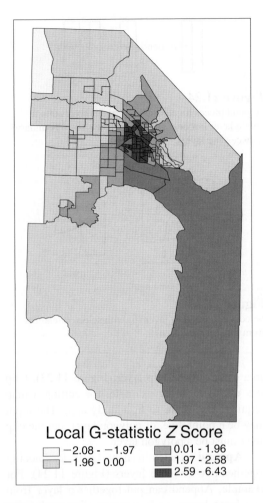

Local G-statistic *Z* Score

☐ −2.08 - −1.97	▦ 0.01 - 1.96
☐ −1.96 - 0.00	▨ 1.97 - 2.58
	▧ 2.59 - 6.43

Figure 11.21

Z scores for the local G-statistics by block group in Ada County, Idaho.

management (Beckler, French, and Chandler 2005), and housing prices (Getis and Ord 1992).

Spatial autocorrelation is useful for analyzing temporal changes of spatial distributions (Goovaerts and Jacquez 2005; Tsai et al. 2006). Likewise, it is useful for quantifying the spatial dependency over distance classes (Overmars, de Koning, and Veldkamp 2003). Spatial autocorrelation is also important for validating the use of standard statistical tests. Statistical inference typically applies to controlled experiments, which are seldom used in geographic research (Goodchild 2009). If the data exhibit significant spatial autocorrelation, it should encourage the researcher to incorporate spatial dependence into the analysis (Legendre 1993; Malczewski and Poetz 2005).

11.5 FEATURE MANIPULATION

Tools are available in a GIS package for manipulating and managing features in one or more feature layers. Like overlay, these feature tools are often needed for data preprocessing and data analysis; however, unlike overlay, these tools do not combine geometries and attributes from input layers into a single layer. Feature manipulation is easy to follow graphically, even though terms describing the various tools may differ between GIS packages. Box 11.5 summarizes map manipulation tools in ArcGIS.

Dissolve aggregates features in a feature layer that have the same attribute value or values (Figure 11.22). For example, we can aggregate roads by highway number or counties by state.

Box 11.5	Feature Manipulation Using ArcGIS

Because of their diverse applications, feature manipulation tools are found in different toolsets in ArcToolbox and in different versions of Arc-GIS. Erase and Update are in the Analysis Tools/ Overlay toolset. Clip, Select, and Split are in the Analysis Tools/Extract toolset. Append is in the Data Management Tools/General toolset. And Dissolve and Eliminate are in the Data Management Tools/Generalization toolset. Erase, Update, Split, and Eliminate require the ArcInfo license, whereas Clip, Select, Append, and Dissolve require only the ArcView license.

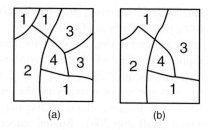

(a)   (b)

Figure 11.22
Dissolve removes boundaries of polygons that have the same attribute value in *(a)* and creates a simplified layer *(b)*.

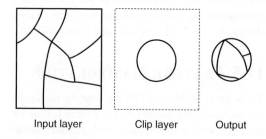

Input layer   Clip layer   Output

Figure 11.23
Clip creates an output that contains only those features of the input layer that fall within the area extent of the clip layer.

An important application of Dissolve is to simplify a classified polygon layer. Classification groups values of a selected attribute into classes and makes obsolete boundaries of adjacent polygons, which have different values initially but are now grouped into the same class. Dissolve can remove these unnecessary boundaries and creates a new, simpler layer with the classification results as its attribute values. Another application is to aggregate both spatial and attribute data of the input layer. For instance, to dissolve a county layer, we can choose state name as the attribute to dissolve and number of high schools to aggregate. The output is a state layer with an attribute showing the total number of high schools by state.

Clip creates a new layer that includes only those features of the input layer that fall within the

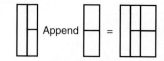

Append

Figure 11.24
Append pieces together two adjacent layers into a single layer but does not remove the shared boundary between the layers.

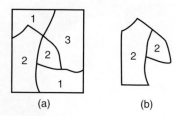

(a)   (b)

Figure 11.25
Select creates a new layer *(b)* with selected features from the input layer *(a)*.

area extent of the clip layer (Figure 11.23). Clip is a useful tool, for example, for cutting a map acquired elsewhere to fit a study area. The input may be a point, line, or polygon layer, but the clip layer must be a polygon layer.

Append creates a new layer by piecing together two or more layers (Figure 11.24). For example, Append can put together a layer from four input layers, each corresponding to the area extent of a USGS 7.5-minute quadrangle. The output can then be used as a single layer for data query or display.

Select creates a new layer that contains features selected from a user-defined query expression (Figure 11.25). For example, we can create a layer showing high-canopy closure by selecting stands that have 60 to 80 percent closure from a stand layer.

Eliminate creates a new layer by removing features that meet a user-defined query expression (Figure 11.26). For example, Eliminate can implement the minimum mapping unit concept by removing polygons that are smaller than the defined unit in a layer.

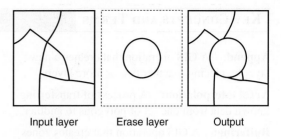

Input layer   Erase layer   Output

Figure 11.28
Erase removes features from the input layer that fall within the area extent of the erase layer.

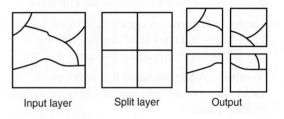

Input layer   Split layer   Output

Figure 11.29
Split uses the geometry of the split layer to divide the input layer into four separate layers.

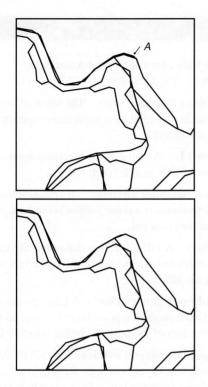

Figure 11.26
Eliminate removes some small slivers along the top boundary (*A*).

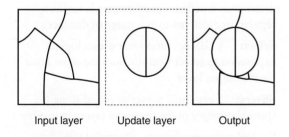

Input layer   Update layer   Output

Figure 11.27
Update replaces the input layer with the update layer and its features.

Update uses a "cut and paste" operation to replace the input layer with the update layer and its features (Figure 11.27). As the name suggests, Update is useful for updating an existing layer with new features in limited areas. It is a better option than redigitizing the entire map.

Erase removes from the input layer those features that fall within the area extent of the erase layer (Figure 11.28). Suppose a suitability analysis stipulates that potential sites cannot be within 300 meters of any stream. A stream buffer layer can be used in this case as the erase layer to remove itself from further consideration.

Split divides the input layer into two or more layers (Figure 11.29). A split layer, which shows area subunits, is used as the template for dividing the input layer. For example, a national forest can split a stand layer by district so that each district office can have its own layer.

In ArcGIS, Clip and Split are also editing tools. These editing tools work with individual features. For example, the editing tool of Split splits a line at a specified location or a polygon along a line sketch. The tool does not work with layers. It is therefore important that we understand the function of a tool before using it.

KEY CONCEPTS AND TERMS

Append: A GIS operation that creates a new layer by piecing together two or more layers.

Areal interpolation: A process of transferring known data from one set of polygons to another.

Buffering: A GIS operation that creates zones consisting of areas within a specified distance of select features.

Clip: A GIS operation that creates a new layer including only those features of the input layer that fall within the area extent of the clip layer.

Cluster tolerance: A distance tolerance that forces points and lines to be snapped together if they fall within the specified distance.

Dissolve: A GIS operation that removes boundaries between polygons that have the same attribute value(s).

Eliminate: A GIS operation that creates a new layer by removing features that meet a user-defined logical expression from the input layer.

Erase: A GIS operation that removes from the input layer those features that fall within the area extent of the erase layer.

Error propagation: The generation of errors in the overlay output that are due to inaccuracies of the input layers.

G-statistic: A spatial statistic that measures the clustering of high and low values in a data set. The G-statistic can be either general or local.

Identity: An overlay method that preserves only features that fall within the area extent defined by the input layer.

Intersect: An overlay method that preserves only those features falling within the area extent common to the input layers.

Line-in-polygon overlay: A GIS operation in which a line layer is dissected by the polygon boundaries on the overlay layer, and each line segment on the output combines attributes from the line layer and the polygon within which it falls.

Local Indicators of Spatial Association (LISA): The local version of Moran's I.

Minimum mapping unit: The smallest area unit that is managed by a government agency or an organization.

Moran's I: A statistic that measures spatial autocorrelation in a data set.

Nearest neighbor analysis: A spatial statistic that determines if a point pattern is random, regular, or clustered.

Overlay: A GIS operation that combines the geometries and attributes of the input layers to create the output.

Point-in-polygon overlay: A GIS operation in which each point of a point layer is assigned the attribute data of the polygon within which it falls.

Polygon-on-polygon overlay: A GIS operation in which the output combines the polygon boundaries from the inputs to create a new set of polygons, each carrying attributes from the inputs.

Ripley's K-function: A spatial statistic that determines whether a point pattern is random, regular, or clustered over a range of distances.

Select: A GIS operation that uses a logical expression to select features from the input layer for the output layer.

Slivers: Very small polygons found along the shared boundary of the two input layers in overlay.

Spatial autocorrelation: A spatial statistic that measures the relationship among values of a variable according to the spatial arrangement of the values. Also called *spatial association* or *spatial dependence.*

Split: A GIS operation that divides the input layer into two or more layers.

Symmetrical Difference: An overlay method that preserves features falling within the area that is common to only one of the input layers.

Union: A polygon-on-polygon overlay method that preserves all features from the input layers.

Update: A GIS operation that replaces the input layer with the update layer and its features.

REVIEW QUESTIONS

1. Define a buffer zone.
2. Describe *three* variations in buffering.
3. Provide an application example of buffering from your discipline.
4. Describe a point-in-polygon overlay operation.
5. A line-in-polygon operation produces a line layer, which typically has more records (features) than the input line layer. Why?
6. Provide an example of a polygon-on-polygon overlay operation from your discipline.
7. Describe a scenario in which Intersect is preferred over Union for an overlay operation.
8. Suppose the input layer shows a county and the overlay layer shows a national forest. Part of the county overlaps the national forest. We can express the output of an Intersect operation as [county] AND [national forest]. How can you express the outputs of a Union operation and an Identity operation?
9. Define slivers from an overlay operation.
10. What is a minimum mapping unit? And, how can a minimum mapping unit be used to deal with the sliver problem?
11. Although many slivers from an overlay operation represent inaccuracies in the digitized boundaries, they can also represent the inaccuracies of attribute data (i.e., identification errors). Provide an example for the latter case.
12. Explain the areal interpolation problem by using an example from your discipline.
13. Both nearest neighbor analysis and Moran's I can apply to point features. How do they differ in terms of input data?
14. Explain spatial autocorrelation in your own words.
15. Both Moran's I and the G-statistic have the global (general) and local versions. How do these two versions differ in terms of pattern analysis?
16. The local G-statistic can be used as a tool for hot spot analysis. Why?
17. What does a Dissolve operation accomplish?
18. Suppose you have downloaded a vegetation map from the Internet. But the map is much larger than your study area. Describe the steps you will follow to get the vegetation map for your study area.
19. Suppose you need a map showing toxic waste sites in your county. You have downloaded a shapefile from the Environmental Protection Agency (EPA) website that shows toxic waste sites in every county of your state. What kind of operation will you use on the EPA map so that you can get only the county you need?

APPLICATIONS: VECTOR DATA ANALYSIS

This applications section has four tasks. Task 1 covers the basic tools of vector data analysis including Buffer, Overlay, and Select. Because ArcGIS does not automatically update the area and perimeter values of an overlay output in shapefile format, Task 1 also uses the Calculate Geometry tool to update the area and perimeter values. Task 2 covers overlay operations with multicomponent polygons. Task 3 introduces two different options for measuring distances between point and line features. Task 4 deals with spatial autocorrelation.

Task 1 Perform Buffering and Overlay

What you need: shapefiles of *landuse, soils,* and *sewers.*

Task 1 simulates GIS analysis for a real-world project. The task is to find a suitable site for a new university aquaculture lab by using the following selection criteria:

- Preferred land use is brushland (i.e., LUCODE = 300 in *landuse.shp*).
- Choose soil types suitable for development (i.e., SUIT $>= 2$ in *soils.shp*).
- Site must be within 300 meters of sewer lines.

1. Start ArcCatalog, and connect to the Chapter 11 database. Launch ArcMap. Add *sewers.shp, soils.shp,* and *landuse.shp* to Layers, and rename Layers Task 1. All three shapefiles are measured in meters.

2. First buffer *sewers.* Click ArcToolbox window to open it. Select Environments from the context menu of ArcToolbox, and set the Chapter 11 database to be the current workspace. Double-click the Buffer tool in the Analysis Tools/Proximity toolset. In the Buffer dialog, select *sewers* for the input features, enter *sewerbuf.shp* for the output feature class, enter 300 (meters) for the distance, select ALL for the dissolve type, and click OK. Open the attribute table of *sewerbuf.* The table has only one record for the dissolved buffer zone.

Q1. What is the definition of Side Type in the Buffer dialog?

3. Next overlay *soils, landuse,* and *sewerbuf.* Double-click the Intersect tool in the Analysis Tools/Overlay toolset. Select *soils, landuse,* and *sewerbuf* for the input features. (If you are using the ArcInfo version of ArcGIS, you can input all three layers; otherwise, overlay two layers at a time.)

Enter *final.shp* for the output feature class. Click OK to run the operation.

Q2. How is the XY Tolerance defined in the Intersect dialog?

Q3. How many records does *final* have?

4. The final step is to select from *final* those polygons that meet the first two criteria. Double-click the Select tool in the Analysis Tools/Extract toolset. Select *final* for the input features, name the output feature class *sites.shp,* and click the SQL button for Expression. In the Query Builder dialog, enter the following expression in the expression box: "SUIT" $>= 2$ AND "LUCODE" = 300. Click OK to dismiss the dialogs.

Q4. How many parcels are included in *sites?*

5. Open the attribute table of *sites.* Notice that the table contains two sets of area and perimeter. Moreover, each field contains duplicate values. This is because ArcGIS Desktop does not automatically update the area and perimeter values of the output shapefile. An easy option to get the updated values is to convert *sites.shp* to a geodatabase feature class. The feature class will have the updated values in the fields shape_area and shape_length. For this task, you will use a simple tool to perform the update. Close the attribute table of *sites.*

6. Double-click the Add Field tool in the Data Management Tools/Fields toolset. Select *sites* for the input table, enter Shape_Area for the field name, select Double for the field type, enter 11 for the field precision, enter 3 for the field scale, and click OK. Use the same tool and the same field definition to add Shape_Leng as a new field to *sites.*

7. Right-click Shape_Area in the attribute table of *sites* and select Calculate Geometry. Click Yes to do a calculate outside of an edit session. In the Calculate Geometry dialog, select Area for the property and square

meters for units. Click OK. Shape_Area is now populated with correct area values.

8. Right-click Shape_Leng and select Calculate Geometry. Click Yes in the next dialog. In the Calculate Geometry dialog, select Perimeter for the property and Meters for units. Click OK. Shape_Leng is now populated with correct perimeter values.

Q5. What is the sum of Shape_Area values in *sites.shp*?

Task 2 Overlay Multicomponent Polygons

What you need: *boise_fire, fire1986,* and *fire1992,* three feature classes in the *regions* feature dataset of *boise_fire.mdb. boise_fire* records forest fires in the Boise National Forest from 1908 to 1996, *fire1986* fires in 1986, and *fire1992* fires in 1992.

Task 2 lets you use multipart polygon features (Chapter 3) in overlay operations. Both *fire1986* and *fire1992* are polygon layers derived from *boise_fire.* An overlay of multipart polygons results in an output with fewer features (records), thus simplifying the data management task.

1. Insert a new data frame in ArcMap, and rename it Task 2. Add the feature dataset *regions* to Task 2. Open the attribute table of *boise_fire.* Historical fires are recorded by year in YEAR1 to YEAR6 and by name in NAME1 to NAME6. The multiple fields for year and name are necessary because a polygon can have multiple fires in the past. Open the attribute table of *fire1986.* It has only one record, although the layer actually contains seven simple polygons. The same is true with *fire1992.*

2. First union *fire1986* and *fire1992* in an overlay operation. Double-click the Union tool in the Analysis Tools/Overlay toolset. Select *fire1986* and *fire1992* for the input features, and enter *fire_union* for the output feature class in the *regions* feature dataset. Click OK to run the operation. Open the attribute table of *fire_union.*

Q6. Explain what each record in *fire_union* represents.

3. Next intersect *fire1986* and *fire1992.* Double-click the Intersect tool in the Analysis Tools/ Overlay toolset. Select *fire1986* and *fire1992* for the input features, and enter *fire_intersect* for the output feature class. Click OK to run the operation.

Q7. Explain what the single record in *fire_intersect* represents.

Task 3 Measure Distances Between Points and Lines

What you need: *deer.shp* and *edge.shp.*

Task 3 asks you to measure each deer location in *deer.shp* to its closest old-growth/clear-cut edge in *edge.shp.* There are two options to complete the task. The first option is to use the join data by location method, which is used for this task. The other option is to use the Near tool in the Analysis Tools/Proximity toolset (available to the ArcInfo version), which is explained at the end of the task.

1. Insert a new data frame in ArcMap, and rename it Task 3. Add *deer.shp* and *edge.shp* to Task 3.

2. Right-click *deer,* point to Joins and Relates, and select Join. Click the first dropdown arrow in the Join Data dialog, and select to join data from another layer based on spatial location. Make sure that *edge* is the layer to join to *deer.* Click the radio button stating that each point will be given all the attributes of the line that is closest to it, and a distance field showing how close that line is. Specify *deer_edge.shp* for the output shapefile. Click OK to run the operation.

3. Right-click *deer_edge* and open its attribute table. The field to the far right of the table is Distance, which lists for each deer location the distance to its closest edge.

Q8. How many deer locations are within 50 meters of their closest edge?

4. The Near tool uses a dialog to get the input features (*deer*), the near features (*edge*), and the optional search radius. The Near tool does not have the option of creating a new output data set. The result of the analysis is stored in the attribute table of *deer,* which may also include location (*x*- and *y*-coordinates of the near feature) and angle (the angle between the input and near features), in addition to distance.

Task 4 Compute General and Local G-Statistics

What you need: *adabg00.shp,* a shapefile containing block groups from Census 2000 for Ada County, Idaho.

In Task 4, you will first determine if a spatial clustering of Latino population exists in Ada County. Then you will test to see if any local "hot spots" of Latino population exist in the county.

1. Insert a new data frame in ArcMap. Rename the new data frame Task 4, and add *adabg00.shp* to Task 4.

2. Right-click *adabg00,* and select Properties. On the Symbology tab, choose Quantities/ Graduated colors to display the field values of Latino. Zoom in to the top center of the map, where Boise is located, and examine the spatial distribution of Latino population. The large block group to the southwest has a high percentage of Latino population (11%) but the block group's population is just slightly over 4600. The visual dominance of large area units is a shortcoming of the choropleth map.

Q9. What is the range of % Latino in Ada County?

3. Open ArcToolbox. You will first compute the general G-statistic. Double-click the High/ Low Clustering (Getis-Ord General G) tool in the Spatial Statistics Tools/Analyzing Patterns toolset. Select *adabg00* for the input feature class, select Latino for the input field, check the box for General Report, and take defaults for the other fields. Click OK to execute the command.

4. After the operation is complete, select Results from the Geoprocessing menu. Under Current Session, expand High/Low Clustering (Getis-Ord General G) and then double-click the HTML Report File. At the top of the window, it lists the observed general G-statistic, the Z score, the probability value, and an interpretation of the result. Close the window.

5. Next you will run the local G-statistic. Double-click the Hot Spot Analysis (Getis-Ord Gi*) tool in the Spatial Statistics Tools/ Mapping Clusters toolset. Select *adabg00* for the input feature class, select Latino for the input field, enter *local_g.shp* for the output feature class in the Chapter 11 database, and specify a distance band of 5000 (meters). Click OK to execute the command.

6. Open the attribute table of *local_g.* The field GiZScore stores the Z Score, and the field GiPValue stores the probability value, for each block group.

Q10. What is the value range of GiZScore?

7. On the map you can "see a 'hot spot'" in Boise and another in the large block group to the southwest.

Challenge Task

What you need: *lochsa.mdb,* a personal geodatabase containing two feature classes for the Lochsa area of the Clearwater National Forest in Idaho.

lochsa_elk in the geodatabase has a field called USE that shows elk habitat use in summer or winter. *lt_prod* has a field called Prod that shows five timber productivity classes derived from the land type data, with 1 being most productive and 5 being least productive. Some polygons in *lt_prod* have the Prod value of -99 indicating absence of data. Also, *lochsa_elk* covers a larger area than *lt_prod* due to the difference in data availability.

This challenge task asks you to prove, or disprove, the statement that "the winter habitat area tends to have a higher area percentage of the productivity classes 1 and 2 than the summer habitat area." In other words, you need to find the answer to the following two questions.

Q1. What is the area percentage of the summer habitat area with the Prod value of 1 or 2?

Q2. What is the area percentage of the winter habitat area with the Prod value of 1 or 2?

REFERENCES

Anselin, L. 1995. Local Indicators of Spatial Association—LISA. *Geographical Analysis* 27: 93–116.

Arbia, G., D. A. Griffith, and R. P. Haining. 1998. Error Propagation Modeling in Raster GIS: Overlay Operations. *International Journal of Geographical Information Science* 12: 145–67.

Bailey, T. C., and A. C. Gatrell. 1995. *Interactive Spatial Data Analysis*. Harlow, England: Longman Scientific & Technical.

Beckler, A. A., B. W. French, and L. D. Chandler. 2005. Using GIS in Areawide Pest Management: A Case Study in South Dakota. *Transactions in GIS* 9: 109–27.

Boots, B. N., and A. Getis. 1988. *Point Pattern Analysis*. Newbury Park, CA: Sage Publications.

Cai, Q., G. Rushton, B. Bhaduri, E. Bright, and P. Coleman. 2006. Estimating Small-Area Populations by Age and Sex Using Spatial Interpolation and Statistical Inference Methods. *Transactions in GIS* 10: 577–98.

Castelle, A. J., A. W. Johnson, and C. Conolly. 1994. Wetland and Stream Buffer Requirements: A Review. *Journal of Environmental Quality* 23: 878–82.

Ceccato, V., R. Haining, and P. Signoretta. 2002. Exploring Offence Statistics in Stockholm City Using Spatial Analysis Tools. *Annals of the Association*

of American Geographers 92: 29–51.

Chainey, S. P., L. Thompson, and S. Uhlig. 2008. The Utility of Hotspot Mapping for Predicting Spatial Patterns of Crime. *Security Journal* 21: 4–28.

Chang, K., D. L. Verbyla, and J. J. Yeo. 1995. Spatial Analysis of Habitat Selection by Sitka Black-Tailed Deer in Southeast Alaska, USA. *Environmental Management* 19: 579–89.

Chrisman, N. R. 1987. The Accuracy of Map Overlays: A Reassessment. *Landscape and Urban Planning* 14: 427–39.

Clark, P. J., and F. C. Evans. 1954. Distance to Nearest Neighbor as a Measure of Spatial Relationships in Populations. *Ecology* 35: 445–53.

Cliff, A. D., and J. K. Ord. 1973. *Spatial Autocorrelation*. New York: Methuen.

Daniels, R. B., and J. W. Gilliam. 1996. Sediment and Chemical Load Reduction by Grass and Riparian Filters. *Soil Science Society of America Journal* 60: 246–51.

Dosskey, M. G. 2002. Setting Priorities for Research on Pollution Reduction Functions of Agricultural Buffers. *Environmental Management* 30: 641–50.

Fisher, P. F., and M. Langford. 1996. Modeling Sensitivity to Accuracy in Classified Imagery:

A Study of Areal Interpolation by Dasymetric Mapping. *The Professional Geographer* 48: 299–309.

Flowerdew, R., and M. Green. 1989. Statistical Methods for Inference Between Incompatible Zonal Systems. In M. F. Goodchild and S. Gopal, eds., *Accuracy of Spatial Databases,* pp. 21–34. London: Taylor and Francis.

Flowerdew, R., and M. Green. 1995. Areal Interpolation and Types of Data. In S. Fotheringham and P. Rogerson, eds., *Spatial Analysis and GIS,* pp. 121–45. London: Taylor and Francis.

Fortney, J., K. Rost, and J. Warren. 2000. Comparing Alternative Methods of Measuring Geographic Access to Health Services. *Health Services & Outcomes Research Methodology* 1: 173–84.

Getis, A., and J. K. Ord. 1992. The Analysis of Spatial Association by Use of Distance Statistics. *Geographical Analysis* 24: 189–206.

Getis, A., and J. K. Ord. 1996. Local Spatial Statistics: An Overview. In P. Longley and M. Batty, eds., *Spatial Analysis: Modelling in a GIS Environment,* pp. 261–77. Cambridge, England: GeoInformation International.

Goodchild, M. F. 2009. What Problem? Spatial Autocorrelation and Geographic Information Science. *Geographical Analysis* 41: 411–17.

Goodchild, M. F., L. Anselin, and U. Deichmann. 1993. A Framework for the Areal Interpolation of Socioeconomic Data. *Environment and Planning A* 25: 383–97.

Goodchild, M. F., and N. S. Lam. 1980. Areal Interpolation: A Variant of the Traditional Spatial Problem. *Geoprocessing* 1: 293–312.

Goovaerts, P., and G. M. Jacquez. 2005. Detection of Temporal Changes in the Spatial Distribution of Cancer Rates Using Local Moran's I and Geostatistically Simulated Spatial Neutral Models. *Journal of Geographical Systems* 7: 137–59.

Greiling, D. A., G. M. Jacquez, A. M. Kaufmann, and R. G. Rommel. 2005. Space-Time Visualization and Analysis in the Cancer Atlas Viewer. *Journal of Geographical Systems* 7: 67–84.

Haining, R. 2003. *Spatial Data Analysis: Theory and Practice.* Cambridge, England: Cambridge University Press.

Hawley, K., and H. Moellering. 2005. A Comparative Analysis of Areal Interpolation Methods. *Cartography and Geographic Information Science* 32: 411–23.

Heuvelink, G. B. M. 1998. *Error Propagation in Environmental Modeling with GIS.* London: Taylor and Francis.

Holt, J. B., C. P. Lo, and T. W. Hodler. 2004. Dasymetric Estimation of Population Density and Areal Interpolation of Census Data. *Cartography and Geographic Information Science* 31: 103–21.

Iverson, L. R., D. L. Szafoni, S. E. Baum, and E. A. Cook. 2001. A Riparian Wildlife Habitat Evaluation Scheme Developed Using GIS. *Environmental Management* 28: 639–54.

Jacquez, G. M., and D. A. Grieling. 2003. Local Clustering in Breast, Lung, and Colorectal Cancer in Long Island, New York. *International Journal of Health Geographics* 2: 3. (Open access at **http://www .ij-healthgeographics.com/ content/2/1/3).**

Lam, N. S., M. Fan, and K. Liu. 1996. Spatial-Temporal Spread of the AIDS Epidemic, 1982–1990: A Correlogram Analysis of Four Regions of the United States. *Geographical Analysis* 28: 93–107.

Legendre, P. 1993. Spatial Autocorrelation: Trouble or New Paradigm? *Ecology* 74: 1659–73.

Li, F., and L. Zhang. 2007. Comparison of Point Pattern Analysis Methods for Classifying the Spatial Distributions of Spruce-Fir Stands in the North-East USA. *Forestry* 80: 337–49.

MacDougall, E. B. 1975. The Accuracy of Map Overlays. *Landscape Planning* 2: 23–30.

Malczewski, J., and A. Poetz. 2005. Residential Burglaries and Neighborhood Socioeconomic Context in London, Ontario: Global and Local Regression Analysis. *The Professional Geographer* 57: 516–29.

Marcon, E., and F. Puech. 2003. Evaluating the Geographic Concentration of Industries Using Distance-Based Methods. *Journal of Economic Geography* 3: 409–28.

Murray, A. T., I. McGuffog, J. S. Western, and P. Mullins.

2001. Exploratory Spatial Data Analysis Techniques for Examining Urban Crime. *British Journal of Criminology* 41: 309–29.

Newcomer, J. A., and J. Szajgin. 1984. Accumulation of Thematic Map Errors in Digital Overlay Analysis. *The American Cartographer* 11: 58–62.

Ord, J. K., and A. Getis. 1995. Local Spatial Autocorrelation Statistics: Distributional Issues and an Application. *Geographical Analysis* 27: 286–306.

Overmars, K. P., G. H. J. de Koning, and A. Veldkamp. 2003. Spatial Autocorrelation in Multi-Scale Land Use Models. *Ecological Modelling* 164: 257–70.

Proctor, S. P., A. Imai, D. Ozonoff, S. Gopal, J. Wolfe, and R. F. White. 2005. Spatial Analysis of 1991 Gulf War Troop Locations in Relationship with Postwar Health Symptom Reports Using GIS Techniques. *Transactions in GIS* 9: 381–96.

Qiu, Z. 2003. A VSA-Based Strategy for Placing Conservation Buffers in Agricultural Watersheds. *Environmental Management* 32: 299–311.

Reibel, M., and A. Agrawal. 2007. Areal Interpolation of Population Counts Using Pre-classified Land Cover Data. *Population Research and Policy Review* 26: 619–633.

Reibel, M., and M. E. Bufalino. 2005. Street-Weighted Interpolation Techniques for Demographic Count Estimation in Incompatible Zone Systems. *Environment and Planning A* 37: 127–39.

Ripley, B. D. 1981. *Spatial Statistics*. New York: Wiley.

Schutt, M. J., T. J. Moser, P. J. Wigington, Jr., D. L. Stevens, Jr., L. S. McAllister, S. S. Chapman, and T. L. Ernst. 1999. Development of Landscape Metrics for Characterizing Riparian-Stream Networks. *Photogrammetric Survey and Remote Sensing* 65: 1157–67.

Thibault, P. A. 1997. Ground Cover Patterns Near Streams for Urban Land Use Categories. *Landscape and Urban Planning* 39: 37–45.

Tsai, B., K. Chang, C. Chang, and C. Chu. 2006. Analyzing Spatial and Temporal Changes of Aquaculture in Yunlin County, Taiwan. *The Professional Geographer* 58: 161–71.

Veregin, H. 1995. Developing and Testing of an Error Propagation Model for GIS Overlay Operations. *International Journal of Geographical Information Systems* 9: 595–619.

Wang, F., and P. Donaghy. 1995. A Study of the Impact of Automated Editing on Polygon Overlay Analysis Accuracy. *Computers & Geosciences* 21: 1177–85.

Wiegand, T., and K. A. Moloney. 2004. Rings, Circles, and Null-Models for Point Pattern Analysis in Ecology. *Oikos* 104: 209–29.

Xie, Y. 1995. The Overlaid Network Algorithms for Areal Interpolation Problem. *Computer, Environment and Urban Systems* 19: 287–306.

Zimmerman, J. K. H., B. Vondracek, and J. Westra. 2003. Agricultural Land Use Effects on Sediment Loading and Fish Assemblages in Two Minnesota (USA) Watersheds. *Environmental Management* 32: 93–105.

CHAPTER 12

RASTER DATA ANALYSIS

The raster data model uses a regular grid to cover the space and the value in each grid cell to represent the characteristic of a spatial phenomenon at the cell location. This simple data structure of a raster with fixed cell locations not only is computationally efficient, but also facilitates a large variety of data analysis operations.

In contrast with vector data analysis, which uses points, lines, and polygons, raster data analysis uses cells and rasters. Raster data analysis can be performed at the level of individual cells, or groups of cells, or cells within an entire raster. Some raster data operations use a single raster; others use two or more rasters. An important consideration in raster data analysis is the type of cell value. Statistics such as mean and standard deviation are designed for numeric values, whereas others such as majority (the most frequent cell value) are designed for both numeric and categorical values.

Various types of data are stored in raster format. Raster data analysis, however, operates only on software-specific raster data such as Esri grids in ArcGIS. Therefore, to use some digital elevation models (DEMs) and other raster data in data analysis, we must process them first and convert them to software-specific raster data.

Chapter 12 covers the basic tools for raster data analysis. Section 12.1 describes the analysis environment including the area for analysis and the output cell size. Section 12.2 through 12.5 cover four common types of raster data analysis: local operations, neighborhood operations, zonal operations, and physical distance measures. Section 12.6 covers operations that do not fit into the common classification of raster data analysis. Section 12.7 introduces map algebra, which allows complex raster data operations. Section 12.8 uses overlay and buffering as examples to compare vector- and raster-based operations.

12.1 DATA ANALYSIS ENVIRONMENT

The analysis environment refers to the area for analysis and the output cell size. The area extent for analysis may correspond to a specific raster, or an area defined by its minimum and maximum *x*-, *y*-coordinates, or a combination of rasters. Given a combination of rasters with different area extents, the area extent for analysis can be based on the union or intersect of the rasters. The union option uses an area extent that encompasses all input rasters, whereas the intersect option uses an area extent that is common to all input rasters.

An **analysis mask** can also determine the area extent for analysis. An analysis mask limits analysis to cells that do not carry the cell value of "no data." No data differs from zero. Zero is a valid cell value, whereas no data means the absence of data. For example, an elevation raster converted from a DEM may contain no-data cells along its border. But in many other cases the user enters no-data cells intentionally to limit the area extent for analysis. For example, one option to limit the analysis of soil erosion to only private lands is to code public lands with no data. An analysis mask can be based on a feature layer or a raster (Box 12.1).

We can define the output cell size at any scale deemed suitable. Typically, the output cell size is set to be equal to, or larger than, the largest cell size among the input rasters. This follows the rationale that the resolution of the output should correspond to that of the lowest-resolution input raster (Chapter 4). For instance, if the input cell sizes range from 10 to 30 meters, the output cell size should be 30 meters or larger. Given the specified output cell size, a GIS package uses a resampling technique to convert all input rasters to that cell size prior to data analysis. Common resampling methods are nearest neighbor, bilinear interpolation, and cubic convolution (Chapter 6).

12.2 LOCAL OPERATIONS

Constituting the core of raster data analysis, **local operations** are cell-by-cell operations. A local operation can create a new raster from either a single input raster or multiple input rasters. The cell values of the new raster are computed by a function relating the input to the output or are assigned by a classification table.

Box 12.1 | **How to Make an Analysis Mask**

The source of an analysis mask can be either a feature layer or a raster. A feature layer showing the outline of a study area can be used as an analysis mask, thus limiting the extent of raster data analysis to the study area. A raster to be used as an analysis mask must have cell values within the area of interest and no data on the outside. If necessary, reclassification can assign no data to the outside area. After an analysis mask is made, it can be applied to raster data analysis by defining it in the Environment Settings menu of ArcToolbox.

12.2.1 Local Operations with a Single Raster

Given a single raster as the input, a local operation computes each cell value in the output raster as a mathematical function of the cell value in the input raster. As shown in Figure 12.1, a large number of mathematical functions are available in a GIS package.

Converting a floating-point raster to an integer raster, for example, is a simple local operation that uses the integer function to truncate the cell value at the decimal point on a cell-by-cell basis. Converting a slope raster measured in percent to one measured in degrees is also a local operation but requires a more complex mathematical expression. In Figure 12.2, the expression [slope_d] = 57.296 × arctan ([slope_p]/100) can convert *slope_p* measured in percent to *slope_d* measured

Arithmetic	+, −, /, *, absolute, integer, floating-point
Logarithmic	exponentials, logarithms
Trigonometric	sin, cos, tan, arcsin, arccos, arctan
Power	square, square root, power

Figure 12.1
Arithmetic, logarithmic, trigonometric, and power functions for local operations.

15.2	16.0	18.5
17.8	18.3	19.6
18.0	19.1	20.2

(a)

8.64	9.09	10.48
10.09	10.37	11.09
10.20	10.81	11.42

(b)

Figure 12.2
A local operation can convert a slope raster from percent (*a*) to degrees (*b*).

in degrees. Because computer packages typically use radian instead of degree in trigonometric functions, the constant 57.296 ($360/2\pi$, $\pi = 3.1416$) changes the angular measure to degrees.

12.2.2 Reclassification

A local operation, **reclassification** creates a new raster by classification. Reclassification is also referred to as recoding, or transforming, through lookup tables (Tomlin 1990). Two reclassification methods may be used. The first method is a one-to-one change, meaning that a cell value in the input raster is assigned a new value in the output raster. For example, irrigated cropland in a land-use raster is assigned a value of 1 in the output raster. The second method assigns a new value to a range of cell values in the input raster. For example, cells with population densities between 0 and 25 persons per square mile in a population density raster are assigned a value of 1 in the output raster and so on. An integer raster can be reclassified by either method, but a floating-point raster can only be reclassified by the second method.

Reclassification serves three main purposes. First, reclassification can create a simplified raster. For example, instead of having continuous slope values, a raster can have 1 for slopes of 0 to 10%, 2 for 10 to 20%, and so on. Second, reclassification can create a new raster that contains a unique category or value such as irrigated cropland or slopes of 10 to 20%. Third, reclassification can create a new raster that shows the ranking of cell values in the input raster. For example, a reclassified raster can show the ranking of 1 to 5, with 1 being least suitable and 5 being most suitable.

12.2.3 Local Operations with Multiple Rasters

Local operations with multiple rasters are also referred to as compositing, overlaying, or superimposing maps (Tomlin 1990). Because local operations can work with multiple rasters, they are the equivalent of vector-based overlay operations.

A greater variety of local operations have multiple input rasters than have a single input raster.

Besides mathematical functions that can be used on individual rasters, other measures that are based on the cell values or their frequencies in the input rasters can also be derived and stored in the output raster. Some of these measures are, however, limited to rasters with numeric data.

Summary statistics, including maximum, minimum, range, sum, mean, median, and standard deviation, are measures that apply to rasters with numeric data. Figure 12.3, for example, shows a local operation that calculates the mean from three input rasters. If a cell contains no data in one of the input rasters, the cell also carries no data in the output raster by default.

Other measures that are suitable for rasters with numeric or categorical data are statistics such as majority, minority, and number of unique values. For each cell, a majority output raster tabulates the most frequent cell value among the input rasters, a minority raster tabulates the least frequent cell value, and a variety raster tabulates the number of different cell values. Figure 12.4, for example, shows the output with the majority statistics from three input rasters.

Some local operations do not involve statistics or computation. A local operation called Combine assigns a unique output value to each unique combination of input values. Suppose a slope raster has three cell values (0 to 20%, 20 to 40%, and greater than 40% slope), and an aspect raster has four cell values (north, east, south, and west aspects). The Combine operation creates an output raster with a value for each unique combination of slope and aspect, such as 1 for greater than 40% slope and the south aspect, 2 for 20 to 40% slope and the south aspect, and so on (Figure 12.5).

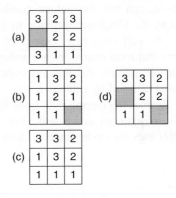

Figure 12.4

The cell value in (d) is the majority statistic derived from three input rasters (a, b, and c) in a local operation. The shaded cells have no data.

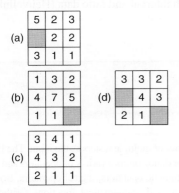

Figure 12.3

The cell value in (d) is the mean calculated from three input rasters (a, b, and c) in a local operation. The shaded cells have no data.

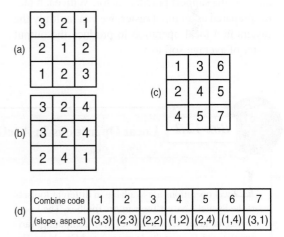

Figure 12.5

Each cell value in (c) represents a unique combination of cell values in (a) and (b). The combination codes and their representations are shown in (d).

Box 12.2 describes local operations available in ArcGIS and the organization of tools and commands for local operations in the software package.

12.2.4 Applications of Local Operations

As the core of raster data analysis, local operations have many applications. A change detection study, for example, can use the unique combinations produced by the Combine operation to trace the change of the cell value (e.g., change of vegetation cover). But local operations are perhaps most useful for GIS models that require mathematical computation on a cell-by-cell basis.

The Revised Universal Soil Loss Equation (RUSLE) (Wischmeier and Smith 1978; Renard et al. 1997) uses six environmental factors in the equation:

(12.1)

$$A = R\,K\,L\,S\,C\,P$$

where A is the average soil loss, R is the rainfall–runoff erosivity factor, K is the soil erodibility factor, L is the slope length factor, S is the slope steepness factor, C is the crop management factor, and P is the support practice factor. With each factor prepared as an input raster, we can multiply the rasters in a local operation to produce the output raster of average soil loss.

A study of favorable wolf habitat uses the following logit model (Mladenoff et al. 1995):

(12.2)

$$\text{logit}\,(p) = -6.5988 + 14.6189\,R,$$

and

$$p = 1/[1 + e^{\text{logit}(p)}]$$

where p is the probability of occurrence of a wolf pack, R is road density, and e is the natural exponent. $\text{logit}\,(p)$ can be calculated in a local operation using a road density raster as the input. Likewise, p can be calculated in another local operation using $\text{logit}\,(p)$ as the input.

Because rasters are superimposed in local operations, error propagation can be an issue in interpreting the output. Raster data do not directly involve digitizing errors. (If raster data are converted from vector data, digitizing errors with vector data are carried into raster data.) Instead, the main source of errors is the quality of the cell value, which in turn can be traced to other data sources. For example, if raster data are converted from satellite images, statistics for assessing the classification accuracy of satellite images can be used to assess the quality of raster data (Congalton 1991; Veregin 1995). But these statistics are based on binary data (i.e., correctly or incorrectly classified). It is more difficult to model error propagation with interval and ratio data (Heuvelink 1998).

Box 12.2   **Local Operations in ArcGIS**

Local operations are available in different toolsets of the Spatial Analyst Tools in ArcToolbox. The Local toolset includes Cell Statistics (for computing various statistics from multiple rasters) and Combine. The Reclass toolset has a number of tools for data reclassification. The Map Algebra toolset has Raster Calculator. Raster Calculator is a versatile tool, which can take a single raster or multiple rasters as the input. The Raster Calculator dialog offers a variety of operators and functions: arithmetic operators, logical operators, Boolean connectors, and mathematical functions (arithmetic, logarithmic, trigonometric, and power functions). We can use these operators and functions directly to prepare a single map algebra expression (Section 12.7).

12.3 NEIGHBORHOOD OPERATIONS

A **neighborhood operation,** also called a focal operation, involves a focal cell and a set of its surrounding cells. The surrounding cells are chosen for their distance and/or directional relationship to the focal cell. Common neighborhoods include rectangles, circles, annuluses, and wedges (Figure 12.6). A rectangle is defined by its width and height in cells, such as a 3-by-3 area centered at the focal cell. A circle extends from the focal cell with a specified radius. An annulus or doughnut-shaped neighborhood consists of the ring area between a smaller circle and a larger circle centered at the focal cell. And a wedge consists of a piece of a circle centered at the focal cell. As shown in Figure 12.6, some cells are only partially covered in the defined neighborhood. The general rule is to include a cell if the center of the cell falls within the neighborhood.

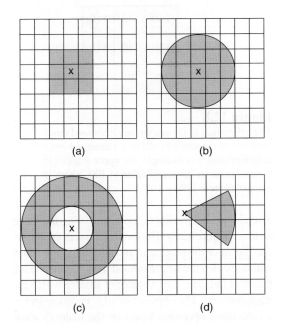

(a)   **(b)**

(c)   **(d)**

Figure 12.6

Four common neighborhood types: rectangle (*a*), circle (*b*), annulus (*c*), and wedge (*d*). The cell marked with an x is the focal cell.

12.3.1 Neighborhood Statistics

A neighborhood operation typically uses the cell values within the neighborhood, with or without the focal cell value, in computation, and then assigns the computed value to the focal cell. To complete a neighborhood operation on a raster, the focal cell is moved from one cell to another until all cells are visited. Different rules devised by GIS software developers are applied to focal cells on the margin of a raster, where a neighborhood such as a 3-by-3 rectangle cannot be used. Although a neighborhood operation works on a single raster, its process is similar to that of a local operation with multiple rasters. Instead of using cell values from different input rasters, a neighborhood operation uses the cell values from a defined neighborhood.

The output from a neighborhood operation can show summary statistics including maximum, minimum, range, sum, mean, median, and standard deviation, as well as tabulation of measures such as majority, minority, and variety. These statistics and measures are the same as those from local operations with multiple rasters.

A **block operation** is a neighborhood operation that uses a rectangle (block) and assigns the calculated value to all block cells in the output raster. Therefore, a block operation differs from a regular neighborhood operation because it does not move from cell to cell but from block to block. Box 12.3 describes neighborhood operations available in ArcGIS and their intended use.

12.3.2 Applications of Neighborhood Operations

An important application of neighborhood operations is data simplification. The moving average method, for instance, reduces the level of cell value fluctuation in the input raster (Figure 12.7). The method typically uses a 3-by-3 or a 5-by-5 rectangle as the neighborhood. As the neighborhood is moved from one focal cell to another, the average of cell values within the neighborhood is computed and assigned to the focal cell. The output raster of moving averages represents a generalization of the original cell values. Another example is a

Box 12.3 | Neighborhood Operations in ArcGIS

ArcToolbox offers a Neighborhood toolset in Spatial Analyst Tools. Among the tools in the toolset, the Focal Statistics tool calculates neighborhood statistics by using the inputs of statistic type and neighborhood. The statistic options include minimum, maximum, mean, median, sum, range, standard deviation, majority, minority, and variety. The neighborhood options include rectangle, circle, doughnut, and wedge. In contrast, the Block Statistics tool calculates statistics by block (e.g., 3-by-3 rectangle).

1	2	2	2	2
1	2	2	2	3
1	2	1	3	3
2	2	2	3	3
2	2	2	2	3

(a)

1.56	2.00	2.22
1.67	2.11	2.44
1.67	2.11	2.44

(b)

Figure 12.7
The cell values in (b) are the neighborhood means of the shaded cells in (a) using a 3-by-3 neighborhood. For example, 1.56 in the output raster is calculated from (1 + 2 + 2 + 1 + 2 + 2 + 1 + 2 + 1)/9.

200	200	110	210	210
200	200	110	210	210
150	150	100	170	170
140	140	130	160	160
140	140	130	160	160

(a)

100	110	110
100	110	110
50	70	70

(b)

Figure 12.8
The cell values in (b) are the neighborhood range statistics of the shaded cells in (a) using a 3-by-3 neighborhood. For example, the upper-left cell in the output raster has a cell value of 100, which is calculated from (200 − 100).

neighborhood operation that uses variety as a measure, tabulates the number of different cell values in the neighborhood, and assigns the number to the focal cell. One can use this method to show, for example, the variety of vegetation types or wildlife species in an output raster.

Neighborhood operations are common in image processing. These operations are variously called filtering, convolution, or moving window operations for spatial feature manipulation (Lillesand et al. 2004). Edge enhancement, for example, can use a range filter,

essentially a neighborhood operation using the range statistic (Figure 12.8). The range measures the difference between the maximum and minimum cell values within the defined neighborhood. A high range value therefore indicates the existence of an edge within the neighborhood. The opposite of edge enhancement is a smoothing operation based on the majority measure (Figure 12.9). The majority operation assigns the most frequent cell value within the neighborhood to the focal cell, thus creating a smoother raster than the original raster.

Another area of study that heavily depends on neighborhood operations is terrain analysis. The slope, aspect, and surface curvature measures

(a)

(b)

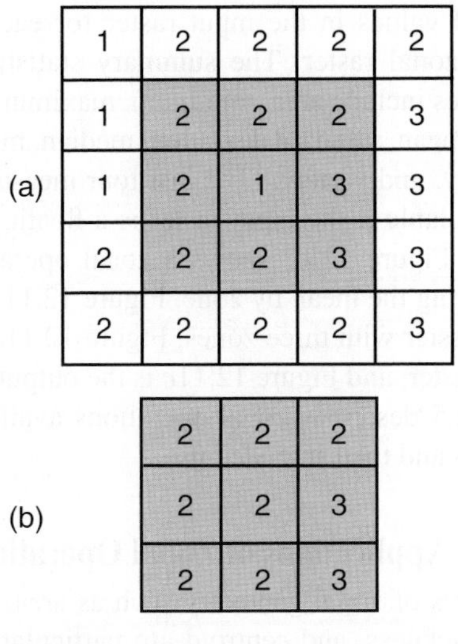

Figure 12.9
The cell values in (b) are the neighborhood majority statistics of the shaded cells in (a) using a 3-by-3 neighborhood. For example, the upper-left cell in the output raster has a cell value of 2 because there are five 2s and four 1s in its neighborhood.

of a cell are all derived from neighborhood operations using elevations from its adjacent neighbors (i.e., a 3-by-3 rectangle) (Chapter 13). For some studies, the definition of a neighborhood can extend far beyond a cell's immediate neighbors (Box 12.4).

Neighborhood operations can also be important to studies that need to select cells by their neighborhood characteristics. For example, installing a gravity sprinkler irrigation system requires information about elevation drop within a circular neighborhood of a cell. Suppose that a system requires an elevation drop of 130 feet within a distance of 0.5 mile to make it financially feasible. A neighborhood operation on an elevation raster can answer the question by using a circle with a radius of 0.5 mile as the neighborhood and (elevation) range as the statistic. A query of the output raster can show which cells meet the criterion.

Because of its ability to summarize statistics within a defined area, a neighborhood operation can be used to select sites that meet a study's specific criteria. An example is that by Crow, Host, and Mladenoff (1999), which uses neighborhood operations to select a stratified random sample of 16 plots, representing two ownerships located within two regional ecosystems.

Box 12.4 | **More Examples of Neighborhood Operations**

Although most neighborhood operations for terrain analysis use a 3-by-3 neighborhood, there are exceptions. For example, a regression model developed by Beguería and Vicente-Serrano (2006) for predicting precipitation in a climatically complex area used the following spatial variables, each to be derived from a neighborhood operation:

• Mean elevation within a circle of 2.5 kilometers and 25 kilometers

• Mean slope within a circle of 2.5 kilometers and 25 kilometers
• Mean relief energy within a circle of 2.5 kilometers and 25 kilometers (maximum elevation – elevation at the focal cell)
• Barrier effect to the four cardinal directions (maximum elevation within a wedge of radius of 2.5/25 kilometers and mean direction north/ south/west/east – elevation at the focal cell)

12.4 ZONAL OPERATIONS

A **zonal operation** works with groups of cells of same values or like features. These groups are called zones. Zones may be contiguous or noncontiguous. A contiguous zone includes cells that are spatially connected, whereas a noncontiguous zone includes separate regions of cells.

12.4.1 Zonal Statistics

A zonal operation may work with a single raster or two rasters. Given a single input raster, zonal operations measure the geometry of each zone in the raster, such as area, perimeter, thickness, and centroid (Figure 12.10). The area is the sum of the cells that fall within the zone times the cell size. The perimeter of a contiguous zone is the length of its boundary, and the perimeter of a noncontiguous zone is the sum of the length of each region. The thickness calculates the radius (in cells) of the largest circle that can be drawn within each zone. And the centroid is the geometric center of a zone located at the intersection of the major axis and the minor axis of an ellipse that best approximates the zone.

Given two rasters in a zonal operation, one input raster and one zonal raster, a zonal operation produces an output raster, which summarizes the cell values in the input raster for each zone in the zonal raster. The summary statistics and measures include area, minimum, maximum, sum, range, mean, standard deviation, median, majority, minority, and variety. (The last four measures are not available if the input raster is a floating-point raster.) Figure 12.11 shows a zonal operation of computing the mean by zone. Figure 12.11*b* is the zonal raster with three zones, Figure 12.11*a* is the input raster, and Figure 12.11*c* is the output raster. Box 12.5 describes zonal operations available in ArcGIS and their intended use.

12.4.2 Applications of Zonal Operations

Measures of zonal geometry such as area, perimeter, thickness, and centroid are particularly useful for studies of landscape ecology (Forman and Godron 1986; McGarigal and Marks 1994). Many other geometric measures can be derived from area

Zone	Area	Perimeter	Thickness
1	36,224	1708	77.6
2	48,268	1464	77.4

Figure 12.10

Thickness and centroid for two large watersheds (zones). Area is measured in square kilometers, and perimeter and thickness are measured in kilometers. The centroid of each zone is marked with an x.

1	2	2	1
1	4	5	1
2	3	7	6
1	3	4	4

(a)

1	1	2	2
1	1	2	2
1	1	3	3
3	3	3	3

(b)

2.17	2.17	2.25	2.25
2.17	2.17	2.25	2.25
2.17	2.17	4.17	4.17
4.17	4.17	4.17	4.17

(c)

Figure 12.11

The cell values in (*c*) are the zonal means derived from an input raster (*a*) and a zonal raster (*b*). For example, 2.17 is the mean of {1, 1, 2, 2, 4, 3} for zone 1.

Box 12.5 | **Zonal Operations in ArcGIS**

ArcToolbox offers the Zonal toolset in Spatial Analyst Tools. The Zonal Statistics tool can perform zonal operations with a zonal data set and a value raster and calculate the summary statistics of area, minimum, maximum, range, mean, standard deviation, sum, variety, majority, minority, and median by zone. The Zonal Histogram tool can plot a series of histograms based on the cell values in the value raster for each zone in the zonal data set. And the Zonal Geometry tool can measure for each zone in a data set the specified geometry measure.

and perimeter. For example, a shape index called the areal roundness is defined as 354 times the square root of a zone's area divided by its perimeter (Tomlin 1990). Essentially, the areal roundness compares the shape of a contiguous zone to a circle. The area roundness value approaches 100 for a circular shape and 0 for a highly distorted shape. The expression for computing areal roundness is: 354 × Sqrt (zonalarea ([raster1]))/zonalperimeter ([raster1]), where raster1 is the input raster and Sqrt is the square root function.

Zonal operations with two rasters can generate useful descriptive statistics for comparison purposes. For example, to compare topographic characteristics of different soil textures, we can use a soil raster that contains the categories of sand, loam, and clay as the zonal raster and slope, aspect, and elevation as the input rasters. By running a series of zonal operations, we can summarize the slope, aspect, and elevation characteristics associated with the three soil textures.

12.5 PHYSICAL DISTANCE MEASURE OPERATIONS

In a GIS project, distances may be expressed as physical distances or cost distances. The **physical distance** measures the straight-line or Euclidean distance, whereas the **cost distance** measures the cost for traversing the physical distance. The distinction between the two types of distance measures is important in real-world applications.

A truck driver, for example, is more interested in the time or the fuel cost for covering a route than in its physical distance. The cost distance in this case is based on not only the physical distance but also the speed limit and road condition. Chapter 12 covers the physical distance measure operations, and Chapter 17 covers the cost distance for least-cost path analysis.

Physical distance measure operations calculate straight-line distances away from cells designated as the source cells. For example, to get the distance between cells (1, 1) and (3, 3) in Figure 12.12, we can use the following formula:

$$\text{cell size} \times \sqrt{(3-1)^2 + (3-1)^2}$$

or cell size × 2.828. If the cell size were 30 meters, the distance would be 84.84 meters.

A physical distance measure operation essentially buffers the source cells with wavelike continuous

(0, 0)

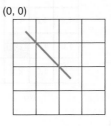

Figure 12.12
A straight-line distance is measured from a cell center to another cell center. This illustration shows the straight-line distance between cell (1,1) and cell (3,3).

distances over the entire raster (Figure 12.13) or to a specified maximum distance. This is why physical distance measure operations are also called extended neighborhood operations (Tomlin 1990) or global (i.e., the entire raster) operations.

Some GIS packages including ArcGIS allow the use of a vector-based data set (e.g., a shapefile) as the source in a physical distance measure operation. This option is based on the consideration of convenience because the data set is converted from vector to raster data before the operation starts.

The continuous distance raster from a physical distance measure operation can be used directly in subsequent operations. But it can also be further

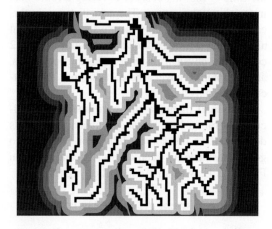

Figure 12.13
Continuous distance measures from a stream network.

processed to create a specific distance zone, or a series of distance zones, from the source cells. Reclassify can convert a continuous distance raster into a raster with one or more discrete distance zones. A variation of Reclassify is an operation called **Slice,** which can divide a continuous distance raster into equal-interval or equal-area distance zones.

12.5.1 Allocation and Direction

Besides calculating straight-line distances, physical distance measure operations can also produce allocation and direction rasters (Figure 12.14). The cell value in an allocation raster corresponds to the closest source cell for the cell. The cell value in a direction raster corresponds to the direction in degrees that the cell is from the closest source cell. The direction values are based on the compass directions: 90° to the east, 180° to the south, 270° to the west, and 360° to the north. (0° is reserved for the source cell.) Box 12.6 describes distance measure operations available in ArcGIS and their intended use.

12.5.2 Applications of Physical Distance Measure Operations

Like buffering around vector-based features, physical distance measure operations have many applications. For example, we can create equal-interval distance zones from a stream network or regional

1.0	**2**	1.0	2.0
1.4	1.0	1.4	2.2
1.0	1.4	2.2	2.8
1	1.0	2.0	3.0

(a)

2	**2**	2	2
2	2	2	2
1	1	1	2
1	1	1	1

(b)

90	**2**	270	270
45	360	315	287
180	225	243	315
1	270	270	270

(c)

Figure 12.14
Based on the source cells denoted as *1* and *2,* (*a*) shows the physical distance measures in cell units from each cell to the closest source cell; (*b*) shows the allocation of each cell to the closest source cell; and (*c*) shows the direction in degrees from each cell to the closest source cell. The cell in a dark shade (row 3, column 3) has the same distance to both source cells. Therefore, the cell can be allocated to either source cell. The direction of 243° is to the source cell 1.

Box 12.6 | Distance Measure Operations in ArcGIS

ArcToolbox has the Distance toolset in Spatial Analyst Tools. The toolset includes tools for measuring distance, direction, and allocation based on either Euclidean or cost distance. Chapter 17 covers cost distance measures for least-cost path analysis.

---

fault lines. Another example is the use of distance measure operations as tools for implementing a model, such as the potential nesting habitat model of greater sandhill cranes in northwestern Minnesota, developed by Herr and Queen (1993). Based on continuous distance zones measured from undisturbed vegetation, roads, buildings, and agricultural land, the model categorizes potentially suitable nesting vegetation as optimal, suboptimal, marginal, and unsuitable.

12.6 OTHER RASTER DATA OPERATIONS

Local, neighborhood, zonal, and distance measure operations cover the majority of raster data operations. Some operations, however, do not fit well into the classification.

12.6.1 Raster Data Management

In the United States, raster data are usually available by U.S. Geological Survey 1:24,000 scale quadrangle. Thus the first step in using these data sets is to clip them, or combine them, to fit the study area for a GIS project. To clip a raster, we can specify an analysis mask or the minimum and maximum x-, y-coordinates of a rectangular area for the analysis environment and then use the larger raster as the input (Figure 12.15). **Mosaic** is a command that can combine multiple input rasters into a single raster. If the input rasters overlap, a GIS package typically provides options for filling in the cell values in the overlapping areas. ArcGIS, for example, lets the user choose the first input raster's data or the blending of data from the input rasters for the overlapping areas. If small gaps exist between the input rasters, one option is

(a)   (b)   (c)

Figure 12.15
An analysis mask (b) is used to clip an input raster (a). The output raster is (c), which has the same area extent as the analysis mask. (The symbology differs between a and c because of different value ranges.)

to use the neighborhood mean operation to fill in missing values.

12.6.2 Raster Data Extraction

Raster data extraction creates a new raster by extracting data from an existing raster. A data set, a graphic object, or a query expression can be used to define the area to be extracted. If the data set is a point feature layer, the extraction tool extracts values at the point locations (e.g., using bilinear interpolation, Chapter 6) and attaches the values to a new field in the point feature attribute table. If the data set is a raster or a polygon feature layer, the extraction tool extracts the cell values within the area defined by the raster or the polygon layer and assigns no data to cells outside the area.

A graphic object for raster data extraction can be a rectangle, a set of points, a circle, or a polygon. The object is input in x-, y-coordinates. For example, a circle can be entered with a pair of x-, y-coordinates for its center and a length for its radius (Figure 12.16). Using graphic objects, we can extract elevation data within a radius of 50 miles from an earthquake epicenter or elevation data at a series of weather stations.

The extract-by-attribute operation creates a new raster that has cell values meeting the query expression. For example, we can create a new raster within a particular elevation zone (e.g., 900 to 1000 meters). Those cells outside the elevation zone carry no data on the output.

12.6.3 Raster Data Generalization

Several operations can generalize or simplify raster data. One such method is resampling, which can build different pyramid levels (different resolutions) for a large raster data set (Chapter 6). **Aggregate** is similar to a resampling technique in that it also creates an output raster that has a larger cell size (i.e., a lower resolution) than the input. But, instead of using nearest neighbor, bilinear interpolation, or cubic convolution, Aggregate calculates each output cell value as the mean, median, sum, minimum, or maximum of the input cells that fall within the output cell (Figure 12.17).

Some data generalization operations are based on zones, or groups of cells of the same value. For example, ArcGIS has a tool called RegionGroup, which identifies for each cell in the output raster the zone that the cell is connected to (Figure 12.18). We may consider RegionGroup to be a classification method that uses both the cell values and the spatial connection of cells as the classification criteria.

Generalizing or simplifying the cell values of a raster can be useful for certain applications. For example, a raster derived from a satellite image or LIDAR (light detection and ranging) data usually has a high degree of local variations. These

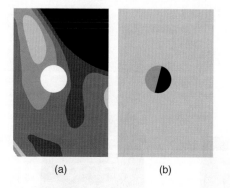

(a)   (b)

Figure 12.16
A circle, shown in white, is used to extract cell values from the input raster (*a*). The output (*b*) has the same area extent as the input raster but has no data outside the circular area. (To highlight the contrast, *b* uses a different symbology than *a*.)

1	3	2	5
1	3	2	7
1	1	2	5
2	4	3	2

(a)

2	4
2	3

(b)

Figure 12.17
An Aggregate operation creates a lower-resolution raster (*b*) from the input (*a*). The operation uses the mean statistic and a factor of 2 (i.e., a cell in *b* covers 2-by-2 cells in *a*). For example, the cell value of 4 in (*b*) is the mean of {2, 2, 5, 7} in (*a*).

(a)   (b)

Figure 12.18
Each cell in the output (*b*) has a unique number that identifies the connected region to which it belongs in the input (*a*). For example, the connected region that has the same cell value of 3 in (*a*) has a unique number of 4 in (*b*).

local variations can become unnecessary noises. To remove them, we can use Aggregate or a resampling technique.

12.7 MAP ALGEBRA

A large variety of data analysis operations are available for raster data. To facilitate data analysis, **map algebra,** an informal language with syntax similar to algebra, can be used for manipulating raster data sets (Tomlin 1990; Pullar 2001). A map algebra expression consists of objects and operations. Using the example of 57.296 × arctan ([slope_p]/100), a local operation to convert a slope raster measured in percent ([slope_p]) into one measured in degrees, [slope_p] (a raster data set), 57.296 (a constant), and 100 (a number) are objects. Operations in the example include multiplication and division (operators) and arctan (a function). Many operations have parameters. For example, the Zonal Mean operation has three parameters: ZonalMean(<zone_grid>, <value_grid>, {DATA | NODATA}), where <zone_grid> is the zonal raster, <value_grid> is the input raster, and {DATA | NODATA} stipulates how no-data cells in the input raster are handled. Parameters provide information on the input, output, and choices for the operation.

Map algebra expressions can be simple or complex. A single expression executes one operation

such as Zonal Mean, whereas a complex expression combines two or more operations. For example, Select(Slope([emidalat], degree), 'value < 20') is a complex expression: it executes the Slope operation on the elevation raster emidalat, Slope([emidalat], degree), before selecting areas with slopes less than 20 degrees. The output is therefore a slope raster, in which areas with slopes less than 20 degrees have cell values but other areas are coded no-data.

12.8 COMPARISON OF VECTOR- AND RASTER-BASED DATA ANALYSIS

Vector data analysis and raster data analysis represent the two basic types of GIS analyses. They are treated separately because a GIS package cannot run them together in the same operation. Although some GIS packages allow the use of vector data in some raster data operations (e.g., extraction operations), the data are converted into raster data before the operation starts.

Each GIS project is different in terms of data sources and objectives. Moreover, vector data can be easily converted into raster data and vice versa. We must therefore choose the type of data analysis that is efficient and appropriate. In the following, overlay and buffering, the two most common operations in GIS, are used as examples to compare vector- and raster-based operations.

12.8.1 Overlay

A local operation with multiple rasters is often compared to a vector-based overlay operation. The two operations are similar in that they both use multiple data sets as inputs. But important differences exist between them.

First, to combine the geometries and attributes from the input layers, a vector-based overlay operation must compute intersections between features and insert points at the intersections. This type of computation is not necessary for a raster-based local operation because the input rasters have the same cell size and area extent. Even if the input rasters have to be first resampled to the same cell size, the computation is still less complicated than calculating line intersections. Second, a raster-based local operation has access to various mathematical functions to create the output whereas a vector-based overlay operation only combines attributes from the input layers. Any computations with the attributes must follow the overlay operation.

Although a raster-based local operation is computationally more efficient than a vector-based overlay operation, the latter has its advantages as well. An overlay operation can combine multiple attributes from each input layer. Once combined into a layer, all attributes can be queried and analyzed individually or in combination. For example, a vegetation stand layer can have the attributes of height, crown closure, stratum, and crown diameter, and a soil layer can have the attributes of depth, texture, organic matter, and pH value. An overlay operation combines the attributes of both layers into a single layer and allows all attributes to be queried and analyzed. In contrast, each input raster in a local operation is associated with a single set of cell values (i.e., a single spatial variable). In other words, to query or analyze the same stand and soil attributes as above, a raster-based local operation would require one raster for each attribute. A vector-based overlay operation is therefore more efficient than a raster-based local operation if the data sets to be analyzed have a large number of attributes that share the same geometry.

12.8.2 Buffering

A vector-based buffering operation and a raster-based physical distance measure operation are similar in that they both measure distances from select features. But they differ in at least two aspects. First, a buffering operation uses x- and y- coordinates in measuring distances, whereas a raster-based operation uses cells in measuring physical distances. A buffering operation can therefore create more accurate buffer zones than a raster-based operation can. This accuracy difference can be important, for example, in implementing riparian zone management programs. Second, a buffering operation is more flexible and offers more options. For example, a buffering operation can create multiple rings (buffer zones), whereas a raster-based operation creates continuous distance measures. Additional data processing (e.g., Reclassify or Slice) is required to define buffer zones from continuous distance measures. A buffering operation has the option of creating separate buffer zones for each select feature or a dissolved buffer zone for all select features. It would be difficult to create and manipulate separate distance measures using a raster-based operation.

KEY CONCEPTS AND TERMS

Aggregate: A generalization operation that produces an output raster with a larger cell size (i.e., a lower resolution) than the input raster.

Analysis mask: A mask that limits raster data analysis to cells that do not carry the cell value of no data.

Block operation: A neighborhood operation that uses a rectangle (block) and assigns the calculated value to all block cells in the output raster.

Cost distance: Distance measured by the cost of moving between cells.

Local operation: A cell-by-cell raster data operation.

Map algebra: A term referring to algebraic operations with raster layers.

Mosaic: A raster data operation that can piece together multiple input rasters into a single raster.

Neighborhood operation: A raster data operation that involves a focal cell and a set of its surrounding cells.

Physical distance: Straight-line distance between cells.

Physical distance measure operation: A raster data operation that calculates straight-line distances away from the source cells.

Raster data extraction: An operation that uses a data set, a graphic object, or a query expression to extract data from an existing raster.

Reclassification: A local operation that reclassifies cell values of an input raster to create a new raster.

Slice: A raster data operation that divides a continuous raster into equal-interval or equal-area classes.

Zonal operation: A raster data operation that involves groups of cells of same values or like features.

REVIEW QUESTIONS

1. How can an analysis mask save time and effort for raster data operations?

2. Why is a local operation also called a cell-by-cell operation?

3. Refer to Figure 12.3. Show the output raster if the local operation uses the minimum statistic.

4. Refer to Figure 12.4. Show the output raster if the local operation uses the variety statistic.

5. Figure 12.5c has two cells with the same value of 4. Why?

6. Neighborhood operations are also called focal operations. What is a focal cell?

7. Describe the common types of neighborhoods used in neighborhood operations.

8. Refer to Figure 12.8. Show the output raster if the neighborhood operation uses the variety measure.

9. Refer to Figure 12.9. Show the output raster if the neighborhood operation uses the minority statistic.

10. What kinds of geometric measures can be derived from zonal operations on a single raster?

11. A zonal operation with two rasters must define one of them as the zonal raster first. What is a zonal raster?

12. Refer to Figure 12.11. Show the output raster if the zonal operation uses the range statistic.

13. Suppose you are asked to produce a raster that shows the average precipitation in each major watershed in your state. Describe the procedure you will follow to complete the task.

14. Explain the difference between the physical distance and the cost distance.

15. What is a physical distance measure operation?

16. A government agency will most likely use a vector-based buffering operation, rather than a raster-based physical distance measure operation, to define a riparian zone. Why?

17. Refer to Box 12.4. Suppose that you are given an elevation raster. How can you prepare a raster showing mean relief energy within a circle of 2.5 kilometers?

18. Write a map algebra expression that will select areas with elevations higher than 3000 feet from an elevation raster (*emidalat*), which is measured in meters.

APPLICATIONS: RASTER DATA ANALYSIS

This applications section covers the basic operations of raster data analysis. Task 1 covers a local operation. Task 2 runs a local operation using the Combine function. Task 3 uses a neighborhood operation. Task 4 uses a zonal operation. And Task 5 includes a physical distance measure operation in data query. All five tasks and the challenge question require the use of the Spatial Analyst extension. Click the Customize menu, point to Extensions, and make sure that the Spatial Analyst extension is checked.

Task 1 Perform a Local Operation

What you need: *emidalat*, an elevation raster with a cell size of 30 meters.

Task 1 lets you run a local operation to convert the elevation values of *emidalat* from meters to feet.

1. Start ArcCatalog, and connect to the Chapter 12 database. Select Properties from the context menu of *emidalat* in the Catalog tree. The Raster Dataset Properties dialog shows that *emidalat* has 186 columns, 214 rows, a cell size of 30 (meters), and a value range of 855 to 1337 (meters). Also, *emidalat* is a floating-point Esri grid. (A floating-point grid allows use of integers for cell values.)

2. Launch ArcMap. Add *emidalat* to Layers, and rename Layers Tasks 1&3. Open ArcToolbox. Right-click ArcToolbox, and select Environment. Set the Chapter 12 database as the current and scratch workspace. Double-click the Times tool in the Spatial Analyst Tools/Math toolset. In the next dialog, select *emidalat* for the input raster or constant value 1, enter 3.28 for the input raster or constant value 2, and save the output raster as *emidaft* in the current workspace. *emidaft* is the same as *emidalat* except that it is measured in feet. Click OK.

Q1. What is the range of cell values in *emidaft*?

Task 2 Perform a Combine Operation

What you need: *slope_gd*, a slope raster with 4 slope classes; and *aspect_gd*, an aspect raster with flat areas and 4 principal directions.

Task 2 covers the use of the Combine function. Combine is a local operation that can work with two or more rasters.

1. Select Data Frame from the Insert menu in ArcMap. Rename the new data frame Task 2, and add *slope_gd* and *aspect_gd* to Task 2.

2. Double-click the Combine tool in the Spatial Analyst Tools/Local toolset. In the next dialog, select *aspect_gd* and *slope_gd* for the input rasters, and enter *slp_asp* for the output raster. Click OK to run the operation. *slp_asp* shows a unique output value for the unique combination of input values. Open the attribute table of *slp_asp* to find the unique combinations and their cell counts.

Q2. How many cells in *combine* have the combination of a slope class of 2 and an aspect class of 4?

Task 3 Perform a Neighborhood Operation

What you need: *emidalat*, as in Task 1.

Task 3 asks you to run a neighborhood mean operation on *emidalat*.

1. Activate Tasks 1&3 in ArcMap. Double-click the Focal Statistics tool in the Spatial Analyst Tools/Neighborhood toolset. In the next dialog, select *emidalat* for the input raster, save the output raster as *emidamean*, accept the default neighborhood of a 3-by-3 rectangle, select mean for the statistic type, and click OK. *emidamean* shows the neighborhood mean of *emidalat*.

Q3. What other neighborhood statistics are available in Spatial Analyst besides the mean?

Task 4 Perform a Zonal Operation

What you need: *precipgd,* a raster showing the average annual precipitation in Idaho; *hucgd,* a watershed raster.

Task 4 asks you to derive annual precipitation statistics by watershed. Both *precipgd* and *hucgd* are projected onto the same projected coordinate system and are measured in meters. The precipitation measurement unit for *precipgd* is 1/100 of an inch; for example, the cell value of 675 means 6.75 inches.

1. Select Data Frame from the Insert menu in ArcMap. Rename the new data frame Task 4, and add *precipgd* and *hucgd* to Task 4.

2. Double-click the Zonal Statistics tool in the Spatial Analyst Tools/Zonal toolset. In the next dialog, select *hucgd* for the input raster, select *precipgd* for the input value raster, save the output raster as *huc_precip,* select mean for the statistics type, and click OK.

3. To show the mean precipitation for each watershed, do the following. Right-click *huc_precip,* and select Properties. Click Unique Values in the Show panel, and click OK.

Q4. Which watershed has the highest average annual precipitation in Idaho?

4. The Zonal Statistics as Table tool in the Spatial Analyst Tools/Zonal toolset can output the summary of the values in each zone into a table.

Task 5 Measure Physical Distances

What you need: *strmgd,* a raster showing streams; and *elevgd,* a raster showing elevation zones.

Task 5 asks you to locate the potential habitat of a plant species. The cell values in *strmgd* are the ID values of streams. The cell values in *elevgd* are elevation zones 1, 2, and 3. Both rasters have the cell resolution of 100 meters. The potential habitat of the plant species must meet the following criteria:

- Elevation zone 2
- Within 200 meters of streams

1. Select Data Frame from the Insert menu in ArcMap. Rename the new data frame Task 5, and add *strmgd* and *elevgd* to Task 5.

2. Double-click the Euclidean Distance tool in the Spatial Analyst Tools/Distance toolset. In the next dialog, select *strmgd* for the input raster, save the output distance raster as *strmdist,* and click OK. *strmdist* shows continuous distance zones away from streams in *strmgd.*

3. This step is to create a new raster that shows areas within 200 meters of streams. Double-click the Reclassify tool from the Spatial Analyst Tools/Reclass toolset. In the Reclassify dialog, select *strmdist* for the input raster, and click the Classify button. In the Classification dialog, change the number of classes to 2, enter 200 for the first break value, and click OK to dismiss the dialog. Back to the Reclassify dialog, save the output raster as *rec_dist,* and click OK. *rec_dist* separates areas that are within 200 meters of streams (1) from areas that are beyond (2).

4. This step is to combine *rec_dist* and *elev_gd.* Double-click the Combine tool in the Spatial Analyst Tools/Local toolset. In the next dialog, select *rec_dist* and *elev_gd* for the input rasters, save the output raster as *habitat1,* and click OK.

5. Now you are ready to extract potential habitat area from *habitat1.* Double-click the Extract by Attributes tool in the Spatial Analyst Tools/Extraction toolset. In the next dialog, select *habitat1* for the input raster, enter the expression, "REC_DIST"=1 AND "ELEVGD"=2, in the Query Builder, save the output raster as *habitat2,* and click OK. *habitat2* shows the selected habitat area.

Q5. Verify that *habitat2* is the correct potential habitat area.

6. As an option, you can use *strmdist* directly without reclassification to complete Task 5. In that case, you will use Raster Calculator and the following map algebra expression: ("strmdist" <= 200) & ("elev_gd" == 2).

Challenge Task

What you need: *emidalat, emidaslope,* and *emidaaspect.*

This challenge task asks you to construct a raster-based model using elevation, slope, and aspect.

1. Use the Reclassify tool in the Spatial Analyst Tools/Reclass toolset to classify *emidalat* into five elevation zones according to the table below and save the classified raster as *rec_emidalat.*

Old Values	New Values
855–900	1
900–1000	2
1000–1100	3
1100–1200	4
>1200	5

2. *emidaslope* and *emidaaspect* have already been reclassified and ranked. Create a model by using the following equation: *emidaelev* + 3 × *emidaslope* + *emidaaspect*. Name the model *emidamodel.*

Q1. What is the range of cell values in *emidamodel*?

Q2. What area percentage of *emidamodel* has cell value > 20?

REFERENCES

Beguería, S., and S. M. Vicente-Serrano. 2006. Mapping the Hazard of Extreme Rainfall by Peaks over Threshold Extreme Value Analysis and Spatial Regression Techniques. *Journal of Applied Meteorology and Climatology* 45: 108–24.

Congalton, R. G. 1991. A Review of Assessing the Accuracy of Classification of Remotely Sensed Data. *Photogrammetric Engineering & Remote Sensing* 37: 35–46.

Crow, T. R., G. E. Host, and D. J. Mladenoff. 1999. Ownership and Ecosystem as Sources of Spatial Heterogeneity in a Forested Landscape, Wisconsin, USA. *Landscape Ecology* 14: 449–63.

Forman, R. T. T., and M. Godron. 1986. *Landscape Ecology.* New York: Wiley.

Herr, A. M., and L. P. Queen. 1993. Crane Habitat Evaluation Using GIS and Remote Sensing. *Photogrammetric Engineering & Remote Sensing* 59: 1531–38.

Heuvelink, G. B. M. 1998. *Error Propagation in Environmental Modelling with GIS.* London: Taylor and Francis.

Lillesand, T. M., R. W. Kiefer, and J. W. Chipman. 2004. *Remote Sensing and Image Interpretation*, 5th ed. New York: Wiley.

McGarigal, K., and B. J. Marks. 1994. *Fragstats: Spatial Pattern Analysis Program for Quantifying Landscape Structure.* Forest Science Department: Oregon State University.

Mladenoff, D. J., T. A. Sickley, R. G. Haight, and A. P. Wydeven. 1995. A Regional Landscape Analysis and Prediction of Favorable Gray Wolf Habitat in the Northern Great Lakes Regions. *Conservation Biology* 9: 279–94.

Pullar, D. 2001. MapScript: A Map Algebra Programming Language Incorporating Neighborhood Analysis. *GeoInformatica* 5: 145–63.

Renard, K. G., G. R. Foster, G. A. Weesies, D. K. McCool, and D. C. Yoder (coordinators).

1997. Predicting Soil Erosion by Water: A Guide to Conservation Planning with the Revised Universal Soil Loss Equation (RUSLE). *Agricultural Handbook 703*. Washington, DC: U.S. Department of Agriculture.

Tomlin, C. D. 1990. *Geographic Information Systems and*

Cartographic Modeling. Englewood Cliffs, NJ: Prentice Hall.

Veregin, H. 1995. Developing and Testing of an Error Propagation Model for GIS Overlay Operations. *International Journal of Geographical Information Systems* 9: 595–619.

Wischmeier, W. H., and D. D. Smith. 1978. Predicting Rainfall Erosion Losses: A Guide to Conservation Planning. *Agricultural Handbook 537*. Washington, DC: U.S. Department of Agriculture.

CHAPTER 13

TERRAIN MAPPING AND ANALYSIS

The terrain with its undulating, continuous surface is a familiar phenomenon to GIS users. The land surface has been the object of mapping and analysis for hundreds of years. Mapmakers have devised various techniques for terrain mapping such as contouring, hill shading, hypsometric tinting, and 3-D perspective view. And geomorphologists have developed measures of the land surface including slope, aspect, and surface curvature.

Terrain mapping and analysis techniques are no longer tools for specialists. GIS has made it relatively easy to incorporate them into a variety of applications. Slope and aspect play a regular role in hydrologic modeling, snow cover evaluation, soil mapping, landslide delineation, soil erosion, and predictive mapping of vegetation communities. Hill shading, perspective view, and even 3-D flyby are common features in presentations and reports.

Most GIS packages treat elevation data (*z* values) as attribute data at point or cell locations rather than as an additional coordinate to *x*- and *y*-coordinates as in a true 3-D model. In raster format, the *z* values correspond to the cell values. In vector format, the *z* values are stored in an attribute field or with the feature geometry. Terrain mapping and analysis can use raster data, vector data, or both as inputs. This is perhaps why GIS vendors typically group terrain mapping and analysis functions into a module or an extension, separate from the basic GIS tools.

Chapter 13 is organized into the following five sections. Section 13.1 covers two common types of input data for terrain mapping and analysis: DEM (digital elevation model) and TIN (triangulated irregular network). Section 13.2 describes different terrain mapping methods. Section 13.3 covers slope and aspect analysis using either a DEM or a TIN. Section 13.4 focuses on deriving surface curvature from a DEM. Section 13.5 compares DEM and TIN for terrain mapping and analysis. Viewshed analysis and watershed analysis, which are closely related to terrain analysis, are covered in Chapter 14.

13.1 DATA FOR TERRAIN MAPPING AND ANALYSIS

Two common types of input data for terrain mapping and analysis are the raster-based DEM and the vector-based TIN. They cannot be used together for analysis, but a DEM can be converted into a TIN and a TIN into a DEM.

13.1.1 DEM

A DEM represents a regular array of elevation points. Most GIS users in the United States use DEMs from the U.S. Geological Survey (USGS). Alternative sources for DEMs come from satellite images, radar data, and LIDAR (light detection and ranging) data (Chapter 4). (LIDAR can be a data source for both DEM and TIN.) Regardless of its origin, a point-based DEM must be converted to raster data before it can be used for terrain mapping and analysis. This conversion simply places each elevation point in a DEM at the center of a cell in an elevation raster. DEM and elevation raster can therefore be used interchangeably.

The quality of a DEM can influence the accuracy of terrain measures including slope and aspect. The USGS offers a nationwide coverage of elevation data through the National Elevation Dataset (NED). NED 1 Arc Second, or the 30-meter DEM, has a vertical accuracy of $+/-$ 7 to 15 meters, whereas NED 1/3 Arc Second, or the 10-meter DEM, has a vertical accuracy of $+/-$ 7 meters (Chapter 4). Based on the root mean square (RMS) error (Chapter 7), a recent report from the USGS suggests that the absolute vertical accuracy of NED 1/3 Arc Second ranges from 2.03 to 4.66 meters, depending

on the production method, with an average of 2.44 meters (Gesch 2007). This statistic was derived by comparing DEM data with more than 13,000 high-precision survey points across the conterminous United States. As the NED is continually upgraded with new acquisitions of high-resolution data, the USGS expects improvement of the vertical accuracy of NED DEMs.

LIDAR data can provide high-quality DEMs with a spatial resolution of 0.5 to 2 meters and a vertical accuracy of about 15 centimeters (Flood 2001; Murphy et al. 2008). The USGS National Geospatial Program's base LIDAR specification cites the RMS of 15 to 18.5 centimeters as minimum requirement for the vertical accuracy of LIDAR data (**http://lidar.cr.usgs.gov/**). LIDAR data are therefore ideal for studies that require detailed topographic data such as floodplain mapping, telecommunications, transportation, ecosystems, and forest structure (Hill et al. 2000; Lefsky et al. 2002; Lim et al. 2003).

13.1.2 TIN

A TIN approximates the land surface with a series of nonoverlapping triangles. Elevation values (*z* values) along with *x-, y-*coordinates are stored at nodes that make up the triangles. In contrast to a DEM, a TIN is based on an irregular distribution of elevation points.

DEMs are usually the primary data source for compiling preliminary TINs through a conversion process. But a TIN can use other data sources (Box 13.1). Additional point data may include surveyed elevation points, GPS (global positioning system) data, and LIDAR data. Line data may include contour lines and breaklines. **Breaklines** are line features that represent changes of the land surface such as streams, shorelines, ridges, and roads. And area data may include lakes and reservoirs.

Because triangles in a TIN can vary in size by the complexity of topography, not every point in a DEM needs to be used to create a TIN. Instead, the process is to select points that are more important in representing the terrain. Several algorithms

Box 13.1 | **Terrain Data Format**

T errain is a data format used in ArcGIS for terrain mapping and analysis. This data format is designed for handling massive point-based data sets such as LIDAR and side-scan sonar (used to image ocean floor) data that have become popular for terrain mapping and analysis in recent years. A terrain is stored in a feature dataset of a geodatabase, along with other feature classes for breaklines, lakes, and study area boundaries. A TIN-based surface can therefore be constructed on the fly by using the feature dataset and its contents.

for selecting significant points from a DEM have been proposed in GIS (Lee 1991; Kumler 1994). The most popular algorithm is the maximum z-tolerance.

The **maximum z-tolerance** algorithm selects points from an elevation raster to construct a TIN such that, for every point in the elevation raster, the difference between the original elevation and the estimated elevation from the TIN is within the specified maximum z-tolerance. The algorithm uses an iterative process. The process begins by constructing a candidate TIN. Then, for each triangle in the TIN, the algorithm computes the elevation difference from each point in the raster to the enclosing triangular facet. The algorithm determines the point with the largest difference. If the difference is greater than a specified z-tolerance, the algorithm flags the point for addition to the TIN. After every triangle in the current TIN is checked, a new triangulation is recomputed with the selected additional points. This process continues until all points in the raster are within the specified maximum z-tolerance.

Elevation points selected by the maximum z-tolerance algorithm, plus additional elevation points from contour lines, survey data, GPS data, or LIDAR data are connected to form a series of non-overlapping triangles in an initial TIN. A common algorithm for connecting points is the **Delaunay triangulation** (Watson and Philip 1984; Tsai 1993). Triangles formed by the Delaunay triangulation have the following characteristics: all nodes

(points) are connected to the nearest neighbors to form triangles; and triangles are as equiangular, or compact, as possible.

Depending on the software design, breaklines may be included in the initial TIN or used to modify the initial TIN. Breaklines provide the physical structure in the form of triangle edges to show changes of the land surface (Figure 13.1). If the z values for each point of a breakline are known, they can be stored in a field. If not, they can be estimated from the underlying DEM or TIN surface.

Triangles along the border of a TIN are sometimes stretched and elongated, thus distorting the topographic features derived from those triangles. The cause of this irregularity stems from the sudden drop of elevation along the edge. One way to solve this problem is to include elevation points beyond the border of a study area for processing and then to clip the study area from the larger coverage. The process of building a TIN is certainly more complex than preparing an elevation raster from a DEM.

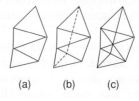

(a)   (b)   (c)

Figure 13.1

A breakline, shown as a dashed line in (*b*), modifies the triangles in (*a*) and creates new triangles in (*c*).

Just as a DEM can be converted to a TIN, a TIN can also be converted to a DEM. The process requires each elevation point of the DEM to be estimated (interpolated) from its neighboring nodes that make up the TIN. Each of these nodes has its x-, y-coordinates as well as a z (elevation) value. Because each triangular facet is supposed to have a constant slope and aspect, converting a TIN to a DEM can be based on local first-order (planar) polynomial interpolation. (Chapter 15 covers local polynomial interpolation as one of the spatial interpolation methods.) TIN to DEM conversion is useful for producing a DEM from LIDAR data. The process first connects LIDAR data (points) to form a TIN and then compiles the DEM by interpolating elevation points from the TIN.

13.2 TERRAIN MAPPING

Common terrain mapping techniques include contouring, vertical profiling, hill shading, hypsometric tinting, and perspective view (Collier et al. 2003).

13.2.1 Contouring

Contouring is a common method for terrain mapping. **Contour lines** connect points of equal elevation, the **contour interval** represents the vertical distance between contour lines, and the **base contour** is the contour from which contouring starts. Suppose a DEM has elevation readings ranging from 743 to 1986 meters. If the base contour were set at 800 and the contour interval at 100, then contouring would create the contour lines of 800, 900, 1000, and so on.

The arrangement and pattern of contour lines reflect the topography. For example, contour lines are closely spaced in steep terrain and are curved in the upstream direction along a stream (Figure 13.2). With some training and experience in reading contour lines, we can visualize, and even judge the accuracy of, the terrain as simulated by digital data.

Automated contouring follows two basic steps: (1) detecting a contour line that intersects a raster cell or a triangle, and (2) drawing the contour line through the raster cell or triangle

(Jones et al. 1986). A TIN is a good example for illustrating automated contouring because it has elevation readings for all nodes from triangulation. Given a contour line, every triangle edge is examined to determine if the line should pass through the edge. If it does, linear interpolation, which assumes a constant gradient between the end nodes of the edge, can determine the contour line's position along the edge. After all the positions are calculated, they are connected to form the contour line (Figure 13.3). The initial contour line consists of straight-line segments, which can be smoothed by fitting a mathematical function such as splines to points that make up the line.

Figure 13.2
A contour line map.

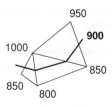

Figure 13.3
The contour line of 900 connects points that are interpolated to have the value of 900 along the triangle edges.

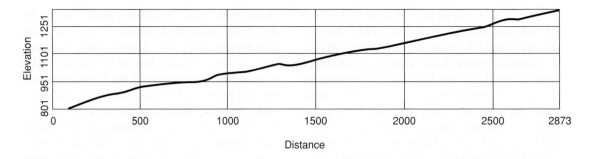

Figure 13.4
A vertical profile.

Contour lines do not intersect one another or stop in the middle of a map, although they can be close together in cases of cliffs or form closed lines in cases of depressions or isolated hills. Contour maps created from a GIS sometimes contain irregularities and even errors. Irregularities are often caused by the use of large cells, whereas errors are caused by the use of incorrect parameter values in the smoothing algorithm (Clarke 1995).

13.2.2 Vertical Profiling

A **vertical profile** shows changes in elevation along a line, such as a hiking trail, a road, or a stream (Figure 13.4). The manual method usually involves the following steps:

1. Draw a profile line on a contour map.

2. Mark each intersection between a contour and the profile line and record its elevation.

3. Raise each intersection point to a height proportional to its elevation.

4. Plot the vertical profile by connecting the elevated points.

Automated profiling follows the same procedure but substitutes the contour map with an elevation raster or a TIN.

13.2.3 Hill Shading

Also known as *shaded relief*, **hill shading** simulates how the terrain looks with the interaction between sunlight and surface features (Figure 13.5). A mountain slope directly facing incoming light

Figure 13.5
An example of hill shading, with the sun's azimuth at 315° (NW) and the sun's altitude at 45°.

will be very bright; a slope opposite to the light will be dark. Hill shading helps viewers recognize the shape of landform features. Hill shading can be mapped alone, such as Thelin and Pike's (1991) digital shaded-relief map of the United States. But often hill shading is the background for terrain or thematic mapping.

Hill shading used to be produced by talented artists. But the computer can now generate high-quality shaded-relief maps. Four factors control the visual effect of hill shading. The sun's azimuth is the direction of the incoming light, ranging from 0° (due north) to 360° in a clockwise direction. Typically, the default for the sun's azimuth is 315°. With the light source located above the upper-left corner of the shaded-relief map, the shadows appear to fall toward the viewer, thus avoiding the pseudoscopic effect (Box 13.2). The sun's altitude is the angle of the incoming light measured above the horizon between 0° and 90°. The other two factors are the surface's slope and

aspect: slope ranges from 0° to 90° and aspect from 0° to 360° (Section 13.3).

Computer-generated hill shading uses the relative radiance value computed for every cell in an elevation raster or for every triangle in a TIN (Eyton 1991). The relative radiance value ranges from 0 to 1; when multiplied by the constant 255, it becomes the illumination value for computer screen display. An illumination value of 255 would be white and a value of 0 would be black on a shaded-relief map. Functionally similar to the relative radiance value, the incidence value can be obtained from multiplying the relative radiance by the sine of the sun's altitude (Franklin 1987). Box 13.3

Box 13.2 The Pseudoscopic Effect

A shaded-relief map looks right when the shadows appear to fall toward the viewer. If the shadows appear to fall away from the viewer, as occurs if we use 135° as the sun's azimuth, the hills on the map look like

depressions and the depressions look like hills in an optical illusion called the pseudoscopic effect (Campbell 1984). Of course, the sun's azimuth at 315° is totally unrealistic for most parts of the Earth's surface.

Box 13.3 A Worked Example of Computing Relative Radiance

The relative radiance value of a raster cell or a TIN triangle can be computed by the following equation:

$$R_f = \cos(A_f - A_s)\sin(H_f)\cos(H_s) + \cos(H_f)\sin(H_s)$$

where R_f is the relative radiance value of a facet (a raster cell or a triangle), A_f is the facet's aspect, A_s is the sun's azimuth, H_f is the facet's slope, and H_s is the sun's altitude.

Suppose a cell in an elevation raster has a slope value of 10° and an aspect value of 297° (W to NW), the sun's altitude is 65°, and the sun's azimuth is 315° (NW). The relative radiance value of the cell can be computed by:

$$R_f = \cos(297 - 315)\sin(10)\cos(65) + \cos(10)\sin(65) = 0.9623$$

The cell will appear bright with an R_f value of 0.9623.

If the sun's altitude is lowered to 25° and the sun's azimuth remains at 315°, then the cell's relative radiance value becomes:

$$R_f = \cos(297 - 315)\sin(10)\cos(25) + \cos(10)\sin(25) = 0.5658$$

The cell will appear in medium gray with an R_f value of 0.5658.

The incidence value of a facet can be computed by:

$$\cos(H_f) + \cos(A_f - A_s)\sin(H_f)\cos(H_s)$$

where the notations are the same as for the relative radiance. One can also derive the incidence value by multiplying the relative radiance value by $\sin(H_s)$.

shows the computation of the relative radiance value and the incidence value.

Besides producing hill shading, both the relative radiance and the incidence can be used in image processing as variables representing the interaction between the incoming radiation and local topography.

13.2.4 Hypsometric Tinting

Hypsometry depicts the distribution of the Earth's mass with elevation. **Hypsometric tinting,** also known as *layer tinting,* applies color symbols to different elevation zones (Figure 13.6). The use of well-chosen color symbols can help viewers see the progression in elevation, especially on a small-scale map. One can also use hypsometric tinting to highlight a particular elevation zone, which may be important, for instance, in a wildlife habitat study.

13.2.5 Perspective View

Perspective views are 3-D views of the terrain: the terrain has the same appearance as it would have if viewed at an angle from an airplane (Figure 13.7).

Figure 13.6
A hypsometric map. Different elevation zones are shown in different gray symbols.

Four parameters can control the appearance of a 3-D view (Figure 13.8):

- **Viewing azimuth** is the direction from the observer to the surface, ranging from 0° to 360° in a clockwise direction.
- **Viewing angle** is the angle measured from the horizon to the altitude of the observer. A viewing angle is always between 0° and 90°. An angle of 90° provides a view from directly above the surface. And an angle of 0° provides a view from directly ahead of the

Figure 13.7
A 3-D perspective view.

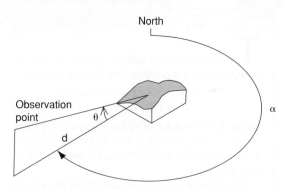

Figure 13.8
Three controlling parameters of the appearance of a 3-D view: the viewing azimuth α is measured clockwise from the north, the viewing angle θ is measured from the horizon, and the viewing distance d is measured between the observation point and the 3-D surface.

surface. Therefore, the 3-D effect reaches its maximum as the angle approaches 0° and its minimum as the angle approaches 90°.

- **Viewing distance** is the distance between the viewer and the surface. Adjustment of the viewing distance allows the surface to be viewed up close or from a distance.
- **z-scale** is the ratio between the vertical scale and the horizontal scale. Also called the *vertical exaggeration factor,* z-scale is useful for highlighting minor landform features.

Because of its visual appeal, 3-D perspective view is a display tool in many GIS packages. The 3-D Analyst extension to ArcGIS, for example, provides the graphical interfaces for manipulating the viewing parameters. Using the extension, we can rotate the surface, navigate the surface, or take a close-up view of the surface. To make perspective views more realistic, we can superimpose these views with layers such as hydrographic features (Figure 13.9), land cover, vegetation, and roads in a process called **3-D draping.** We can also add clouds and change the color of the sky. ArcGIS is not the only package for making 3-D views; for example, 3-D Nature offers World Construction Set and Visual Nature Studio **(http://3dnature.com/).**

Typically, DEMs and TINs provide the surfaces for 3-D perspective views and 3-D draping. But as long as a feature layer has a field that stores z values

or a field that can be used for calculating z values, the layer can be viewed in 3-D perspective. Figure 13.10, for example, is a 3-D view based on elevation zones. Likewise, to show building features in 3-D, we can extrude them by using the building height as the z value. Figure 13.11 is a view of 3-D buildings created by Google Maps.

Figure 13.10
A 3-D perspective view of elevation zones.

Figure 13.11
A view of 3-D buildings in Boston, Massachusetts.

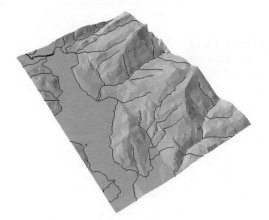

Figure 13.9
Draping of streams and shorelines on a 3-D surface.

13.3 SLOPE AND ASPECT

Slope measures the rate of change of elevation at a surface location. Slope may be expressed as percent slope or degree slope. Percent slope is 100 times the ratio of rise (vertical distance) over run (horizontal distance), whereas degree slope is the arc tangent of the ratio of rise over run (Figure 13.12).

Aspect is the directional measure of slope. Aspect starts with 0° at the north, moves clockwise, and ends with 360° also at the north. Because it is a circular measure, an aspect of 10° is closer to 360° than to 30°. We often have to manipulate aspect measures before using them in data analysis. A common method is to classify aspects into the four principal directions (north, east, south, and west) or eight principal directions (north, northeast, east, southeast, south, southwest, west, and northwest) and to treat aspects as categorical data (Figure 13.13). Rather than converting aspects to categorical data, Chang and Li (2000) have proposed a method for capturing the principal direction while retaining the numeric measure. For instance, to capture the N–S principal direction, we can set 0° at north, 180° at south, and 90° at both west and east (Figure 13.14). Perhaps the most common method for converting aspect measures to linear measures is to use their sine or cosine values, which range from −1 to 1 (Zar 1984).

As basic elements for analyzing and visualizing the terrain, slope and aspect are important in studies of watershed units, landscape units, and morphometric measures (Moore et al. 1991). When used with other variables, slope and aspect can assist in solving problems in forest inventory

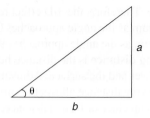

Figure 13.12
Slope θ, either measured in percent or degrees, can be calculated from the vertical distance *a* and the horizontal distance *b*.

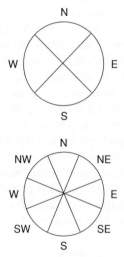

Figure 13.13
Aspect measures are often grouped into the four principal directions or eight principal directions.

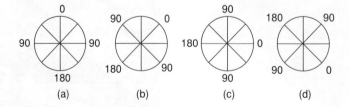

(a)   (b)   (c)   (d)

Figure 13.14
Transformation methods to capture the N–S direction (*a*), the NE–SW direction (*b*), the E–W direction (*c*), and the NW–SE direction (*d*).

estimates, soil erosion, wildlife habitat suitability, site analysis, and many other fields (Lane et al. 1998; Wilson and Gallant 2000).

13.3.1 Computing Algorithms for Slope and Aspect Using Raster

Slope and aspect used to be derived manually from a contour map. Fortunately, this practice has become rare with the use of GIS. Although conceptually slope and aspect vary continuously over space, a GIS does not compute them at points. Instead, it computes slope and aspect for discrete units such as cells of an elevation raster or triangles of a TIN. The resolution of the input raster or TIN can therefore influence the computational results.

The slope and aspect for an area unit (i.e., a cell or triangle) are measured by the quantity and direction of tilt of the unit's normal vector—a directed line perpendicular to the unit (Figure 13.15). Given a normal vector (n_x, n_y, n_z), the formula for computing the unit's slope is:

(13.1)

$$\sqrt{n_x^2 + n_y^2}/n_z$$

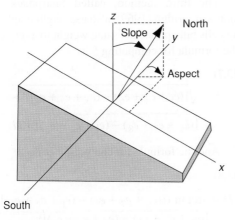

Figure 13.15
The normal vector to the area unit is the directed line perpendicular to the unit. The quantity and direction of tilt of the normal vector determine the slope and aspect of the unit. (Redrawn from Hodgson, 1998, *CaGIS* 25, (3): pp. 173–85; reprinted with the permission of the American Congress on Surveying and Mapping.)

And the formula for computing the unit's aspect is:

(13.2)

$$\arctan(n_y/n_x)$$

Different approximation (finite difference) methods have been proposed for calculating slope and aspect from an elevation raster. Here we will examine three common methods. All three methods use a 3-by-3 moving window to estimate the slope and aspect of the center cell, but they differ in the number of neighboring cells used in the estimation and the weight applying to each cell.

The first method, which is attributed to Fleming and Hoffer (1979) and Ritter (1987), uses the four immediate neighbors of the center cell. The slope (S) at C_0 in Figure 13.16 can be computed by:

(13.3)

$$S = \sqrt{(e_1 - e_3)^2 + (e_4 - e_2)^2}/2d$$

where e_i are the neighboring cell values, and d is the cell size. The n_x component of the normal vector to C_0 is $(e_1 - e_3)$, or the elevation difference in the x dimension. The n_y component is $(e_4 - e_2)$, or the elevation difference in the y dimension. To compute the percent slope at C_0, we can multiply S by 100.

S's directional angle D can be computed by:

(13.4)

$$D = \arctan((e_4 - e_2)/(e_1 - e_3))$$

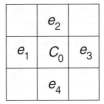

Figure 13.16
Ritter's algorithm for computing slope and aspect at C_0 uses the four immediate neighbors of C_0.

Box 13.4 | **Conversion of D to Aspect**

The notations used here are the same as in Eq. (13.1) to (13.4). In the following, the text after an apostrophe is an explanatory note.

If $S <> 0$ then
  $T = D \times 57.296$
  If $n_x = 0$
    If $n_y < 0$ then
      Aspect $= 180$
    Else

Aspect $= 360$
ElseIf $n_x > 0$ then
  Aspect $= 90 - T$
Else '$n_x < 0$
  Aspect $= 270 - T$
Else '$S = 0$
  Aspect $= -1$ 'undefined aspect for flat surface
End If

D is measured in radians and is with respect to the x-axis. Aspect, on the other hand, is measured in degrees and from a north base of $0°$. Box 13.4 shows an algorithm for converting D to aspect (Ritter 1987; Hodgson 1998).

The second method for computing slope and aspect is called Horn's algorithm (1981), an algorithm used in ArcGIS. Horn's algorithm uses eight neighboring cells and applies a weight of 2 to the four immediate neighbors and a weight of 1 to the four corner cells. Horn's algorithm computes slope at C_0 in Figure 13.17 by:

(13.5)

$$S = \sqrt{[(e_1 + 2e_4 + e_6) - (e_3 + 2e_5 + e_8)]^2 + [(e_6 + 2e_7 + e_8) - (e_1 + 2e_2 + e_3)]^2}/8d$$

e_1	e_2	e_3
e_4	C_0	e_5
e_6	e_7	e_8

Figure 13.17
Both Horn's algorithm and Sharpnack and Akin's algorithm for computing slope and aspect at C_0 use the eight neighboring cells of C_0.

And the D value at C_0 is computed by:

(13.6)

$$D = \arctan([(e_7 + 2e_6 + e_8) - (e_1 + 2e_2 + e_3)]/ [(e_1 + 2e_4 + e_6) - (e_3 + 2e_5 + e_8)])$$

D can be converted to aspect by using the same algorithm as for the first method except that $n_x = (e_1 + 2e_4 + e_6)$ and $n_y = (e_3 + 2e_5 + e_8)$. Box 13.5 shows a worked example using Horn's algorithm.

The third method, called Sharpnack and Akin's algorithm (1969), also uses eight neighboring cells but applies the same weight to every cell. The formula for computing S is:

(13.7)

$$S = \sqrt{[(e_1 + e_4 + e_6) - (e_3 + e_5 + e_8)]^2 + [(e_6 + e_7 + e_8) - (e_1 + e_2 + e_3)]^2}/6d$$

And the formula for computing D is:

(13.8)

$$D = \arctan([(e_6 + e_7 + e_8) - (e_1 + e_2 + e_3)]/ [(e_1 + e_4 + e_6) - (e_3 + e_5 + e_8)])$$

13.3.2 Computing Algorithms for Slope and Aspect Using TIN

Suppose a triangle is made of the following three nodes: $A (x_1, y_1, z_1)$, $B (x_2, y_2, z_2)$, and $C (x_3, y_3, z_3)$ (Figure 13.18). The normal vector is a cross product

Box 13.5 | **A Worked Example of Computing Slope and Aspect Using Raster**

The following diagram shows a 3-by-3 window of an elevation raster. Elevation readings are measured in meters, and the cell size is 30 meters.

1006	1012	1017
1010	1015	1019
1012	1017	1020

This example computes the slope and aspect of the center cell using Horn's algorithm:

$n_x = (1006 + 2 \times 1010 + 1012) - (1017 + 2 \times 1019 + 1020) = -37$

$n_y = (1012 + 2 \times 1017 + 1020) - (1006 + 2 \times 1012 + 1017) = 19$

$S = \sqrt{(-37)^2 + (19)^2}/(8 \times 30) = 0.1733$

$S_p = 100 \times 0.1733 = 17.33$

$D = \arctan(n_y/n_x) = \arctan(19/-37) = -0.4744$

$T = -0.4744 \times 57.296 = -27.181$

Because $S <> 0$ and $n_x < 0$,

Aspect $= 270 - (-27.181) = 297.181$.

For comparison, S_p has a value of 17.16 using the Fleming and Hoffer algorithm and a value of 17.39 using the Sharpnack and Akin algorithm. Aspect has a value of 299.06 using the Fleming and Hoffer algorithm and a value of 296.56 using the Sharpnack and Akin algorithm.

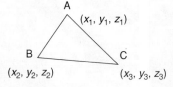

Figure 13.18

The algorithm for computing slope and aspect of a triangle in a TIN uses the x, y, and z values at the three nodes of the triangle.

of vector AB, $[(x_2 - x_1), (y_2 - y_1), (z_2 - z_1)]$, and vector AC, $[(x_3 - x_1), (y_3 - y_1), (z_3 - z_1)]$. And the three components of the normal vector are:

(13.9)

$$n_x = (y_2 - y_1)(z_3 - z_1) - (y_3 - y_1)(z_2 - z_1)$$

$$n_y = (z_2 - z_1)(x_3 - x_1) - (z_3 - z_1)(x_2 - x_1)$$

$$n_z = (x_2 - x_1)(y_3 - y_1) - (x_3 - x_1)(y_2 - y_1)$$

The S and D values of the triangle can be derived from Eq. (13.1) and (13.2) and the D value can then be converted to the aspect measured in degrees and

from a north base of $0°$. Box 13.6 shows a worked example using a TIN.

13.3.3 Factors Influencing Slope and Aspect Measures

The accuracy of slope and aspect measures can influence the performance of models that use slope and aspect as inputs (Srinivasan and Engel 1991). Therefore, it is important to examine several factors that can influence slope and aspect measures.

The first and perhaps the most important factor is the resolution of DEM used for deriving slope and aspect. Slope and aspect layers created from the 7.5-minute DEM are expected to contain greater amounts of details than those from the 1-degree DEM (Isaacson and Ripple 1990). Reports from experimental studies also show that the accuracy of slope and aspect estimates decreases with a decreasing DEM resolution (Chang and Tsai 1991; Gao 1998; Deng, Wilson, and Bauer, 2007).

Figure 13.19 shows hill-shaded maps of a small area north of Tacoma, Washington, created from three different DEMs. The 30-meter and 10-meter DEMs are USGS DEMs. The 1.83-meter

| *Box 13.6* | **A Worked Example of Computing Slope and Aspect Using TIN** |

Suppose a triangle in a TIN is made of the following nodes with their *x, y,* and *z* values measured in meters.

Node 1: $x_1 = 532260$, $y_1 = 5216909$, $z_1 = 952$
Node 2: $x_2 = 531754$, $y_2 = 5216390$, $z_2 = 869$
Node 3: $x_3 = 532260$, $y_3 = 5216309$, $z_3 = 938$

The following shows the steps for computing the slope and aspect of the triangle.

$n_x = (5216390 - 5216909)(938 - 952)$
$\quad - (5216309 - 5216909)(869 - 952)$
$\quad = -42534$

$n_y = (869 - 952)(532260 - 532260)$
$\quad - (938 - 952)(531754 - 532260) = -7084$

$n_z = (531754 - 532260)(5216309 - 5216909)$
$\quad - (532260 - 532260)(5216390 - 5216909)$
$\quad = 303600$

$S_p = 100 \times \left[\sqrt{(-42534)^2 + (-7084)^2}/303600 \right]$
$\quad = 14.20$

$D = \arctan(-7084 / -42534) = 0.165$

$T = 0.165 \times 57.296 = 9.454$

Because $S <> 0$ and $n_x < 0$, aspect $= 270 - T = 260.546$.

(a)   (b)   (c)

Figure 13.19
DEMs at three different resolutions: USGS 30-meter DEM (*a*), USGS 10-meter DEM (*b*), and 1.83-meter DEM derived from LIDAR data (*c*).

DEM is a bare-ground DEM compiled from LIDAR data. It is not difficult to see an increase of topographic details with an increasing DEM resolution. Slope maps in Figure 13.20 are derived from the three DEMs, with darker symbols for steeper slopes. Slope measures range from 0° to 77.82° for the 30-meter DEM, 0° to 83.44°

for the 10-meter DEM, and 0° to 88.45° for the 1.83-meter DEM. Details on the slope maps therefore increase with an increasing DEM resolution.

The quality of DEM can also influence slope and aspect measures. We can now extract DEMs from satellite imagery (e.g., SPOT images) on the personal computer (Chapter 4). But the quality of

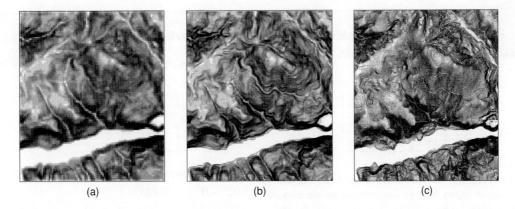

(a)   (b)   (c)

Figure 13.20
Slope layers derived from the three DEMs in Figure 13.19. The darkness of the symbol increases as the slope becomes steeper.

such DEMs varies depending on the software package and the quality of input data including ground control points. Comparing a USGS 7.5-minute DEM and a DEM produced from a SPOT panchromatic stereopair, Bolstad and Stowe (1994) report that slope and aspect errors for the SPOT DEM are significantly different from zero, whereas those for the USGS DEM are not.

Slope and aspect measures can vary by the computing algorithm. Skidmore (1989) reports that Horn's algorithm and Sharpnack and Akin's algorithm, both involving eight neighboring cells, are the best for estimating both slope and aspect in an area of moderate topography. Ritter's algorithm involving four immediate neighbors turns out to be better than other algorithms in studies by Hodgson (1998) and Jones (1998). But Kienzle (2004) reports no significant statistical difference between Ritter's algorithm and Horn's algorithm for slope or aspect measures. Therefore, there is no consensus as to which algorithm is better overall.

Finally, local topography can be a factor in estimating slope and aspect. Errors in slope estimates tend to be greater in areas of higher slopes, but errors in aspect estimates tend to be greater in areas of lower relief (Chang and Tsai 1991; Zhou, Liu, and Sun 2006). Data precision problems (i.e., the rounding of elevations to the nearest whole number) can be the cause for aspect and

slope errors in areas of low relief (Carter 1992; Florinsky 1998). Slope errors on steeper slopes may be due in part to difficulties in stereocorrelation in forested terrain (Bolstad and Stowe 1994).

13.4 SURFACE CURVATURE

GIS applications in hydrological studies often require computation of surface curvature to determine if the surface at a cell location is upwardly convex or concave (Gallant and Wilson 2000). Similar to slope and aspect, different algorithms are available for calculating surface curvature (Schmidt, Evans, and Brinkmann 2003). A common algorithm is to fit a 3-by-3 window with a quadratic polynomial equation (Zevenbergen and Thorne 1987; Moore et al. 1991):

(13.10)

$$z = Ax^2y^2 + Bx^2y + Cxy^2 + Dx^2 + Ey^2 + Fxy + Gx + Hy + I$$

The coefficients A–I can be estimated by using the elevation values in the 3-by-3 window and the raster cell size. The measures of profile curvature, plan curvature, and curvature, in addition to slope and aspect, can then be computed from the coefficients (Box 13.7).

Profile curvature is estimated along the direction of maximum slope. Plan curvature is estimated

Box 13.7   **A Worked Example of Computing Surface Curvature**

1017	1010	1017
1012	1006	1019
1015	1012	1020

e_1	e_2	e_3
e_4	e_0	e_5
e_6	e_7	e_8

The diagram above represents a 3-by-3 window of an elevation raster, with a cell size of 30 meters. This example shows how to compute the profile curvature, plan curvature, and surface curvature at the center cell. The first step is to estimate the coefficients D–H of the quadratic polynomial equation that fits the 3-by-3 window:

$$D = [(e_4 + e_5)/2 - e_0]/L^2$$
$$E = [(e_2 + e_7)/2 - e_0]/L^2$$
$$F = (-e_1 + e_3 + e_6 - e_8)/4L^2$$
$$G = (-e_4 + e_5)/2L$$
$$H = (e_2 - e_7)/2L$$

where e_0 to e_8 are elevation values within the 3-by-3 window according to the diagram that follows, and L is the cell size.

Profile curvature $= -2[(DG^2 + EH^2 + FGH)/(G^2 + H^2)] = -0.0211$

Plan curvature $= 2[(DH^2 + EG^2 - FGH)/(G^2 + H^2)] = 0.0111$

Curvature $= -2(D + E) = -0.0322$

All three measures are based on 1/100 (z units). The negative curvature value means the surface at the center cell is upwardly concave in the form of a shallow basin surrounded by higher elevations in the neighboring cells. Incidentally, the slope and aspect at the center cell can also be derived by using the coefficients G and H in the same way they are used by Fleming and Hoffer (1979) and Ritter (1987) (Section 13.3.1):

$$\text{Slope} = \sqrt{G^2 + H^2}$$
$$\text{Aspect} = \arctan(-H/-G)$$

across the direction of maximum slope. And curvature measures the difference between the two: profile curvature—plan curvature. A positive curvature value at a cell means that the surface is upwardly convex at the cell location. A negative curvature value means that the surface is upwardly concave. And a 0 value means that the surface is flat.

13.5 RASTER VERSUS TIN

We can often choose either elevation rasters or TINs for terrain mapping and analysis. GIS packages, such as ArcGIS, can also convert a raster to a TIN or a TIN to a raster (Box 13.8). Given these options, we might ask which data model to use. There is no easy answer to the question.

Essentially, rasters and TINs differ in data flexibility and computational efficiency.

A main advantage of using a TIN lies in the flexibility with input data sources. We can construct a TIN using inputs from DEM, contour lines, GPS data, LIDAR data, and survey data. We can add elevation points to a TIN at their precise locations and add breaklines, such as streams, roads, ridgelines, and shorelines, to define surface discontinuities. In comparison, a DEM or an elevation raster cannot indicate a stream in a hilly area and the accompanying topographic characteristics if the stream width is smaller than the DEM resolution.

An elevation raster is fixed with a given cell size. We cannot add new sample points to an elevation raster to increase its surface accuracy. Assuming

| Box 13.8 | **Terrain Mapping and Analysis Using ArcGIS** |

Tools for terrain mapping and analysis are available through the Spatial Analyst and 3-D Analyst extensions to ArcGIS. The Spatial Analyst toolbar has Raster Analysis Layer List, Contour, and Histogram. Starting with ArcGIS 10, many commands that used to be on the Spatial Analyst toolbar are only available in ArcToolbox. The Spatial Analyst Tools/Surface toolset includes tools for aspect, contour, curvature, cut fill, hillshade, and slope.

The 3-D Analyst toolbar has tools for creating and editing TINs; converting between TIN, raster, and feature layer; interpolating point, line, and polygon; and creating profile. Additionally, the 3-D Analyst toolbar offers 3-D visualization environments through ArcGlobe and ArcScene. ArcGlobe is functionally similar to ArcScene except that it can work with large and varied data sets such as high-resolution satellite images, high-resolution DEMs, and vector data. ArcScene, on the other hand, is designed for small local projects. ArcToolbox, through the 3-D Analyst Tools, offers many tools similar to those on the 3-D Analyst toolbar. Additionally, ArcToolbox has tools for deriving aspect, contour, and slope feature layers directly from TINs.

that the production method is the same, the only way to improve the accuracy of a raster is to increase its resolution, for example, from 30 meters to 10 meters. Researchers, especially those working with small watersheds, have in fact advocated DEMs with a 10-meter resolution (Zhang and Montgomery 1994) and even higher (Gertner et al. 2002). But increasing DEM resolution is a costly operation because it requires the recompiling of elevation data.

Besides providing flexibility of data sources, TIN is also an excellent data model for terrain mapping and 3-D display. Compared with the fuzziness of a DEM, the triangular facets of a TIN better define the land surface and create a sharper image. Most GIS users seem to prefer the look of a map based on a TIN than on a DEM (Kumler 1994).

Computational efficiency is the main advantage of using rasters for terrain analysis. The simple

data structure makes it relatively easy to perform neighborhood operations on an elevation raster. Therefore, using an elevation raster to compute slope, aspect, surface curvature, relative radiance, and other topographic variables is fast and efficient. In contrast, the computational load from using a TIN can increase significantly as the number of triangles increases. For some terrain analysis operations, a GIS package may in fact convert a TIN to an elevation raster prior to data analysis.

Finally, which data model is more accurate in measuring elevation, slope, aspect, and other topographic parameters? The answer depends on the production method. If a TIN is made from sampling a DEM, the TIN cannot be as accurate as the full DEM; likewise, a DEM cannot be as accurate as a TIN if the DEM is interpolated from the TIN (Kumler 1994).

KEY CONCEPTS AND TERMS

3-D draping: The method of superimposing thematic layers such as vegetation and roads on 3-D perspective views.

Aspect: The directional measure of slope.

Base contour: The contour from which contouring starts.

Breaklines: Line features that represent changes of the land surface such as streams, shorelines, ridges, and roads.

Contour interval: The vertical distance between contour lines.

Contour lines: Lines connecting points of equal elevation.

Delaunay triangulation: An algorithm for connecting points to form triangles such that all points are connected to their nearest neighbors and triangles are as compact as possible.

Hill shading: A graphic method that simulates how the land surface looks with the interaction between sunlight and landform features. The method is also known as *shaded relief.*

Hypsometric tinting: A mapping method that applies color symbols to different elevation zones. The method is also known as *layer tinting.*

Maximum *z*-tolerance: A TIN construction algorithm, which ensures that, for each elevation point selected, the difference between the original elevation and the estimated elevation from the TIN is within the specified tolerance.

Perspective view: A graphic method that produces 3-D views of the land surface.

Slope: The rate of change of elevation at a surface location, measured as an angle in degrees or as a percentage.

Vertical profile: A chart showing changes in elevation along a line such as a hiking trail, a road, or a stream.

Viewing angle: A parameter for creating a perspective view, measured by the angle from the horizon to the altitude of the observer.

Viewing azimuth: A parameter for creating a perspective view, measured by the direction from the observer to the surface.

Viewing distance: A parameter for creating a perspective view, measured by the distance between the viewer and the surface.

***z*-scale:**   The ratio between the vertical scale and the horizontal scale in a perspective view. Also called the *vertical exaggeration factor.*

REVIEW QUESTIONS

1. Describe the two common types of data for terrain mapping and analysis.
2. Go to the USGS NED website (**http://ned.usgs.gov/**) and read the information about the three types of DEMs available for download.
3. List the types of data that can be used to compile an initial TIN.
4. List the types of data that can be used to modify a TIN.
5. The maximum *z*-tolerance algorithm is an algorithm used by ArcGIS for converting a DEM into a TIN. Explain how the algorithm works.
6. Suppose you are given a DEM to make a contour map. The elevation readings in the DEM range from 856 to 1324 meters. If you were to use 900 as the base contour and 100 as the contour interval, what contour lines would be on the map?

7. Describe factors that can influence the visual effect of hill shading.
8. Explain how the viewing azimuth, viewing angle, viewing distance, and *z*-scale can change a 3-D perspective view.
9. Describe in your own words (no equation) how a computing algorithm derives the slope of the center cell by using elevations of its four immediate neighbors.
10. Describe in your own words (no equation) how ArcGIS derives the slope of the center cell by using elevations of its eight surrounding neighbors.
11. What factors can influence the accuracy of slope and aspect measures from a DEM?
12. Suppose you need to convert a raster from degree slope to percent slope. How can you complete the task in ArcGIS?

13. Suppose you have a polygon layer with a field showing degree slope. What kind of data processing do you have to perform in ArcGIS so that you can use the polygon layer in percent slope?

14. What are the advantages of using an elevation raster for terrain mapping and analysis?

15. What are the advantages of using a TIN for terrain mapping and analysis?

APPLICATIONS: TERRAIN MAPPING AND ANALYSIS

This applications section includes three tasks. Task 1 lets you create a contour layer, a vertical profile, a shaded-relief layer, and a 3-D perspective view from a DEM. In Task 2, you will derive a slope layer, an aspect layer, and a surface curvature layer from DEM data. Task 3 lets you build and modify a TIN. You will use Spatial Analyst, 3-D Analyst, and ArcToolbox to perform terrain mapping and analysis in this section.

Task 1 Use DEM for Terrain Mapping

What you need: *plne,* an elevation raster; and *streams.shp,* a stream shapefile.

The elevation raster *plne* is imported from a USGS 7.5-minute DEM. The shapefile *streams .shp* shows major streams in the study area.

1.1 Create a contour layer

1. Start ArcCatalog, and connect to the Chapter 13 database. Launch ArcMap. Add *plne* to Layers, and rename Layers Tasks 1&2. Click the Customize menu, point to Extensions, and make sure that both the Spatial Analyst and 3-D Analyst extensions are checked.

2. Open ArcToolbox. Set the Chapter 13 database for the current workspace. Double-click the Contour tool in the Spatial Analyst Tools/Surface toolset. In the Contour dialog, select *plne* for the input raster, save the output polyline features as *ctour.shp,* enter 100 (meters) for the contour interval, enter 800 (meters) for the base contour, and click OK.

3. *ctour* appears in the map. Select Properties from the context menu of *ctour.* On the Labels tab, check the box to label features in

this layer and select CONTOUR for the label field. Click OK. The contour lines are now labeled. (To remove the contour labels, right-click *ctour* and uncheck Label Features.)

1.2 Create a vertical profile

1. Add *streams.shp* to Tasks 1&2. This step selects a stream for the vertical profile. Open the attribute table of *streams.* Click the Select By Attributes button and enter the following SQL statement in the expression box: "USGH_ ID" = 167. Click Apply. Close the *streams* attribute table. Zoom in on the selected stream.

2. Click the Customize menu, point to Toolbars, and check the 3-D Analyst toolbar. *plne* should appear as a layer in the toolbar. Click the Interpolate Line tool on the 3-D Analyst toolbar. Use the mouse pointer to digitize points along the selected stream. Double-click the last point to finish digitizing. A rectangle with handles appears around the digitized stream.

3. Click the Create Profile Graph tool on the 3-D Analyst toolbar. A vertical profile appears with a default title and footer. Right-click the title bar of the graph and select Properties. The Graph Properties dialog allows you to enter a new title and footer and to choose other advanced design options.

Q1. What is the elevation range along the vertical profile? Does the range correspond to the readings on *ctour* from Task 1.1?

4. The digitized stream becomes a graphic element on the map. You can delete it by using the Select Elements tool to first select

it. To unselect the stream, choose Clear Selected Features from the Selection menu.

1.3 Create a hillshade layer

1. Double-click the Hillshade tool in the Spatial Analyst Tools/Surface toolset. In the Hillshade dialog, select *plne* for the Input surface, save the output raster as *hillshade*. Notice that the default azimuth is 315 and the default altitude is 45. Click OK to run the tool.

2. Try different values of azimuth and altitude to see how these two parameters affect hill shading.

Q2. Does the hillshade layer look darker or lighter with a lower altitude?

1.4 Create a perspective view

1. Click the ArcScene tool on the 3-D Analyst toolbar to open the ArcScene application. Add *plne* and *streams.shp* to view. By default, *plne* is displayed in a planimetric view, without the 3-D effect. Select Properties from the context menu of *plne*. On the Base Heights tab, click the radio button for floating on a custom surface, and select *plne* for the surface. Click OK to dismiss the dialog.

2. *plne* is now displayed in a 3-D perspective view. The next step is to drape *streams* on the surface. Select Properties from the context menu of *streams*. On the Base Heights tab, click the radio button for floating on a custom surface, and select *plne* for the surface. Click OK.

3. Using the properties of *plne* and *streams,* you can change the look of the 3-D view. For example, you can change the color scheme for displaying *plne*. Select Properties from the context menu of *plne*. On the Symbology tab, right-click the Color Ramp box and uncheck Graphic View. Click the Color Ramp dropdown arrow and select Elevation #1. Click OK. Elevation #1 uses the conventional color scheme to display the 3-D view of *plne*. Click the symbol for *streams* in

the table of contents. Select the River symbol from the Symbol Selector, and click OK.

4. You can tone down the color symbols for *plne* so that *streams* can stand out more. Select Properties from the context menu of *plne*. On the Display tab, enter 40 (%) transparency and click OK.

5. Click the View menu and select Scene Properties. The Scene Properties dialog has four tabs: General, Coordinate System, Extent, and Illumination. The General tab has options for the vertical exaggeration factor (the default is none), background color, and a check box for enabling animated rotation. The Illumination tab has options for azimuth and altitude.

6. ArcScene has standard tools to navigate, zoom in or out, center on target, zoom to target, and to perform other 3-D manipulations. For example, the Navigate tool allows you to rotate the 3-D surface.

7. Besides the preceding standard tools, ArcScene has additional toolbars for perspective views. Click the Customize menu, point to Toolbars, and check the boxes for 3-D Effects and Animation. The 3-D Effects toolbar has tools for adjusting transparency, lighting, and shading. The Animation toolbar has tools for making animations. For example, you can save an animation as an .avi file and use it in a PowerPoint presentation. Close ArcScene.

Task 2 Derive Slope, Aspect, and Curvature from DEM

What you need: *plne,* an elevation raster, same as Task 1.

Task 2 covers slope, aspect, and surface curvature.

2.1 Derive a slope layer

1. Double-click the Slope tool in the Spatial Analyst Tools/Surface toolset. Select *plne* for the input raster, specify *plne_slope* for the output raster, select PERCENT_RISE for the

output measurement, and click OK to execute the command.

Q3. What is the range of percent slope values in *plne_slope*?

2. *plne_slope* is a continuous raster. You can divide *plne_slope* into slope classes. Double-click the Reclassify tool in the Spatial Analyst Tools/Reclass toolset. In the Reclassify dialog, select *plne_slope* for the input raster and click on Classify. In the next dialog, use the number of classes of 5, enter 10, 20, 30, and 40 as the first four break values, save the output raster as *rec_slope,* and click OK.

2.2 Derive an aspect layer

1. Double-click the Aspect tool in the Spatial Analyst Tools/Surface toolset. Select *plne* for the input raster, specify *plne_aspect* for the output raster, and click OK.

2. *plne_aspect* shows an aspect layer with the eight principal directions and flat area. But it is actually a continuous raster. You can verify the statement by checking the layer properties. To create an aspect raster with the eight principal directions, you need to reclassify *plne_aspect*.

3. Double-click the Reclassify tool in the Spatial Analyst Tools/Reclass toolset. Select *plne_aspect* for the input raster and click on Classify. In the Classification dialog, make sure that the number of classes is 10. Then click the first cell under Break Values and enter −1. Enter 22.5, 67.5, 112.5, 157.5, 202.5, 247.5, 292.5, 337.5, and 360 in the following nine cells. Click OK to dismiss the Classification dialog.

4. The old values in the Reclassify dialog are now updated with the break values you have entered. Now you have to change the new values. Click the first cell under new values and enter −1. Click and enter 1, 2, 3, 4, 5, 6, 7, 8, and 1 in the following nine cells. The last cell has the value 1 because the cell

(337.5° to 360°) and the second cell (−1° to 22.5°) make up the north aspect. Save the output raster as *rec_aspect,* and click OK. The output is an integer aspect raster with the eight principal directions and flat (−1).

Q4. The value with the largest number of cells on the reclassed aspect raster is −1. Can you speculate why?

2.3 Derive a surface curvature layer

1. Double-click the Curvature tool in the Spatial Analyst Tools/Surface toolset. Select *plne* for the input raster, specify *plne_curv* for the output raster, and click OK.

2. A positive cell value in *plne_curv* indicates that the surface at the cell location is upwardly convex. A negative cell value indicates that the surface at the cell location is upwardly concave. The ArcGIS Desktop Help further suggests that the curvature output value should be within −0.5 to 0.5 in a hilly area and within −4 to 4 in rugged mountains. The elevation data set *plne* is from the Priest Lake area in North Idaho, a mountainous area with steep terrain. Therefore, it is no surprise that *plne_curv* has cell values ranging from −6.89 to 6.33.

3. Right-click *plne_curv* and select Properties. On the Symbology tab, change the show type to Classified, and then click on Classify. In the Classification dialog, first select 6 for the number of classes and then enter the following break values: −4, −0.5, 0, 0.5, 4, and 6.34. Return to the Properties dialog, select a diverging color ramp (e.g., red to green diverging, bright), and click OK.

4. Through the color symbols, you can now tell upwardly convex cells in *plne_curv* from upwardly concave cells. Priest Lake on the west side carries the symbol for the −0.5 to 0 class; its cell value is actually 0 (flat surface). Add *streams.shp* to Tasks 1&2, if necessary. Check cells along the stream channels. Many of these cells should have symbols indicating upwardly concave.

Task 3 Build and Display a TIN

What you need: *emidalat,* an elevation raster; and *emidastrm.shp,* a stream shapefile.

Task 3 shows you how to construct a TIN from an elevation raster and to modify the TIN with *emidastrm.shp* as breaklines. You will also display different features of the TIN.

1. Select Data Frame from the Insert menu in ArcMap. Rename the new data frame Task 3, and add *emidalat* and *emidastrm.shp* to Task 3.

2. Double-click the Raster to TIN tool in the 3-D Analyst Tools/Conversion/From Raster toolset. Select *emidalat* for the input raster, specify *emidatin* for the output TIN, and change the *Z* Tolerance value to 10. Click OK to run the command.

Q5. The default *Z* tolerance in the Raster to TIN dialog is 48.2. What happens when you change the tolerance to 10?

3. *emidatin* is an initial TIN converted from *emidalat.* This step is to modify *emidatin* with *emidastrm,* which contains streams. Double-click the Edit TIN tool in the 3-D Analyst Tools/TIN Management toolset. Select *emidatin* for the input TIN, and select *emidastrm* for the input feature class. Notice that the default for SF_type (surface feature type) is hardline. Click OK to edit the TIN.

4. You can view the edited *emidatin* in a variety of ways. Select Properties from the context menu of *emidatin.* Click the Symbology tab. Click the Add button below the Show frame. An Add Renderer scroll list appears with choices related to the display of edges, faces, or nodes that make up *emidatin.* Click "Faces with the same symbol" in the list, click Add,

and click Dismiss. Uncheck all the boxes in the Show frame except Faces. Make sure that the box to show hillshade illumination effect in 2-D display is checked. Click OK on the Layer Properties dialog. With its faces in the same symbol, *emidatin* can be used as a background in the same way as a shaded-relief map for displaying map features such as streams, vegetation, and so on.

Q6. How many nodes and triangles are in the edited *emidatin*?

Challenge Task

What you need: *lidar, usgs10,* and *usgs30.*

This challenge task lets you work with DEMs at three different resolutions: *lidar* at 1.83 meters, *usgs10* at 10 meters, and *usgs30* at 30 meters.

1. Insert a new data frame in ArcMap and rename the data frame Challenge. Add *lidar, usgs10,* and *usgs30* to Challenge.

Q1. How many rows and columns are in each DEM?

2. Create a hillshade layer from each DEM and compare the layers in terms of the coverage of topographic details.

3. Create a degree slope layer from each DEM and reclassify the slope layer into nine classes: 0–10°, 10–20°, 20–30°, 30–40°, 40–50°, 50–60°, 60–70°, 70–80°, and 80–90°.

Q2. List the area percentage of each slope class from each DEM.

Q3. Summarize the effect of DEM resolution on the slope layers.

REFERENCES

Bolstad, P. V., and T. Stowe. 1994. An Evaluation of DEM Accuracy: Elevation, Slope, and Aspect. *Photogrammetric Engineering and Remote Sensing* 60: 1327–32.

Campbell, J. 1984. *Introductory Cartography.* Englewood Cliffs, NJ: Prentice Hall.

Carter, J. R. 1992. The Effect of Data Precision on the Calculation of Slope and Aspect Using Gridded DEMs. *Cartographica* 29: 22–34.

Chang, K., and Z. Li. 2000. Modeling Snow Accumulation with a Geographic Information System. *International Journal of Geographical Information Science* 14: 693–707.

Chang, K., and B. Tsai. 1991. The Effect of DEM Resolution on Slope and Aspect Mapping. *Cartography and Geographic Information Systems* 18: 69–77.

Clarke, K. C. 1995. *Analytical and Computer Cartography,* 2d ed. Englewood Cliffs, NJ: Prentice Hall.

Collier, P., D. Forrest, and A. Pearson. 2003. The Representation of Topographic Information on Maps: The Depiction of Relief. *The Cartographic Journal* 40: 17–26.

Deng, Y., J. P. Wilson, and B. O. Bauer. 2007. DEM Resolution Dependencies of Terrain Attributes across a Landscape. *International Journal of Geographical Information Science* 21: 187–213.

Eyton, J. R. 1991. Rate-of-Change Maps. *Cartography and*

Geographic Information Systems 18: 87–103.

Fleming, M. D., and R. M. Hoffer. 1979. *Machine Processing of Landsat MSS Data and DMA Topographic Data for Forest Cover Type Mapping.* LARS Technical Report 062879. Laboratory for Applications of Remote Sensing, Purdue University, West Lafayette, IN.

Flood, M. 2001. Laser Altimetry: From Science to Commercial LIDAR Mapping. *Photogrammetric Engineering and Remote Sensing* 67: 1209–17.

Florinsky, I. V. 1998. Accuracy of Local Topographic Variables Derived from Digital Elevation Models. *International Journal of Geographical Information Systems* 12: 47–61.

Franklin, S. E. 1987. Geomorphometric Processing of Digital Elevation Models. *Computers & Geosciences* 13: 603–9.

Gallant J. C., and J. P. Wilson. 2000. Primary Topographic Attributes. In J. P. Wilson and J. C. Gallant, eds., *Terrain Analysis: Principles and Applications,* pp. 51–85. New York: Wiley.

Gao, J. 1998. Impact of Sampling Intervals on the Reliability of Topographic Variables Mapped from Raster DEMs at a Micro-Scale. *International Journal of Geographical Information Systems* 12: 875–90.

Gertner, G., G. Wang, S. Fang, and A. B. Anderson. 2002.

Effect and Uncertainty of Digital Elevation Model Spatial Resolutions on Predicting the Topographical Factor for Soil Loss Estimation. *Journal of Soil and Water Conservation* 57: 164–74.

Gesch, D. B. 2007. The National Elevation Dataset. In D. Maune, ed., *Digital Elevation Model Technologies and Applications: The DEM Users Manual,* 2d ed., pp. 99–118. Bethesda, MD: American Society for Photogrammetry and Remote Sensing.

Hill, J. M., L. A. Graham, and R. J. Henry. 2000. Wide-Area Topographic Mapping and Applications Using Airborne Light Detection and Ranging (LIDAR) Technology. *Photogrammetric Engineering and Remote Sensing* 66: 908–14.

Hodgson, M. E. 1998. Comparison of Angles from Surface Slope/Aspect Algorithms. *Cartography and Geographic Information Systems* 25: 173–85.

Horn, B. K. P. 1981. Hill Shading and the Reflectance Map. *Proceedings of the IEEE* 69 (1): 14–47.

Isaacson, D. L., and W. J. Ripple. 1990. Comparison of 7.5-Minute and 1-Degree Digital Elevation Models. *Photogrammetric Engineering and Remote Sensing* 56: 1523–27.

Jones, K. H. 1998. A Comparison of Algorithms Used to Compute Hill Slope as a Property of the DEM. *Computers & Geosciences* 24: 315–23.

Jones, T. A., D. E. Hamilton, and C. R. Johnson. 1986. *Contouring Geologic Surfaces with the Computer.* New York: Van Nostrand Reinhold.

Kienzle, S. 2004. The Effect of DEM Raster Resolution on First Order, Second Order and Compound Terrain Derivatives. *Transactions in GIS* 8: 83–111.

Kumler, M. P. 1994. An Intensive Comparison of Triangulated Irregular Networks (TINs) and Digital Elevation Models (DEMs). *Cartographica* 31 (2): 1–99.

Lane, S. N., K. S. Richards, and J. H. Chandler, eds., 1998. *Landform Monitoring, Modelling and Analysis.* Chichester, England: Wiley.

Lee, J. 1991. Comparison of Existing Methods for Building Triangular Irregular Network Models of Terrain from Raster Digital Elevation Models. *International Journal of Geographical Information Systems* 5: 267–85.

Lefsky, M. A., W. B. Cohen, G. G. Parker, and D. J. Harding. 2002. Lidar Remote Sensing for Ecosystem Studies. *BioScience* 52: 19–30.

Lim, K., P. Treitz, M. Wulder, B. St-Onge, and M. Flood. 2003. Lidar Remote Sensing of Forest Structure. *Progress in Physical Geography* 27: 88–106.

Moore, I. D., R. B. Grayson, and A. R. Ladson. 1991. Digital Terrain Modelling: A Review of Hydrological, Geomorphological, and Biological Applications. *Hydrological Process* 5: 3–30.

Murphy, P. N. C., J. Ogilvie, F. Meng, and P. Arp. 2008. Stream Network Modelling Using Lidar and Photogrammetric Digital Elevation Models: A Comparison and Field Verification. *Hydrological Processes* 22: 1747–54.

Ritter, P. 1987. A Vector-Based Slope and Aspect Generation Algorithm. *Photogrammetric Engineering and Remote Sensing* 53: 1109–11.

Schmidt, J., I. Evans, and J. Brinkmann. 2003. Comparison of Polynomial Models for Land Surface Curvature Calculation. *International Journal of Geographical Information Science* 17: 797–814.

Sharpnack, D. A., and G. Akin. 1969. An Algorithm for Computing Slope and Aspect from Elevations. *Photogrammetric Engineering* 35: 247–48.

Skidmore, A. K. 1989. A Comparison of Techniques for Calculating Gradient and Aspect from a Gridded Digital Elevation Model. *International Journal of Geographical Information Systems* 3: 323–34.

Srinivasan, R., and B. A. Engel. 1991. Effect of Slope Prediction Methods on Slope and Erosion Estimates. *Applied Engineering in Agriculture* 7: 779–83.

Thelin, G. P., and R. J. Pike. 1991. *Landforms of the Conterminous United States: A Digital Shaded-Relief Portrayal,* Map I-2206, scale 1:3,500,000. Washington, DC: U.S. Geological Survey.

Tsai, V. J. D. 1993. Delaunay Triangulations in TIN Creation: An Overview and Linear Time Algorithm. *International Journal of Geographical Information Systems* 7: 501–24.

Watson, D. F., and G. M. Philip. 1984. Systematic Triangulations. *Computer Vision, Graphics, and Image Processing* 26: 217–23.

Wilson, J. P., and J. C. Gallant, eds. 2000. *Terrain Analysis: Principles and Applications.* New York: Wiley.

Zar, J. H. 1984. *Biostatistical Analysis,* 2d ed. Englewood Cliffs, NJ: Prentice Hall.

Zevenbergen, L. W., and C. R. Thorne. 1987. Quantitative Analysis of Land Surface Topography. *Earth Surface Processes and Landforms* 12: 47–56.

Zhang, W., and D. R. Montgomery. 1994. Digital Elevation Model Raster Size, Landscape Representation, and Hydrologic Simulations. *Water Resources Research* 30: 1019–28.

Zhou, Q., X. Liu, and Y. Sun. 2006. Terrain Complexity and Uncertainty in Grid-Based Digital Terrain Analysis. *International Journal of Geographical Information Science* 20: 1137–47.

VIEWSHEDS AND WATERSHEDS

Terrain analysis covers basic topographic parameters such as slope, aspect, and surface curvature (Chapter 13) as well as more specific applications. Chapter 14 focuses on two such applications: viewshed analysis and watershed analysis.

A viewshed is an area that is visible from a viewpoint and, in GIS, viewshed analysis refers to the derivation and accuracy assessment of viewsheds. Examples of viewsheds range from the vast area visible from the observation deck of the Empire State Building to the service area of a communication tower. Studies of viewsheds, in addition to delineating visible areas, may also analyze the visual impact

or the "value" that the view offers. They have found, for example, people are willing to pay more for a hotel room with a view (Lange and Schaeffer 2001) or a high-rise apartment with a view of parks and water (Bishop, Lange, and Mahbubul 2004).

A watershed is the area that drains surface water to a common outlet, and watershed analysis traces the flow and channeling of surface water on the terrain so that watersheds can be correctly delineated. Studies of watersheds rarely stop at the mapping of watershed boundaries. As an example, a watershed delineation program in Wisconsin was designed for assessing the impact of agricultural non-point source pollution on water quality and aquatic ecosystems (Maxted, Diebel, and Vander Zanden 2009).

Chapter 14 is organized into six sections. Section 14.1 introduces viewshed analysis using a digital elevation model (DEM) or a triangulated irregular network (TIN). Section 14.2 examines various parameters such as viewpoint, viewing angle, search distance, and tree height that can affect viewshed analysis. Section 14.3 provides an overview of viewshed applications. Section 14.4 introduces watershed analysis and the analysis procedure using a DEM. Section 14.5 discusses factors such as methods for determining flow directions that can influence the outcome of a watershed analysis. Section 14.6 covers applications of watershed analysis.

14.1 VIEWSHED ANALYSIS

A **viewshed** refers to the portion of the land surface that is visible from one or more viewpoints (Figure 14.1). The process for deriving viewsheds is called viewshed or visibility analysis. A viewshed analysis requires two input data sets. The first is usually a point layer containing one or more viewpoints such as a layer containing communication towers. If a line layer such as a layer containing historical trails is used, the viewpoints are points that make up the linear features. The second input is a DEM (i.e., an elevation raster) or a TIN, which represents the land surface. Using these two inputs, viewshed analysis can derive visible areas, representing, for example, service areas of communication towers or scenic views from historical trails.

14.1.1 Line-of-Sight Operation

The **line-of-sight** operation is the basis for viewshed analysis. The line of sight, also called *sightline,* connects the viewpoint and the target. If any

land, or any object on the land, rises above the line, then the target is invisible to the viewpoint. If no land or object blocks the view, then the target is visible to the viewpoint. Rather than just marking the target point as visible or not, a GIS can display a sightline with symbols for the visible and invisible portions along the sightline.

Figure 14.2 illustrates a line-of-sight operation over a TIN. Figure 14.2*a* shows the visible (white) and invisible (black) portions of the sightline connecting the viewpoint and the target point.

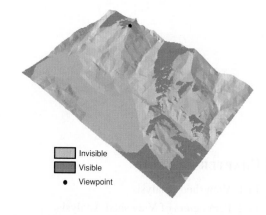

Figure 14.1
A viewshed example.

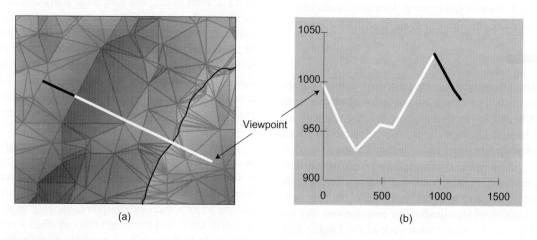

Figure 14.2
A sightline connects two points on a TIN in (*a*). The vertical profile of the sightline is depicted in (*b*). In both diagrams, the visible portion is shown in white and the invisible portion in black.

Figure 14.2*b* shows the vertical profile along the sightline. In this case, the viewpoint is at an elevation of 994 meters on the east side of a stream. The visible portion of the sightline follows the downslope, crosses the stream at the elevation of 932 meters, and continues uphill on the west side of the stream. The ridgeline at an elevation of 1028 meters presents an obstacle and marks the beginning of the invisible portion.

Viewshed analysis expands the line-of-sight operation to cover every possible cell or every possible TIN facet in the study area. Because viewshed analysis can be a time-consuming operation, various algorithms have been developed for computing viewsheds (De Floriani and Magillo 2003). Some algorithms are designed for elevation rasters, and others are for TINs. A commercial GIS package usually does not provide choices of algorithms or information on the adopted algorithm (Riggs and Dean 2007). ArcGIS, for example, takes an elevation raster as the data source and saves the output of a viewshed analysis in raster format to take advantage of the computational efficiency of raster data.

14.1.2 Raster-Based Viewshed Analysis

Deriving a viewshed from an elevation raster follows a series of steps. First, a sightline is set up between the viewpoint and a target location (e.g., the center of a cell). Second, a set of intermediate points is derived along the sightline. Typically, these intermediate points are chosen from the intersections between the sightline and the grid lines of the elevation raster, disregarding areas inside the grid cells (De Floriani and Magillo 2003). Third, the elevations of the intermediate points are estimated (e.g., by linear interpolation). Finally, the computing algorithm examines the elevations of the intermediate points and determines if the target is visible or not.

This procedure can be repeated using each cell in the elevation raster as a target (Clarke 1995). The result is a raster that classifies cells into the visible and invisible categories. An algorithm proposed by Wang, Robinson, and White. (1996) takes a different approach to save the computer processing time.

Before running the line-of-sight operations, their algorithm first screens out invisible cells by analyzing the local surface at the viewpoint and the target.

14.1.3 TIN-Based Viewshed Analysis

Deriving viewsheds from a TIN is not as well defined as from an elevation raster. Different rules can be applied. The first rule determines whether a TIN triangle can be divided into visible and invisible parts (De Floriani and Magillo 1994, 1999) or whether an entire triangle can be defined as either visible or invisible (Goodchild and Lee 1989; Lee 1991). The latter is simpler than the former in computer processing time. Assuming that an entire triangle is to be either visible or invisible, the second rule determines whether the visibility is to be based on one, two, or all three points that make up the triangle or the center point (label point) of the triangle (Riggs and Dean 2007). The one-point rule is not as stringent as the two- or three-point rule.

14.1.4 Cumulative Viewshed

The output of a viewshed analysis, using either an elevation raster or a TIN as the data source, is a binary map showing visible and not-visible areas. Given one viewpoint, a viewshed map has the value of 1 for visible and 0 for not visible. Given two or more viewpoints, a viewshed map becomes a **cumulative viewshed** map. Two options are common for presenting a cumulative viewshed map. The first option uses counting operations. For example, a cumulative viewshed map based on two viewpoints has three possible values: 2 for visible from both points, 1 for visible from one point, and 0 for not visible (Figure 14.3*a*). The number of possible values in such a viewshed map is $n + 1$, where n is the number of viewpoints. The second option uses Boolean operations. Suppose two viewpoints for a viewshed analysis are labeled J and K. Using the viewsheds derived for each viewpoint and the Combine local operation (Chapter 12), we can divide the visible areas of a cumulative viewshed map into visible to J only, visible to K only, or visible to both J and K (Figure 14.3*b*).

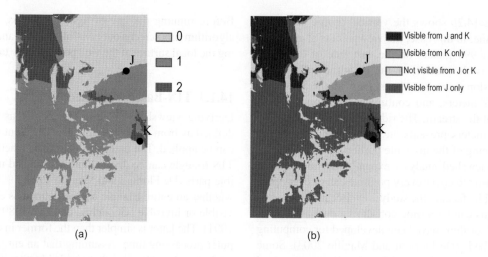

Figure 14.3
Two options for presenting a cumulative viewshed map: the counting option (*a*) and the Boolean option (*b*).

14.1.5 Accuracy of Viewshed Analysis

The accuracy of viewshed analysis depends on the accuracy of the surface data, the data model (i.e., TIN versus DEM), and the rule for judging visibility (Fisher 1991, 1993; Riggs and Dean 2007). According to Maloy and Dean (2001), the average level of agreement is only slightly higher than 50 percent between GIS-predicted raster-based viewsheds and field-surveyed viewsheds. A more recent study (Riggs and Dean 2007) finds the level of agreement ranging between 66 and 85 percent, depending on the GIS software and the DEM resolution. These findings suggest that it may be better to express visibility in probabilistic terms (Fisher 1996). Although a probabilistic visibility map is currently not available from commercial GIS packages, it is possible to produce one by incorporating an error source such as DEM errors in Monte Carlo simulations (Nackaerts, Govers, and Van Orshoven 1999).

14.2 PARAMETERS OF VIEWSHED ANALYSIS

A number of parameters can influence the result of a viewshed analysis. The first parameter is the viewpoint. A viewpoint located along a ridge line would

have a wider view than a viewpoint located in a narrow valley. There are at least two scenarios in dealing with the viewpoint in GIS. The first scenario assumes that the location of the point is fixed. If the elevation at the point is known, it can be entered directly in a field. If not, it can be estimated from an elevation raster or a TIN. ArcGIS, for example, uses bilinear interpolation (Chapter 6) to interpolate the elevation of a viewpoint from an elevation raster. The second scenario assumes that the viewpoint is to be selected. If we further assume that the objective is to gain maximum visibility, then we should select the viewpoint at a high elevation with open views. A GIS provides various tools that can help locate suitable viewpoints (Box 14.1).

After the viewpoint is determined, its elevation should be increased by the height of the observer and, in some cases, the height of a physical structure. For instance, a forest lookout station is usually 15 to 20 meters high. The height of the observation station, when added as an offset value to the elevation at the station, makes the viewpoint higher than its immediate surroundings, thus increasing its viewshed (Figure 14.4).

The second parameter is the **viewing azimuth,** which sets horizontal angle limits to the view.

ArcGIS has various tools that we can use to select viewpoints at high elevations with open views. Tools such as contouring and hill shading can provide an overall view of the topography in a study area. Tools for data query can narrow the selection to a specific elevation range such as within 100 meters of the highest elevation in the data set. Data extraction tools can extract elevation readings from an underlying surface (i.e., a raster or a TIN) at point locations, along a line, or within a circle, a box, or a polygon. These extraction tools are useful for narrowing the selection of a viewpoint to a small area with high elevations. However, it will be difficult to find the specific elevation of a viewpoint because that is estimated using four closest cell values if the underlying surface is a raster, and elevations at three nodes of a triangle if the underlying surface is a TIN.

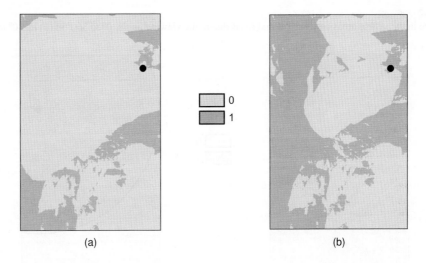

(a)   (b)

Figure 14.4
The increase of the visible areas from (*a*) to (*b*) is a direct result of adding 20 meters to the height of the viewpoint.

Figure 14.5, for example, uses a viewing angle from 0° to 180°. The default is a full 360° sweep, which is unrealistic in many instances. To simulate the view from the window of a property (e.g., a home or an office), a 90° viewing azimuth (45° either side of the perpendicular to the window) is more realistic than a 360° sweep (Lake et al. 1998).

Viewing radius is the third parameter, which sets the search distance for deriving visible areas. Figure 14.6, for example, shows viewable areas within a radius of 8000 meters around the viewpoint. The default view distance is typically infinity. The setting of the search radius can vary with project. A 5-km radius has been used, for example, to simulate the view from an observation tower that guards the access to an ancient city (Nackaerts, Govers, and Van Orshoven 1999).

Other parameters include vertical viewing angle limits, the Earth's curvature, tree height, and building height. Vertical viewing angles can

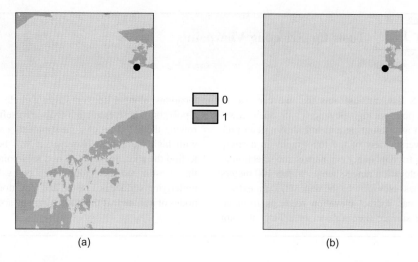

(a)   (b)

Figure 14.5
The difference in the visible areas between (a) and (b) is due to the viewing angle: 0° to 360° in (a) and 0° to 180° in (b).

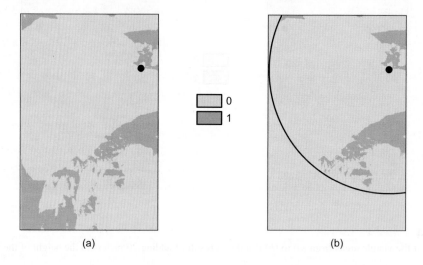

(a)   (b)

Figure 14.6
The difference in the visible areas between (a) and (b) is due to the search radius: infinity in (a) and 8000 meters from the viewpoint in (b).

range from 90° above the horizontal plane to −90° below. The Earth's curvature can be either ignored or corrected in deriving a viewshed. Tree height can be an important factor for a viewshed analysis involving forested lands. Estimated tree heights can be added to ground elevations to create forest (canopy) elevations (Wing and Johnson 2001). But tree height is not an issue if the DEM represents surface elevation as with an SRTM (Shuttle Radar Topography Mission) DEM (Chapter 4). Similar to tree heights, building heights and locations can be incorporated into DEMs for viewshed analysis in urban areas (Sander and Monson 2007).

Box 14.2 | **Parameters for Viewshed Analysis**

To set up the parameters for viewshed analysis in ArcGIS, we must work with the attribute table of the viewpoint data set. Each parameter has a predefined field name. OFFSETA and OFFSETB define the heights to be added to the viewpoint and the target respectively. AZIMUTH1 and AZIMUTH2 define the view's horizontal angle limits. For example, to limit the viewing angle from 0° to 180°, we can assign 0 to AZIMUTH1 and 180 to AZIMUTH2. RADIUS1 and RADIUS2 define the search distance. For example, to limit the search distance to 2 miles from the viewpoint, we can assign 0 to RADIUS1 and 2 miles to RADIUS2. VERT1 and VERT2 define the view's vertical angle limits. The vertical angle values range from 90° to −90°. Task 1 in the applications section shows how to define and add an OFFSETA value to a viewpoint.

A GIS package handles the parameters for viewshed analysis as attributes in the viewpoint data set. Therefore, we must set up the parameters before running viewshed analysis. Box 14.2 describes the setup of viewshed analysis parameters in ArcGIS.

14.3 APPLICATIONS OF VIEWSHED ANALYSIS

Viewshed analysis is useful for the site selection of facilities such as forest lookout stations, wireless telephone base stations, and microwave towers for radio and television. The location of these facilities is chosen to maximize the viewable (service) areas without having too much overlap. Viewshed analysis can help locate these facilities, especially at the preliminary stage.

Viewshed analysis can be useful for evaluating housing and resort area developments, although the objective of the analysis can differ between the developer and current residents. New facilities can easily intrude on residents in rural areas (Davidson, Watson, and Selman 1993). Visual intrusion and noise associated with road development can also affect property values in an urban environment (Lake et al. 1998). And in rural areas, the clustering of large greenhouses for horticulture can impact the visual quality (Rogge, Nevens, and Gulinck 2008).

Viewshed analysis is closely related to landscape analysis of visual quality and visual impact (Bishop 2003). The U.S. Forest Service, for example, uses viewshed analysis for delineating landscape view areas and for reducing the visual impact of clear-cuts and logging activities. A national park or a nature reserve may want to monitor and assess the impacts of land cover change on the visual landscape (Miller 2001). And measures derived from visibility analysis may be useful for quantifying the perceptual characteristics of landscapes (O'Sullivan and Turner 2001). Viewshed analysis can also be integrated with least-cost path calculations (Chapter 17) to provide scenic paths for hikers and others (Lee and Stucky 1998). Finally, viewshed analysis can be a tool for preparing 3-D visualization (Kumsap, Borne, and Moss 2005).

14.4 WATERSHED ANALYSIS

A **watershed** refers to an area, defined by topographic divides, that drains surface water to a common outlet. A watershed is often used as a unit area for the management and planning of water and other natural resources. **Watershed analysis** refers to the process of using DEMs and following water flows to delineate stream networks and watersheds.

Traditionally, watershed boundaries are drawn manually onto a topographic map. The person who draws the boundaries uses topographic features on the map to determine where a divide is

Box 14.3 **Watershed Boundary Dataset (WBD)**

For most purposes, hydrologic units are synonymous with watersheds. A hierarchical four-level hydrologic unit code (HUC), including region, subregion, accounting units, and cataloging unit, was developed in the United States during the 1970s. The new Watershed Boundary Dataset (WBD) adds two finer levels of watershed and subwatershed to the four-level HUC. Watersheds and subwatersheds are delineated and georeferenced to the USGS 1:24,000-scale quadrangle maps based on a common set of guidelines documented in the Federal Standards for Delineation of Hydrologic Unit Boundaries (**http://acwi.gov/spatial/index.html**). On average, a watershed covers 40,000 to 250,000 acres and a subwatershed covers 10,000 to 40,000 acres. The term watershed therefore refers to a specific hydrologic unit in the WBD.

located. Today computer programs are also used to derive watersheds from DEMs. Using computer technology, we can generate preliminary watershed boundaries in a fraction of the time needed for the traditional method.

Delineation of watersheds can take place at different spatial scales (Band et al. 2000; Whiteaker et al. 2007). A large watershed may cover an entire stream system and, within the watershed, there may be smaller watersheds, one for each tributary in the stream system. This is why the Watershed Boundary Dataset that is being developed by a number of federal agencies in the United States includes six levels (Box 14.3).

Delineation of watersheds can also be area-based or point-based. An area-based method divides a study area into a series of watersheds, one for each stream section. A point-based method, on the other hand, derives a watershed for each select point. The select point may be an outlet, a gauge station, or a dam. Whether area- or point-based, the automated method for delineating watersheds follows a series of steps, starting with a filled DEM.

14.4.1 Filled DEM

A **filled DEM** is void of depressions or sinks. A depression is a cell or cells surrounded by higher-elevation values, thus representing an area of internal drainage. Although some depressions are real, such as quarries or glaciated potholes, many are imperfections in the DEM. Therefore depressions must be removed from an elevation raster. A common method for removing a depression is to increase its cell value to the lowest overflow point out of the sink (Jenson and Domingue 1988). The flat surface resulting from sink filling still needs to be interpreted to define the drainage flow. One approach is to impose two shallow gradients and to force flow away from higher terrain surrounding the flat surface toward the edge bordering lower terrain (Garbrecht and Martz 2000).

14.4.2 Flow Direction

A **flow direction raster** shows the direction water will flow out of each cell of a filled elevation raster. Flow directions are commonly determined using single or multiple flow direction methods. D8 is a popular single flow direction method. Used by ArcGIS, the D8 method assigns a cell's flow direction to the one of its eight surrounding cells that has the steepest distance-weighted gradient (Figure 14.7) (O'Callaghan and Mark 1984). Multiple flow direction methods allow flow divergence or flow bifurcation (Freeman 1991; Gallant and Wilson 2000; Endreny and Wood 2003). An example is the D∞ (D Infinity) method, which partitions flow from a cell into two adjacent cells (Tarboton 1997). The D∞ method first forms eight

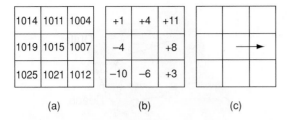

(a)   (b)   (c)

Figure 14.7

The flow direction of the center cell in (a) is determined by first calculating the distance-weighted gradient to each of its eight neighbors. For the four immediate neighbors, the gradient is calculated by dividing the elevation difference (b) between the center cell and the neighbor by 1. For the four corner neighbors, the gradient is calculated by dividing the elevation difference (b) by 1.414. The results show that the steepest gradient, and therefore the flow direction, is from the center cell to its right.

triangles by connecting the centers of the cell and its eight surrounding cells. It selects the triangle with the maximum downhill slope as the flow direction. The two neighboring cells that the triangle intersects receive the flow in proportion to their closeness to the aspect of the triangle. The D∞ method is available as an extension to ArcGIS.

14.4.3 Flow Accumulation

A **flow accumulation raster** tabulates for each cell the number of cells that will flow to it. The tabulation is based on the flow direction raster (Figure 14.8). With the appearance of a spanning tree (Figure 14.9), a flow accumulation raster records how many upstream cells will contribute drainage to each cell (the cell itself is not counted).

A flow accumulation raster can be interpreted in two ways. First, cells having high accumulation values generally correspond to stream channels, whereas cells having an accumulation value of zero generally correspond to ridge lines. Second, if multiplied by the cell size, the accumulation value equals the drainage area.

14.4.4 Stream Network

A stream network can be derived from a flow accumulation raster. The derivation is based on

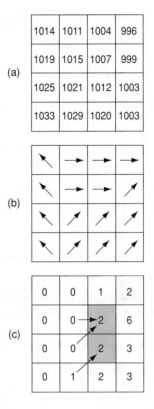

(a)

(b)

(c)

Figure 14.8

This illustration shows a filled elevation raster (a), a flow direction raster (b), and a flow accumulation raster (c). Both shaded cells in (c) have the same flow accumulation value of 2. The top cell receives its flow from its left and lower-left cells. The bottom cell receives its flow from its lower-left cell, which already has a flow accumulation value of 1.

the *channel initiation threshold,* which represents the amount of discharge needed to maintain a channel head, with contributing cells serving as a surrogate for discharge (Lindsay 2006). A threshold value of 500, for example, means that each cell of the drainage network has a minimum of 500 contributing cells.

14.4.5 Stream Links

After a stream network is derived from a flow accumulation raster, each section of the stream raster line is assigned a unique value and is associated

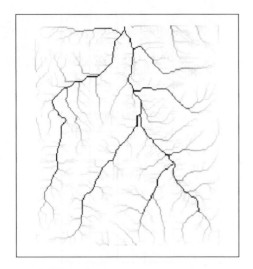

Figure 14.9
A flow accumulation raster, with darker symbols representing higher flow accumulation values.

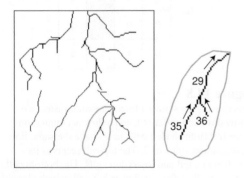

Figure 14.10
To derive the stream links, each section of the stream network is assigned a unique value and a flow direction. The inset map on the right shows three stream links.

with a flow direction (Figure 14.10). A stream link raster therefore resembles a topology-based stream layer: the intersections or junctions are like nodes, and the stream sections between junctions are like arcs or reaches (Figure 14.11).

14.4.6 Areawide Watersheds

The final step is to delineate a watershed for each stream section (Figure 14.12). This operation uses the flow direction raster and the stream link raster

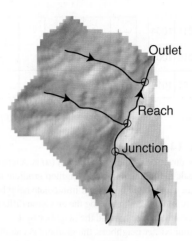

Figure 14.11
A stream link raster includes reaches, junctions, flow directions, and an outlet.

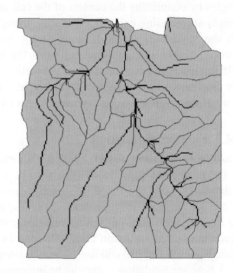

Figure 14.12
Areawide watersheds.

as the inputs. A denser stream network (i.e., based on a smaller threshold value) will have more, but smaller, watersheds. Figure 14.12 does not cover the entire area of the original DEM. The missing areas around the rectangular border are areas that do not have flow accumulation values higher than the specified threshold value.

14.4.7 Point-Based Watersheds

Instead of deriving a watershed for each identified stream section, the task for some projects is to delineate specific watersheds based on points of interest (Figure 14.13). These points of interest may be stream gauge stations, dams, or water quality monitoring stations. In watershed analysis, these points are called **pour points** or *outlets*.

Delineation of individual watersheds based on pour points follows the same procedure as for delineation of areawide watersheds. The only difference is to substitute a point raster for a stream link raster. In the point raster, a cell representing a pour point must be located over a cell that is part of the stream link. If a pour point is not located directly over a stream link, it will result in a small, incomplete watershed for the outlet (Lindsay, Rothwell, and Davies 2008).

Figure 14.14 illustrates the importance of the location of a pour point. The pour point in the example represents a USGS gauge station on a river. The geographic location of the station is recorded in longitude and latitude values at the USGS website. When plotted on the stream link raster, the station is about 50 meters away from the stream. The location error results in a very small watershed for the station, as shown in Figure 14.14. By moving the station to be on the stream, it results in a large water-

shed spanning beyond the length of a 1:24,000 scale quadrangle map (Figure 14.15). ArcGIS has a command (Snap Pour Point) that can snap a pour point to a stream cell within a user-defined search radius

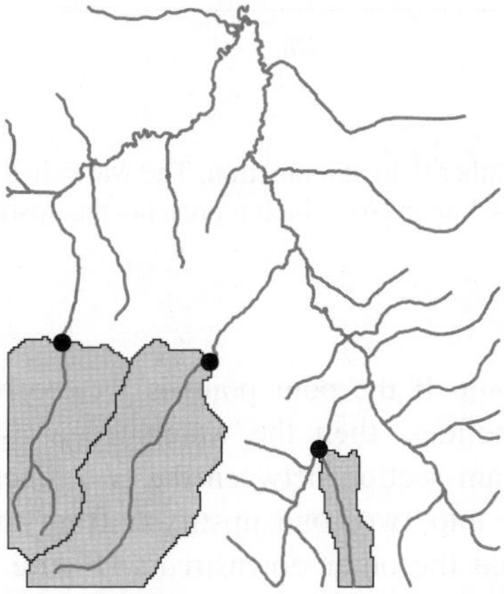

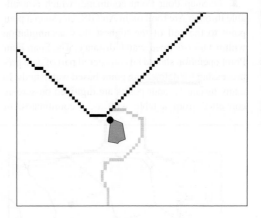

Figure 14.14

If a pour point (black circle) is not snapped to a cell with a high flow accumulation value (dark cell symbol), it usually has a small number of cells (shaded area) identified as its watershed.

Figure 14.15

When the pour point in Figure 14.14 is snapped to a cell with a high flow accumulation value (i.e., a cell representing a stream channel), its watershed extends to the border of a USGS 1:24,000-scale quadrangle map and beyond.

Figure 14.13

Point-based watersheds (shaded area).

Box 14.4 | Snapping Pour Points

T he Snap Pour Point command, which is available through ArcToolbox in ArcGIS, can snap a pour point to the cell of the highest flow accumulation within a user-defined search distance. The Snap Pour Point operation should be considered part of data preprocessing for delineating point-based watersheds. In many instances, pour points are digitized on-screen, converted from a table with x-, y-coordinates, or selected from an existing data set. These points are rarely right on top of a computer-generated stream channel. The discrepancy can be caused by the poor data quality of the pour points, the inaccuracy of the computer-generated stream channel, or both. Task 4 of the applications section covers use of the Snap Pour Point command.

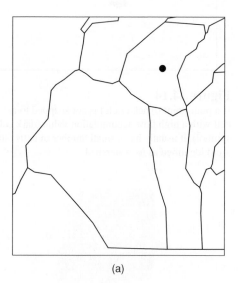

(a)

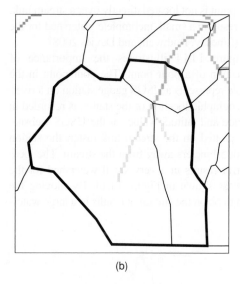

(b)

Figure 14.16
The pour point (black circle) in (*a*) is located along a stream section rather than at a junction. The watershed derived for the pour point is a merged watershed, shown in thick black line in (*b*), which represents the upstream contributing area at the pour point.

(Box 14.4). Lindsay, Rothwell, and Davies (2008) have offered a method that uses water body names to better reposition outlet points.

The relative location of the pour point to a stream network determines the size of a point-based watershed. If the pour point is located at a junction, then the watersheds upstream from the junction are merged to form the watershed for the pour point. If the pour point is located between two junctions, then the watershed assigned to the stream section between the two junctions is divided into two, one upstream from the pour point and the other downstream (Figure 14.16). The upstream portion of the watershed is then merged with watersheds further upstream to form the watershed for the pour point.

14.5 FACTORS INFLUENCING WATERSHED ANALYSIS

The outcome of watershed analysis including a stream network and watersheds can be influenced by DEM resolution, flow direction method, and flow accumulation threshold.

14.5.1 DEM Resolution

A higher-resolution DEM can better define the topographic features and produce a more detailed stream network than a lower-resolution DEM. This is illustrated in Figures 14.17 and 14.18, comparing a 30-meter DEM with a 10-meter DEM. In a study involving field verification, Murphy et al. (2008) report that a 1-meter LIDAR DEM can predict more first-order streams (smallest tributaries) and produce flow channels that extend further upslope into the landscape than a 10-meter DEM.

A stream network delineated automatically from watershed analysis can deviate from that on a USGS 1:24,000 scale digital line graph (DLG), especially in areas of low relief. To get a better match, a method called "stream burning" can integrate a vector-based stream layer into a DEM for watershed analysis (Kenny and Matthews 2005). For example, to produce "hydro enforced" DEMs, the USGS adjusts elevation values of cells immediately adjacent to a vector stream by assuming a continuous gradient from the larger to smaller contour values intersected by the stream (Simley 2004). Likewise, Murphy et al. (2008) subtract a constant value from the elevation of those cells classified as surface water. By "burning" streams into a DEM, it means that water flows on the DEM will more logically accumulate into streams identified on the topographic map.

14.5.2 Flow Direction Method

Commercial GIS packages, including ArcGIS, use the D8 method mainly because it is simple and can produce good results in mountainous topography with convergent flows (Freeman 1991). But it tends to produce flow in parallel lines along principal directions (Moore 1996). And it cannot represent adequately divergent flow over convex slopes and ridges (Freeman 1991) and does not do well in highly variable topography with floodplains and wetlands (Liang and Mackay 2000). As an example, Figure 14.19 shows that the D8 method

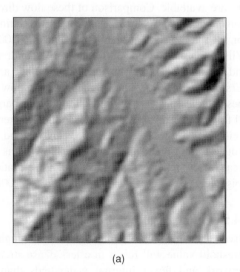

(a)

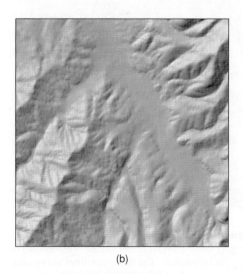

(b)

Figure 14.17
DEMs at a 30-meter resolution (*a*) and a 10-meter resolution (*b*).

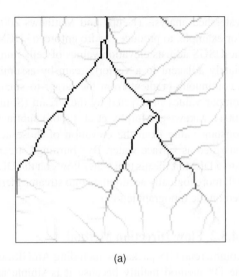
(a)

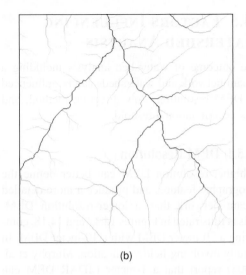
(b)

Figure 14.18
Stream networks derived from the DEMs in Figure 14.17. The stream network derived from the 30-meter DEM (*a*) has fewer details than that from the 10-meter DEM (*b*).

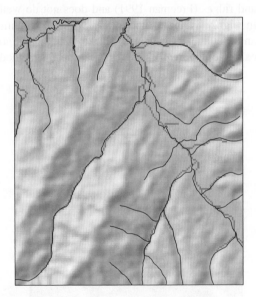

Figure 14.19
The gray raster lines represent stream segments derived using the D8 method. The thin black lines are stream segments from the 1:24,000-scale DLG. The two types of lines correspond well in well-defined valleys but poorly on the bottomlands.

performs well in well-defined valleys but poorly in relatively flat areas.

D8 is a single-flow direction method. A variety of multiple-flow direction methods including D∞ are available. Comparison of these flow direction methods have been made in controlled settings for specific catchment area (Zhou and Liu 2002; Wilson, Lam, and Deng 2007), flow path (Endreny and Wood 2003), and upslope contributing area (Erskine et al. 2006). The D8 method ranks low in these studies because it yields straight and parallel flow paths and does not represent flow patterns well along ridges and side slopes. A method to reduce the problem of straight-line flow paths in areas of low relief is to add to the DEM elevation noises (e.g., 0 to 5 centimeters) (Murphy et al. 2008).

14.5.3 Flow Accumulation Threshold

Given the same flow accumulation raster, a higher threshold value will result in a less dense stream network and fewer internal watersheds than a lower threshold value. Figure 14.20 illustrates the effect of the threshold value. Figure 14.20*a* shows

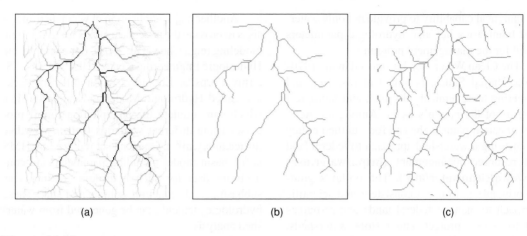

(a)   (b)   (c)

Figure 14.20

(*a*) A flow accumulation raster; (*b*) a stream network based on a threshold value of 500 cells; and (*c*) a stream network based on a threshold value of 100 cells.

the flow accumulation raster, Figure 14.20*b* the stream network based on a threshold of 500 cells, and Figure 14.20*c* the stream network based on a threshold of 100 cells. Ideally, the resulting stream network from a threshold value should correspond to a network obtained from traditional methods such as from high-resolution topographic maps or field mapping (Tarboton, Bras, and Rodrigues-Iturbe 1991).

Figure 14.21 shows the hydrography from the 1:24,000 scale digital line graph (DLG) for the same area as Figure 14.20. A threshold value between 100 and 500 cells seems to best capture the stream network in the area. Other researchers have also suggested that, instead of using a constant value, one should vary the threshold by slope and other terrain attributes (Montgomery and Foufoula-Georgiou 1993; Heine, Lant, and Sengupta 2004).

Figure 14.21

The stream network from the 1:24,000-scale digital line graph for the same area as Figure 14.11.

14.6 APPLICATIONS OF WATERSHED ANALYSIS

A watershed is a hydrologic unit often used for the management and planning of natural resources. Therefore, an important application of watershed analysis is in the area of watershed management. **Watershed management** approaches the organization and planning of human activities on a watershed by recognizing the interrelationships among land use, soil, and water as well as the linkage between uplands and downstream areas (Brooks et al. 2003). The state of Washington in the United States, for example, has enacted the Watershed Planning Act, which addresses such issues as water resource, water quality, and salmon habitat needs (Ryan and Klug 2005). One requirement for implementing watershed management programs is the analytical capability to provide not only watershed boundaries but also hydrologic parameters useful for the management programs.

The Clean Water Act, introduced in the 1970s in the United States, is aimed at restoring and protecting the chemical, physical, and biological integrity of the nation's water. Among the Act's current action plans is the call for a unified policy for ensuring a watershed approach to federal land and resource management (**http://water.usgs .gov/owq/cleanwater/ufp/**). The policy's guiding principle is to use a consistent and scientific approach to manage federal lands and resources and to assess, protect, and restore watersheds. A multiagency effort has been organized to create the Watershed Boundary Dataset (WBD) and the Natural Resources Conservation Service (NRCS) has been responsible for certification of the WBD (**http://www.ncgc.nrcs.usda.gov/products/ datasets/watershed/**). As of March 2009, the data sets for the conterminous 48 states, Hawaii, and Puerto Rico have been certified.

Another major application of watershed analysis is to provide the necessary inputs for hydrologic modeling (e.g., Chen et al. 2005). For example, the Hydrologic Engineering Center (HEC) of the U.S. Army Corps of Engineers has a software package called Hydrologic Modeling System (HMS), which can simulate the precipitation runoff processes using different scenarios (**http://www.hec .usace.army.mil/**). One data set required by HMS is the basin model including parameter and connectivity data for such hydrologic elements as subbasin, reach, junction, source, and sink. These hydrologic elements can be generated from watershed analysis.

Flood prediction models and snowmelt runoff models are other examples that can use topographic features generated from watershed analysis as inputs. A flood prediction model requires such variables as contributing drainage area, channel slope, stream length, and basin elevation; and a snowmelt runoff model requires the snow-covered area of the watershed and its topographic features.

KEY CONCEPTS AND TERMS

Cumulative viewshed: A viewshed based on two or more viewpoints.

Filled DEM: A digital elevation model that is void of depressions.

Flow accumulation raster: A raster that shows for each cell the number of cells that flow to it.

Flow direction raster: A raster that shows the direction water will flow out of each cell of a filled elevation raster.

Line-of-sight: A line connecting the viewpoint and the target in viewshed analysis. Also called *sightline*.

Pour points: Points used for deriving contributing watersheds.

Viewing azimuth: A parameter that sets horizontal angle limits to the view from a viewpoint.

Viewing radius: A parameter that sets the search distance for viewshed analysis.

Viewshed: Area of the land surface visible from one or more viewpoints.

Watershed: An area that drains water and other substances to a common outlet.

Watershed analysis: An analysis that involves derivation of flow direction, flow accumulation, watershed boundaries, and stream networks.

Watershed management: A practice of managing human activities on a watershed by recognizing the interrelationships among land use, soil, and water as well as the linkage between uplands and downstream areas.

1. Describe the two types of input data required for viewshed analysis.

2. What is a cumulative viewshed map?

3. Some researchers have advocated probabilistic visibility maps. Why?

4. What parameters can we choose for viewshed analysis?

5. Suppose you are asked by the U.S. Forest Service to run viewshed analysis along a scenic highway. You have chosen a number of points along the highway as viewpoints. You want to limit the view to within 2 miles from the highway and within a horizontal viewing angle from due west to due east. What parameters will you use? What parameter values will you specify?

6. What does the parameter of OFFSETA do for viewshed analysis in ArcGIS?

7. Besides the examples cited in Chapter 14, could you think of another viewshed application from your discipline?

8. Draw a diagram that shows the elements of watershed, topographic divide, stream section, stream junction, and outlet.

9. What is a filled DEM? Why is a filled DEM needed for a watershed analysis?

10. The example in Figure 14.7 shows an eastward flow direction. Suppose the elevation of the lower-left cell is changed from 1025 to 1028. Will the flow direction remain the same?

11. What kinds of criticisms has the D8 method received?

12. How do you interpret a flow accumulation raster (Figure 14.9)?

13. Deriving a drainage network from a flow accumulation raster requires the use of a channel initiation threshold value. Explain how the threshold value can alter the outcome of the drainage network.

14. To generate areawide watersheds from a DEM, we must create several intermediate rasters. Draw a flow chart that starts with a DEM, followed by the intermediate rasters, and ends with a watershed raster.

15. A watershed identified for a pour point is often described as a merged watershed. Why?

16. Describe the effect of DEM resolution on watershed boundary delineation.

17. Explain the difference between D8 and D∞ in calculation of flow direction.

18. Describe an application example of watershed analysis in your discipline.

APPLICATIONS: VIEWSHEDS AND WATERSHEDS

This applications section includes four tasks. Task 1 covers viewshed analysis and the effect of the viewpoint's height offset on the viewshed. Task 2 creates a cumulative viewshed by using two viewpoints, one of which is added through on-screen digitizing. Task 3 covers the steps for deriving areawide watersheds from a DEM. Task 4 uses the output from Task 3 to derive point-based watersheds. Task 4 also shows the importance of snapping points of interest to the stream channel.

Task 1 Perform Viewshed Analysis

What you need: *plne*, an elevation raster; and *lookout.shp*, a lookout point shapefile.

The lookout point shapefile contains a viewpoint. In Task 1, you first create a hillshade map of *plne* to better visualize the terrain. Next you run a viewshed analysis without specifying any parameter value. Then you add 15 meters to the height of the viewpoint to increase the viewshed coverage.

1. Launch ArcCatalog and connect to the Chapter 14 database. Start ArcMap, and rename the data frame Tasks 1&2. Add *plne* and *lookout.shp* to Tasks 1&2. First, create a hillshade map of *plne*. Open ArcToolbox, and set the Chapter 14 database as the current workspace. Double-click the Hillshade tool in the Spatial Analyst Tools/Surface toolset. Select *plne* for the input raster, save the output raster as *hillshade*, and click OK. *hillshade* is added to the map. Right-click *hillshade* and select Properties. On the Display tab, enter 30% transparent. The 30% transparency allows *hillshade* to be superimposed with other layers.

2. Now run a viewshed analysis. Double-click the Viewshed tool in the Spatial Analyst Tools/Surface toolset. Select *plne* for the input raster, select *lookout* for the input point or polyline observer features, save the output raster as *viewshed*, and click OK.

3. *viewshed* separates visible areas from not-visible areas. Open the attribute table of *viewshed*. The table shows the cell counts for the visibility classes of 0 (not visible) and 1 (visible).

Q1. What area percentage of *plne* is visible from the viewpoint?

4. Suppose the viewpoint has a physical structure that adds a height of 15 meters. You can use the field OFFSETA to include this height in viewshed analysis. Double-click the Add Field tool in the Data Management Tools/Fields toolset. Select *lookout* for the input table, enter OFFSETA for the field name, and click OK. Double-click the Calculate Field tool in the Data Management Tools/Fields toolset. Select *lookout* for the input table, select OFFSETA for the field name, enter 15 for the expression, and click

OK. Open the attribute table of *lookout* to make sure that the offset is set up correctly.

5. Follow Step 2 to run another viewshed analysis with the added height of 15 meters to the viewpoint. Save the output raster as *viewshed15*. *viewshed15* should show an increase of visible areas.

Q2. What area percentage of *plne* is visible from the viewpoint with the added height?

Task 2 Create a New Lookout Shapefile for Viewshed Analysis

What you need: *plne* and *lookout.shp*, same as in Task 1.

Task 2 asks you to digitize one more lookout location before running a viewshed analysis. The output from the analysis represents a cumulative viewshed.

1. Select Copy from the context menu of *lookout* in the table of contents. Select Paste Layer(s) from the context menu of Tasks 1&2. The copied shapefile is also named *lookout*. Right-click the top *lookout*, and select Properties. On the General tab, change the layer name from *lookout* to *newpoints*.

2. Click the Editor Toolbar button to open the toolbar. Click Editor's dropdown arrow, select Start Editing, and choose *newpoints* to edit. In the Create Features window, click *newpoints* and make sure that the construction tool is point.

3. Next add a new viewpoint. To find suitable viewpoint locations, you can use *hillshade* as a guide and the Zoom In tool for close-up looks. You can also use *plne* and the Identify tool to find elevation data. When you are ready to add a viewpoint, click the Point Tool on the Editor toolbar first and then click the intended location of the point. The new viewpoint has an OFFSETA value of 0. Open the attribute table of *newpoints*. For the new point, enter 15 for OFFSETA and 2 for the ID. Click the Editor menu and select Stop

Editing. Save the edits. You are ready to use *newpoints* for viewshed analysis.

4. Double-click the Viewshed tool in the Spatial Analyst Tools/Surface toolset. Select *plne* for the input raster, select *newpoint* for the input point or polyline observer features, save the output raster as *newviewshed*, and click OK.

5. *newviewshed* shows visible and not-visible areas. The visible areas represent the cumulative viewshed. Portions of the viewshed are visible to only one viewpoint, whereas others are visible to both viewpoints. The attribute table of *newviewshed* provides the cell counts of visible from one point and visible from two points.

Q3. What area percentage of *plne* is visible from *newpoints*? Report the increase in viewshed from one to two viewpoints.

6. To save *newpoints* as a shapefile, right-click *newpoints,* point to Data, and select Export Data. In the Export Data dialog, specify the path and name of the output shapefile.

Task 3 Delineate Areawide Watersheds

What you need: *emidalat,* an elevation raster; and *emidastrm.shp,* a stream shapefile.

Task 3 shows you the process of delineating areawide watersheds using an elevation raster as the data source. *emidastrm.shp* serves as a reference.

1. Insert a new data frame in ArcMap. Rename the new data frame Task 3, and add *emidalat* and *emidastrm.shp* to Task 3. If necessary, open the ArcToolbox window.

2. First check to see if there are any sinks in *emidalat*. Double-click the Flow Direction tool in the Spatial Analyst Tools/Hydrology toolset. Select *emidalat* for the input surface raster, enter *temp_flowd* for the output flow direction raster, and click OK. Double-click the Sink tool. Select *temp_flowd* for the input flow direction raster, specify *sinks* for the output raster, and click OK.

Q4. How many sinks does *emidalat* have? Describe where these sinks are located.

3. This step fills the sinks in *emidalat*. Double-click the Fill tool. Select *emidalat* for the input surface raster, specify *emidafill* for the output surface raster, and click OK.

4. You will use *emidafill* for the rest of Task 3. Double-click the Flow Direction tool. Select *emidafill* for the input surface raster, and specify *flowdirection* for the output flow direction raster. Run the command.

Q5. If a cell in *flowdirection* has a value of 64, what is the cell's flow direction? (Use the index of Flow Direction tool/command in ArcGIS Desktop Help to get the answer.)

5. Next create a flow accumulation raster. Double-click the Flow Accumulation tool. Select *flowdirection* for the input flow direction raster, enter *flowaccumu* for the output accumulation raster, and click OK.

Q6. What is the range of cell values in *flowaccumu*?

6. Next create a source raster, which will be used as the input later for watershed delineation. Creating a source raster involves two steps. First select from *flowaccumu* those cells that have more than 500 cells (threshold) flowing into them. Double-click the Con tool in the Spatial Analyst Tools/Conditional toolset. Select *flowaccumu* for the input conditional raster, enter Value > 500 for the expression, enter 1 for the input true raster or constant value, specify *net* for the output raster, and click OK to run the command. Second, assign a unique value to each section of *net* between junctions (intersections). Go back to the Hydrology toolset. Double-click the Stream Link tool. Select *net* for the input stream raster, select *flowdirection* for the input flow direction raster, and specify *source* for the output raster. Run the command.

7. Now you have the necessary inputs for watershed delineation. Double-click the Watershed tool. Select *flowdirection* for the

input flow direction raster, select *source* for the input raster, specify *watershed* for the output raster, and click OK. Change the symbology of *watershed* to that of unique values so that you can see individual watersheds.

Q7. How many watersheds are in *watershed*?

Q8. If the flow accumulation threshold were changed from 500 to 1000, would it increase, or decrease, the number of watersheds?

8. You can also complete Task 3 using a Python script in ArcMap. To use this option, first click the Python window in ArcMap to open it. Assuming that the workspace is d:/chap14 and the workspace contains *emidalat,* the following lists the statements (without >>>) you need to type in the Python window to complete Task 3:

>>> import arcpy

>>> from arcpy import env

>>> from arcpy.sa import *

>>> env.workspace = "d:/chap14"

>>> arcpy.CheckExtension("Spatial")

>>> outflowdirection = FlowDirection("emidalat")

>>> outsink = Sink("outflowdirection")

>>> outfill = Fill("emidalat")

>>> outfd = FlowDirection("outfill")

>>> outflowac = FlowAccumulation("outfd")

>>> outnet2 = Con("outflowac", 1, 0, "VALUE > 500")

>>> outstreamlink = StreamLink("net","outfd")

>>> outwatershed = Watershed("outfd", "outstreamlink")

>>> outwatershed.save("outwatershed")

The first five statements of the script import arcpy and Spatial Analyst tools, and define the Chapter 14 database as the workspace. This is followed by statements that use the tools of FlowDirection, Sink,

Fill, Flowdirection (on the filled DEM), FlowAccumulation, Con, StreamLink, and Watershed. Each time you enter a statement, you will see its output in ArcMap. The last statement saves *outwatershed,*the watershed output, in the Chapter 14 database.

Task 4 Derive Upstream Contributing Areas at Pour Points

What you need: *flowdirection, flowaccumu,* and *source,* all created in Task 3; and *pourpoints.shp,* a shapefile with two points.

In Task 4, you will derive a specific watershed (i.e., upstream contributing area) for each point in *pourpoints.shp.*

1. Insert a data frame in ArcMap and rename it Task 4. Add *flowdirection, flowaccumu, source,* and *pourpoints.shp* to Task 4.

2. Zoom in on a pour point. The pour point is not right on *source,* the stream link raster created in Task 3. It is the same with the other point. If these pour points were used in watershed analysis, they would generate no or very small watersheds. ArcGIS has a SnapPour command, which can snap a pour point to the cell with the highest flow accumulation value within a search distance. Use the Measure tool to measure the distance between the pour point and the nearby stream segment. A snap distance of 90 meters (3 cells) should place the pour points onto the stream channel.

3. Double-click the Snap Pour Point tool in the Spatial Analyst Tools/Hydrology toolset. Select *pourpoints* for the input raster or feature pour point data, select *flowaccumu* for the input accumulation raster, save the output raster as *snappour,* enter 90 for the snap distance, and click OK. Now check *snappour;* the cells should be aligned with *flowaccumu.*

4. Double-click the Watershed tool. Select *flowdirection* for the input flow direction raster, select *snappour* for the input raster or

feature pour point data, save the output raster as *pourshed*, and click OK.

Q9. How many cells are associated with each of the new pour points?

Challenge Task

What you need: access to the Internet.

1. From the National Elevation Dataset (NED) website, **http://seamless.usgs.gov/,** download a USGS DEM for an area, preferably a mountainous area, near your university.

You can refer to Task 1 of Chapter 5 for the download information.

2. Use the DEM and a threshold value of 500 to run an area-wide watershed analysis. Save the output watershed as *watershed500*. Then use the same DEM and a threshold value of 250 to run another area-wide watershed analysis, and save the output as *watershed250*.

3. Compare *watershed500* with *watershed250* and explain the difference between them.

REFERENCES

Band, L. E., C. L. Tague, S. E. Brun, D. E. Tenenbaum, and R. A. Fernandes. 2000. Modelling Watersheds as Spatial Object Hierarchies: Structure and Dynamics. *Transactions in GIS* 4: 181–96.

Bishop, I. D. 2003. Assessment of Visual Qualities, Impacts, and Behaviours, in the Landscape, by Using Measures of Visibility. *Environment and Planning B: Planning and Design* 30: 677–88.

Bishop, I. D., E. Lange, and A. M. Mahbubul. 2004. Estimation of the influence of view components on high-rise apartment pricing using a public survey and GIS modeling. *Environment and Planning B: Planning and Design* 31: 439–52.

Brooks, K. N., P. F. Ffolliott, H. M. Gregersen, and L. F. DeBano. 2003. *Hydrology and the Management of Watersheds,* 3d ed. Ames, IA: Iowa State Press.

Chen, C. W., J. W. Herr, R. A. Goldstein, G. Ice, and T. Cundy. 2005. Retrospective Comparison of Watershed Analysis Risk Management Framework and Hydrologic Simulation Program Fortran Applications to Mica Creek Watershed. *Journal of Environmental Engineering* 131: 1277–84.

Clarke, K. C. 1995. *Analytical and Computer Cartography,* 2d ed. Englewood Cliffs, NJ: Prentice Hall.

Davidson, D. A., A. I. Watson, and P. H. Selman. 1993. An Evaluation of GIS as an Aid to the Planning of Proposed Developments in Rural Areas. In P. M. Mather, ed., *Geographical Information Handling: Research and Applications,* pp. 251–59. London: Wiley.

De Floriani, L., and P. Magillo. 1994. Visibility Algorithms on Triangulated Terrain Models. *International Journal of Geographical Information Systems* 8: 13–41.

De Floriani, L., and P. Magillo. 1999. Intervisibility on Terrains. In P. A. Longley, M. F. Goodchild, D. J. Maguire, and D. W. Rhind, eds., *Geographical Information Systems, Vol. 1: Principles and Technical Issues,* 2d ed., pp. 543–56. New York: Wiley.

De Floriani, L., and P. Magillo. 2003. Algorithms for Visibility Computation on Terrains: A Survey. *Environment and Planning B: Planning and Design* 30: 709–28.

Endreny, T. A., and E. F. Wood. 2003. Maximizing Spatial Congruence of Observed and DEM-Delineated Overland Flow Networks. *International Journal of Geographical Information Science* 17: 699–713.

Erskine, R. H., T. R. Green, J. A. Ramirez, and L. H. MacDonald. 2006. Comparison of Grid-Based Algorithms for Computing Upslope Contributing Area. *Water Resources Research* 42, W09416, doi:10.1029/2005WR004648.

Fisher, P. F. 1991. First Experiments in Viewshed Uncertainty: The Accuracy of the Viewshed Area. *Photogrammetric Engineering and Remote Sensing* 57: 1321–27.

Fisher, P. F. 1993. Algorithm and Implementation Uncertainty in Viewshed Analysis. *International Journal of Geographical Information Systems* 7: 331–47.

Fisher, P. R. 1996. Extending the Applicability of Viewsheds in Landscape Planning. *Photogrammetric Engineering and Remote Sensing* 62: 1297–1302.

Freeman, T. G. 1991. Calculating Catchment Area with Divergent Flow Based on a Regular Grid. *Computers and Geosciences* 17: 413–22.

Gallant J. C., and J. P. Wilson. 2000. Primary Topographic Attributes. In J. P. Wilson and J. C. Gallant, eds., *Terrain Analysis: Principles and Applications,* pp. 51–85. New York: Wiley.

Garbrecht, J., and L. W. Martz. 2000. Digital Elevation Model Issues in Water Resources Modeling. In D. Maidment and D. Djokic, eds., *Hydrologic and Hydraulic Modeling Support with Geographic Information Systems,* pp. 1–27. Redlands, CA: ESRI Press.

Goodchild, M. F., and J. Lee. 1989. Coverage Problems and Visibility Regions on Topographic Surfaces. *Annals of Operations Research* 18: 175–86.

Heine, R. A., C. L. Lant, and R. R. Sengupta. 2004. Development and Comparison of Approaches for Automated Mapping of Stream Channel Networks. *Annals of the Association of American Geographers* 94: 477–90.

Jenson, S. K., and J. O. Domingue. 1988. Extracting Topographic Structure from Digital Elevation Data for Geographic Information System Analysis. *Photogrammetric Engineering and Remote Sensing* 54: 1593–1600.

Kenny, F., and B. Matthews. 2005. A Methodology for Aligning Raster Flow Direction Data with Photogrammetrically Mapped Hydrology. *Computers & Geosciences* 31: 768–79.

Kumsap, C., F. Borne, and D. Moss. 2005. The Technique of Distance Decayed Visibility for Forest Landscape Visualization. *International Journal of Geographical Information Science* 19: 723–44.

Lake, I. R., A. A. Lovett, I. J. Bateman, and I. H. Langford. 1998. Modelling Environmental Influences on Property Prices in an Urban Environment. *Computers, Environment and Urban Systems* 22: 121–36.

Lange, E., and P. V. Schaeffer. 2001. A comment on the market value of a room with a view. *Landscape and Urban Planning* 55: 113–120.

Lee, J. 1991. Analyses of Visibility Sites on Topographic Surfaces. *International Journal of Geographical Information Systems* 5: 413–29.

Lee, J., and D. Stucky. 1998. On Applying Viewshed Analysis for Determining Least-Cost Paths on Digital Elevation Models. *International Journal of Geographical Information Science* 12: 891–905.

Liang, C., and D. S. Mackay. 2000. A General Model of Watershed Extraction and Representation Using Globally Optimal Flow Paths and Up-Slope Contributing Areas. *International Journal of Geographical Information Science* 14: 337–58.

Lindsay, J. B. 2006. Sensitivity of Channel Mapping Techniques to Uncertainty in Digital Elevation Data. *International Journal of Geographical Information Science* 20: 669–92.

Lindsay, J. B., J. J. Rothwell, and H. Davies. 2008. Mapping Outlet Points Used for Watershed Delineation onto DEM-Derived Stream Networks. *Water Resources Research* 44, W08442, doi:10.1029/2007WR006507.

Maloy, M. A., and D. J. Dean. 2001. An Accuracy Assessment of Various GIS-Based Viewshed Delineation Techniques. *Photogrammetric Engineering and Remote Sensing* 67: 1293–98.

Maxted, J. T., M. W. Diebel, and M. J. Vander Zanden. 2009. Landscape Planning for Agricultural Non-Point Source Pollution Reduction. II. Balancing Watershed Size, Number of Watersheds, and Implementation Effort. *Environmental Management* 43: 60–68.

Miller, D. 2001. A Method for Estimating Changes in the Visibility of Land Cover. *Landscape and Urban Planning* 54: 93–106.

Montgomery, D. R., and E. Foufoula-Georgiou. 1993. Channel Network Source Representation Using Digital Elevation Models. *Water Resources Research* 29: 3925–34.

Moore, I. D. 1996. Hydrological Modeling and GIS. In M. F. Goodchild, L. T. Steyaert, B. O. Parks, C. Johnston, D. Maidment, M. Crane, and S. Glendinning, eds., *GIS and Environmental Modeling: Progress and Research Issues,* pp. 143–48. Fort Collins, CO: GIS World Books.

Murphy, P. N. C., J. Ogilvie, F. Meng, and P. Arp. 2008. Stream Network Modelling using Lidar and Photogrammetric Digital Elevation Models: A Comparison and Field Verification. *Hydrological Processes* 22: 1747–1754.

Nackaerts, K., G. Govers, and J. Van Orshoven. 1999. Accuracy Assessment of Probabilistic Visibilities. *International Journal of Geographical Information Science* 13: 709–21.

O'Callaghan, J. F., and D. M. Mark. 1984. The Extraction of Drainage Networks from Digital Elevation Data. *Computer Vision, Graphics and Image Processing* 28: 323–44.

O'Sullivan, D., and A. Turner. 2001. Visibility Graphs and Landscape Visibility Analysis.

International Journal of Geographical Information Science 15: 221–37.

Riggs, P. D., and D. J. Dean. 2007. An Investigation into the Causes of Errors and Inconsistencies in Predicted Viewsheds. *Transactions in GIS* 11: 175–96.

Rogge, E., F. Nevens, and H. Gulinck. 2008. Reducing the Visual Impact of "Greenhouse Parks" in Rural Landscapes. *Landscape and Urban Planning* 87: 76–83.

Ryan, C. M., and J. S. Klug. 2005. Collaborative Watershed Planning in Washington State: Implementing the Watershed Planning Act. *Journal of Environmental Planning and Management* 48: 491–506.

Sander, H. A., and S. M. Manson. 2007. Heights and Locations of Artificial Structures in Viewshed Calculation: How Close is Close Enough? *Landscape and Urban Planning* 82: 257–70.

Simley J. 2004. *The Geodatabase Conversion.* USGS National Hydrography Newsletter 3(4). USGS. Available at **http://nhd.usgs.gov/newsletter_list.html**

Tarboton, D. G. 1997. A New Method for the Determination of Flow Directions and Upslope Areas in Grid Digital Elevation Models. *Water Resources Research* 32: 309–19.

Tarboton, D. G., R. L. Bras, and I. Rodrigues-Iturbe. 1991. On the Extraction of Channel Networks from Digital Elevation Data. *Water Resources Research* 5: 81–100.

Wang, J. J., G. J. Robinson, and K. White. 1996. A Fast Solution to Local Viewshed Computation Using Grid-Based Digital Elevation Models. *Photogrammetric Engineering and Remote Sensing* 62: 1157–64.

Whiteaker, T. L., D. R. Maidment, H. Gopalan, C. Patino, and D. C. McKinney. 2007. Raster-Network Regionalization for Watershed Data Processing. *International Journal of Geographical Information Science* 21: 341–53.

Wilson, J. P., C. S. Lam, and Y. Deng. 2007. Comparison of the Performance of Flow-Routing Algorithms Used in GIS-Based Hydrologic Analysis. *Hydrological Processes* 21: 1026–1044.

Wing, M. G., and R. Johnson. 2001. Quantifying Forest Visibility with Spatial Data. *Environmental Management* 27: 411–20.

Zhou, Q., and X. Liu. 2002. Error Assessment of Grid-Based Flow Routing Algorithms Used in Hydrological Models. *International Journal of Geographical Information Science* 16: 819–42.

CHAPTER 15

SPATIAL INTERPOLATION

CHAPTER OUTLINE

15.1 Elements of Spatial Interpolation
15.2 Global Methods
15.3 Local Methods
15.4 Kriging
15.5 Comparison of Spatial Interpolation
      Methods

The terrain is one type of surface that is familiar to us. In GIS we also work with another type of surface, which may not be physically present but can be visualized in the same way as the land surface. Cartographers call this type of surface the statistical surface (Robinson et al. 1995). Examples of the statistical surface include precipitation, snow accumulation, water table, and population density.

How can one construct a statistical surface? The answer is similar to that for the land surface except that input data are typically limited to a sample of point data. To make a precipitation map, for example, we will not find a regular array of weather stations like a digital elevation model (DEM). A process of filling in data between the sample points is therefore required.

Spatial interpolation is the process of using points with known values to estimate values at other points. Through spatial interpolation, we can estimate the precipitation value at a location with no recorded data by using known precipitation readings at nearby weather stations. In GIS applications, spatial interpolation is typically applied to a raster with estimates made for all cells. Spatial interpolation is therefore a means of creating surface data from sample points so that the surface data can be used for analysis and modeling.

Chapter 15 has five sections. Section 15.1 reviews the elements of spatial interpolation including control points and type of spatial interpolation. Section 15.2 covers global methods including trend surface and regression models. Section 15.3 covers local methods including Thiessen polygons, density estimation, inverse distance weighted, and splines. Section 15.4 examines kriging, a widely used stochastic local method. Section 15.5 compares interpolation methods. Perhaps more than any other topic in GIS, spatial interpolation depends on the computing algorithm. Worked examples are included in Chapter 15 to show how spatial interpolation is carried out mathematically.

15.1 ELEMENTS OF SPATIAL INTERPOLATION

Spatial interpolation requires two basic inputs: known points and an interpolation method. The requirement of known points distinguishes spatial interpolation from isopleth mapping, which uses assigned points such as polygon centroids for interpolation (Robinson et al. 1995).

15.1.1 Control Points

Control Points are points with known values. Also called *known points, sample points,* or *observations,* control points provide the data necessary for developing an interpolator (e.g., a mathematical equation) for spatial interpolation. The number and distribution of control points can greatly influence the accuracy of spatial interpolation. A basic assumption in spatial interpolation is that the value to be estimated at a point is more influenced by nearby known points than those farther away. To be effective for estimation, control points should be well distributed within the study area. But this ideal situation is rare in real-world applications because a study area often contains data-poor areas.

Figure 15.1 shows 130 weather stations in Idaho and 45 additional stations from the surrounding states. The map clearly shows data-poor areas in Clearwater Mountains, Salmon River Mountains, Lemhi Range, and Owyhee Mountains. These 175 stations, and their 30-year (1970–2000) average annual precipitation data, are used as sample data throughout Chapter 15. As will be shown later, the data-poor areas can cause problems for spatial interpolation.

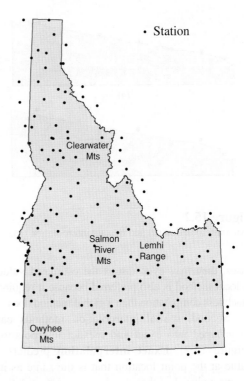

Figure 15.1
A map of 175 weather stations in and around Idaho.

15.1.2 Type of Spatial Interpolation

Spatial interpolation methods can be categorized in several ways. First, they can be grouped into global and local methods. A **global interpolation** method uses every known point available to estimate an unknown value. A **local interpolation** method, on the other hand, uses a sample of known points to estimate an unknown value. Because the difference between the two groups lies in the number of control points used in estimation, one may view the scale from global to local as a continuum.

Conceptually, a global interpolation method is designed to capture the general trend of the surface and a local interpolation method the local or short-range variation. For many phenomena, it is more efficient to estimate the unknown value at a point using a local method than a global method. Faraway points have little influence on the estimated value; in some

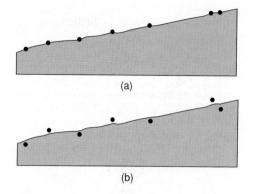

(a)

(b)

Figure 15.2

Exact interpolation (*a*) and inexact interpolation (*b*).

cases, they may even distort the estimated value. A local method is also preferred because it requires much less computation than a global method does.

Second, spatial interpolation methods can be grouped into exact and inexact interpolation (Figure 15.2). **Exact interpolation** predicts a value at the point location that is the same as its known value. In other words, exact interpolation generates a surface that passes through the control points. In contrast, **inexact interpolation,** or approximate interpolation, predicts a value at the point location that differs from its known value.

Third, spatial interpolation methods may be deterministic or stochastic. A **deterministic interpolation** method provides no assessment of errors with predicted values. A **stochastic interpolation** method, on the other hand, considers the presence of some randomness in its variable and offers assessment of prediction errors with estimated variances.

Table 15.1 shows a classification of spatial interpolation methods covered in Chapter 15. Notice that the two global methods can also be used for local operations.

15.2 Global Methods

Global methods include trend surface models and regression models.

15.2.1 Trend Surface Models

An inexact interpolation method, **trend surface analysis** approximates points with known values with a polynomial equation (Davis 1986; Bailey and Gatrell 1995). The equation or the interpolator can then be used to estimate values at other points. A linear or first-order trend surface uses the equation:

(15.1)

$$z_{x,y} = b_0 + b_1 x + b_2 y$$

where the attribute value z is a function of x and y coordinates. The b coefficients are estimated from the known points (Box 15.1). Because the trend surface model is computed by the least-squares method, the "goodness of fit" of the model can be measured and tested. Also, the deviation or the residual between the observed and the estimated values can be computed for each known point.

The distribution of most natural phenomena is usually more complex than an inclined plane

Table 15.1	**A Classification of Spatial Interpolation Methods**			
Global			*Local*	
Deterministic	**Stochastic**	**Deterministic**		**Stochastic**
Trend surface (inexact)*	Regression (inexact)	Thiessen (exact)		Kriging (exact)
		Density estimation (inexact)		
		Inverse distance weighted (exact)		
		Splines (exact)		

*Given some required assumptions, trend surface analysis can be treated as a special case of regression analysis and thus a stochastic method (Griffith and Amrhein 1991).

Box 15.1 | A Worked Example of Trend Surface Analysis

Figure 15.3 shows five weather stations with known values around point 0 with an unknown value. The table below shows the x-, y-coordinates of the points, measured in row and column of a raster with a cell size of 2000 meters, and their known values. This example shows how we can use Eq. (15.1), or a linear trend surface, to interpolate the unknown

Point	x	y	Value
1	69	76	20.820
2	59	64	10.910
3	75	52	10.380
4	86	73	14.600
5	88	53	10.560
0	69	67	?

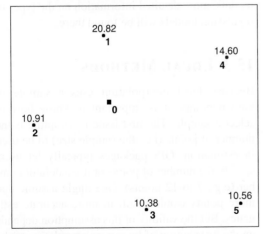

Figure 15.3
Estimation of the unknown value at point 0 from five surrounding known points.

value at point 0. The least-squares method is commonly used to solve for the coefficients of b_0, b_1, and b_2 in Eq. (15.1). Therefore, the first step is to set up three normal equations:

$$\Sigma z = b_0 n + b_1 \Sigma x + b_2 \Sigma y$$
$$\Sigma xz = b_0 \Sigma x + b_1 \Sigma x^2 + b_2 \Sigma xy$$
$$\Sigma yz = b_0 \Sigma y + b_1 \Sigma xy + b_2 \Sigma y^2$$

The equations can be rewritten in matrix form as:

$$\begin{bmatrix} n & \Sigma x & \Sigma y \\ \Sigma x & \Sigma x^2 & \Sigma xy \\ \Sigma y & \Sigma xy & \Sigma y^2 \end{bmatrix} \cdot \begin{bmatrix} b_0 \\ b_1 \\ b_2 \end{bmatrix} = \begin{bmatrix} \Sigma z \\ \Sigma xz \\ \Sigma yz \end{bmatrix}$$

Using the values of the five known points, we can calculate the statistics and substitute the statistics into the equation:

$$\begin{bmatrix} 5 & 377 & 318 \\ 377 & 29007 & 23862 \\ 318 & 23862 & 20714 \end{bmatrix} \cdot \begin{bmatrix} b_0 \\ b_1 \\ b_2 \end{bmatrix} = \begin{bmatrix} 67.270 \\ 5043.650 \\ 4445.800 \end{bmatrix}$$

We can then solve the b coefficients by multiplying the inverse of the first matrix on the left (shown with four decimal digits because of the very small numbers) by the matrix on the right:

$$\begin{bmatrix} 23.2102 & -0.1631 & -0.1684 \\ -0.1631 & 0.0018 & 0.0004 \\ -0.1684 & 0.0004 & 0.0021 \end{bmatrix} \cdot \begin{bmatrix} 67.270 \\ 5043.650 \\ 4445.800 \end{bmatrix} = \begin{bmatrix} -10.094 \\ 0.020 \\ 0.347 \end{bmatrix}$$

Using the coefficients, the unknown value at point 0 can be estimated by:

$$z_0 = -10.094 + (0.020)(69) + (0.347)(67) = 14.535$$

surface from a first-order model. Higher-order trend surface models are required to approximate more complex surfaces. A cubic or a third-order model, for example, includes hills and valleys. A cubic trend surface is based on the equation:

(15.2)

$$z_{x,y} = b_0 + b_1 x + b_2 y + b_3 x^2 + b_4 xy + b_5 y^2 + b_6 x^3 + b_7 x^2 y + b_8 xy^2 + b_9 y^3$$

A third-order trend surface requires estimation of 10 coefficients (i.e., b_i), compared to three coefficients for a first-order surface. A higher-order trend surface model therefore requires more computation than a lower-order model does. A GIS package may offer up to 12th-order trend surface models.

Figure 15.4 shows an isoline (isohyet) map derived from a third-order trend surface of annual precipitation in Idaho created from 175 data points with a cell size of 2000 meters. An isoline map is like a contour map, useful for visualization as well as measurement.

There are variations of trend surface analysis. Logistic trend surface analysis uses known points with binary data (i.e., 0 and 1) and produces a probability surface. **Local polynomial interpolation** uses

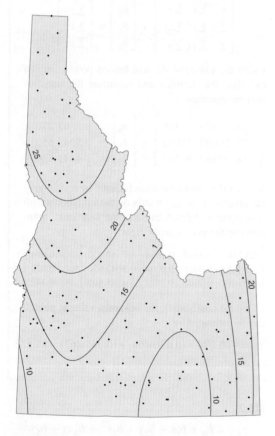

Figure 15.4

An isohyet map in inches from a third-order trend surface model. The point symbols represent known points within Idaho.

a sample of known points to estimate the unknown value of a cell. Local polynominal interpolation can be used for converting a triangulated irregular network (TIN) to a DEM and deriving topographic measures from a DEM (Chapter 13).

15.2.2 Regression Models

A **regression model** relates a dependent variable to a number of independent variables. A regression model can be used for estimation, using the regression equation as the interpolator, or for exploring the relationships between the dependent variable and independent variables. Many regression models use nonspatial attributes and are not considered methods for spatial interpolation. But exceptions can be made for regression models that use spatial variables such as distance to a river or location-specific elevation (Burrough and McDonnell 1998; Rogerson 2001; Beguería and Vicente-Serrano 2006). Because Chapter 18 also covers regression models, more detailed information on the types of regression models will be found there.

15.3 LOCAL METHODS

Because local interpolation uses a sample of known points, it is important to know how to select a sample. The first issue in sampling is the number of points (i.e., the sample size) to be used in estimation. GIS packages typically let users specify the number of points or use a default number (e.g., 7 to 12 points). One might assume that more points would result in more accurate estimates. But the validity of this assumption depends on the distribution of known points relative to the cell to be estimated, the extent of spatial autocorrelation (Chapter 11), and the quality of data (Yang and Hodler 2000). More points usually lead to more generalized estimations, and fewer points may actually produce more accurate estimations (Zimmerman et al. 1999).

After the number of points is determined, the next task is to search for those known points (Figure 15.5). A simple option is to use the closest known points to the point to be estimated.

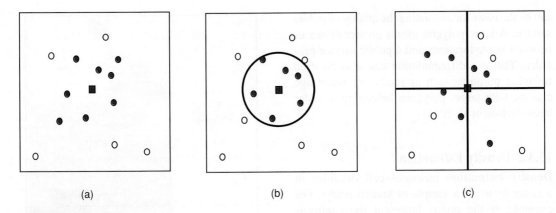

(a)   (b)   (c)

Figure 15.5
Three search methods for sample points: (*a*) find the closest points to the point to be estimated, (*b*) find points within a radius, and (*c*) find points within each quadrant.

An alternative is to select known points within a circle, the size of which depends on the sample size. Some search options may incorporate a quadrant or octant requirement. A quadrant requirement means selecting known points from each of the four quadrants around a cell to be estimated. An octant requirement means using eight sectors. Other search options may consider the directional component by using an ellipse, with its major axis corresponding to the principal direction.

15.3.1 Thiessen Polygons

Thiessen polygons assume that any point within a polygon is closer to the polygon's known point than any other known points. Thiessen polygons were originally proposed to estimate areal averages of precipitation by making sure that any point within a polygon is closer to the polygon's weather station than any other station (Tabios and Salas 1985). Thiessen polygons, also called *Voronoi polygons,* are used in a variety of applications, especially for service area analysis of public facilities such as hospitals (Schuurman et al. 2006).

Thiessen polygons do not use an interpolator but require initial triangulation for connecting known points. Because different ways of connecting points can form different sets of triangles, the Delaunay triangulation—the same method for constructing a TIN—is often used in preparing Thiessen

polygons (Davis 1986). The Delaunay triangulation ensures that each known point is connected to its nearest neighbors, and that triangles are as equilateral as possible. After triangulation, Thiessen polygons can be easily constructed by connecting lines drawn perpendicular to the sides of each triangle at their midpoints (Figure 15.6).

Thiessen polygons are smaller in areas where points are closer together and larger in areas where points are farther apart. This size differentiation

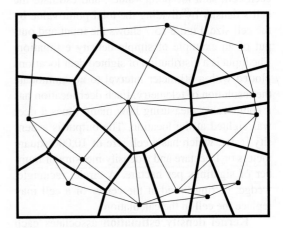

Figure 15.6
Thiessen polygons (in thicker lines) are interpolated from the known points and the Delaunay triangulation (in thinner lines).

can be the basis for evaluating the quality of public service. A large polygon means greater distances between home locations and a public service provider. The size differentiation can also be used for other purposes such as predicting forest age classes, with larger polygons belonging to older trees (Nelson et al. 2004).

15.3.2 Density Estimation

Density estimation measures cell densities in a raster by using a sample of known points. For example, if the points represent the centroids of census tracts and the known values represent reported burglaries, then density estimation can produce a surface map showing the high and low burglary rates within a city. For some applications, density estimation provides an alternative to point pattern analysis, which describes a pattern in terms of random, clustered, and dispersed (Chapter 11).

There are simple and kernel density estimation methods. The simple method is a counting method, whereas the kernel method is based on a probability function and offers options in terms of how density estimation is made. To use the simple density estimation method, we can place a raster on a point distribution, tabulate points that fall within each cell, sum the point values, and estimate the cell's density by dividing the total point value by the cell size. Figure 15.7 shows the input and output of an example of simple density estimation. The input is a distribution of sighted deer locations plotted with a 50-meter interval to accommodate the resolution of telemetry. Each deer location has a count value measuring how many times a deer was sighted at the location. The output is a density raster, which has a cell size of 10,000 square meters or 1 hectare and a density measure of number of sightings per hectare. A circle, rectangle, wedge, or ring based at the center of a cell may replace the cell in the calculation.

Kernel density estimation associates each known point with a kernel function for the purpose of estimation (Silverman 1986; Scott 1992; Bailey and Gatrell 1995). Expressed as a bivariate probability density function, a kernel function looks like

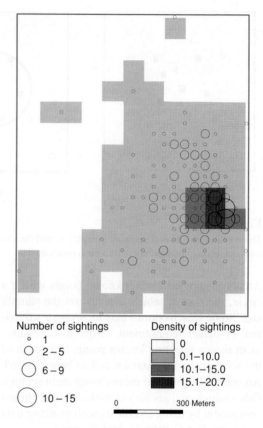

Number of sightings

○	1
○	2 – 5
○	6 – 9
○	10 – 15

Density of sightings

☐	0
▨	0.1–10.0
▨	10.1–15.0
■	15.1–20.7

0   300 Meters

Figure 15.7

Deer sightings per hectare calculated by the simple density estimation method.

a "bump," centering at a known point and tapering off to 0 over a defined bandwidth or window area (Silverman 1986) (Figure 15.8). The kernel function and the bandwidth determine the shape of the bump, which in turn determines the amount of smoothing in estimation. The kernel density estimator at point x is then the sum of bumps placed at the known points x_i within the bandwidth:

(15.3)

$$\hat{f}(x) = \frac{1}{nh^d} \sum_{i=1}^{n} K(\frac{1}{h}(x - x_i))$$

where $K(\ )$ is the kernel function, h is the bandwidth, n is the number of known points within the bandwidth, and d is the data dimensionality. For

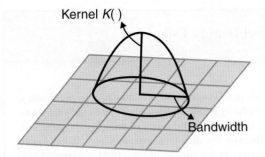

Figure 15.8
A kernel function, which represents a probability density function, looks like a "bump" above a grid.

two-dimensional data ($d = 2$), the kernel function is usually given by:

(15.4)

$$K(x) = 3\pi^{-1}(1 - X^T X)^2, \text{ if } X^T X < 1$$
$$K(x) = 0, \text{ otherwise}$$

By substituting Eq. (15.4) for $K(\ )$, Eq. (15.3) can be rewritten as:

(15.5)

$$\hat{f}(x) = \frac{3}{nh^2\pi} \sum_{i=1}^{n} \{1 - \frac{1}{h^2}[(x - x_i)^2 + (y - y_i)^2]\}^2$$

where π is a constant, and $(x - x_i)$ and $(y - y_i)$ are the deviations in x-, y-coordinates between point x and known point x_i that is within the bandwidth.

Using the same input as for the simple estimation method, Figure 15.9 shows the output raster from kernel density estimation. Density values in the raster are expected values rather than probabilities (Box 15.2). Kernel density estimation usually produces a smoother surface than the simple estimation method does. As a surface interpolation method, kernel density estimation has been applied to a wide variety of fields such as forest resources (Lundquist and Beatty 1999), public health (Reader 2001; Chung, Yang, and Bell 2004), crime (Ackerman and Murray 2004), and natural hazards (van der Veen and Logtmeijer 2005).

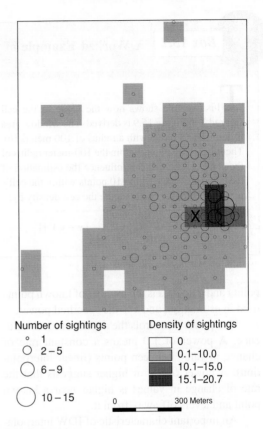

Number of sightings
- ○ 1
- ○ 2 – 5
- ○ 6 – 9
- ○ 10 – 15

Density of sightings
- 0
- 0.1 – 10.0
- 10.1 – 15.0
- 15.1 – 20.7

0   300 Meters

Figure 15.9
Deer sightings per hectare calculated by the kernel density estimation method. The letter X marks the cell, which is used as an example in Box 15.2.

15.3.3 Inverse Distance Weighted Interpolation

Inverse distance weighted (IDW) interpolation is an exact method that enforces the condition that the estimated value of a point is influenced more by nearby known points than by those farther away. The general equation for the IDW method is:

(15.6)

$$z_0 = \frac{\sum_{i=1}^{s} z_i \frac{1}{d_i^k}}{\sum_{i=1}^{s} \frac{1}{d_i^k}}$$

where z_0 is the estimated value at point 0, z_i is the z value at known point i, d_i is the distance between

Box 15.2 | A Worked Example of Kernel Density Estimation

This example shows how the value of the cell marked X in Figure 15.9 is derived. The window area is defined as a circle with a radius of 100 meters (h). Therefore, only points within the 100-meter radius of the center of the cell can influence the estimation of the cell density. Using the 10 points within the cell's neighborhood, we can compute the cell density by:

$$\frac{3}{\pi} \sum_{i=1}^{10} n_i \{1 - \frac{1}{h^2}[(x - x_i)^2 + (y - y_i)^2]\}^2$$

where n_i is the number of sightings at point i, x_i and y_i are the x-, y-coordinates of point i, and x and y are the x-, y-coordinates of the center of the cell to be estimated. Because the density is measured per 10,000 square meters or hectare, h^2 in Eq. (15.5) is canceled out. Also, because the output shows an expected value rather than a probability, n in Eq. (15.5) is not needed. The computation shows the cell density to be 11.421.

point i and point 0, s is the number of known points used in estimation, and k is the specified power.

The power k controls the degree of local influence. A power of 1.0 means a constant rate of change in value between points (linear interpolation). A power of 2.0 or higher suggests that the rate of change in values is higher near a known point and levels off away from it.

An important characteristic of IDW interpolation is that all predicted values are within the range of maximum and minimum values of the known points. Figure 15.10 shows an annual precipitation surface created by the IDW method with a power of 2 (Box 15.3). Figure 15.11 shows an isoline map of the surface. Small, enclosed isolines are typical of IDW interpolation. The odd shape of the 10-inch isoline in the southwest corner is due to the absence of known points.

15.3.4 Thin-Plate Splines

Splines for spatial interpolation are conceptually similar to splines for line smoothing (Chapter 7) except that in spatial interpolation they apply to surfaces rather than lines. **Thin-plate splines** create a surface that passes through the control points and has the least possible change in slope at all points (Franke 1982). In other words, thin-plate splines fit the control points with a minimum curvature surface. The approximation of thin-plate splines is of the form:

(15.7)

$$Q(x, y) = \sum A_i d_i^2 \log d_i + a + bx + cy$$

where x and y are the x-, y-coordinates of the point to be interpolated, $d_i^2 = (x - x_i)^2 + (y - y_i)^2$, and x_i and y_i are the x-, y-coordinates of control point i. Thin-plate splines consist of two components: $(a + bx + cy)$ represents the local trend function, which has the same form as a linear or first-order trend surface, and $d_i^2 \log d_i$ represents a basis function, which is designed to obtain minimum curvature surfaces (Watson 1992). The coefficients A_i, and a, b, and c are determined by a linear system of equations (Franke 1982):

(15.8)

$$\sum_{i=1}^{n} A_i d_i^2 \log d_i + a + bx + cy = f_i$$
$$\sum_{i=1}^{n} A_i = 0$$
$$\sum_{i=1}^{n} A_i x_i = 0$$
$$\sum_{i=1}^{n} A_i y_i = 0$$

where n is the number of control points and f_i is the known value at control point i. The estimation of the coefficients requires $n + 3$ simultaneous equations.

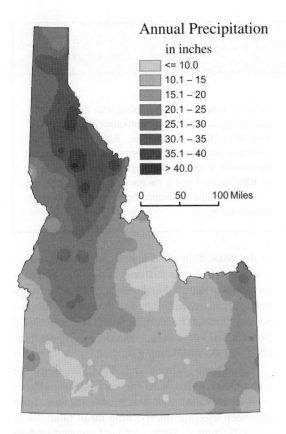

Annual Precipitation

in inches

░	<= 10.0
▒	10.1 – 15
▒	15.1 – 20
▓	20.1 – 25
▓	25.1 – 30
▓	30.1 – 35
▓	35.1 – 40
█	> 40.0

0   50   100 Miles

Figure 15.10
An annual precipitation surface created by the inverse
distance squared method.

Figure 15.11
An isohyet map created by the inverse distance squared
method.

Box 15.3 | **A Worked Example of Inverse Distance Weighted Estimation**

This example uses the same data set as in
Box 15.1, but interpolates the unknown value at
point 0 by the IDW method. The table below shows
the distances in thousands of meters between point 0
and the five known points:

Between Points	Distance
0,1	18.000
0,2	20.880
0,3	32.310
0,4	36.056
0,5	47.202

We can substitute the parameters in Eq. (15.6) by the
known values and the distances to estimate z_0:

$$\Sigma z_i(1/d_i^2) = (20.820)(1/18.000)^2$$
$$+ (10.910)(1/20.880)^2 + (10.380)(1/32.310)^2$$
$$+ (14.600)(1/36.056)^2 + (10.560)(1/47.202)^2$$
$$= 0.1152$$

$$\Sigma(1/d_i^2) = (1/18.000)^2 + (1/20.880)^2$$
$$+ (1/32.310)^2 + (1/36.056)^2 + (1/47.202)^2$$
$$= 0.0076$$
$$z_0 = 0.1152/0.0076 = 15.158$$

Box 15.4 | **Radial Basis Functions**

Radial basis functions (RBF) refer to a large group of interpolation methods. All of them are exact interpolators. The selection of a basis function or equation determines how the surface will fit between the control points. ArcToolbox in ArcGIS offers thin-plate splines with tension and regularized splines. The Geostatistical Analyst extension to ArcGIS, on the other hand, has a menu choice of five RBF

methods: thin-plate spline, spline with tension, completely regularized spline, multiquadric function, and inverse multiquadric function. Each RBF method also has a parameter that controls the smoothness of the generated surface. Although each combination of an RBF method and a parameter value can create a new surface, the difference between the surfaces is usually small.

Unlike the IDW method, the predicted values from thin-plate splines are not limited within the range of maximum and minimum values of the known points. In fact, a major problem with thin-plate splines is the steep gradients in data-poor areas, often referred to as overshoots. Different methods for correcting overshoots have been proposed. Thin-plate splines with tension, for example, allow the user to control the tension to be pulled on the edges of the surface (Franke 1985; Mitas and Mitasova 1988). Other methods include regularized splines (Mitas and Mitasova 1988) and regularized splines with tension (Mitasova and Mitas 1993). All these methods belong to a diverse group called **radial basis functions (RBF)** (Box 15.4).

The **thin-plate splines with tension** method has the following form:

(15.9)

$$a + \sum_{i=1}^{n} A_i R(d_i)$$

where a represents the trend function, and the basis function $R(d)$ is:

(15.10)

$$-\frac{1}{2\pi\varphi^2} \left[\ln\left(\frac{d\varphi}{2}\right) + c + K_0\left(d\varphi\right) \right]$$

where φ is the weight to be used with the tension method. If the weight φ is set close to 0, then the approximation with tension is similar

to the basic thin-plate splines method. A larger φ value reduces the stiffness of the plate and thus the range of interpolated values, with the interpolated surface resembling the shape of a membrane passing through the control points (Franke 1985). Box 15.5 shows a worked example using the thin-plate splines with tension method.

Thin-plate splines and their variations are recommended for smooth, continuous surfaces such as elevation and water table. Splines have also been used for interpolating mean rainfall surface (Hutchinson 1995) and land demand surface (Wickham, O'Neill, and Jones 2000). Figures 15.12 and 15.13 show annual precipitation surfaces created by the regularized splines method and the splines with tension method, respectively. The isolines in both figures are smoother than those generated by the IDW method. Also noticeable is the similarity between the two sets of isolines.

15.4 KRIGING

Kriging is a geostatistical method for spatial interpolation. Kriging differs from other local interpolation methods because kriging can assess the quality of prediction with estimated prediction errors. Originated in mining and geologic engineering in the 1950s, kriging has since been adopted in a wide variety of disciplines. In GIS, kriging has also become a popular method for coverting LIDAR (light detection and ranging) point data into DEMs (e.g., Zhang et al. 2003).

Box 15.5 A Worked Example of Thin-Plate Splines with Tension

This example uses the same data set as in Box 15.1 but interpolates the unknown value at point 0 by splines with tension. The method first involves calculation of $R(d)$ in Eq. (15.10) using the distances between the point to be estimated and the known points, distances between the known points, and the φ value of 0.1. The following table shows the $R(d)$ values along with the distance values.

Points	0,1	0,2	0,3	0,4	0,5
Distance	18.000	20.880	32.310	36.056	47.202
$R(d)$	−7.510	−9.879	−16.831	−18.574	−22.834
Points	1,2	1,3	1,4	1,5	2,3
Distance	31.240	49.476	34.526	59.666	40.000
$R(d)$	−16.289	−23.612	−17.879	−26.591	−20.225
Points	2,4	2,5	3,4	3,5	4,5
Distance	56.920	62.032	47.412	26.076	40.200
$R(d)$	−25.843	−27.214	−22.868	−13.415	−20.305

The next step is to solve for A_i in Eq. (15.9). We can substitute the calculated $R(d)$ values into Eq. (15.9)

and rewrite the equation and the constraint about A_i in matrix form:

$$\begin{bmatrix} 1 & 0 & -16.289 & -23.612 & -17.879 & -26.591 \\ 1 & -16.289 & 0 & -20.225 & -25.843 & -27.214 \\ 1 & -23.612 & -20.225 & 0 & -22.868 & -13.415 \\ 1 & -17.879 & -25.843 & -22.868 & 0 & -20.305 \\ 1 & -26.591 & -27.214 & -13.415 & -20.305 & 0 \\ 0 & 1 & 1 & 1 & 1 & 1 \end{bmatrix} \cdot \begin{bmatrix} a \\ A_1 \\ A_2 \\ A_3 \\ A_4 \\ A_5 \end{bmatrix} = \begin{bmatrix} 20.820 \\ 10.910 \\ 10.380 \\ 14.600 \\ 10.560 \\ 0 \end{bmatrix}$$

The matrix solutions are:

$$a = 13.203 \quad A_1 = 0.396 \quad A_2 = -0.226$$
$$A_3 = -0.058 \quad A_4 = -0.047 \quad A_5 = -0.065$$

Now we can calculate the value at point 0 by:

$$z_0 = 13.203 + (0.396)(-7.510)$$
$$+ (-0.226)(-9.879) + (-0.058)(-16.831)$$
$$+ (-0.047)(-18.574) + (-0.065)(-22.834)$$
$$= 15.795$$

Using the same data set, the estimations of z_0 by other splines are as follows: 16.350 by thin-plate splines and 15.015 by regularized splines (using the τ value of 0.1).

Kriging assumes that the spatial variation of an attribute such as changes in grade within an ore body is neither totally random (stochastic) nor deterministic. Instead, the spatial variation may consist of three components: a spatially correlated component, representing the variation of the regionalized variable; a "drift" or structure, representing a trend; and a random error term. The interpretation of these components has led to development of different kriging methods for spatial interpolation.

15.4.1 Semivariogram

Kriging uses the **semivariance** to measure the spatially correlated component, a component that is also called *spatial dependence* or *spatial autocorrelation*. The semivariance is computed by:

(15.11)

$$\gamma(h) = \frac{1}{2}[z(x_i) - z(x_j)]^2$$

where $\gamma(h)$ is the semivariance between known points, x_i and x_j, separated by the distance h; and z is the attribute value.

Figure 15.14 is a semivariogram cloud, which plots $\gamma(h)$ against h for all pairs of known points in a data set. (Because every known point contributes to the semivariogram, kriging is sometimes described as having *global support*.) If spatial dependence does exist in a data set, known points that are close to each other are expected to have small semivariances, and known points that are farther apart are expected to have larger semivariances.

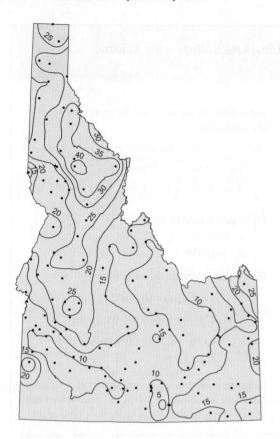

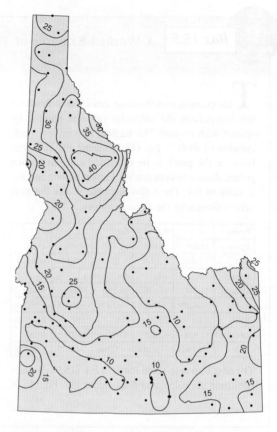

Figure 15.12

An isohyet map created by the regularized splines method.

Figure 15.13

An isohyet map created by the splines with tension method.

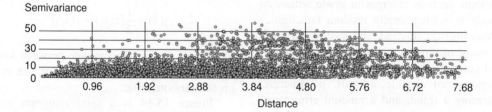

Figure 15.14

A semivariogram cloud.

A semivariogram cloud is an important tool for investigating the spatial variability of the phenomenon under study (Gringarten and Deutsch 2001). But because it has all pairs of known points, a semivariogram cloud is difficult to manage and

use. A process called **binning** is typically used in kriging to average semivariance data by distance and direction. The first part of the binning process is to group pairs of sample points into lag classes. For example, if the lag size (i.e., distance interval)

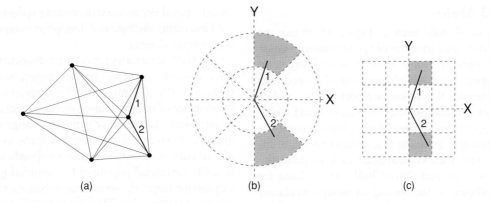

Figure 15.15

A common method for binning pairs of sample points by direction, such as 1 and 2 in (*a*), is to use the radial sector (*b*). Geostatistical Analyst in ArcGIS uses grid cells instead (*c*).

is 2000 meters, then pairs of points separated by less than 2000 meters are grouped into the lag class of 0–2000, pairs of points separated between 2000 and 4000 meters are grouped into the lag class of 2000–4000, and so on. The second part of the binning process is to group pairs of sample points by direction. A common method is to use radial sectors; the Geostatistical Analyst extension to ArcGIS, on the other hand, uses grid cells (Figure 15.15).

The result of the binning process is a set of bins (e.g., grid cells) that sort pairs of sample points by distance and direction. The next step is to compute the average semivariance by:

(15.12)

$$\gamma(h) = \frac{1}{2n} \sum_{i=1}^{n} [z(x_i) - z(x_i + h)]^2$$

where $\gamma(h)$ is the average semivariance between sample points separated by lag h; n is the number of pairs of sample points sorted by direction in the bin; and z is the attribute value.

A **semivariogram** plots the average semivariance against the average distance (Figure 15.16). Because of the directional component, one or more average semivariances may be plotted at the same distance. We can examine the semivariogram in Figure 15.16. If spatial dependence exists among the sample points, then pairs of points

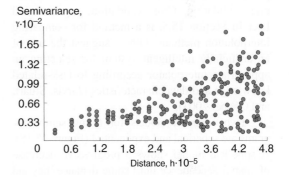

Figure 15.16

A semivariogram after binning by distance.

that are closer in distance will have more similar values than pairs that are farther apart. In other words, the semivariance is expected to increase as the distance increases in the presence of spatial dependence.

A semivariogram can also be examined by direction. If spatial dependence has directional differences, then the semivariance values may change more rapidly in one direction than another. **Anisotropy** is the term describing the existence of directional differences in spatial dependence (Eriksson and Siska 2000). Isotropy represents the opposite case in which spatial dependence changes with the distance but not the direction.

15.4.2 Models

A semivariogram such as Figure 15.16 may be used alone as a measure of spatial autocorrelation in the data set. But to be used as an interpolator in kriging, the semivariogram must be fitted with a mathematical function or model (Figure 15.17). The fitted semivariogram can then be used for estimating the semivariance at any given distance.

Fitting a model to a semivariogram is a difficult and often controversial task in geostatistics (Webster and Oliver 2001). One reason for the difficulty is the number of models to choose from. For example, Geostatistical Analyst offers 11 models. The other reason is the lack of a standardized procedure for comparing the models. Webster and Oliver (2001) recommend a procedure that combines visual inspection and cross-validation. Cross-validation, as discussed later in Section 15.5, is a method for comparing interpolation methods. Others suggest the use of an artificially intelligent system for selecting an appropriate interpolator according to task-related knowledge and data characteristics (Jarvis, Stuart, and Cooper 2003).

Two common models for fitting semivariograms are spherical and exponential (Figure 15.18). A spherical model shows a progressive decrease of spatial dependence until some distance, beyond which spatial dependence levels off. An exponential model exhibits a less gradual pattern than a spherical

model: spatial dependence decreases exponentially with increasing distance and disappears completely at an infinite distance.

A fitted semivariogram can be dissected into three possible elements: nugget, range, and sill (Figure 15.19). The **nugget** is the semivariance at the distance of 0, representing measurement error, or microscale variation, or both. The **range** is the distance at which the semivariance starts to level off. In other words, the range corresponds to the spatially correlated portion of the semivariogram. Beyond the range, the semivariance becomes a relatively constant value. The semivariance at which the leveling takes place is called the **sill.** The sill comprises two components: the nugget and the

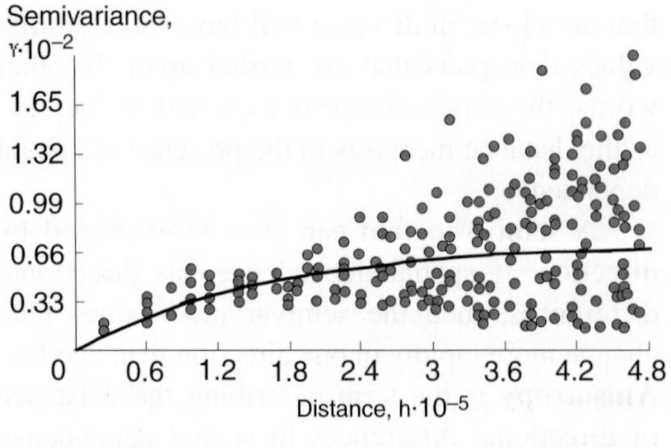

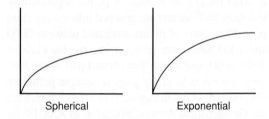

Figure 15.18
Two common models for fitting semivariograms: spherical and exponential.

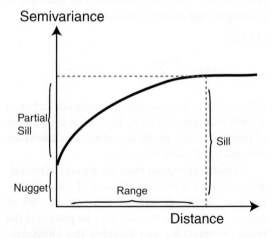

Figure 15.19
Nugget, range, sill, and partial sill.

Figure 15.17
Fitting a semivariogram with a mathematical function or a model.

partial sill. To put it another way, the partial sill is the difference between the sill and the nugget.

15.4.3 Ordinary Kriging

Assuming the absence of a drift, **ordinary kriging** focuses on the spatially correlated component and uses the fitted semivariogram directly for interpolation. The general equation for estimating the z value at a point is:

(15.13)

$$z_0 = \sum_{i=1}^{s} z_x W_x$$

where z_0 is the estimated value, z_x is the known value at point x, W_x is the weight associated with point x, and s is the number of sample points used in estimation. The weights can be derived from solving a set of simultaneous equations. For example, the following equations are needed for a point (0) to be estimated from three known points (1, 2, 3):

(15.14)

$$W_1\gamma(h_{11}) + W_2\gamma(h_{12}) + W_3\gamma(h_{13}) + \lambda = \gamma(h_{10})$$
$$W_1\gamma(h_{21}) + W_2\gamma(h_{22}) + W_3\gamma(h_{23}) + \lambda = \gamma(h_{20})$$
$$W_1\gamma(h_{31}) + W_2\gamma(h_{32}) + W_3\gamma(h_{33}) + \lambda = \gamma(h_{30})$$
$$W_1 + W_2 + W_3 + 0 = 1.0$$

where $\gamma(h_{ij})$ is the semivariance between known points i and j, $\gamma(h_{i0})$ is the semivariance between the ith known point and the point to be estimated, and λ is a Lagrange multiplier, which is added to ensure the minimum possible estimation error. Once the weights are solved, Eq. (15.13) can be used to estimate z_0

$$z_0 = z_1W_1 + z_2W_2 + z_3W_3$$

The preceding example shows that weights used in kriging involve not only the semivariances between the point to be estimated and the known points but also those between the known points. This differs from the IDW method, which uses only weights applicable to the point to be estimated and the known points. Another important difference between kriging and other local methods is that kriging produces a variance measure for each estimated point to indicate the reliability of the estimation. For the example, the variance estimation can be calculated by:

(15.15)

$$s^2 = W_1\gamma(h_{10}) + W_2\gamma(h_{20}) + W_3\gamma(h_{30}) + \lambda$$

Figure 15.20 shows an annual precipitation surface created by ordinary kriging with the exponential model. Figure 15.21 shows the distribution of the standard error of the predicted surface. As

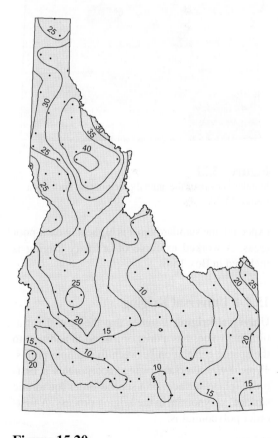

Figure 15.20
An isohyet map created by ordinary kriging with the exponential model.

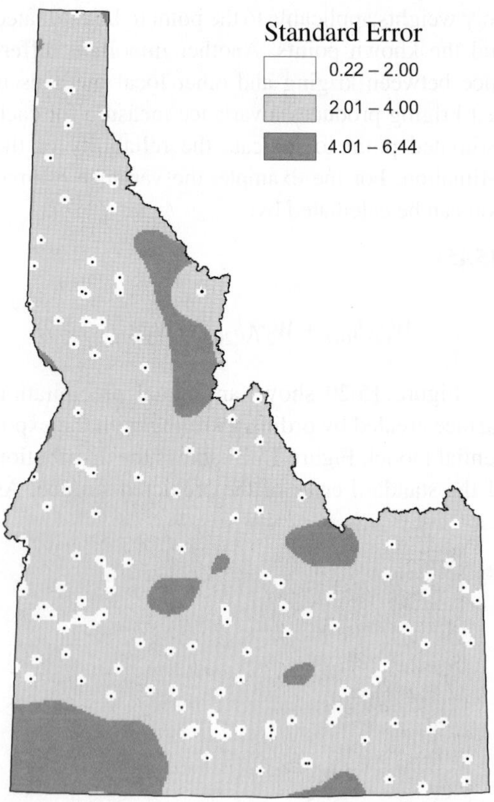
Standard Error
0.22 – 2.00
2.01 – 4.00
4.01 – 6.44

Figure 15.21
Standard errors of the annual precipitation surface in Figure 15.20.

expected, the standard error is highest in data-poor areas. A worked example of ordinary kriging is included in Box 15.6.

15.4.4 Universal Kriging

Universal kriging assumes that the spatial variation in z values has a drift or a trend in addition to the spatial correlation between the sample points. Typically, universal kriging incorporates a first-order (plane surface) or a second-order (quadratic surface) polynomial in the kriging process. A first-order polynomial is:

(15.16)

$$M = b_1x_i + b_2y_i$$

where M is the drift, x_i and y_i are the x-, y-coordinates of sampled point i, and b_1 and b_2 are the drift coefficients. A second-order polynomial is:

(15.17)

$$M = b_1x_i + b_2y_i + b_3x_i^2 + b_4x_iy_i + b_5y_i^2$$

Higher-order polynomials are usually not recommended for two reasons. First, kriging is performed on the residuals after the trend is removed. A higher-order polynomial will leave little variation in the residuals for assessing uncertainty. Second, a higher-order polynomial means a larger number of the b_i coefficients, which must be estimated along with the weights, and a larger set of simultaneous equations to be solved.

Figure 15.22 shows an annual precipitation surface created by universal kriging with the linear (first-order) drift, and Figure 15.23 shows the distribution of the standard error of the predicted surface. Universal kriging produces less reliable estimates than ordinary kriging in this case. A worked example of universal kriging is included in Box 15.7.

15.4.5 Other Kriging Methods

The three basic kriging methods are ordinary kriging, universal kriging, and simple kriging. Simple kriging assumes that the mean of the data set is known. This assumption, however, is unrealistic in most cases.

Other kriging methods include indicator kriging, disjunctive kriging, and block kriging (Bailey and Gatrell 1995; Burrough and McDonnell 1998; Lloyd and Atkinson 2001; Webster and Oliver 2001). Indicator kriging uses binary data (i.e., 0 and 1) rather than continuous data. The interpolated values are therefore between 0 and 1, similar to probabilities. Disjunctive kriging uses a function of the attribute value for interpolation and is more complicated than other kriging methods computationally. Block kriging estimates the average value of a variable over some small area or block rather than at a point.

Cokriging uses one or more secondary variables, which are correlated with the primary

Box 15.6 — A Worked Example of Ordinary Kriging Estimation

This worked example uses ordinary kriging for spatial interpolation. To keep the computation simpler, the semivariogram is fitted with the linear model, which is defined by:

$$\gamma(h) = C_0 + C(h/a), 0 < h \leq a$$
$$\gamma(h) = C_0 + C, h > a$$
$$\gamma(0) = 0$$

where $\gamma(h)$ is the semivariance at distance h, C_0 is the semivariance at distance 0, a is the range, and C is the sill, or the semivariance at a. The output from ArcGIS shows:

$$C_0 = 0, C = 112.475, \text{ and } a = 458,000.$$

Now we can use the model for spatial interpolation. The scenario is the same as in Box 15.1: using five points with known values to estimate an unknown value. The estimation begins by computing distances between points (in thousands of meters) and the semivariances at those distances based on the linear model:

Points ij	0,1	0,2	0,3	0,4	0,5
h_{ij}	18.000	20.880	32.310	36.056	47.202
$\gamma(h_{ij})$	4.420	5.128	7.935	8.855	11.592
Points ij	1,2	1,3	1,4	1,5	2,3
h_{ij}	31.240	49.476	34.526	59.666	40.000
$\gamma(h_{ij})$	7.672	12.150	8.479	14.653	9.823
Points ij	2,4	2,5	3,4	3,5	4,5
h_{ij}	56.920	62.032	47.412	26.076	40.200
$\gamma(h_{ij})$	13.978	15.234	11.643	6.404	9.872

Using the semivariances, we can write the simultaneous equations for solving the weights in matrix form:

$$\begin{bmatrix} 0 & 7.672 & 12.150 & 8.479 & 14.653 & 1 \\ 7.672 & 0 & 9.823 & 13.978 & 15.234 & 1 \\ 12.150 & 9.823 & 0 & 11.643 & 6.404 & 1 \\ 8.479 & 13.978 & 11.643 & 0 & 9.872 & 1 \\ 14.653 & 15.234 & 6.404 & 9.872 & 0 & 1 \\ 1 & 1 & 1 & 1 & 1 & 0 \end{bmatrix} \cdot \begin{bmatrix} W_1 \\ W_2 \\ W_3 \\ W_4 \\ W_5 \\ \lambda \end{bmatrix} = \begin{bmatrix} 4.420 \\ 5.128 \\ 7.935 \\ 8.855 \\ 11.592 \\ 1 \end{bmatrix}$$

The matrix solutions are:

$$W_1 = 0.397 \quad W_2 = 0.318 \quad W_3 = 0.182$$
$$W_4 = 0.094 \quad W_5 = 0.009 \quad \lambda = -1.161$$

Using Eq. (15.13), we can estimate the unknown value at point 0 by:

$$z_0 = (0.397)(20.820) + (0.318)(10.910)$$
$$+ (0.182)(10.380) + (0.094)(14.600)$$
$$+ (0.009)(10.560) = 15.091$$

We can also estimate the variance at point 0 by:

$$s^2 = (4.420)(0.397) + (5.128)(0.318)$$
$$+ (7.935)(0.182) + (8.855)(0.094)$$
$$+ (11.592)(0.009) - 1.161 = 4.605$$

In other words, the standard error of estimate at point 0 is 2.146.

variable of interest, in interpolation. It assumes that the correlation between the variables can improve the prediction of the value of the primary variable. For example, better results in precipitation interpolation have been reported by including elevation as an additional variable in cokriging (Martinez-Cob 1996). Cokriging can be ordinary cokriging, universal cokriging, and so on, depending on the kriging method that is applied to each data set.

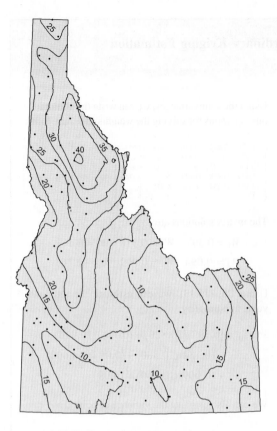

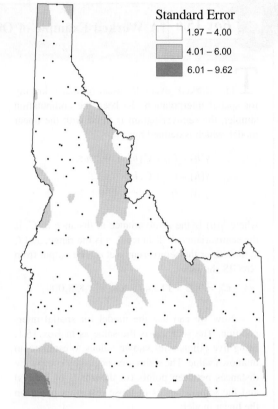

Figure 15.22
An isohyet map created by universal kriging with the linear drift and the spherical model.

Figure 15.23
Standard errors of the annual precipitation surface in Figure 15.22.

15.5 Comparison of Spatial Interpolation Methods

GIS packages such as ArcGIS offer a large number of spatial interpolation methods (Box 15.8). Using the same data but different methods, we can expect to find different interpolation results. For example, Figure 15.24 shows the difference of the interpolated surfaces between IDW and ordinary kriging. The difference ranges from −8 to 3.4 inches: A negative value means that IDW produces a smaller estimate than ordinary kriging, and a positive value means a reverse pattern. Data-poor areas clearly have the largest difference

(i.e., more than 3 inches either positively or negatively), suggesting that spatial interpolation can never substitute for observed data. However, if adding more known points is not feasible, how can we tell which interpolation method is better?

Cross-validation and validation are two common statistical techniques for comparing interpolation methods (Phillips, Dolph, and Marks 1992; Carroll and Cressie 1996; Zimmerman et al. 1999; Lloyd 2005), although some studies have also suggested the importance of the visual quality of generated surfaces such as preservation of distinct spatial pattern and visual pleasantness and faithfulness (Laslett 1994; Declercq 1996; Yang and Hodler 2000).

Box 15.7 A Worked Example of Universal Kriging Estimation

This example uses universal kriging to estimate the unknown value at point 0 (Box 15.1) and assumes that (1) the drift is linear and (2) the semivariogram is fitted with a linear model. Because of the additional drift component, this example uses eight simultaneous equations:

$$W_1\gamma(h_{11}) + W_2\gamma(h_{12}) + W_3\gamma(h_{13}) + W_4\gamma(h_{14})$$
$$+ W_5\gamma(h_{15}) + \lambda + b_1x_1 + b_2y_1 = \gamma(h_{10})$$

$$W_1\gamma(h_{21}) + W_2\gamma(h_{22}) + W_3\gamma(h_{23}) + W_4\gamma(h_{24})$$
$$+ W_5\gamma(h_{25}) + \lambda + b_1x_2 + b_2y_2 = \gamma(h_{20})$$

$$W_1\gamma(h_{31}) + W_2\gamma(h_{32}) + W_3\gamma(h_{33}) + W_4\gamma(h_{34})$$
$$+ W_5\gamma(h_{35}) + \lambda + b_1x_3 + b_2y_3 = \gamma(h_{30})$$

$$W_1\gamma(h_{41}) + W_2\gamma(h_{42}) + W_3\gamma(h_{43}) + W_4\gamma(h_{44})$$
$$+ W_5\gamma(h_{45}) + \lambda + b_1x_4 + b_2y_4 = \gamma(h_{40})$$

$$W_1\gamma(h_{51}) + W_2\gamma(h_{52}) + W_3\gamma(h_{53}) + W_4\gamma(h_{54})$$
$$+ W_5\gamma(h_{55}) + \lambda + b_1x_5 + b_2y_5 = \gamma(h_{50})$$

$$W_1 + W_2 + W_3 + W_4 + W_5 + 0 + 0 + 0 = 1$$
$$W_1x_1 + W_2x_2 + W_3x_3 + W_4x_4 + W_5x_5 + 0 + 0 + 0 = x_0$$
$$W_1y_1 + W_2y_2 + W_3y_3 + W_4y_4 + W_5y_5 + 0 + 0 + 0 = y_0$$

where x_0 and y_0 are the x-, y-coordinates of the point to be estimated, and x_i and y_i are the x-, y-coordinates of known point i; otherwise, the notations are the same as in Box 15.6. The x-, y-coordinates are actually rows and columns in the output raster with a cell size of 2000 meters.

Similar to ordinary kriging, semivariance values for the equations can be derived from the semivariogram and the linear model. The next step is to rewrite the equations in matrix form:

$$\begin{bmatrix} 0 & 7.672 & 12.150 & 8.479 & 14.653 & 1 & 69 & 76 \\ 7.672 & 0 & 9.823 & 13.978 & 15.234 & 1 & 59 & 64 \\ 12.150 & 9.823 & 0 & 11.643 & 6.404 & 1 & 75 & 52 \\ 8.479 & 13.978 & 11.643 & 0 & 9.872 & 1 & 86 & 73 \\ 14.653 & 15.234 & 6.404 & 9.872 & 0 & 1 & 88 & 53 \\ 1 & 1 & 1 & 1 & 1 & 0 & 0 & 0 \\ 69 & 59 & 75 & 86 & 88 & 0 & 0 & 0 \\ 76 & 64 & 52 & 73 & 53 & 0 & 0 & 0 \end{bmatrix} \bullet \begin{bmatrix} W_1 \\ W_2 \\ W_3 \\ W_4 \\ W_5 \\ \lambda \\ b_1 \\ b_2 \end{bmatrix} = \begin{bmatrix} 4.420 \\ 5.128 \\ 7.935 \\ 11.592 \\ 8.855 \\ 1 \\ 69 \\ 67 \end{bmatrix}$$

The solutions are:

$$W_1 = 0.387 \ W_2 = 0.311 \ W_3 = 0.188 \ W_4 = 0.093$$
$$W_1 = 0.021 \ \lambda = -1.154 \ b_1 = 0.009 \ b_2 = -0.010$$

The estimated value at point 0 is:

$$z_0 = (0.387)(20.820) + (0.311)(10.910)$$
$$+ (0.188)(10.380) + (0.093)(14.600)$$
$$+ (0.021)(10.560) = 14.981$$

And, the variance at point 0 is:

$$s^2 = (4.420)(0.387) + (5.128)(0.311)$$
$$+ (7.935)(0.188) + (8.855)(0.093)$$
$$+ (11.592)(0.021) - 1.154 + (0.009)(69)$$
$$- (0.010)(67) = 4.661$$

The standard error (s) at point 0 is 2.159. These results from universal kriging are very similar to those from ordinary kriging.

Cross-validation compares the interpolation methods by repeating the following procedure for each interpolation method to be compared:

1. Remove a known point from the data set.
2. Use the remaining points to estimate the value at the point previously removed.

3. Calculate the predicted error of the estimation by comparing the estimated with the known value.

After completing the procedure for each known point, one can calculate diagnostic statistics to assess the accuracy of the interpolation method.

Box 15.8 | **Spatial Interpolation Using ArcGIS**

ArcToolbox offers the Density and Interpolation toolsets in Spatial Analyst Tools. The Density toolset has Kernel Density, Point Density, and Line Density. The Interpolation toolset has tools for interpolation using trend surface, inverse distance weighted, spline, and kriging (ordinary and universal). Additionally, Ordinary Least Squares and Geographically Weighted Regression are included in the Spatial Statistics Tools/Modeling Spatial Relationships toolset.

Most users, however, will opt to run spatial interpolation using the Geostatistical Analyst extension. Geostatistical Analyst's main menu has the selections of explore data, geostatistical wizard, and create subsets. The explore data selection offers histogram, semivariogram, QQ plot, and others for exploratory data analysis (Chapter 10). The geostatistical wizard offers a large variety of global and local interpolation methods including IDW, trend surface, local polynomial, radial basis function, kriging, and cokriging. The wizard also provides cross-validation measures. The create subsets selection is designed for model validation.

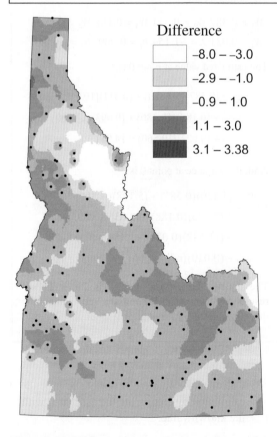

Difference

☐	–8.0 – –3.0
☐	–2.9 – –1.0
☐	–0.9 – 1.0
☐	1.1 – 3.0
☐	3.1 – 3.38

Figure 15.24
Differences between the interpolated surfaces from ordinary kriging and IDW.

Two common diagnostic statistics are the root mean square (RMS) error and the standardized RMS error:

(15.18)

$$\text{RMS} = \sqrt{\frac{1}{n}\sum_{i=1}^{n}(z_{i,\text{act}} - z_{i,\text{est}})^2}$$

(15.19)

$$\text{Standardized RMS} = \sqrt{\frac{1}{n}\sum_{i=1}^{n}\frac{(z_{i,\text{act}} - z_{i,\text{est}})^2}{s^2}}$$

$$= \frac{\text{RMS}}{s}$$

where n is the number of points, $z_{i,\text{act}}$ is the known value of point i, $z_{i,\text{est}}$ is the estimated value of point i, s^2 is the variance, and s is the standard error.

A common measure of accuracy, RMS quantifies the differences between the known and estimated values at sample points. The RMS statistic is available for all exact local methods. Because the standardized RMS requires the variance of the estimated values for the computation, it is only available for kriging. The interpretation of the statistics is as follows:

- A better interpolation method should yield a smaller RMS. By extension, an optimal method should have the smallest RMS, or

the smallest average deviation between the estimated and known values at sample points.
- A better kriging method should yield a smaller RMS and a standardized RMS closer to 1.

If the standardized RMS is 1, it means that the RMS statistic equals *s*. Therefore the estimated standard error is a reliable or valid measure of the uncertainty of predicted values.

The **validation** technique compares the interpolation methods by first dividing known points into two samples: one sample for developing the model for each interpolation method to be compared and the other sample for testing the accuracy of the models. The diagnostic statistics of RMS and standardized RMS derived from the test sample can then be used to compare the methods. Validation may not be a feasible option if the number of known points is too small to be split into two samples.

KEY CONCEPTS AND TERMS

Anisotropy: A term describing the existence of directional differences in spatial dependence.

Binning: A process used in kriging to average semivariance data by distance and direction.

Control points: Points with known values in spatial interpolation. Also called *known points, sample points,* or *observations.*

Cross-validation: A technique for comparing different interpolation methods.

Density estimation: A local interpolation method that measures densities in a raster based on a distribution of points and point values.

Deterministic interpolation: A spatial interpolation method that provides no assessment of errors with predicted values.

Exact interpolation: An interpolation method that predicts the same value as the known value at the control point.

Global interpolation: An interpolation method that uses every control point available in estimating an unknown value.

Inexact interpolation: An interpolation method that predicts a different value from the known value at the control point.

Inverse distance weighted (IDW) interpolation: A local interpolation method that enforces the condition that the unknown value of a point is influenced more by nearby points than by those farther away.

Kernel density estimation: A local interpolation method that associates each known point with a kernel function in the form of a bivariate probability density function.

Kriging: A stochastic interpolation method that assumes that the spatial variation of an attribute includes a spatially correlated component.

Local interpolation: An interpolation method that uses a sample of known points in estimating an unknown value.

Local polynomial interpolation: A local interpolation method that uses a sample of points with known values and a polynomial equation to estimate the unknown value of a point.

Local regression analysis: A local interpolation method that uses the information for each known point to derive a local regression model. Also called *geographically weighted regression analysis.*

Nugget: The semivariance value at the distance 0 in a semivariogram.

Ordinary kriging: A kriging method that assumes the absence of a drift or trend and focuses on the spatially correlated component.

Partial sill: The difference between the sill and the nugget in a semivariogram.

Radial basis functions (RBF): A diverse group of methods for spatial interpolation including thin-plate splines, thin-plate splines with tension, and regularized splines.

Range: The distance at which the semivariance starts to level off in a semivariogram.

Regression model: A global interpolation method that uses a number of independent variables to estimate a dependent variable.

Semivariance: A measure of the degree of spatial dependence among points used in kriging.

Semivariogram: A diagram relating the semivariance to the distance between sample points used in kriging.

Sill: The semivariance at which the leveling starts in a semivariogram.

Spatial interpolation: The process of using points with known values to estimate unknown values at other points.

Stochastic interpolation: A spatial interpolation method that offers assessment of prediction errors with estimated variances.

Thiessen polygons: A local interpolation method that ensures that every unsampled point within a polygon is closer to the polygon's known point than any other known points. Also called *Voronoi polygons.*

Thin-plate splines: A local interpolation method that creates a surface passing through points with the least possible change in slope at all points.

Thin-plate splines with tension: A variation of thin-plate splines for spatial interpolation.

Trend surface analysis: A global interpolation method that uses points with known values and a polynomial equation to approximate a surface.

Universal kriging: A kriging method that assumes that the spatial variation of an attribute has a drift or a structural component in addition to the spatial correlation between sample points.

Validation: A technique for comparing interpolation methods, which splits control points into two samples, one for developing the model and the other for testing the accuracy of the model.

REVIEW QUESTIONS

1. What is spatial interpolation?
2. What kinds of inputs are required for spatial interpolation?
3. Explain the difference between a global method and a local method.
4. How does an exact interpolation method differ from an inexact interpolation method?
5. What are Thiessen polygons?
6. Given a sample size of 12, illustrate the difference between a sampling method that uses the closest points and a quadrant sampling method.
7. Describe how cell densities are derived using the kernel density estimation method.
8. The power k in inverse distance weighted interpolation determines the rate of change

in values from the sample points. Can you think of a spatial phenomenon that should be interpolated with a k value of 2 or higher?

9. Describe how the semivariance can be used to quantify the spatial dependence in a data set.
10. Binning is a process for creating a usable semivariogram from empirical data. Describe how binning is performed.
11. A semivariogram must be fitted with a mathematical model before it can be used in kriging. Why?
12. Both IDW and kriging use weights in estimating an unknown value. Describe the difference between the two interpolation methods in terms of derivation of the weights.

13. Explain the main difference between ordinary kriging and universal kriging.

14. The root mean square (RMS) statistic is commonly used for selecting an optimal interpolation method. What does the RMS statistic measure?

15. Explain how one can use the validation technique for comparing different interpolation methods.

16. Which local interpolation method can usually give you a smooth contour map?

APPLICATIONS: SPATIAL INTERPOLATION

This applications section has five tasks. Task 1 covers trend surface analysis. Task 2 deals with kernel density estimation. Task 3 uses IDW for local interpolation. Tasks 4 and 5 cover kriging: Task 4 uses ordinary kriging and Task 5 universal kriging. Except for Task 2, you will run spatial interpolation in Geostatistical Analyst so that you can use the cross-validation statistics such as the root mean square (RMS) statistic to compare models.

Task 1 Use Trend Surface Model for Interpolation

What you need: *stations.shp,* a shapefile containing 175 weather stations in and around Idaho; and *idoutlgd,* an Idaho outline raster.

In Task 1 you will first explore the average annual precipitation data in *stations.shp,* before running a trend surface analysis.

1. Start ArcCatalog, and connect to the Chapter 15 database. Launch ArcMap. Add *stations.shp* and *idoutlgd* to Layers and rename the data frame Task 1. Make sure that both the Geostatistical Analyst and Spatial Analyst extensions are checked in the Customize menu and the Geostatistical Analyst toolbar is checked in the Customize menu.

2. Click the Geostatistical Analyst dropdown arrow, point to Explore Data, and select Trend Analysis. At the bottom of the Trend Analysis dialog, click the dropdown arrow to select *stations* for the layer and ANN_PREC for the attribute.

3. Maximize the Trend Analysis dialog. The 3-D diagram shows two trend projections: The YZ plane dips from north to south, and the XZ plane dips initially from west to east and then rises slightly. The north–south trend is much stronger than the east–west trend, suggesting that the general precipitation pattern in Idaho decreases from north to south. Close the dialog.

4. Click the Geostatistical Analyst dropdown arrow, and select Geostatistical Wizard. The opening panel lets you choose a geostatistical method. In the Methods frame, click Global Polynomial Interpolation. Click Next.

5. Step 2 lets you choose the order of polynomial. The order of polynomial dropdown menu provides the order from 0 to 10. Select 1 for the order. Step 3 shows scatter plots (Predicted versus Measured values, and Error versus Measured values) and statistics related to the first-order trend surface model. The RMS statistic measures the overall fit of the trend surface model. In this case, it has a value of 6.073. Click Back and change the power to 2. The RMS statistic for the power of 2 has a value of 6.085. Repeat the same procedure with other power numbers. The trend surface model with the lowest RMS statistic is the best overall model for this task. For ANN_PREC, the best overall model has the power of 5. Change the power to 5, and click Finish. Click OK in the Method Summary dialog.

Q1. What is the RMS statistic for the power of 5?

6. *Global Polynomial Interpolation Prediction Map* is a Geostatistical Analyst (ga) output layer and has the same area extent as *stations*. Right-click *Global Polynomial Interpolation Prediction Map* and select Properties. The Symbology tab has four Show options: Hillshade, Contours, Grid, and Filled Contours. Uncheck all Show boxes except Filled Contours, click Filled Contours, and then click on Classify. In the Classification dialog, select the Manual method and 7 classes. Then enter the class breaks of 10, 15, 20, 25, 30, and 35 between the Min and Max values. Click OK to dismiss the dialogs. The contour (isohyet) intervals are color-coded.

7. To clip *Global Polynomial Interpolation Prediction Map* to fit Idaho, first convert the ga layer to a raster. Right-click *Global Polynomial Interpolation Prediction Map,* point to Data, and select Export to Raster. In the Export to Raster dialog, specify *trend5_temp* for the output surface raster, enter 2000 (meters) for the cell size, and click OK to export the data set. (The GA Layer to Grid tool in Geostatistical Analyst Tools/Working with Geostatistical Layers toolset can also perform the conversion.) *trend5_temp* is added to the map. (Extreme cell values in *trend5_temp* are located outside the state border.)

8. Now you are ready to clip *trend5_temp*. Click ArcToolbox window to open it. Set the Chapter 15 database as the workspace. Double-click the Extract by Mask tool in the Spatial Analyst Tools/Extraction toolset. In the next dialog, select *trend5_temp* for the input raster, select *idoutlgd* for the input raster or feature mask data, specify *trend5* for the output raster, and click OK. *trend5* is the clipped *trend5_temp*.

9. You can generate contours from *trend5* for data visualization. Double-click the Contour tool in the Spatial Analyst Tools/Surface toolset. In the Contour dialog, select *trend5*

for the input raster, save the output polyline features as *trend5ctour.shp*, enter 5 for the contour interval and 10 for the base contour, and click OK. To label the contour lines, right-click *trend5ctour* and select Properties. On the Labels tab, check the box to label features in this layer, select CONTOUR from the Label Field dropdown list, and click OK. The map now shows contour labels.

Task 2 Use Kernel Density Estimation Method

What you need: *deer.shp,* a point shapefile showing deer locations.

Task 2 uses the kernel density estimation method to compute the average number of deer sightings per hectare from *deer.shp*. Deer location data have a 50-meter minimum discernible distance; therefore, some locations have multiple sightings.

1. Insert a new data frame in ArcMap and rename it Task 2. Add *deer.shp* to Task 2. Select Properties from the context menu of *deer*. On the Symbology tab, select Quantities and Graduated symbols in the Show box and select SIGHTINGS from the Value dropdown list. Click OK. The map shows deer sightings at each location in graduated symbols.

Q2. What is the value range of SIGHTINGS?

2. Double-click the Kernel Density tool in the Spatial Analyst Tools/Density toolset. Select *deer* for the input point or polyline features, select SIGHTINGS for the population field, specify *kernel_d* for the output raster, enter 100 for the output cell size, enter 100 for the search radius, and select HECTARES for the area units. Click OK to run the command. *kernel_d* shows deer sighting densities computed by the kernel density estimation method.

Q3. What is the value range of deer sighting densities?

3. To view *kernel_d* on top of *deer,* you can use the transparency option. Right-click *kernel_d,* and select Properties. On the Display tab, enter 30% transparency. You can now see two layers superimposed on top of one another.

Task 3 Use IDW for Interpolation

What you need: *stations.shp* and *idoutlgd,* same as in Task 1.

This task lets you create a precipitation raster using the IDW method.

1. Insert a new data frame in ArcMap and rename it Task 3. Add *stations.shp* and *idoutlgd* to Task 3.

2. Click the Geostatistical Analyst dropdown arrow and select Geostatistical Wizard. Click Inverse Distance Weighting in the Methods frame. Click Next.

3. The Step 2 panel includes a graphic frame and a method frame for specifying IDW parameters. The default IDW method uses a power of 2, a maximum of 15 neighbors (control points), a minimum of 10 neighbors, and 1 sector area from which control points are selected. The graphic frame shows *stations* and the points and their weights (You can click on more in the General Properties frame to see the explanation) used in deriving the estimated value for a test location. You can use the Identify Value tool to click any point within the graphic frame and see how the point's predicted value is derived.

4. The Click to optimize Power value button is included in the method frame of Step 2. Because a change of the power value will change the estimated value at a point location, you can click the button and ask Geostatistical Wizard to find the optimal power value while holding other parameter values constant. Geostatistical Wizard employs the cross-validation technique to find the optimal power value. Click the

button to optimize power value, and the Power field shows a value of 3.191. Click Next.

5. Step 3 lets you examine the cross-validation results including the RMS statistic.

Q4. What is the RMS statistic when you use the default parameters including the optimal power value?

Q5. Change the power to 2 and the maximum number of neighbors to 10 and the minimum number to 6. What RMS statistic do you get?

6. Set the parameters back to the default including the optimal power value. Click Finish. Click OK in the Method Summary dialog. You can follow the same steps as in Task 1 to convert *Inverse Distance Weighting Prediction Map* to a raster, to clip the raster by using *idoutlgd* as the analysis mask, and to create isolines from the clipped raster.

Task 4 Use Ordinary Kriging for Interpolation

What you need: *stations.shp* and *idoutlgd.*

In Task 4, you will first examine the semi-variogram cloud from 175 points in *stations.shp.* Then you will run ordinary kriging on *stations.shp* to generate an interpolated precipitation raster and a standard error raster.

1. Select Data Frame from the Insert menu in ArcMap. Rename the new data frame Tasks 4&5, and add *stations.shp* and *idoutlgd* to Tasks 4&5. First explore the semivariogram cloud. Click the Geostatistical Analyst dropdown arrow, point to Explore Data, and select Semivariogram/Covariance Cloud. Select *stations* for the layer and ANN_PREC for the attribute. To view all possible pairs of control points in the cloud, enter 82,000 for the lag size and 12 for the number of lags. Use the mouse pointer to drag a box around the point to the far right of the cloud. Check *stations* in the ArcMap window. The highlighted pair consists of the control

points that are farthest apart in *stations*. The semivariogram shows a typical pattern of spatially correlated data: the semivariance increases rapidly to a distance of about 200,000 meters (2.00×10^5) and then gradually decreases.

2. To zoom in on the distance range of 200,000 meters, change the lag size to 10,000 and the number of lags to 20. The semivariance actually starts to level off at about 125,000 meters. To see if the semivariance may have the directional influence, check the box to show search direction. You can change the search direction by either entering the angle direction or using the direction controller in the graphic. Drag the direction controller in the counterclockwise direction from 0° to 180° but pause at different angles to check the semivariogram. The fluctuation in the semivariance tends to increase from northwest (315°) to southwest (225°). This indicates that the semivariance may have the directional influence. Close the Semivaraince/Covariance Cloud window. Clear selected features.

3. Select Geostatistical Wizard from the Geostatistical Analyst menu. Make sure that the source dataset is *stations* and the data field is ANN_PREC. Click Kriging/CoKriging in the Methods frame. Click Next. Step 2 lets you select the kriging method. Select Ordinary for the Kriging Type and Prediction for the Output Type. Click Next.

4. The Step 3 panel shows a semivariogram, which is similar to the semivariogram/covariance cloud except that the semivariance data have been averaged by distance (i.e., binned). The Models frame lets you choose a mathematical model to fit the empirical semivariogram. A common model is spherical. In the Model #1 frame, click the dropdown arrow for Type and select Spherical. Change the lag size to 40,000 and

the number of lags to 12. Change Anisotropy to be true. Click Next.

5. Step 4 lets you choose the number of neighbors (control points), and the sampling method. Take defaults, and click Next.

6. The Step 5 panel shows the cross-validation results. The Chart frame offers four types of scatter plots (Predicted versus Measured values, Error versus Measured values, Standardized Error versus Measured values, and Quantile-Quantile plot for Standardized Error against Normal values). The Prediction Errors frame lists cross-validation statistics, including the RMS statistic. Record the RMS and standardized RMS statistics.

Q6. What is the RMS value from Step 5?

7. Use the Back button to go back to the Step 3 panel. Notice that a button to optimize entire model is available at the top of the Model frame. To try the model optimization option, select none for Type under Model #1. Then click the optimize button. Click yes to proceed. Now the Model frame shows the parameters for the optimal model. Check the RMS statistic for the optimal model.

Q7. Does the optimal model have a lower RMS statistic than your answer to Q6?

8. Use the optimal model, and click Finish in the Step 4 panel. Click OK in the Method Summary dialog. *Ordinary Kriging Prediction Map* is added to the map. To derive a prediction standard error map, you will click the Ordinary Kringing/Prediction Standard Error Map in Step 2 and repeat Steps 3 to 5.

9. You can follow the same steps as in Task 1 to convert *Ordinary Kriging Prediction Map* and *Ordinary Kriging Prediction Standard Error Map* to rasters, to clip the rasters by using *idoutlgd* as the analysis mask, and to create isolines from the clipped rasters.

Task 5 Use Universal Kriging for Interpolation

What you need: *stations.shp* and *idoutlgd*.

In Task 5 you will run universal kriging on *stations.shp*. The trend to be removed from the kriging process is the first-order trend surface.

1. Click the Geostatistical Analyst dropdown arrow and select Geostatistical Wizard. Select *stations* for the source dataset and ANN_PREC for the data field. Click Kriging/Cokriging in the Methods frame. Click Next.

2. In the Step 2 panel, click Universal for the kriging type and Prediction for the output type. Select First from the Order of trend removal list. Click Next.

3. The Step 3 panel shows the first-order trend that will be removed from the kriging process. Click Next.

4. In the Step 4 panel, click the button to optimize model. Click OK to proceed. Click Next.

5. Take the default values for the number of neighbors and the sampling method. Click Next.

6. The Step 6 panel shows the cross-validation results. The RMS value is slightly higher than ordinary kriging in Task 4, and the standardized RMS value is farther away from 1 than ordinary kriging. This means that the estimated standard error from universal kriging is not as reliable as that from ordinary kriging.

Q8. What is the standardized RMS value from Step 6?

7. Click Finish in the Step 6 panel. Click OK in the Method Summary dialog. *Universal Kriging Prediction Map* is an interpolated map from universal kriging. To derive a prediction standard error map, you will click Universal Kriging/Prediction Standard Error Map in the Step 2 panel and repeat Steps 3 to 6.

8. You can follow the same steps as in Task 1 to convert *Universal Kriging Prediction Map* and *Universal Kriging Prediction Standard Error Map* to rasters, to clip the rasters by

using *idoutlgd* as the analysis mask, and to create isolines from the clipped rasters.

Challenge Task

What you need: *stations.shp* and *idoutlgd*.

This challenge task asks you to compare the interpolation results from two spline methods in Geostatistical Analyst. Except for the interpolation method, you will use the default values for the challenge task. The task has three parts: one, create an interpolated raster using completely regularized spline; two, create an interpolated raster using spline with tension; and three, use a local operation to compare the two rasters. The result can show the difference between the two interpolation methods.

1. Create a Radial Basis Functions Prediction map by using the kernel function of completely regularized spline. Convert the map to a raster, and save the raster as *regularized* with a cell size of 2000.

2. Create a Radial Basis Functions Prediction map by using the kernel function of spline with tension. Convert the map to a raster, and save the raster as *tension* with a cell size of 2000.

3. Select Raster Analysis from the Environment Settings menu of ArcToolbox. Select *idoutlgd* for the analysis mask.

4. Use Raster Calculator in the Spatial Analyst Tools/Map Algebra toolset to subtract *tension* from *regularized*.

5. The result shows the difference in cell values between the two rasters within *idoutlgd*. Display the difference raster in three classes: lowest value to -0.5, -0.5 to 0.5, and 0.5 to highest value.

Q1. What is the range of cell values in the difference raster?

Q2. What does a positive cell value in the difference raster mean?

Q3. Is there a pattern in terms of where high cell values, either positive or negative, are distributed in the difference raster?

REFERENCES

Ackerman, W. V., and A. T. Murray. 2004. Assessing Spatial Patterns of Crime in Lima, Ohio. *Cities* 21: 423–37.

Bailey, T. C., and A. C. Gatrell. 1995. *Interactive Spatial Data Analysis.* Harlow, England: Longman Scientific & Technical.

Beguería, S., and S. M. Vicente-Serrano. 2006. Mapping the Hazard of Extreme Rainfall by Peaks over Threshold Extreme Value Analysis and Spatial Regression Techniques. *Journal of Applied Meteorology and Climatology* 45: 108–24.

Burrough, P. A., and R. A. McDonnell. 1998. *Principles of Geographical Information Systems.* Oxford, England: Oxford University Press.

Carroll, S. S., and N. Cressie. 1996. A Comparison of Geostatistical Methodologies Used to Estimate Snow Water Equivalent. *Water Resources Bulletin* 32: 267–78.

Chung, K., D. Yang, and R. Bell. 2004. Health and GIS: Toward Spatial Statistical Analyses. *Journal of Medical Systems* 28: 349–60.

Davis, J. C. 1986. *Statistics and Data Analysis in Geology,* 2d ed. New York: Wiley.

Declercq, F. A. N. 1996. Interpolation Methods for Scattered Sample Data: Accuracy, Spatial Patterns, Processing Time. *Cartography and Geographic Information Science* 23: 128–44.

Eriksson, M., and P. P. Siska. 2000. Understanding Anisotropy Computations. *Mathematical Geology* 32: 683–700.

Franke, R. 1982. Smooth Interpolation of Scattered Data by Local Thin Plate Splines. *Computers and Mathematics with Applications* 8: 273–81.

Franke, R. 1985. Thin Plate Splines with Tension. *Computer-Aided Geometrical Design* 2: 87–95.

Griffith, D. A., and C. G. Amrhein. 1991. *Statistical Analysis for Geographers.* Englewood Cliffs, NJ: Prentice Hall.

Gringarten E., and C. V. Deutsch. 2001. Teacher's Aide: Variogram Interpretation and Modeling. *Mathematical Geology* 33: 507–34.

Hutchinson, M. F. 1995. Interpolating Mean Rainfall Using Thin Plate Smoothing Splines. *International Journal of Geographical Information Systems* 9: 385–403.

Jarvis, C. H., N. Stuart, and W. Cooper. 2003. Infometric and Statistical Diagnostics to Provide Artificially-Intelligent Support for Spatial Analysis: the Example of Interpolation. *International Journal of Geographical Information Science* 17: 495–516.

Laslett, G. M. 1994. Kriging and Splines: An Empirical Comparison of Their Predictive Performance in Some Applications. *Journal of the American Statistical Association* 89: 391–409.

Lloyd, C. D. 2005. Assessing the Effect of Integrating Elevation Data into the Estimation of Monthly Precipitation in Great Britain. *Journal of Hydrology* 308: 128–50.

Lloyd, C. D., and P. M. Atkinson. 2001. Assessing Uncertainty in Estimates with Ordinary and Indicator Kriging. *Computers & Geosciences* 27: 929–37.

Lundquist, J. E., and J. S. Beatty. 1999. A Conceptual Model for Defining and Assessing Condition of Forest Stands. *Environmental Management* 23: 519–25.

Martinez-Cob, A. 1996. Multivariate Geostatistical Analysis of Evapotranspiration and Precipitation in Mountainous Terrain. *Journal of Hydrology* 174: 19–35.

Mitas, L., and H. Mitasova. 1988. General Variational Approach to the Interpolation Problem. *Computers and Mathematics with Applications* 16: 983–92.

Mitasova, H., and L. Mitas. 1993. Interpolation by Regularized Spline with Tension: I. Theory and Implementation. *Mathematical Geology* 25: 641–55.

Nelson, T., B. Boots, M. Wulder, and R. Feick. 2004. Predicting Forest Age Classes from High Spatial Resolution Remotely Sensed Imagery Using Voronoi Polygon Aggregation. *GeoInformatica* 8: 143–55.

Phillips, D. L., J. Dolph, and D. Marks. 1992. A Comparison of Geostatistical Procedures for Spatial Analysis of Precipitation in Mountainous Terrain. *Agricultural and Forest Meteorology* 58: 119–41.

Reader, S. 2001. Detecting and Analyzing Clusters of

Low–Birth Weight Incidence Using Exploratory Spatial Data Analysis. *GeoJournal* 53: 149–59.

Robinson, A. H., J. L. Morrison, P. C. Muehrcke, A. J. Kimerling, and S. C. Guptill. 1995. *Elements of Cartography,* 6th ed. New York: Wiley.

Rogerson, P. A. 2001. *Statistical Methods for Geography.* London: Sage Publications.

Schuurman, N., R. S. Fiedler, S. C. W. Grzybowski, and D. Grund. 2006. Defining Rational Hospital Catchments for Non-Urban Areas Based on Travel-Time. *International Journal of Health Geographics* 5: 43 doi: 10. 1186/1476-072X-5-43.

Scott, D. W. 1992. *Multivariate Density Estimation: Theory, Practice, and Visualization.* New York: Wiley.

Silverman, B. W. 1986. *Density Estimation.* London: Chapman and Hall.

Tabios, G. Q., III, and J. D. Salas. 1985. A Comparative Analysis of Techniques for Spatial Interpolation of Precipitation. *Water Resources Bulletin* 21: 365–80.

Van der Veen, A., and C. Logtmeijer. 2005. Economic Hotspots: Visualizing Vulnerability to Flooding. *Natural Hazards* 36: 65–80.

Watson, D. F. 1992. *Contouring: A Guide to the Analysis and Display of Spatial Data.* Oxford, England: Pergamon Press.

Webster, R., and M. A. Oliver. 2001. *Geostatistics for Environmental Scientists.* Chichester, England: Wiley.

Wickham, J. D., R. V. O'Neill, and K. B. Jones. 2000. A Geography of Ecosystem Vulnerability.

Landscape Ecology 15: 495–504.

Yang, X., and T. Hodler. 2000. Visual and Statistical Comparisons of Surface Modeling Techniques for Point-Based Environmental Data. *Cartography and Geographic Information Science* 27: 165–75.

Zhang, K., S. C. Chen, D. Whitman, M. L. Shyu, J. Yan, and C. Zhang. 2003. A Progressive Morphological Filter for Removing Nonground Measurement from Airborne LIDAR Data. *IEEE Transaction on Geoscience and Remote Sensing* 41: 872–82.

Zimmerman, D., C. Pavlik, A. Ruggles, and M. P. Armstrong. 1999. An Experimental Comparison of Ordinary and Universal Kriging and Inverse Distance Weighting. *Mathematical Geology* 31: 375–90.

CHAPTER 16

GEOCODING AND DYNAMIC SEGMENTATION

Dr. Snow's map was introduced in Chapter 10 as an example of data exploration, but it can also be used as an introduction to geocoding. To locate the culprit for the outbreak of cholera, Snow first mapped (geocoded) the locations of the homes of those who had died from cholera. The geocoding was done manually then, but now it is performed regularly on the Internet. For example, how can we find a nearby bank in an unfamiliar city? With a cell phone, we can go online, use a browser such as Google Maps, zoom in to the closest street intersection, and search for nearby banks. After a bank is selected, we can even have a view of the bank and

its surrounding. Although not apparent to us, this routine activity involves geocoding, the process of plotting street addresses or intersections as point features on a map. Geocoding has become perhaps the most commercialized GIS-related operation.

Like geocoding, dynamic segmentation can also locate spatial features from a data source that lacks x-, y-coordinates. Dynamic segmentation works with data such as locations of traffic accidents, which are typically reported in linear distances from some known points (e.g., mileposts). Dynamic segmentation is popular with government agencies that produce and distribute geospatial data. Many transportation departments use dynamic segmentation to manage data such as speed limits, rest areas, bridges, and pavement conditions along highway routes. Natural resource agencies also use dynamic segmentation to store and analyze stream reach data and aquatic habitat data.

Chapter 16 covers geocoding and dynamic segmentation in five sections. Section 16.1 explains the basics of geocoding including the reference database and the process and options of address matching.

Section 16.2 discusses other types of geocoding besides address matching. Section 16.3 reviews applications of geocoding. Section 16.4 covers the basic elements of routes and events in dynamic segmentation. Section 16.5 describes the applications of dynamic segmentation in data management, data query, data display, and data analysis.

16.1 GEOCODING

Geocoding refers to the process of assigning spatial locations to data that are in tabular format but have fields that describe their locations. According to Cooke (1998), geocoding started during the 1960s—the same time as GIS—when the U.S. Census Bureau was looking for ways of mapping survey data gathered across the country, address by address.

The most common type of geocoding is **address geocoding,** or *address matching,* which plots street addresses as point features on a map. Address geocoding requires two sets of data. The first data set contains individual street addresses in a table, one record per address (Figure 16.1). The second is a reference database that consists of a street map and attributes for each street segment such as the street name, address ranges, and ZIP codes. Address geocoding interpolates the location of a street address by comparing it with data in the reference database.

16.1.1 Geocoding Reference Database

A reference database must have a road network with appropriate attributes for geocoding. Many GIS users in the United States derive a geocoding reference database from the TIGER/Line files, which are extracts of geographic/cartographic information from the U.S. Census Bureau's MAF/TIGER (Master Address File/Topologically Integrated Geographic Encoding and Referencing) database (**http://www.census.gov/geo/www/tiger/index.html**). The TIGER/Line files contain legal and statistical area boundaries such as counties, census tracts, and block groups, as well as streets, roads, streams, and water bodies. More importantly, the TIGER/Line attributes also include for each street segment the street name, the beginning and end address numbers on each side, and the ZIP code on each side (Figure 16.2).

Geocoding reference databases are also available from commercial companies such as Tele Atlas (**http://www.teleatlas.com/**), NAVTEQ (**http://www.navteq.com/**), and Pitney Bowes (**http://www.pbinsight.com/**). These companies advertise their

```
Name,    Address,    ZIP
Iron Horse, 407 E Sherman Ave, 83814
Franlin's Hoagies, 501 N 4th St, 83814
McDonald's, 208 W Appleway, 83814
Rockin Robin Cafe, 3650 N Government way, 83815
Olive Garden, 525 W Canfield Ave, 83815
Fernan Range Station, 2502 E Sherman Ave, 83814
FBI, 250 Northwest Blvd, 83814
ID Fish & Game, 2750 W Kathleen Ave, 83814
ID Health & Welfare, 1120 W Ironwood Dr, 83814
ID Transportation Dept, 600 W Prairie Ave, 83815
```

Figure 16.1

A sample address table records name, address, and ZIP code.

```
FEDIRP: A direction that precedes a street name.
FENAME: The name of a street.
FETYPE: The street name type such as St, Rd, and Ln.
FRADDL: The beginning address number on the left side of a street segment.
TOADDL: The ending address number on the left side of a street segment.
FRADDR: The beginning address number on the right side of a street segment.
TOADDR: The ending address number on the right side of a street segment.
ZIPL: The zip code for the left side of a street segment.
ZIPR: The zip code for right side of a street segment.
```

Figure 16.2

The TIGER/Line files include the attributes of FEDIRP, FENAME, FETYPE, FRADDL, TOADDL, FRADDR, TOADDR, ZIPL, and ZIPR, which are important for geocoding.

street and address data products to be current, verified, and updated.

16.1.2 The Address Matching Process

The geocoding process uses a geocoding engine, which can be embedded in a GIS package. (In ArcGIS, the geocoding engine is called the Address Locator.) In general, the geocoding process consists of three phases: preprocessing, matching, and plotting.

The preprocessing phase involves parsing and address standardization (Yang et al. 2004). Parsing breaks down an address into a number of components. Using an example from the United States, the address "630 S. Main Street, Moscow, Idaho 83843-3040" has the following components:

- street number (630)
- prefix direction (S or South)
- street name (Main)
- street type (Street)
- city (Moscow)
- state (Idaho)
- ZIP+4 code (83843-3040)

The result of the parsing process is a record in which there is a value for each of the address components to be matched. Variations from the example can occur. Some addresses have an apartment number in addition to a street number, whereas others have suffixes such as NE following the street name and type.

Not all street addresses are as complete or as well structured as the preceding example. Address standardization identifies and places each address component in order. It also standardizes variations of an address component to a consistent form. For example, "Avenue" is standardized as "Ave," "North" as "N," and "Third" as "3rd." If a geocoding engine uses the Soundex system (a system that codes together names of the same or similar sounds but of different spellings) to check the spelling, then names such as Smith and Smythe are treated the same.

Next, the geocoding engine matches the address against a reference database. A variety of mismatches can occur. According to Harries (1999),

common errors in address recording (for crime mapping) include misspelling of the street name, incorrect address number, incorrect direction prefix or suffix, incorrect street type, and abbreviation not recognized by the geocoding engine. Other errors are incorrect or missing ZIP codes, post office box addresses, and rural route addresses (Hurley et al. 2003; Yang et al. 2004). A reference database can have its own set of problems as well. A reference database can be outdated because it does not have information on new streets, street name changes, street closings, and ZIP code changes. In some cases, a reference database may even have missing address ranges, gaps in address ranges, or incorrect ZIP codes.

If an address is judged to be matched, the final step is to plot it as a point feature. Suppose the reference database is derived from the TIGER/Line files. The geocoding engine first locates the street segment in the reference database that contains the address in the input table. Then it interpolates where the address falls within the address range. For example, if the address is 620 and the address range is from 600 to 700 in the database, the address will be located about one-fifth of the street segment from 600 (Figure 16.3). This process is linear interpolation. Figure 16.4 shows a geocoded map in which street addresses are converted into point features.

An alternative to linear interpolation is the use of a "location of address" database, which has been developed in some countries. In such a database, the location of an address is denoted by a pair of x- and y-coordinates, corresponding to the centroid of a building base or footprint. For

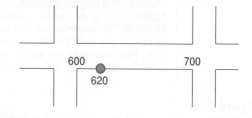

Figure 16.3
Linear interpolation for address geocoding.

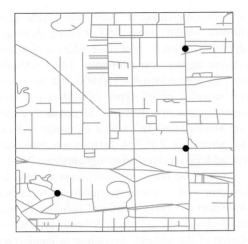

Figure 16.4
Address geocoding plots street addresses as points on a map.

example, GeoDirectory is an address database that has the coordinates of the center point of every building in Ireland **(http://www.geodirectory.ie/).** In the United States, Sanborn offers CitySets, a database that includes building footprints for the core downtown areas of major cities **(http://www .sanborn.com/).** Address geocoding using a

location of address database simply converts *x*- and *y*-coordinates into points.

16.1.3 Address Matching Options

A geocoding engine must have options to deal with possible errors in the address table, the reference database, or both. Typically, a geocoding engine has provisions for relaxing the matching conditions. ArcGIS, for example, offers the minimum candidate score and the minimum match score. The former determines the likely candidates in the reference database for matching, and the latter determines whether an address is matched or not. Adopting a lower match score can result in more matches but can also introduce more possible geocoding errors. ArcGIS, however, does not explain exactly how the scores are tabulated (Box 16.1).

Given the various matching options, we can expect to run the geocoding process more than once. The first time through, the geocoding engine reports the number of matched addresses as well as the number of unmatched addresses. For each unmatched address, the geocoding engine lists available candidates based on the minimum candidate score. We can either accept a candidate for

Box 16.1 **Scoring System for Geocoding**

A GIS package such as ArcGIS does not explain how the scoring system works in geocoding services. Take the example of the spelling sensitivity setting, which has a possible value between 0 and 100. ArcGIS sets the default at 80. But the help document suggests that we use a higher spelling sensitivity if we are sure that the input addresses are spelled correctly and a lower setting if we think that the input addresses may contain spelling errors. Although the suggestion makes sense, it does not say exactly how to set spelling sensitivity with a numeric value.

There are probably too many scoring rules used by a commercial geocoding engine to be listed in a help document. But we may occasionally discover how the

scoring system works in a specific case. For example, if the street name "Appleway Ave" is spelled "Appleway" for a street address, the scoring system in ArcGIS deducts 15 from the match score. According to Yang et al. (2004), ArcGIS uses the preprogrammed weights for each address element, weighting the house number the most and the street suffix the least. The match score we see is the sum of separate scores. In their article on the development of a geocoding certainty indicator, Davis and Fonseca (2007) describe the procedure they follow in assessing the degree of certainty for the three phases of parsing, matching, and locating in geocoding. The article gives a glimpse of how the scoring system may be like in a commercial geocoding system.

a match or modify the unmatched address before running the gecoding process again.

16.1.4 Offset Plotting Options

The side offset and the end offset are options that allow a geocoded address to be plotted away from its interpolated location along a street segment (Figure 16.5). The side offset places a geocoded point at a specified distance from the side of a street segment. This option is useful for point-in-polygon overlay analysis (Chapter 11), such as linking addresses to census tracts or land parcels (Ratcliffe 2001). The end offset places a point feature at a distance from the end point of a street segment, thus preventing the geocoded point from falling on top of a cross street. The end offset uses a distance that is given as a specified percentage of the length of a street segment.

16.1.5 Quality of Geocoding

The quality of geocoding is often expressed by the match rate or "hit" rate. What is an acceptable hit rate? For crime mapping and analysis, one researcher has derived statistically a minimum acceptable hit rate of 85 percent (Ratcliffe 2004). Using a GIS, a recent study reports the hits rate of 87 percent of geocoding the addresses of over 700 sex offenders (Zandberger and Hart 2009). Various tools are available for improving the accuracy of address matching. They include standardizing addresses to the USPS format (e.g., changing "4345 North 73 Street" to "4345 N 73rd St.") and using Internet mapping engines such as Google

Earth Pro (Bigham et al. 2009). The USPS offers software vendors the CASS (Coding Accuracy Support System) to test their address-matching software against test addresses. The CASS file contains approximately 150,000 test addresses with samples of all types of addressing used around the country. To be CASS certified, a software vendor must pass with a minimum score of 98.5 percent for ZIP + 4 (**http://www.usps.com/ncsc/addressservices/certprograms/cass.htm**).

Besides hit rates, positional accuracy has also been proposed to assess the quality of geocoding (Whitsel et al. 2004). Positional accuracy is measured by how close each geocoded point is to the true location of the address. According to Zandbergen and Hart (2009), typical positional errors for residential addresses range from 25 to 168 meters. Positional errors are often caused by errors in the street number ranges in the street network data.

Because of geocoding errors, Davis and Fonseca (2007) have proposed a geocoding certainty indicator based on the degree of certainty during the three stages of the geocoding process (parsing, matching, and plotting). According to their study, the indicator can be used either as a threshold beyond which the geocoded result should be left out of any statistical analysis or as a weight that can be incorporated into spatial analysis.

16.2 VARIATIONS OF GEOCODING

Intersection matching, also called *corner matching,* matches address data with street intersections on a map (Figure 16.6). An address entry for intersection matching must list two streets such as "E Sherman Ave & N 4th St." A geocoding engine finds the location of the point where the two streets intersect. Intersection matching is a common geocoding method for police collision report data (Levine and Kim 1998; Bigham et al. 2009). Like address matching, intersection matching can run into problems. A street intersection may not exist, and the reference database may not cover new or renamed streets. A winding street crossing another street more than once can also present a unique problem. Additional data such as address numbers

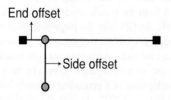

Figure 16.5
The end offset moves a geocoded point away from the end point of a street segment, and the side offset places a geocoded point away from the side of a street segment.

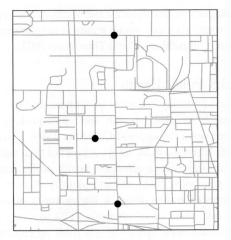

Figure 16.6
An example of intersection matching.

or adjacent ZIP codes are required to determine which street intersection to plot.

ZIP code geocoding matches a ZIP code to the code's centroid location. It differs from address matching or intersection matching in two ways. First, it is not street-level geocoding. Second, it uses a reference database that contains the x-, y-coordinates, either geographic or projected, of ZIP code centroids, rather than a street network.

Parcel-level geocoding matches a parcel number to the parcel's centroid location and, if a parcel database is available, plots the parcel boundary.

Reverse geocoding is the reverse of address geocoding; it converts point locations into descriptive addresses.

Place name alias geocoding matches a place name such as a well-known restaurant or a museum with a street address, locates the street address, and plots it as a point feature. It requires a place name alias table, complete with place names and their street addresses.

Photo geocoding attaches location information to photographs. Photographs taken with a digital camera with built-in GPS (global positioning system) can have associated longitude and latitude readings. Photo geocoding uses these geographic coordinates to plot the point locations. (Examples of "geotagged" photographs can be found on Flickr, **http://www .flickr.com/map**).

16.3 APPLICATIONS OF GEOCODING

Geocoding is perhaps the most commercialized GIS-related operation; it plays an important role in location-based services and other business applications. Geocoding is also a tool for wireless emergency service, crime mapping and analysis, and public health monitoring.

16.3.1 Location-Based Services

A location-based service refers to any service or application that extends spatial information processing to end users via the Internet and/or wireless network. Early examples of location-based services relied on the computer access to the Internet. To find a street address and the directions to the address, one would visit MapQuest and get the results **(http://www.mapquest.com).** MapQuest is now only one of many websites that provide this kind of service. Others include Google, Yahoo!, and Microsoft. Many U.S. government agencies such as the Census Bureau also combine address matching with online interactive mapping at their websites.

The current popularity of location-based services ties to the use of GPS and mobile devices of all kinds. Mobile devices allow users to access the Internet and location-based services virtually anywhere. A mobile phone user can now be located and can in turn receive location information such as nearby ATMs or restaurants. Other types of services include tracking people, tracking commercial vehicles (e.g., trucks, pizza delivery vehicles), routing workers, meeting customer appointments, and measuring and auditing mobile workforce productivity.

16.3.2 Business Applications

For business applications, geocoding is most useful in matching the ZIP codes of customers and prospects to the census data. Census data such as income, percent population in different age groups, and education level can help businesses prepare promotional mailings that are specifically targeted at their intended recipients. For example, Tapestry

Segmentation, a database developed by Esri, connects ZIP codes to 65 types of neighborhoods in the United States based on their socioeconomic and demographic characteristics. The database is designed for mail houses, list brokers, credit card companies, and any business that regularly sends large promotional mailings.

Parcel-level geocoding links parcel IDs to parcel boundaries and allows property and insurance companies to use the information for a variety of applications such as determining the insurance rate based on the distance of a parcel to areas prone to floods, brush fires, or earthquakes.

Other business applications include site analysis and market area analysis. For example, a spatial pattern analysis of real estate prices can be based on point features geocoded from home purchase transactions (Basu and Thibodeau 1998). Telecommunication providers can also use geocoded data to determine appropriate infrastructure placements (e.g., cell phone towers) for expanding customer bases.

16.3.3 Wireless Emergency Services

A wireless emergency service uses a built-in GPS receiver to locate a mobile phone user in need of emergency dispatch services (i.e., fire, ambulance, or police). This application is enhanced by a 2001 mandate by the Federal Communications Commission (FCC) commonly called *automatic location identification,* which requires that all wireless carriers in the United States provide a certain degree of accuracy in locating mobile phone users who dial 911.

16.3.4 Crime Mapping and Analysis

Crime mapping and analysis typically starts with geocoding. Crime records almost always have street addresses or other locational attributes (Ratcliffe 2004; Harries 2006; Grubesic 2010). Geocoded crime location data are input into software packages such as CrimeStat **(http://www .icpsr.umich.edu/NACJD/crimestat.htm)** for "hot spot" analysis (Chapter 11) and space time analysis. Using historical crime data, software

packages can also forecast the numbers and types of crimes that can occur within a neighborhood (Gorr, Olligschlaeger, and Thompson 2003).

16.3.5 Public Health

Geocoding has become an important tool for research in public health and epidemiology in recent years (Moore and Carpenter 1999; Krieger 2003; Uhlmann et al. 2009). As part of their public health surveillance activities, health professionals use geocoding to locate and identify changes in patterns of human diseases. For example, one can geocode cases of tuberculosis (TB) to study the spatial-temporal spread of TB from an epidemic focus into the surrounding regions. Geocoding can also be used to derive neighborhood socioeconomic data for cross-sectional analysis of public health surveillance data (Krieger et al. 2003) and to provide input data for measuring geographic access to health services based on the travel times and distances between the subjects and the medical providers (Fortney, Rost, and Warren 2000).

16.4 DYNAMIC SEGMENTATION

Vector data used in GIS are two-dimensional data measured in x-, y-coordinates (Chapter 3). There are, however, data that are linearly measured. For example, highway accidents are typically reported in linear distances from known points such as mileposts (e.g., on Highway 95 two miles south of postmile marker 13). Unless the location of linearly measured data is provided, the data cannot be used in a GIS environment. Another example is pavement conditions along highways. Intuitively, highways can be divided into segments so that each segment corresponds to a pavement condition. But this will require a large amount of data editing in segmentation. And the segmented highways can only be used with pavement conditions. This section discusses dynamic segmentation, a method designed for the use of linearly referenced data with GIS layers measured in x-, y-coordinates.

To understand dynamic segmentation, route and event must be introduced first. A **route** is a

Box 16.2 | Linear Location Referencing System

Transportation agencies normally use a linear location referencing system to locate events such as accidents and potholes, and facilities such as bridges and culverts, along roads and transit routes. To locate events, a linear referencing system uses distance measures from known points, like the beginning of a route, a milepost, or a road intersection. The address of an accident location, for example, contains a route name and a distance offset from a milepost. The address for a linear event, on the other hand, uses two distance offsets. A projected coordinate system uses x-, y-coordinates. The dynamic segmentation model uses routes, measures, and events to bring a projected coordinate system together with a linear referencing system, two fundamentally different measuring systems.

linear feature, such as a street, highway, or stream used in a GIS, which also has a linear measurement system stored with its geometry. **Events** are linearly referenced data, such as speed limits, traffic accidents, or fishery habitat conditions, which occur along routes. An example of a linearly referencing system is the milepost referencing, which uses distances from a point of origin (e.g., mile 0) to locate mile points along a highway (Box 16.2). **Dynamic segmentation** can be defined formally as the process of computing the location of events along a route (Nyerges 1990). Thus, through dynamic segmentation, traffic accident locations can be computed at run time (dynamically) from the linearly referenced data and plotted along a highway route.

16.4.1 Routes

To be used with linearly referenced event data, a route must have a built-in measurement system. Figure 16.7 shows a route, along which each point has a pair of x- and y-coordinates and an m value. The x and y values locate the linear feature in a two-dimensional coordinate system, and the m value is a linear measure, which can be based on the geometric length of line segments or interpolated from postmile or other reference markers. This kind of route has been described as a "route dynamic location object" (Sutton and Wyman 2000). In ArcGIS, it is called "route feature class," which has gained acceptance by government agencies (Box 16.3).

16.4.2 Creating Routes

A route links a series of line segments together. In a GIS, routes can be created interactively or through data conversion (Box 16.4).

Using the interactive method, we must first digitize a route or select existing lines from a layer that make up a route (Figure 16.8). Then we can apply a measuring command to the route to compute the route measures. If necessary, route measures can further be calibrated based on points with known distances.

The data conversion method can create routes at once from all linear features or from features selected by a data query. For example, we can create a route system for each numbered highway in a

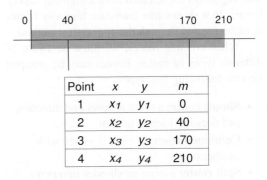

Point	x	y	m
1	x_1	y_1	0
2	x_2	y_2	40
3	x_3	y_3	170
4	x_4	y_4	210

Figure 16.7

An example of a geodatabase route feature class.

Box 16.3 | **Route Feature Classes**

T he National Hydrography Dataset (NHD) is a program that has migrated from the coverage (NHD-inARC) to the geodatabase (NHDinGEO) **(http://nhd.usgs.gov/geodatabase_review.html)**. NHDinARC data use a route subclass to store the transport and coastline reaches (route.rch). In contrast, NHDin-GEO data has a new NHDFlowline feature class, which stores *m* values with its feature geometry. These *m* values apply to other feature classes that use the reach reference.

Route feature classes have also been adopted by transportation agencies for data delivery. An example is the Washington State Department of Transportation (WSDOT) **(http://www.wsdot.wa.gov/mapsdata/geodatacatalog/default.htm)**. WSDOT has built state highway routes with linear measures. These highway routes provide the basis for locating features such as rest areas, speed limits, and landscape type along the routes. Tasks 2 and 3 in the applications section use data sets downloaded from the WSDOT website.

Box 16.4 | **Create Routes Using ArcGIS**

A rcGIS offers both the conversion method and the interactive method for creating route feature classes from existing lines. The Create Routes tool in ArcToolbox can convert all linear features or selected features from a coverage, a shapefile, or a geodatabase feature class into routes. The output routes can be saved in a shapefile or a geodatabase feature

class. ArcToolbox also has other tools for calibrating routes, dissolving route events, locating features along routes, making route event layers, overlaying route events, and transforming route events. ArcMap has the Route Editing toolbar, which offers tools for creating and calibrating routes interactively.

state or for each interstate highway (after the interstate highways are selected from a highway layer). Figure 16.9 shows five interstate highway routes in Idaho created through the conversion method.

When creating routes, we must be aware of different types of routes. Routes may be grouped into the following four types:

- **Simple route:** a route follows one direction and does not loop or branch.
- **Combined route:** a route is joined with another route.
- **Split route:** a route subdivides into two routes.
- **Looping route:** a route intersects itself.

Simple routes are straightforward but the other three types require special handling. An example of a combined route is an interstate highway, which has different traffic conditions depending on the traffic direction. In this case, two routes can be built for the same interstate highway, one for each direction. An example of a split route is the split of a business route from an interstate highway (Figure 16.10). At the point of the split, two separate linear measures begin: one continues on the interstate highway, and the other stops at the end of the business route.

A looping route is a route with different parts. Unless the route is dissected into parts, the route measures will be off. A bus route in Figure 16.11

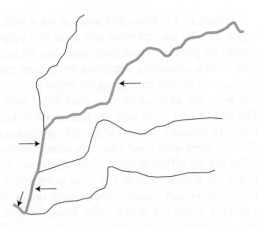

Figure 16.8
The interactive method requires the selection or
digitizing of the line segments that make up a route
(shown in a thicker line symbol).

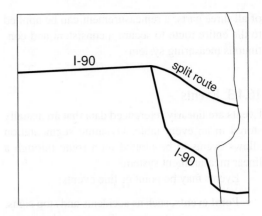

Figure 16.10
An example of a split route.

Figure 16.11
A looping route divided into three parts for the purpose
of route measuring.

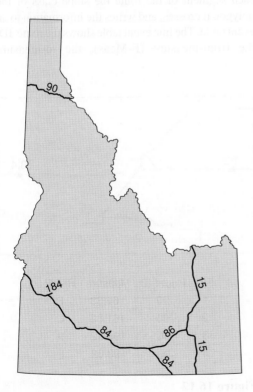

Figure 16.9
Interstate highway routes in Idaho.

serves as an example. The bus route has two
loops. We can dissect the bus route into three parts:
(1) from the origin on the west side of the city to
the first crossing of the route, (2) between the first
and the second crossing, and (3) from the second
crossing back to the origin. Each part is created and
measured separately. Following the completion

of all three parts, a remeasurement can be applied to the entire route to secure a consistent and continuous measuring system.

16.4.3 Events

Events are linearly referenced data that are usually stored in an event table. Dynamic segmentation allows events to be plotted on a route through a linear measurement system.

Events may be point or line events:

- **Point events,** such as accidents and stop signs, occur at point locations. To relate point events to a route, a point event table must have the route ID, the location measures of the events, and attributes describing the events.
- **Line events,** such as pavement conditions, cover portions of a route. To relate line events to a route, a line event table must have the route ID and the from- and to-measures.

A route can be associated with different events, both point and line. Therefore, pavement conditions, traffic volumes, and traffic accidents can all be linked to the same highway route.

16.4.4 Creating Event Tables

There are two common methods for creating event tables. The first method creates an event table from an existing table that already has data on route IDs and linear measures through milepost, river mile, or other data. For example, the Washington State Department of Transportation's website has event tables of rest areas, speed limits, landscape types, and other data (**http://www.wsdot.wa.gov/ mapsdata/geodatacatalog/default.htm**). These tables are in dBASE format. Other formats such as comma delimited text format and Excel are also acceptable.

The second method creates an event table by locating point or polygon features along a route, similar to a vector-based overlay operation (Chapter 11). The input layers consist of a route layer and a point or polygon layer. Instead of creating an output layer, the procedure creates an event table.

Figure 16.12 shows rest areas along a highway route. To convert these rest areas into point events, we prepare a point layer containing the rest areas, use a measuring command to measure the locations of the rest areas along the highway route, and write the information to a point event table. The point event table lists the feature ID (FID), route-ID, measure value, and the side of the road (RDLR). Rest areas 3 and 4 have the same measure value but on different sides of the highway. To determine whether a point is a point event or not, the computer uses a user-defined search radius. If a point is within the search radius from the route, then the point is a point event.

Figure 16.13 shows a stream network and a slope layer with four slope classes. To create a line event table showing slope classes along the stream route, the overlay method calculates the intersection between the route and the slope layer, assigns each segment of the route the slope class of the polygon it crosses, and writes the information to an event table. The line event table shows the route-ID, the from-measure (F-Meas), the to-measure

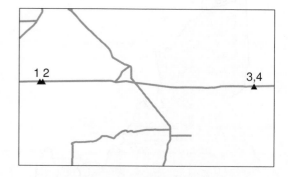

FID	Route-ID	Measure	RDLR
1	90	161.33	R
2	90	161.82	L
3	90	198.32	R
4	90	198.32	L

Figure 16.12

An example of converting point features to point events. See Section 16.4.4 for explanation.

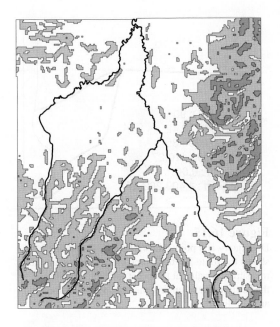

Route-ID	F-Meas	T-Meas	Slope-Code
1	0	7638	1
1	7638	7798	2
1	7798	7823	1
1	7823	7832	2
1	7832	8487	1
1	8487	8561	1
1	8561	8586	2
1	8586	8639	1
1	8639	8643	2

Figure 16.13
An example of creating a line event table by overlaying a route layer and a polygon layer. See Section 16.4.4 for explanation.

(T-Meas), and the slope-code for each stream segment. For example, the slope-code is 1 for the first 7638 meters of the stream route 1, changes to the slope-code of 2 for the next 160 meters (7798−7638), and so on.

Whether prepared from an existing table or generated from locating features along a route, an event table can be easily edited and updated. For example, after a point event table is created for rest areas along a highway route, we can add other attributes such as highway district and management unit to the table. And, if there are new rest areas, we can add them to the same event table so long as the measure values of these new rest areas are known.

16.5 APPLICATIONS OF DYNAMIC SEGMENTATION

Dynamic segmentation can convert linearly referenced data stored in a tabular report into events along routes. Once these data are associated with routes, they can be displayed, queried, and analyzed in a GIS environment.

16.5.1 Data Management

From the data management perspective, dynamic segmentation is useful in two ways. First, different routes can be built on the same linear features. A transit network can be developed from the same street database for modeling transit demand (Choi and Jang 2000). Similarly, multiple trip paths can be built on the same street network, each with a starting and end point, to provide disaggregate travel data for travel demand modeling (Shaw and Wang 2000).

Second, different events can reference the same route or, more precisely, the same linear measurement system stored with the route. Fishery biologists can use dynamic segmentation to store various environmental data with a stream network for habitat studies. And, as shown in Box 16.3, dynamic segmentation is an efficient method for delivering hydrologic data sets and for locating features along highway routes.

16.5.2 Data Display

Once an event table is linked to a route, the table is georeferenced and can be used as a feature layer. This conversion is similar to geocoding an address table or displaying a table with x-, y-coordinates.

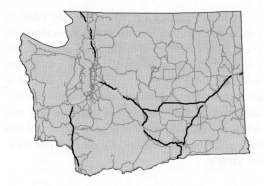

Figure 16.14
The thicker, solid line symbol represents those portions of Washington State's highway network that have the legal speed limit of 70 miles per hour.

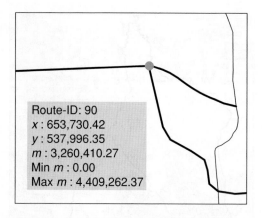

Figure 16.15
Data query at a point, shown here by the small circle, shows the route-ID, the *x*- and *y*-coordinates, and the measure (*m*) value at the point location. Additionally, the beginning and ending measure values of the route are also listed.

For data display, an event layer is the same as a stream layer or a city layer. We can use point symbols to display point events and line symbols to display line events (Figure 16.14).

16.5.3 Data Query

We can perform both attribute data query and spatial data query on an event table and its associated event layer. For example, we can query a point event table to select recent traffic accidents along a highway route. The attribute query result can be displayed in the event table as well as in the event layer. To determine if these recent accidents follow a pattern similar to those of past accidents, we can perform a spatial query and select past accidents within 10 miles of recent accidents for investigation.

Data query can also be performed on a route to find the measure value of a point. This is similar to clicking a feature to get the attribute data about the feature (Figure 16.15).

16.5.4 Data Analysis

Both routes and events can be used as inputs for data analysis. For example, to study the efficiency or equity of public transit routes, we can first buffer the routes with a distance of one-half mile and

overlay the buffer zone with census blocks or block groups. We can then use demographic data from the buffer zone (e.g., population density, income levels, number of vehicles owned per family, and commuting time to work) for the planning and operation of the public transit system.

Events, after they are converted to an event layer, can be analyzed in the same way as any point or linear feature layer. For example, road accidents can be located by dynamic segmentation and then analyzed in terms of their concentration patterns (Steenberghen, Aerts, and Thomas 2009).

Data analysis can also be performed between two event layers. Thus we can analyze the relationship between traffic accidents and speed limits along a highway route. The analysis is similar to a vector-based overlay operation: the output layer shows the location of each accident and its associated legal speed limit. A query on the output layer can determine whether accidents are more likely to happen on portions of the highway route with higher speed limits. However, data analysis can become complex when more than two event layers are involved and each layer has a large number of sections (Huang 2003).

KEY CONCEPTS AND TERMS

Address geocoding: A process of plotting street addresses in a table as point features on a map. Also called *address matching*.

Combined route: A route that is joined with another route.

Dynamic segmentation: The process of computing the location of events along a route.

Events: Attributes occurring along a route.

Geocoding: A process of assigning spatial locations to data that are in tabular format but have fields that describe their locations.

Intersection matching: A process of plotting street intersections as point features on a map. Also called *corner matching*.

Line events: Events that occur along a portion of a route, such as pavement conditions.

Looping route: A route that intersects itself.

Parcel-level geocoding: A process of matching a parcel number to the parcel's centroid location.

Photo geocoding: A process of attaching location information to photographs.

Place name alias geocoding: A process of plotting place names such as well-known restaurants as point features on a map.

Point events: Events that occur at point locations along a route, such as accidents and stop signs.

Reverse geocoding: A process of converting location data in latitude and longitude into descriptive addresses.

Route: A linear feature that has a linear measurement system stored with its geometry.

Simple route: A route that follows one direction and does not loop or branch.

Split route: A route that subdivides into two routes.

ZIP code geocoding: A process of matching ZIP codes to their centroid locations.

REVIEW QUESTIONS

1. Describe the two required inputs for address geocoding.

2. List attributes in the TIGER/Line files that are important for geocoding.

3. Go to the MapQuest website (**http://www .mapquest.com/).** Type the street address of your bank for search. Does it work correctly?

4. Explain linear interpolation as a process for plotting an address along a street segment.

5. Describe the three phases of the address geocoding process.

6. What address matching options are usually available in a GIS package?

7. What factors can cause low hit rates in address geocoding?

8. Explain the difference between the side offset and end offset options in geocoding.

9. What is ZIP code geocoding?

10. What is photo geocoding?

11. Geocoding is one of the most commercialized GIS-related activities. Can you think of a commercial application example of geocoding besides those that have already been mentioned in Chapter 16?

12. Explain in your own words how events are located along a route using dynamic segmentation.

13. Go to Esri's geodatabase Web page (**http:// www.support.esri.com/datamodels).**

Download the custom template of the transportation data model. Use the search function in Adobe Reader to look for the keyword "route." How many route feature classes are included in the data model?

14. Explain how a route's linear measure system is stored in the geodatabase.

15. Describe methods for creating routes from existing linear features.

16. How can you tell a point event table from a line event table?

17. Suppose you are asked to prepare an event table that shows portions of Interstate 5 in California crossing earthquake-prone zones (polygon features). How can you complete the task?

18. Check whether the transportation department in your state maintains a website for distributing GIS data. If it does, what kinds of data are available for dynamic segmentation applications?

APPLICATIONS: GEOCODING AND DYNAMIC SEGMENTATION

This applications section covers five tasks. Task 1 asks you to geocode ten street addresses by using a reference database derived from the 2000 TIGER/Line files. Task 2 lets you display and query highway routes and events downloaded from the Washington State Department of Transportation website. Again using data from the same website, you will analyze the spatial relationship between two event layers in Task 3. In Tasks 4 and 5, you will build routes from existing line features and locate features along routes. Task 4 locates slope classes along a stream route, and Task 5 locates cities along Interstate 5.

Task 1 Geocode Street Addresses

What you need: *Streets,* a street feature class of Kootenai County, Idaho, derived from the 2000 TIGER/Line files; and *cda_add,* a table containing street addresses of 5 restaurants and 5 government offices in Coeur d'Alene, the largest city in Kootenai County. Both *streets* and *cda_add* reside in *Kootenai.gdb.*

In Task 1, you will learn how to create point features from street addresses. Address geocoding requires an address table and a reference dataset. The address table contains a list of street addresses to be located. The reference dataset has the address information for locating street addresses. Task 1 includes the following

four steps: view the input data; create an address locator; run geocoding; and rerun geocoding for unmatched addresses.

1. Launch ArcMap. Click the Catalog window to open Catalog, and connect the Catalog tree to the Chapter 16 database. Add *streets* and *cda_add* to Layers and rename Layers Task 1. Right-click *streets* and open its attribute table. Because *streets* is derived from the TIGER/Line files, it has all the attributes from the original data. Figure 16.2 has the description of some of these attributes. Right-click *cda_add,* and open the table. The table contains the fields of Name, Address, and Zip. Close both tables.

2. Click ArcToolbox window to open ArcToolbox. Right-click ArcToolbox and set the current workspace of the environment settings to the Chapter 16 database. Double-click the Create Address Locator tool in the Geocoding Tools toolset. In the Create Address Locator dialog, select US Address – Dual Ranges for the address locator style, and select *streets* for the reference data. The Field Map window shows the field name and alias name. The field names with an asterisk are required, and you must correlate them to the right fields in *streets.* Start with

the field name of From Left, click on its corresponding alias and select FRADDL from the dropdown menu. For the field name of To Left, select TOADDL; for the field name of From Right, select FRADDR; and for the field name of To Right, select TOADDR. For the field name of Street Name, the alias name of FENAME is correct. Save the output address locator as *Task 1* in the Chapter 16 database, and click OK. The process of creating *Task 1* will take a while.

3. You will now work with the Geocoding toolbar. Click the Customize menu, point to Toolbars, and check Geocoding. Click Geocode Addresses on the toolbar. Choose Task 1 as the address locator. In the next dialog, select *cda_add* for the address table, save the output feature class as *cda_geocode* in *Kootenai.gdb*, and click OK. The completion message shows 9 (90%) are matched and 1 (10%) is unmatched.

4. To deal with the unmatched record, click on Rematch on the completion message window. This opens the Interactive Rematch dialog. Scroll down the records at the top, and you will find the unmatched record is that of 2750 W Kathleen Ave 83814. The reason for the unmatch is the wrong ZIP code. Change the ZIP from 83814 to 83815 and enter the new ZIP code. The Candidates window shows the geocoding candidates. Highlight the top candidate with a score of 100 and click Match. All ten records are now geocoded. Click Close to dismiss the dialog.

5. You can view *cda_geocode* on top of *streets*. All ten geocoded points are located within the city of Coeur d'Alene.

6. Before you leave Task 1, you can take a look at the address locator properties. Double-click *Task 1* in the Catalog tree. The properties include the input address fields, matching options, output options, and the output fields.

Q1. The default spelling sensitivity value is 80. If you were to change it to 60, how would the change affect the geocoding process?

Q2. What output options are available?

Task 2 Display and Query Routes and Events

What you need: *decrease24k.shp,* a shapefile showing Washington State highways; and *Speed-LimitsDecAll.dbf,* a dBASE file showing legal speed limits on the highways.

Originally in geographic coordinates, *decrease 24k.shp* has been projected onto the Washington State Plane, South Zone, NAD83, and Units feet. The linear measurement system is in miles. You will learn how to display and query routes and events in Task 2.

1. Insert a new data frame in ArcMap and rename it Task 2. Add *decrease24k.shp* and *SpeedLimitsDecAll.dbf* to Task 2. Open the attribute table of *decrease24k.shp*. The table shows state route identifiers (SR) and route measure attributes (Polyline M). Close the table.

2. This step adds the Identify Route Locations tool. The tool does not appear on any toolbar by default. You need to add it. Select Customize Mode from the Customize menu. On the Commands tab, select the category of Linear Referencing. The Commands frame shows five commands. Drag and drop the Identify Route Locations command to a toolbar in ArcMap. Close the Customize dialog.

3. Use the Select By Attribute tool to select "SR" = '026' in *decrease24k*. Highway 26 should now be highlighted in the map. Zoom in on Highway 26. Click the Identify Route Locations tool, and then click a point along Highway 26. This opens the Identify Route Location Results dialog and shows the measure value of the point you clicked as well as other information.

Q3. What is the total length of Highway 26 in miles?

Q4. In which direction is the route mileage accumulated?

4. Clear the selected features. Now you will work with the speed limits event table. Double-click the Make Route Event Layer tool in the Linear Referencing Tools toolset. In the next dialog, from top to bottom, select *decrease24k* for the input route features, SR for the route identifier field, *SpeedLimitsDecAll* for the input event table, SR for the route identifier field (ignore the error message), line event type, B_ARM (beginning accumulated route mileage) for the from-measure field, and E_ARM (end accumulated route mileage) for the to-measure field. Click OK to dismiss the dialog. A new layer, *SpeedLimitsDecAll Events,* is added to Task 2.

5. Right-click *SpeedLimitsDecAll Events* and select Properties. On the Symbology tab, choose Quantities/Graduate colors in the Show frame and LEGSPDDEC (legal speed description) for the value in the Fields frame. Choose a color ramp and a line width that can better distinguish the speed limits classes. Click OK to dismiss the dialog. Task 2 now shows the speed limits data on top of the state highway routes.

Q5. How many records of *SpeedLimitsDecAll Events* have speed limits > 60?

Task 3 Analyze Two Event Layers

What you need: *decrease24k.shp,* same as Task 2; *RoadsideAll.dbf,* a dBASE file showing classification of roadside landscape along highways; and *RestAreasAll.dbf,* a dBASE file showing rest areas.

In Task 3, you will use ArcToolbox to overlay the event tables of rest areas and the roadside landscape classes. The output event table can then be added to ArcMap as an event layer.

1. Insert a new data frame in ArcMap and rename it Task 3. Add *decrease24k.shp, RoadsideAll.dbf,* and *RestAreasAll.dbf*

to Task 3. Open *RoadsideAll.* The CLASSIFICA field stores five landscape classes: forested, open, rural, semi-urban, and urban. (Ignore the coincident features.) Open *RestAreasAll.* The table has a large number of attributes including the name of the rest area (FEATDESCR) and the county (COUNTY). Close the tables.

2. Double-click the Overlay Route Events tool in the Linear Reference Tools toolset. The Overlay Route Events dialog consists of three parts: the input event table, the overlay event table, and the output event table. Select *RestAreasAll* for the input event table, SR for the route identifier field (ignore the warning message), POINT for the event type, and ARM for the measure field. Select *RoadsideAll* for the overlay event table, SR for the route identifier field, LINE for the event type, BEGIN_ARM for the from-measure field, and END_ARM for the to-measure field. Select INTERSECT for the type of overlay. Enter *Rest_Roadside.dbf* for the output event table, SR for the route identifier field, and ARM for the measure field. Click OK to perform the overlay operation.

3. Open *Rest_Roadside.* The table combines attributes from *RestAreasAll* and *RoadsideAll.*

Q6. How many rest areas are located in forested areas?

Q7. Are any rest areas located in urban areas?

4. Similar to Task 2, you can use the Add Route Events tool to add *Rest_Roadside* as an event layer. The event layer can display the rest areas and their attributes such as the landscape classification.

Task 4 Create a Stream Route and Analyze Slope Along the Route

What you need: *plne,* an elevation raster; and *streams.shp,* a stream shapefile.

Task 4 lets you analyze slope classes along a stream. Task 4 consists of several subtasks: (1) use

ArcToolbox to create a slope polygon shapefile from *plne*; (2) import a select stream from *streams.shp* as a feature class to a new geodatabase; (3) use ArcToolbox to create a route from the stream feature class; and (4) run an overlay operation to locate slope classes along the stream route.

1. Insert a new data frame in ArcMap and rename the data frame Task 4. Add *plne* to Task 4. Use the Slope tool in the Spatial Analyst Tools/Surface toolset to create a percent slope raster from *plne* andname it *plne_slp*. Use the Reclassify tool in the Spatial Analyst Tools/Reclass toolset to reclassify the percent slope raster into the following five classes: <10%, 10–20%, 20–30%, 30–40%, and >40%. Name the reclassified raster *reclass_slp*. Then use the Raster to Polygon tool in the Conversion Tools/From Raster toolset to convert the reclassified slope raster to a polygon shapefile. Name the shapefile *slope*. The field GRIDCODE in *slope* shows the five slope classes.

2. Right-click the Chapter 16 database in the Catalog tree, point to New, and select Personal Geodatabase. Name the geodatabase *stream.mdb*.

3. Next import one of the streams in *streams.shp* as a feature class in *stream.mdb*. Right-click *stream.mdb*, point to Import, and select Feature Class (single). Select *streams.shp* for the input features, enter *stream165* for the output feature class name, and click on the SQL button for the expression. In the Query Builder dialog, enter the following expression: "USGH_ID" = 165 and click OK. Click OK in the Feature Class to Feature Class dialog to run the import operation. Expand *stream.mdb*, and *stream165* should appear as a feature class in the database.

4. Double-click the Create Routes tool in the Linear Referencing Tools toolset. Select *stream165* for the input line features, select USGH_ID for the route identifier field, enter *StreamRoute* for the output route feature class in *stream.mdb*, and click OK.

5. Now you will run an operation to locate slope classes along *StreamRoute*. Double-click the Locate Features Along Routes tool in the Linear Referencing Tools toolset. Select *slope* for the input features, select *StreamRoute* for the input route features, enter USGH_ID for the route identifier field, enter *Stream_Slope.dbf* for the output event table in the Chapter 16 database, uncheck the box for keeping zero length line events, and click OK to run the overlay operation.

6. Double-click the Make Route Event Layer tool in the Linear Referencing Tools toolset. In the next dialog, make sure that *StreamRoute* is the input route features, USGH_ID is the route identifier field, *Stream_Slope* is the input event table, RID is the route identifier field and Line is the event type. Then click OK to add the event layer.

7. Turn off all layers except *Stream_Slope Events* in Task 4 in the table of contents. Right-click *Stream_Slope Events* and select Properties. On the Symbology tab, select Categories and Unique Values for the Show option, select GRIDCODE for the value field, click on Add All Values, and click OK. Zoom in on *Stream_Slope Events* to view the changes of slope classes along the route.

Q8. How many records are in the *Stream_Slope Events* layer?

Q9. How many records in *Stream_Slope Events* have the GRIDCODE value of 5 (i.e., slope >40%)?

Task 5 Locate Cities Along U.S. Interstate 5

What you need: *interstates.shp*, a line shapefile containing interstate highways in the conterminous United States; and *uscities.shp*, a point shapefile containing cities in the conterminous United States. Both shapefiles are based on projected coordinates in meters.

In Task 5, you will locate cities along Interstate 5 that runs from Washington State to California. The task consists of three subtasks: extract Interstate 5 from *interstates.shp* to create a new shapefile; create a route from the Interstate 5 shapefile; and locate cities in *uscities.shp* that are within 10 miles of Interstate 5.

1. Insert a new data frame in ArcMap and rename it Task 5. Add *interstates.shp* and *uscities.shp* to Task 5. Double-click the Select tool in the Analysis Tools/Extract toolset. Select *interstates.shp* for the input features, specify *I5.shp* for the output feature class, and click on the SQL button for the expression. Enter the following expression in the Query Builder: "RTE_NUM1" = '5'. (There are two spaces before 5.) Click OK to dismiss the dialogs. Next add a numeric field to *I5* for the route identifier. Double-click the Add Field tool in the Data Management Tools/Fields toolset. Select *I5* for the input table, enter RouteNum for the field name, select DOUBLE for the field type, and click OK. Double-click the Calculate Field tool. Select *I5* for the input table, RouteNum for the field name, enter 5 for the expression, and click OK.

2. Double-click the Create Routes tool in the Linear Referencing Tools toolset. Select *I5* for the input line features, select RouteNum for the route identifier field, specify *Route5.shp* for the output route feature class, select LENGTH for the measure source, enter 0.00062137119 for the measure factor, and click OK. The measure factor converts the linear measure units from meters to miles.

3. This step locates cities within 10 miles of *Route5*. Double-click the Locate Features Along Routes tool in the Linear Referencing Tools toolset. Select *uscities* for the input features, select *Route5* for the input route features, select RouteNum for the route identifier field, enter 10 miles for the search radius, specify *Route5_cities.dbf* for the output event table, and click OK.

4. This step adds *Route5_cities.dbf* to Task 5. Double-click the Make Route Event Layer tool in the Linear Referencing Tools toolset. Make sure that *Route5* is the input route features, *Route5_cities* is the event table, and the type of events is point events. Click OK to add the event layer.

Q10. How many cities in Oregon are within 10 miles of *Route5*?

Challenge Task

What you need: access to the Internet.

In Task 1, you have used the Address Locator in ArcGIS to geocode ten addresses in Coeur d'Alene, Idaho. This challenge question asks you to use two Internet browsers of your choice to geocode the same addresses. (The choices include Google, Yahoo!, Microsoft, and MapQuest.) Then you can compare the search results with Task 1 based on: (1) the hit rate and (2) the positional accuracy.

Q1. Which geocoding engine has the best performance?

References

Basu, S., and T. G. Thibodeau. 1998. Analysis of Spatial Autocorrelation in House Prices. 1998. *The Journal of Real Estate Finance and Economics* 17: 61–85.

Bigham, J. M., T. M. Rice, S. Pande, J. Lee, S. H. Park, N. Gutierrez, and D. R. Ragland.

2009. Geocoding Police Collision Report Data from California: A Comprehensive Approach. *International Journal of Health Geographics* 8: 72 doi:10.1186/1476-072X-8-72.

Choi, K., and W. Jang. 2000. Development of a Transit

Network from a Street Map Database with Spatial Analysis and Dynamic Segmentation. *Transportation Research Part C: Emerging Technologies* 8: 129–46.

Cooke, D. F. 1998. Topology and TIGER: The Census

Bureau's Contribution. In T. W. Foresman, ed., *The History of Geographic Information Systems: Perspectives from the Pioneers,* pp. 47–57. Upper Saddle River, NJ: Prentice Hall.

Davis, Jr., C. A., and F. T. Fonseca. 2007. Assessing the Certainty of Locations Produced by an Address Geocoding System. *Geoinformatica* 11: 103–29.

Fortney, J., K. Rost, and J. Warren. 2000. Comparing Alternative Methods of Measuring Geographic Access to Health Services. *Health Services & Outcomes Research Methodology* 1: 173–84.

Gorr, W., A. Olligschlaeger, and Y. Thompson. 2003. Short-Term Forecasting of Crime. *International Journal of Forecasting* 19: 579–94.

Grubesic, T. H. 2010. Sex Offender Clusters. *Applied Geography* 30: 2–18.

Harries, K. 1999. *Mapping Crime: Principles and Practice.* Washington, DC: U.S. Department of Justice.

Harries, K. 2006. Extreme Spatial Variations in Crime Density in Baltimore County, MD. *Geoforum* 37: 404–16.

Huang, B. 2003. An Object Model with Parametric Polymorphism for Dynamic Segmentation. *International Journal of Geographical Information Science* 17: 343–60.

Hurley, S. E., T. M. Saunders, R. Nivas, A. Hertz, and P. Reynolds. 2003. Post Office Box Addresses: A Challenge for Geographic Information System-Based Studies. *Epidemiology* 14: 386–91.

Krieger, N. 2003. Place, Space, and Health: GIS and Eepidemiology. *Epidemiology* 14: 384–85.

Krieger, N., P. D. Waterman, J. T. Chen, M. Soobader, and S. V. Subramanian. 2003. Monitoring Socioeconomic Inequalities in Sexually Transmitted Infections, Tuberculosis, and Violence: Geocoding and Choice of Area-Based Socioeconomic Measures—The Public Health Disparities Geocoding Project (US). *Public Health Reports* 118: 240–60.

Levine, N., and K. E. Kim. 1998. The Location of Motor Vehicle Crashes in Honolulu: A Methodology for Geocoding Intersections. *Computers, Environment and Urban Systems* 22: 557–76.

Moore, D. A., and T. E. Carpenter. 1999. Spatial Analytical Methods and Geographic Information Systems: Use in Health Research and Epidemiology. *Epidemiologic Reviews* 21: 143–61.

Nyerges, T. L. 1990. Locational Referencing and Highway Segmentation in a Geographic Information System. *ITE Journal* 60: 27–31.

Ratcliffe, J. H. 2001. On the Accuracy of TIGER-Type Geocoded Address Data in Relation to Cadastral and Census Areal Units. *International Journal of Geographical Information Science* 15: 473–85.

Ratcliffe, J. H. 2004. Geocoding Crime and a First Estimate of a Minimum Acceptable Hit Rate. *International Journal of Geographical Information Science* 18: 61–72.

Shaw, S., and D. Wang. 2000. Handling Disaggregate Spatiotemporal Travel Data in GIS. *GeoInformatica* 4: 161–78.

Steenberghen, T., K. Aerts, and I. Thomas. 2009. Spatial Clustering of Events on a Network. *Journal of Transport Geography* doi:10.1016/j.jtrangeo.2009.08.005.

Sutton, J. C., and M. M. Wyman. 2000. Dynamic Location: An Iconic Model to Synchronize Temporal and Spatial Transportation Data. *Transportation Research Part C* 8: 37–52.

Uhlmann, S., E. Galanis, T. Takaro, S. Mak, L. Gustafson, G. Embree, N. Bellack, K. Corbett, J. Isaac-Renton. 2009. Where's the Pump? Associating Sporadic Enteric Disease with Drinking Water Using a *Geographic Information System,* in British Columbia, Canada, 1996–2005. *Journal of Water & Health* 7: 692–98.

Whitsel, E. A., K. M. Rose, J. L. Wood, A. C. Henley, D. Liao, and G. Heiss. 2004. Accuracy and Repeatability of Commercial Geocoding. *American Journal of Epidemiology* 160: 1023–1029.

Yang, D., L. M. Bilaver, O. Hayes, and R. Goerge. 2004. Improving Geocoding Practices: Evaluation of Geocoding Tools. *Journal of Medical Systems* 28: 361–70.

Zandbergen, P. A., and T. C. Hart. 2009. Geocoding Accuracy Considerations in Determining Residency Restrictions for Sex Offenders.*Criminal Justice Policy Review* 20: 62–90.

CHAPTER 7

LEAST-COST PATH ANALYSIS AND NETWORK ANALYSIS

Chapter 17 covers least-cost path analysis and network analysis, both dealing with movement and linear features. Least-cost path analysis is raster-based and has a narrower focus. Using a cost raster that defines the cost of moving through each cell, it finds the least accumulated cost path between cells. Least-cost path analysis is useful as a planning tool for locating a new road or a new pipeline that is least costly (optimal) in terms of the construction costs as well as the potential costs of environmental impacts.

Network analysis requires a network that is vector-based and topologically connected (Chapter 3). Perhaps the most common network

analysis is shortest path analysis, which is used, for example, in in-vehicle navigation systems to help drivers find the shortest route between an origin and a destination. Network analysis also includes the traveling salesman problem, vehicle routing problem, closest facility, allocation, and location-allocation.

Least-cost path analysis and shortest path analysis share some common terms and concepts. But they differ in data format and data analysis environment. Least-cost path analysis uses raster data to locate a "virtual" least cost path. In contrast, shortest path analysis finds the shortest path between stops on an existing network. By having both analyses in the same chapter, we can compare raster data and vector data for GIS applications.

Chapter 17 includes the following five sections. Section 17.1 introduces least-cost path analysis and its basic elements. Section 17.2 covers applications of least-cost path analysis. Section 17.3 examines the basic structure of a road network. Section 17.4 describes how to put together a road network with appropriate attributes for network analysis. Section 17.5 provides an overview of network analysis.

17.1 LEAST-COST PATH ANALYSIS

A least-cost path analysis requires a source raster, a cost raster, cost distance measures, and an algorithm for deriving the least accumulative cost path.

17.1.1 Source Raster

A **source raster** defines the source cell(s). Only the source cell has a cell value in the source raster; all other cells are assigned no data. Similar to physical distance measure operations (Chapter 12), cost distance measures spread from the source cell. But in the context of least-cost path analysis, one can consider the source cell as an end point of a path, either the origin or the destination. The analysis derives for a cell the least accumulated cost path to the source cell or to the closest source cell if two or more source cells are present.

17.1.2 Cost Raster

A **cost raster** defines the cost or impedance to move through each cell. A cost raster has three characteristics. First, the cost for each cell is usually the

sum of different costs. As an example, Box 17.1 summarizes the cost for constructing a pipeline, which may include the construction and operational costs as well as the potential costs of environmental impacts.

Second, the cost may represent the actual or relative cost. Relative costs are expressed in ranked values. For example, costs may be ranked from 1 to 5, with 5 being the highest cost value. A project such as a pipeline project typically involves a wide variety of cost factors. Some factors can be measured in actual costs but others such as aesthetics, wildlife habitats, and cultural resources are difficult to measure in actual costs. Relative costs are therefore a means of standardizing different cost factors for least-cost path analysis.

Third, the cost factors may be weighted depending on the relative importance of each factor. Thus, if factor A is deemed to be twice as important as factor B, factor A can be assigned a weight of 2 and factor B a weight of 1.

To put together a cost raster, we start by compiling and evaluating a list of cost factors. We then make a raster for each cost factor, multiply each cost factor by its weight, and use a local operation to sum the individual cost rasters. The local sum is the total cost necessary to traverse each cell.

Box 17.1 | **Cost Raster for a Site Analysis of Pipelines**

A site analysis of a pipeline project must consider the construction and operational costs. Some of the variables that can influence the costs include the following:

- Distance from source to destination
- Topography, such as slope and grading
- Geology, such as rock and soils
- Number of stream, road, and railroad crossings
- Right-of-way costs
- Proximity to population centers

In addition, the site analysis should consider the potential costs of environmental impacts during construction and liability costs that may result from accidents after the project has been completed. Environmental impacts of a proposed pipeline project may involve the following:

- Cultural resources
- Land use, recreation, and aesthetics
- Vegetation and wildlife
- Water use and quality
- Wetlands

17.1.3 Cost Distance Measures

The cost distance measure in a path analysis is based on the node-link cell representation (Figure 17.1). A node represents the center of a cell, and a link—either a lateral link or a diagonal line—connects the node to its adjacent cells. A lateral link connects a cell to one of its four immediate neighbors, and a diagonal link connects the cell to one of the corner neighbors. The distance is 1.0 cell for a lateral link and 1.414 cells for a diagonal link.

The cost distance to travel from one cell to another through a lateral link is 1.0 cell times the average of the two cost values:

$$1 \times [(C_i + C_j)/2]$$

where C_i is the cost value at cell i, and C_j the cost value at neighboring cell j. The cost distance to

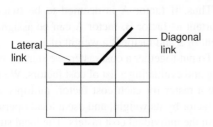

Figure 17.1
Cost distance measures follow the node-link cell representation: a lateral link connects two direct neighbors, and a diagonal link connects two diagonal neighbors.

travel from one cell to another through a diagonal link, on the other hand, is 1.414 cells times the average of the two cost values (Figure 17.2).

The node-link cell representation limits the movement to between neighboring cells and the direction of movement to the eight principal directions. This is why a least-cost path often displays a zigzag pattern. It is possible to connect a cell to more neighbors (i.e., beyond the eight neighboring cells). But this can create nonintuitive intersecting paths, in addition to requiring more computer processing (Miller and Shaw 2001).

17.1.4 Deriving the Least Accumulative Cost Path

Given a cost raster, we can calculate the accumulative cost between two cells by summing the costs associated with each link that connects the two cells (Figure 17.3). But finding the least accumulative cost path is more challenging. This is because many different paths can connect two cells that are not immediate neighbors. The least accumulative cost path can only be derived after each possible path has been evaluated.

Finding the least accumulative cost path is an iterative process based on Dijkstra's algorithm (1959). The process begins by activating cells adjacent to the source cell and by computing costs to the cells. The cell with the lowest cost distance is chosen from the active cell list, and its value is

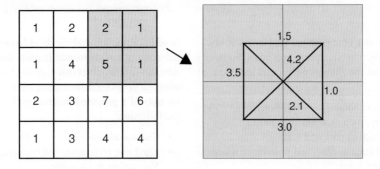

Figure 17.2
The cost distance of a lateral link is the average of the costs in the linked cells, for example, $(1 + 2)/2 = 1.5$. The cost distance of a diagonal link is the average cost times 1.414, for example, $1.414 \times [(1 + 5)/2] = 4.2$.

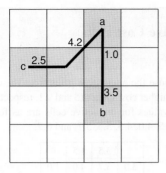

Figure 17.3

The accumulative cost from cell *a* to cell *b* is the sum of 1.0 and 3.5, the costs of two lateral links. The cumulative cost from cell *a* to cell *c* is the sum of 4.2 and 2.5, the costs of a diagonal link and a lateral link.

assigned to the output raster. Next, cells adjacent to the chosen cell are activated and added to the active cell list. Again, the lowest cost cell is chosen from the list and its neighboring cells are activated. Each time a cell is reactivated, meaning that the cell is accessible to the source cell through a different path, its accumulative cost must be recomputed. The lowest accumulative cost is then assigned to the reactivated cell. This process continues until all cells in the output raster are assigned with their least accumulative costs to the source cell.

Figure 17.4 illustrates the cost distance measure operation. Figure 17.4a shows a raster with the source cells at the opposite corners. Figure 17.4b represents a cost raster. To simplify the computation,

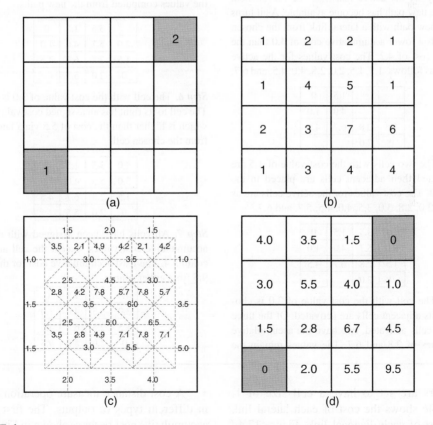

Figure 17.4

The cost distance for each link (*c*) and the least accumulative cost distance from each cell (*d*) are derived using the source cells (*a*) and the cost raster (*b*). See Box 17.2 for the derivation.

Box 17.2 | Derivation of the Least Accumulative Cost Path

Step 1. Activate cells adjacent to the source cells, place the cells in the active list, and compute cost values for the cells. The cost values for the active cells are as follows: 1.0, 1.5, 1.5, 2.0, 2.8, and 4.2.

		1.5	0
		4.2	1.0
1.5	2.8		
0	2.0		

Step 2. The active cell with the lowest value is assigned to the output raster, and its adjacent cells are activated. The cell at row 2, column 3, which is already on the active list, must be reevaluated, because a new path has become available. As it turns out, the new path with a lateral link from the chosen cell yields a lower accumulative cost of 4.0 than the previous cost of 4.2. The cost values for the active cells are as follows: 1.5, 1.5, 2.0, 2.8, 4.0, 4.5, and 6.7.

		1.5	0
		4.0	1.0
1.5	2.8	6.7	4.5
0	2.0		

Step 3. The two cells with the cost value of 1.5 are chosen, and their adjacent cells are placed in the active list. The cost values for the active cells are as follows: 2.0, 2.8, 3.0, 3.5, 4.0, 4.5, 5.7, and 6.7.

	3.5	1.5	0
3.0	5.7	4.0	1.0
1.5	2.8	6.7	4.5
0	2.0		

Step 4. The cell with the cost value of 2.0 is chosen and its adjacent cells are activated. Of the three adjacent cells activated, two have the accumulative cost values of 2.8 and 6.7. The values remain the

same because the alternative paths from the chosen cell yield higher cost values (5 and 9.1, respectively). The cost values for the active cells are as follows: 2.8, 3.0, 3.5, 4.0, 4.5, 5.5, 5.7, and 6.7.

	3.5	1.5	0
3.0	5.7	4.0	1.0
1.5	2.8	6.7	4.5
0	2.0	5.5	

Step 5. The cell with the cost value of 2.8 is chosen. Its adjacent cells all have accumulative cost values assigned from the previous steps. These values remain unchanged because they are all lower than the values computed from the new paths.

	3.5	1.5	0
3.0	5.7	4.0	1.0
1.5	2.8	6.7	4.5
0	2.0	5.5	

Step 6. The cell with the cost value of 3.0 is chosen. The cell to its right has an assigned cost value of 5.7, which is higher than the cost of 5.5 via a lateral link from the chosen cell.

4.0	3.5	1.5	0
3.0	5.5	4.0	1.0
1.5	2.8	6.7	4.5
0	2.0	5.5	

Step 7. All cells have been assigned with the least accumulative cost values, except the cell at row 4, column 4. The least accumulative cost for the cell is 9.5 from either source.

4.0	3.5	1.5	0
3.0	5.5	4.0	1.0
1.5	2.8	6.7	4.5
0	2.0	5.5	9.5

both rasters are set to have a cell size of 1. Figure 17.4c shows the cost of each lateral link and the cost of each diagonal link. Figure 17.4d shows for each cell the least accumulative cost. Box 17.2 explains how Figure 17.4d is derived.

A cost distance measure operation can result in different types of outputs. The first is a least accumulative cost raster as shown in Figure 17.4d. The second is a direction raster, showing the direction of the least-cost path for each cell. The third is an allocation raster, showing the assignment of each cell to a source cell on the basis of cost distance measures. The fourth type is a shortest path raster, which shows the least cost path from each cell to a source cell. Using the same data as in Figure 17.4, Figure 17.5a shows two examples of the least-cost path and Figure 17.5b shows the assignment of each cell to a source cell. The darkest cell in Figure 17.5b can be assigned to either one of the two sources.

17.1.5 Options for Least-Cost Path Analysis

The outcome of a least-cost path analysis is directly influenced by the selection of cost factors and, perhaps more importantly, the weighting of each factor. This is why, in recent studies (e.g., Atkinson et al. 2005; Choi et al. 2009), least-cost path analysis has often been integrated with multicriteria evaluation (Malczewski 2006), a topic to be discussed in detail in Chapter 18.

When the terrain is used for deriving the least-cost path, the surface is typically assumed to be uniform for all directions. But in reality, the terrain changes in elevation, slope, and aspect in different directions. To provide a more realistic analysis of how we traverse the terrain, the "surface distance" can be used (Collischonn and Pilar 2000; Yu, Lee, and Munro-Stasiuk 2003). Calculated from an elevation raster or a digital elevation model (DEM), the surface distance measures the ground or actual distance that must be covered from one cell to another. The surface distance increases when the elevation difference (gradient) between two cells increases. Is it accurate to calculate distances from a DEM? According to Rasdorf et al. (2004), the surface distances calculated from a DEM along highway centerlines compare favorably with distances obtained by driving cars equipped with a distance measurement instrument. In addition to the surface distance, the vertical factor (i.e., the difficulty of overcoming the vertical elements such as uphill or downhill) and the horizontal factor (i.e., the difficulty of overcoming the horizontal elements such as crosswinds) can also be considered.

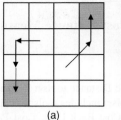 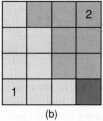

(a)   (b)

Figure 17.5
The least-cost path (a) and the allocation raster (b) are derived using the same input data as in Figure 17.4.

ArcGIS uses the term **path distance** to describe the cost distance based on the surface distance, vertical factor, and horizontal factor. Box 17.3 describes least-cost path analysis and path distance analysis tools that are available in ArcGIS.

17.2 APPLICATIONS OF LEAST-COST PATH ANALYSIS

Least-cost path analysis is useful for planning roads, pipelines, canals, transmission lines, and trails. As examples, least-cost path analysis has been used by Rees (2004) to locate footpaths in mountainous areas, by Atkinson et al. (2005) to derive an arctic all-weather road, and by Snyder et al. (2008) to find trail location for all-terrain vehicles.

Least-cost path analysis can also be applied to wildlife movements. A common application in wildlife management is corridor or connectivity study (Kindall and van Manen 2007; Falcucci et al. 2008). In such studies, the source cells represent habitat concentration areas and the cost factors typically include vegetation, topography, and human activities such as roads. Analysis results can show the least costly routes for wildlife movement.

A more unusual application of least-cost path analysis is a model of gas dispersion built by Hepner and Finco (1995). In the study, the cost raster is the product of surface distance, slope impedance, and wind impedance, and the accumulative cost represents the difficulty of gas travel between the release point (source cell) and any given cell in the study area.

Box 17.3 | **Cost Distance Measure Operations in ArcGIS**

ArcToolbox has the Distance toolset in Spatial Analyst Tools. The toolset includes tools for calculating Cost Distance, Cost Back Link, Cost Path, and Cost Allocation, all based on the inputs of a source data set and a cost raster. It also has tools for path distance measures including Path Distance, Path Distance Allocation, and Path Distance Back Link. The cost distance in a path distance analysis is calculated by multiplying the surface distance, vertical factor, and horizontal factor. The surface distance is calculated from an elevation raster. The vertical factor and the horizontal factor are user-defined and optional. Path distance analysis is designed to provide a more realistic analysis of how we traverse the terrain.

17.3 NETWORK

A **network** is a system of linear features that has the appropriate attributes for the flow of objects. A road system is a familiar network. Other networks include railways, public transit lines, bicycle paths, and streams. A network is typically topology-based: lines meet at intersections, lines cannot have gaps, and lines have directions.

Because many network applications involve road systems, a discussion of a road network is presented here. It starts by describing the geometry and attribute data of a road network including link impedance, turns, one-way streets, and overpasses/underpasses. Then it shows how these data can be put together to form a street network for a real-world example.

17.3.1 Link and Link Impedance

A **link** refers to a road segment defined by two end points. Also called edges or arcs, links are the basic geometric features of a network. **Link impedance** is the cost of traversing a link. A simple measure of the cost is the physical length of the link. But the length may not be a reliable measure of cost, especially in cities where speed limits and traffic conditions vary significantly along different streets. A better measure of link impedance is the travel time estimated from the length and the speed limit of a link. For example, if the speed limit is 30 miles per hour and the length is 2 miles, then the link travel time is 4 minutes (2/30 × 60 minutes).

There are variations in measuring the link travel time. The travel time may be directional: the travel time in one direction may be different from that in the other direction. In that case, the directional travel time can be entered separately in two fields (e.g., from-to and to-from). The travel time may also vary by the time of the day and by the day of the week, thus requiring the setup of different network attributes for different applications.

17.3.2 Junction and Turn Impedance

A **junction** refers to a street intersection. A junction is also called a node. A turn is a transition from one street segment to another at a junction. **Turn impedance** is the time it takes to complete a turn, which is significant in a congested street network (Ziliaskopoulos and Mahmassani 1996). Turn impedance is directional. For example, it may take 5 seconds to go straight, 10 seconds to make a right turn, and 30 seconds to make a left turn at a stoplight. A negative turn impedance value means a prohibited turn, such as turning the wrong way onto a one-way street.

Because a network typically has many turns with different conditions, a table can be used to assign the turn impedance values in the network. Each record in a turn table shows the street segments involved in a turn and the turn impedance in minutes or seconds. A driver approaching a street intersection can go straight, turn right, turn left, and, in some cases, make a U-turn. Assuming the intersection involves four street segments, as most intersections do, this means at least 12 possible turns at the intersection, excluding U-turns. Depending on the level of details for a study, we may not have to include every intersection and every possible turn in a turn table. A partial turn table listing only turns at intersections with stoplights may be sufficient for a network application.

17.3.3 One-Way or Closed Streets

The value of a designated field in a network's attribute table can show one-way or closed streets. For example, the value T can mean that a street segment is one-way; F, it is not one-way; and N, it does not allow traffic in either direction. The direction of a one-way street is determined by the beginning point and the end point of the line segment.

17.3.4 Overpasses and Underpasses

There are at least two methods for including overpasses and underpasses in a network. The first method uses nonplanar features: an overpass and

the street underneath it are both represented as continuous lines without a node at their intersection (Figure 17.6). The second method treats overpasses and underpasses as planar features and uses an elevation field to separate the two intersecting lines. In Figure 17.7, a higher "elevation" value (1) is assigned to the junction used by the overpass and a lower value (0) is assigned to the street underneath the overpass.

17.4 PUTTING TOGETHER A NETWORK

Putting together a road network involves three tasks: gather the linear features of the network, build the necessary topology for the network, and attribute the network features. Here we use an example from Moscow, Idaho. A university town of 25,000 people, Moscow, Idaho, has more or less the same network features as other larger cities except for overpasses or underpasses.

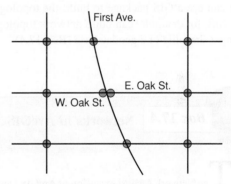

Street name	F-elev	T-elev
First Ave.	0	1
First Ave.	1	0
W. Oak St.	0	0
E. Oak St.	0	0

Figure 17.7

First Ave. crosses Oak St. with an overpass.
A planar representation with two nodes is used at the intersection: one for First Ave., and the other for Oak St. First Ave has 1 for the T-elev and F-elev values, indicating that the overpass is on First Ave.

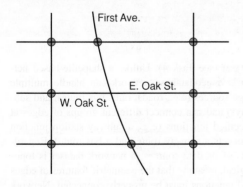

Figure 17.6

First Ave. crosses Oak St. with an overpass.
A nonplanar representation with no nodes is used at the intersection of Oak St. and First Ave.

17.4.1 Gathering Linear Features

The TIGER/Line files from the U.S. Census Bureau are a common data source for making preliminary street networks in the United States. They are available for download at the Census Bureau website (**http://www.census.gov/geo/www/tiger/**). The 2009 TIGER/Line files, for example, is in shapefile format and is measured in longitude and latitude values based on NAD83 (North American Datum of 1983) (Chapter 2). To use it for real-world applications, a preliminary network compiled from the TIGER/Line files must be converted to projected coordinates. Road network data can also be digitized or purchased from commercial companies.

17.4.2 Editing and Building Network

A network converted from the TIGER/Line files has the built-in topology: The streets are connected at nodes and the nodes are designated as either from-nodes or to-nodes. If topology is not available in a data set (e.g., a shapefile or a CAD file), one can use a GIS package to build the topology. ArcGIS, for example, can build network topology using a shapefile or a geodatabase (Box 17.4).

Editing and updating the road network is the next step. When superimposing the road network over orthophotos or high-resolution satellite images, one may find that the street centerlines from the TIGER/Line files deviate from the streets. Such mistakes must be corrected. Also, new streets must be added. In some cases, pseudo nodes (i.e., nodes that are not required topologically) must be removed so that a street segment between two intersections is not unnecessarily broken up. But a pseudo node is needed at the location where one street changes into another along a continuous link (rare but does happen). Without the pseudo node, the continuous link will be treated as the same street. After the TIGER/Line shapefile has been edited and updated, it can be converted to a street network in a GIS.

17.4.3 Attributing the Network Features

Network attributes in this example include the link impedance, one-way streets, speed limits, and turns (Box 17.5). The link impedance value can be the physical distance or the travel time. The physical distance is the length of a street segment. The travel time can be derived from the length and the speed limit of a street segment. Roads con-

---

Box 17.4 | **Networks in ArcGIS**

The Network Analyst extension of ArcGIS Desktop stores networks as network datasets. A network dataset combines network elements and data sources for the elements. Network elements refer to edges, junctions, and turns. Network sources can be shapefiles or geodatabase feature classes. A shapefile-based network consists of a network dataset and a junction shapefile, both created from a polyline shapefile (e.g., a street network shapefile). Task 3 in the applications section covers a shapefile-based network. A geodatabase network includes both network elements and sources as feature classes in a feature dataset (see Task 4). Unlike a shapefile-based network, a geodatabase network can handle multiple edge sources (e.g., roads, rails, bus routes, and subways) and can connect different groups of edges at specified junctions (e.g., a subway station junction connects a subway route and a bus route). Regardless of its data sources, a network dataset is topological, meaning that the geometric features of edges and junctions must be properly connected. Network Analyst has the Build tool to build the topology (i.e., connectivity) between the network elements.

Box 17.5 | **Network Attributes**

N etwork Analyst groups network attributes into cost (e.g., travel time), descriptors (e.g., speed limit), restrictions (e.g., one-way streets), and hierarchy (e.g., primary, secondary, and local roads). These network attributes are created when the network dataset is built. The attribute values are based on a field or fields in the network sources. For example, a field called Minutes in an edge source can assign travel time to a cost network attribute.

verted from the TIGER/Line files have the field MTFCC (MAF/TIGER feature class code), classifying roads as primary road, secondary road, and so on, which can be used to assign speed limits. Moscow, Idaho, has three speed limits: 35 miles/hour for principal arterials, 30 miles/hour for minor arterials, and 25 miles/hour for all other city streets. With speed limits in place, the travel time for each street segment can be computed from the segment's length and the speed limit.

Moscow, Idaho, has two one-way streets serving as the northbound and the southbound lanes of a state highway. These one-way streets are denoted by the value T in a direction field. The direction of all street segments that make up a one-way street must be consistent and pointing in the right direction. Street segments that are incorrectly oriented must be flipped.

Making a turn table for street intersections with traffic signals is next. Each turn is defined by the edge that the turn starts, the edge that the turn ends, and the turn impedance value in minutes or seconds.

Figure 17.8 shows a street intersection at node 341 with no restrictions for all possible turns except U-turns. This example uses two turn impedance values: 30 seconds or 0.5 minute for a left turn, and 15 seconds or 0.25 minute for either a right turn or going straight. Some street intersections do not allow certain types of turns. For example, Figure 17.9 shows a street intersection at node 265 with stop signs posted only for the east–west traffic. Therefore, the turn impedance values are assigned only to turns to be made

from edge 342 and edge 340. Box 17.6 provides more information on how to prepare a turn table in ArcGIS.

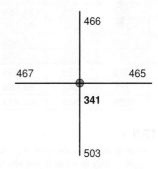

Node#	Arc1#	Arc2#	Angle	Minutes
341	503	467	90	0.500
341	503	466	0	0.250
341	503	465	−90	0.250
341	467	503	−90	0.250
341	467	466	90	0.500
341	467	465	0	0.250
341	466	503	0	0.250
341	466	467	−90	0.250
341	466	465	90	0.500
341	465	503	90	0.500
341	465	467	0	0.250
341	465	466	−90	0.250

Figure 17.8

Possible turns at node 341. This turn table was created using ArcInfo Workstation. Box 17.6 provides more information on turn tables.

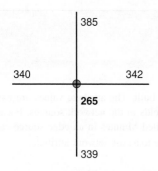

Node#	Arc1#	Arc2#	Angle	Minutes
265	339	342	–87.412	0.000
265	339	340	92.065	0.000
265	339	385	7.899	0.000
265	342	339	87.412	0.500
265	342	340	–0.523	0.250
265	342	385	–84.689	0.250
265	340	339	–92.065	0.250
265	340	342	0.523	0.250
265	340	385	95.834	0.500
265	385	339	–7.899	0.000
265	385	342	84.689	0.000
265	385	340	–95.834	0.000

Figure 17.9
Node 265 has stop signs for the east–west traffic. Turn impedance applies only to turns in the shaded rows.

17.5 NETWORK ANALYSIS

A network with the appropriate attributes can be used for a variety of applications. Some applications are directly accessible through commands in a GIS package. Others require the integration of GIS and specialized software.

17.5.1 Shortest Path Analysis

Shortest path analysis finds the path with the minimum cumulative impedance between nodes on a network. Because the link impedance can be measured in distance or time, a shortest path may represent the shortest route or fastest route.

Shortest path analysis typically begins with an impedance matrix in which a value represents the impedance of a direct link between two nodes on a network and an ∞ (infinity) means no direct connection. The problem is to find the shortest distances (least cost) from a node to all other nodes. A variety of algorithms can be used to solve the problem (Zhan and Noon 1998; Zeng and Church 2009); among them, the most commonly used algorithm is the Dijkstra algorithm (1959).

To illustrate how the Dijkstra algorithm works, Figure 17.10 shows a road network with six nodes and eight links and Table 17.1 shows the travel time measured in minutes between nodes. The value of ∞ above and below the principal diagonal in the impedance matrix in Table 17.1 means no direct path between two nodes. To find the shortest path from node 1 to all other nodes in Figure 17.10, we can solve the problem by using an iterative procedure (Lowe and Moryadas 1975). At each step, we choose the shortest path from a list of candidate paths and place the node of the shortest path in the solution list.

The first step chooses the minimum among the three paths from node 1 to nodes 2, 3, and 4, respectively:

$$\min (p_{12}, p_{13}, p_{14}) = \min (20, 53, 58)$$

We choose p_{12} because it has the minimum impedance value among the three candidate paths. We then place node 2 in the solution list with node 1.

A new candidate list of paths that are directly or indirectly connected to nodes in the solution list (nodes 1 and 2) is prepared for the second step:

$$\min (p_{13}, p_{14}, p_{12} + p_{23}) = \min (53, 58, 59)$$

We choose p_{13} and add node 3 to the solution list. To complete the solution list with other nodes on the network we go through the following steps:

$$\min (p_{14}, p_{13} + p_{34}, p_{13} + p_{36}) = \min (58, 78, 72)$$
$$\min (p_{13} + p_{36}, p_{14} + p_{45}) = \min (72, 71)$$
$$\min (p_{13} + p_{36}, p_{14} + p_{45} + p_{56}) = \min (72, 84)$$

Table 17.2 summarizes the solution to the shortest path problem from node 1 to all other nodes.

Box 17.6 | Turn Table and Turn Feature Class

The Turntable command in ArcInfo Workstation creates a turn table for every possible turn in a coverage. Each turn is limited to moving from one arc (the from arc) to another arc (the to arc). The turn angle is the angle between the two arcs. Although turn angles are supposed to be 90 for a left turn, −90 for a right turn, and 0 for going straight, deviations often occur in reality as well as in digital files. Given a turn table, one can use a Network Analyst tool to convert it into a turn feature class (see Task 4). Introduced in ArcGIS 9.3, the global turn delay evaluator is similar to the turn table. The evaluator assigns turn impedances to left, right, U-, and straight turns. Each turn is also limited to moving from one edge to another.

To model turns and turn impedances more accurately, Network Analyst uses a turn feature class. A turn feature class differs from a turn table or the global turn delay evaluator in two important aspects. First, Network Analyst allows multi-edge turns. A multi-edge turn connects one edge to another through a sequence of connected intermediate edges. In contrast, turns recorded in a turn table are two-edge turns. A two-edge turn, which is most common in a street network, refers to a simple movement from one edge to another edge. Second, turn angle is no longer a default field in a turn feature class. Instead, there are fields describing the positions along the line features involved in turns. A turn feature class can be created interactively using the Editor tools in ArcMap.

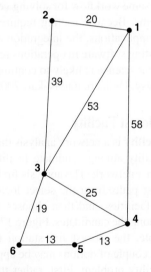

Figure 17.10
Link impedance values between cities on a road network.

TABLE 17.1 | The Impedance Matrix Among Six Nodes in Figure 17.10

	(1)	(2)	(3)	(4)	(5)	(6)
(1)	∞	20	53	58	∞	∞
(2)	20	∞	39	∞	∞	∞
(3)	53	39	∞	25	∞	19
(4)	58	∞	25	∞	13	∞
(5)	∞	∞	∞	13	∞	13
(6)	∞	∞	19	∞	13	∞

A shortest-path problem with six nodes and eight links is simple to solve. A real road network, however, has many more nodes and links. Zhan and Noon (1998), for example, lists 2878 nodes and 8428 links in the State of Georgia network with three levels of roads (interstate, principal arterial, and major arterial). This is why, over the years, researchers have continued to propose new shortest-path algorithms to reduce the computation time (e.g., Zhen and Church 2009).

Shortest-path analysis has many applications. Perhaps the common application is to help a driver find the shortest route between an origin and a destination. This can be done via a navigation system, either in a vehicle or on a cell phone. Shortest routes are also useful as measures of accessibility.

TABLE 17.2	Shortest Paths from Node 1 to All Other Nodes in Figure 17.11		
From-Node	To-Node	Shortest Path	Minimum Cumulative Impedance
1	2	p_{12}	20
1	3	p_{13}	53
1	4	p_{14}	58
1	5	$p_{14} + p_{45}$	71
1	6	$p_{13} + p_{36}$	72

Thus they can be used as input data to a wide range of accessibility studies connecting users to park-and-ride facility (Farhan and Murray 2005), urban trail system (Krizek, El-Geneidy, and Thompson 2007), food supermarkets (Apparicio, Cloutier, and Shearmur 2007), and community resources (Witten, Exeter, and Field 2003).

17.5.2 Traveling Salesman Problem

The **traveling salesman problem** is a routing problem, which stipulates that the salesman must visit each of the select stops only once, and the salesman may start from any stop but must return to the original stop. The objective is to determine which route, or tour, the salesman can take to minimize the total impedance value. A common solution to the traveling salesman problem uses a heuristic method (Lin 1965). Beginning with an initial random tour, the method runs a series of locally optimal solutions by swapping stops that yield reductions in the cumulative impedance. The iterative process ends when no improvement can be found by swapping stops. This heuristic approach can usually create a tour with a minimum, or near minimum, cumulative impedance. Similar to shortest-path analysis, a number of algorithms are available for the local search procedure, with Tabu Search being the best-known algorithm. For some applications, a time window constraint can also be added to the traveling salesman problem so that the tour must

be completed with a minimum amount of time delay.

17.5.3 Vehicle Routing Problem

The vehicle routing problem is an extension of the traveling salesman problem. Given a fleet of vehicles and customers, the main objective of the vehicle routing problem is to schedule vehicle routes and visits to customers in such a way that the total travel time is minimized. Additional constraints such as time windows, vehicle capacity, and dynamic conditions (e.g., traffic congestion) may also exist.

In an early application of GIS to the vehicle routing problem, Weigel and Cao (1999) used a GIS (ARC/INFO) to prepare an origin-destination cost matrix connecting customers and Sears stores as an input to external routing software. After routes were assigned, route maps were generated for dispatch drivers. ArcGIS Network Analyst adopts the same workflow for solving vehicle routing problems. Because routing requires complex modeling applications, the integration of GIS and special routing software in operations research and management science is likely to continue (Keenan 2008; Bräysy, Dullaert, and Nakari 2009).

17.5.4 Closest Facility

Closest facility is a network analysis that finds the closest facility among candidate facilities to any location on a network. The analysis first computes the shortest paths from the select location to all candidate facilities, and then chooses the closest facility among the candidates. Figure 17.11 shows, for example, the closest fire station to a street address. A couple of options may be applied to the closest facility problem. First, rather than getting a single facility, the user may ask for a number of closest facilities. Second, the user may specify a search radius in distance or travel time, thus limiting the candidate facilities.

Closest facility analysis is regularly performed in to location-based services (LBS), an application heavily promoted by GIS companies in recent years (Chapter 16). An LBS user can use

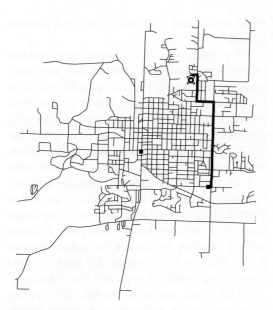

Figure 17.11
Shortest path from a street address to its closest fire station, shown by the square symbol.

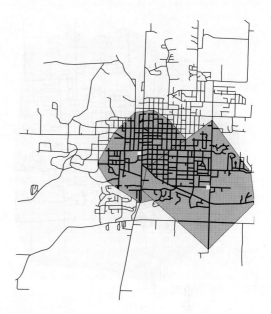

Figure 17.12
The service areas of two fire stations within a 2-minute response time.

a Web-enabled cell phone and a browser such as Google Maps to find the closest hospital, restaurant, or ATM. Closest facility is also important as a measure of quality of service such as health care (Schuurman, Leight, and Berube 2008).

17.5.5 Allocation

Allocation is the study of the spatial distribution of resources through a network. Resources in allocation studies often refer to public facilities, such as fire stations, schools, hospitals, or even open spaces (in case of earthquakes) (Tarabanis and Tsionas 1999). Because the distribution of the resources defines the extent of the service area, the main objective of spatial allocation analysis is to measure the efficiency of these resources.

A common measure of efficiency in the case of emergency services is the response time—the time it takes for a fire truck or an ambulance to reach an incident. Figure 17.12, for example, shows areas of a small city covered by two existing fire stations within a 2-minute response time. The map shows

a large portion of the city is outside the 2-minute response zone. The response time to the city's outer zone is about 5 minutes (Figure 17.13). If residents of the city demand that the response time to any part of the city be 2 minutes or less, then the options are either to relocate the fire stations or, more likely, to build new fire stations. A new fire station should be located to cover the largest portion of the city unreachable in 2 minutes by the existing fire stations. The problem then becomes a location and allocation problem, which is covered in Section 17.5.6.

For health care, the efficiency of allocation can be measured by the number or percentage of population served by hospitals. In their study of health care in non-urban areas of British Columbia, Canada, Schuurman et al. (2006) first defined a hospital's catchment (service area) as within one-hour travel time of the hospital on a road network. They then used spatial query (Chapter 10) to select census blocks within the catchment for analysis of variations in service coverage and population served.

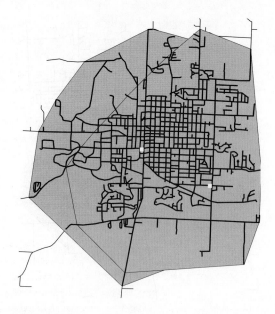

Figure 17.13
The service areas of two fire stations within a 5-minute response time.

17.5.6 Location–Allocation

Location–allocation solves problems of matching the supply and demand by using sets of objectives and constraints. The private sector offers many location–allocation examples. Suppose a company operates soft-drink distribution facilities to serve supermarkets. The objective in this example is to minimize the total distance traveled, and a constraint, such as a 2-hour drive distance, may be imposed on the problem. A location–allocation analysis matches the distribution facilities and the supermarkets while meeting both the objective and the constraint.

Location–allocation is also important in the public sector. For instance, a local school board may decide that all school-age children should be within 1 mile of their schools and the total distance traveled by all children should be minimized. In this case, schools represent the supply, and school-age children represent the demand. The objective of this location–allocation analysis is to provide equitable service to a population

while maximizing efficiency in the total distance traveled.

The setup of a location–allocation problem requires inputs in supply, demand, and impedance measures. The supply consists of facilities at point locations. The demand may consist of individual points, or aggregate points representing line or polygon data. For example, the locations of school-age children may be represented as individual points or aggregate points (e.g., centroids) in unit areas such as census block groups. Impedance measures between the supply and demand can be expressed as travel distance or travel time. They can be measured along the shortest path between two points on a road network or along the straight line connecting two points. Shortest path distances are likely to yield more accurate results than straight-line distances.

Two most common models for solving location-allocation problems are minimum impedance (time or distance) and maximum coverage (Ghosh, McLafferty, and Craig 1995; Church 1999). The minimum impedance model, also called the p-median location model, minimizes the total distance or time traveled from all demand points to their nearest supply centers (Hakimi 1964). In contrast, the maximum coverage model maximizes the demand covered within a specified time or distance (Church and ReVelle 1974). Both models may take on added constraints or options. A maximum distance constraint may be imposed on the minimum impedance model so that the solution, while minimizing the total distance traveled, ensures that no demand is beyond the specified maximum distance. A desired distance option may be used with the maximum covering model to cover all demand points within the desired distance. Box 17.7 summarizes the location-allocation algorithms and other network analysis functions available in ArcGIS.

Here we will examine location–allocation problems in matching ambulance service and nursing homes. Suppose (1) there are two existing fire stations to serve seven nursing homes, and (2) an emergency vehicle should reach any nursing home in 4 minutes or less. Figure 17.14 shows that this objective cannot be achieved with the two existing fire stations as two of the seven nursing homes are outside the 4-minute range on a road network using either the minimum impedance or maximum coverage model. One solution is to increase the number of fire stations from two to three. Figure 17.14 shows three candidates for the additional fire station. Based on either the minimum impedance or maximum coverage model, Figure 17.15 shows the selected candidate and the matching of nursing homes to the facilities by line symbols. One nursing home, however, is still outside the 4-minute range in Figure 17.15. There are two options to have a complete coverage of all nursing homes: one is to increase the facilities from three to four, and the other is to relax the constraint of the response time from 4 to 5 minutes. Figure 17.16 shows the result of the change of the response time from 4 to 5 minutes. All seven nursing homes can now be served by the three facilities. Notice that the selected candidate in Figure 17.16 differs from that in Figure 17.15.

Figure 17.15
The map shows the result of matching three fire stations, two existing ones and one candidate, with six nursing homes based on the minimum impedance model and an impedance cutoff of 4 minutes on the road network.

Figure 17.14
The two solid squares represent existing fire stations, the three gray squares candidate facilities, and the seven circles nursing homes. The map shows the result of matching two existing fire stations with nursing homes based on the minimum impedance model and an impedance cutoff of 4 minutes on the road network.

Figure 17.16
The map shows the result of matching three fire stations, two existing ones and one candidate, with seven nursing homes based on the minimum impedance model and an impedance cutoff of 5 minutes on the road network.

Box 17.7 | **Network Analysis Using ArcGIS**

Network Analyst can solve shortest path, closest facilities, service area (allocation), and origin–destination cost matrix problems. These analytical functions are found on the Network Analyst toolbar and in the Analysis toolset of Network Analyst Tools in ArcToolbox. The Turn Feature Class toolset of Network Analyst Tools has tools for creating and editing turn feature classes as well as a tool for converting a turn table into a turn feature class.

ArcGIS 10 has introduced location–allocation analysis as part of network analysis. The algorithms for location–allocation analysis include minimum impedance and maximum coverage, two common algorithms discussed in this chapter. Other algorithms are minimum facilities, maximum attendance, maximum market share, and target market share.

KEY CONCEPTS AND TERMS

Allocation: A study of the spatial distribution of resources on a network.

Closest facility: A network analysis that finds the closest facility among candidate facilities.

Cost raster: A raster that defines the cost or impedance to move through each cell.

Junction: A street intersection.

Link: A segment separated by two nodes in a road network.

Link impedance: The cost of traversing a link, which may be measured by the physical length or the travel time.

Location–allocation: A spatial analysis approach that matches the supply and demand by using sets of objectives and constraints.

Network: A system of linear features that has the appropriate attributes for the flow of objects such as traffic flow.

Path distance: A term used in ArcGIS to describe the cost distance that is calculated from the surface distance, the vertical factor, and the horizontal factor.

Shortest path analysis: A network analysis approach that finds the path with the minimum cumulative impedance between nodes on a network.

Source raster: A raster that defines the source to which the least-cost path from each cell is calculated.

Traveling salesman problem: A network analysis scenario that finds the best route with the conditions of visiting each stop only once, and returning to the node where the journey starts.

Turn impedance: The cost of completing a turn on a road network, which is usually measured by the time delay.

REVIEW QUESTIONS

1. What is a source raster for least-cost path analysis?

2. Define a cost raster.

3. Box 17.1 summarizes the various costs for a pipeline project. Use Box 17.1 as a reference and list the types of costs that can be associated with a new road project.

4. Cost distance measure operations are based on the node-link cell representation. Explain the representation by using a diagram.

5. Refer to Figure 17.4d. Explain how the cell value of 5.5 (row 2, column 2) is derived. Is it the least accumulative cost possible?

6. Refer to Figure 17.4d. Show the least accumulative cost path for the cell at row 2, column 3.

7. Refer to Figure 17.5b. The cell at row 4, column 4 can be assigned to either one of the two source cells. Show the least-cost path from the cell to each source cell.

8. What is an allocation raster?

9. How does the surface distance differ from the regular (planimetric) cost distance?

10. Explain the difference between a network and a line shapefile.

11. What is link impedance?

12. What is turn impedance?

13. What fields are normally included in a turn table?

14. Suppose the impedance value between nodes 1 and 4 is changed from 58 to 40 (e.g., because of lane widening) in Figure 17.10. Will it change the result in Table 17.2?

15. The result of an allocation analysis is typically presented as service areas. Why?

16. Define location–allocation analysis.

17. Explain the difference between the minimum distance model and the maximum covering model in location–allocation analysis.

APPLICATIONS: PATH ANALYSIS AND NETWORK APPLICATIONS

This applications section has six tasks. Tasks 1 and 2 cover path analysis. You will work with the least accumulative cost distance in Task 1 and the path distance in Task 2. Tasks 3 to 6 require the use of the Network Analyst extension. Task 3 runs a shortest path analysis. Task 4 lets you build a geodatabase network dataset. Task 5 runs a closest facility analysis. And Task 6 runs an allocation analysis.

Task 1 Compute the Least Accumulative Cost Distance

What you need: *sourcegrid* and *costgrid,* the same rasters as in Figure 17.4; and *pathgrid,* a raster to be used with the shortest path function. All three rasters are sample rasters and do not have the projection file.

In Task 1, you will use the same inputs as in Figures 17.4a and 17.4b to create the same outputs as in Figures 17.4d, 17.5a, and 17.5b.

1. Connect to the Chapter 17 database in ArcCatalog. Launch ArcMap. Rename the data frame Task 1, and add *sourcegrid, costgrid,* and *pathgrid* to Task 1. Ignore the warning message about the spatial reference.

2. Click ArcToolbox window to open ArcToolbox. Set the current workspace as the Chapter 17 database. Double-click the Cost Distance tool in the Spatial Analyst Tools/ Distance toolset. In the next dialog, select *sourcegrid* for the input raster, select *costgrid* for the input cost raster, save the output

distance raster as *CostDistance*, and save the output backlink raster as *CostDirection*. Click OK to run the command.

3. *CostDistance* shows the least accumulative cost distance from each cell to a source cell. You can use the Identify tool to click a cell and find its accumulative cost.

Q1. Are the cell values in *CostDistance* the same as those in Figure 17.4*d*?

4. *CostDirection* shows the least cost path from each cell to a source cell. The cell value in the raster indicates which neighboring cell to traverse to reach a source cell. The directions are coded 1 to 8, with 0 representing the cell itself (Figure 17.17).

5. Double-click the Cost Allocation tool in the Spatial Analyst Tools/Distance toolset. Select *sourcegrid* for the input raster, select *costgrid* for the input cost raster, save the output allocation raster as *Allocation*, and click OK. *Allocation* shows the allocation of cells to each source cell. The output raster is the same as Figure 17.5*b*.

6. Double-click the Cost Path tool in the Spatial Analyst Tools/Distance toolset. Select *pathgrid* for the input raster, select *costgrid* for the input cost distance raster, select *CostDirection* for the input cost backlink raster, save the output raster as *ShortestPath*, and click OK. *ShortestPath* shows the path

from each cell in *pathgrid* to its closest source. *ShortestPath* is the same as Figure 17.5*a*.

Task 2 Compute the Path Distance

What you need: *emidasub,* an elevation raster; *peakgrid,* a source raster with one source cell; and *emidapathgd,* a path raster that contains two cell values. All three rasters are projected onto UTM coordinates in meters.

In Task 2, you will find the least-cost path from each of the two cells in *emidapathgd* to the source cell in *peakgrid*. The least-cost path is based on the path distance. Calculated from an elevation raster, the path distance measures the ground or actual distance that must be covered between cells. The source cell is at a higher elevation than the two cells in *emidapathgd*. Therefore, you can imagine the objective of Task 2 is to find the least-cost hiking path from each of the two cells in *emidapathgd* to the source cell in *peakgrid*.

1. Insert a new data frame in ArcMap, and rename it Task 2. Add *emidasub, peakgrid,* and *emidapathgd* to Task 2. Select Properties from the context menu of *emidasub*. On the Symbology tab, right-click the Color Ramp box to uncheck Graphic View. Then select Elevation #1. As shown in the map, the source cell in *peakgrid* is located near the summit of the elevation surface and the two cells in *emidapathgd* are located in low elevation areas.

2. Double-click the Path Distance tool in the Spatial Analyst Tools/Distance toolset. Select *peakgrid* for the input raster, specify *pathdist1* for the output distance raster, select *emidasub* for the input surface raster, specify *backlink1* for the output backlink raster, and click OK to run the command.

Q2. What is the range of cell values in *pathdist1*?

Q3. If a cell value in *pathdist1* is 900, what does the value mean?

3. Double-click Cost Path in the Spatial Analyst Tools/Distance toolset. Select *emidapathgd*

6	7	8
5	0	1
4	3	2

Figure 17.17
Direction measures in a direction raster are numerically coded. The focal cell has a code of 0. The numeric codes 1 to 8 represent the direction measures of 90°, 135°, 180°, 225°, 315°, 360°, and 45° in a clockwise direction.

for the input raster, select *pathdist1* for the input cost distance raster, select *backlink1* for the input cost backlink raster, specify *path1* for the output raster, and click OK.

4. Open the attribute table of *path1*. Click the left cell of the first record. As shown in the map, the first record is the cell in *peakgrid*. Click the second record, which is the least-cost path from the first cell in *emidapathgd* (in the upper-right corner) to the cell in *peakgrid*.

Q4. What does the third record in the *path1* attribute table represent?

Task 3 Run Shortest Path Analysis

What you need: *uscities.shp,* a point shapefile containing cities in the conterminous United States; and *interstates.shp,* a line shapefile containing interstate highways in the conterminous United States. Both shapefiles are based on the Albers Conic Equal Area projection in meters.

The objective of Task 3 is to find the shortest route between two cities in *uscities.shp* on the interstate network. The shortest route is defined by the link impedance of travel time. The speed limit for calculating the travel time is 65 miles/hour. Helena, Montana, and Raleigh, North Carolina, are two cities for this task.

1. Right-click *interstates.shp* in the Catalog tree and select Item Descriptions. On the Preview tab of the Item Descriptions window, preview the table of *interstates. shp* has several attributes that are important for network analysis. The field MINUTES shows the travel time in minutes for each line segment. The field NAME lists the interstate number. And the field METERS shows the physical length in meters for each line segment.

2. Select Extensions from the Customize menu. Make sure that Network Analyst is checked. Select Toolbars from the Customize menu, and make sure that Network Analyst is checked.

3. This step uses *interstates.shp* in the Catalog tree to set up a network dataset. Right-click *interstates.shp* and select New Network Dataset. The New Network Dataset dialog allows you to set up various parameters for the network dataset. Accept the default name *interstates_ND* for the name of the network dataset. Opt not to model turns. Click on the Connectivity button in the next dialog. The Connectivity dialog shows interstates as the source, end point for connectivity, and 1 connectivity group. Click OK to exit the Connectivity dialog. In the New Network Dataset dialog, opt not to model the elevation of your network features. The next window shows Meters and Minutes as attributes for the network dataset. Click Next. Select yes to establish driving directions settings, and click the Directions button. The Network Directions Properties dialog shows that the display length units will be in miles and the length attribute is in meters. NAME in *interstates.shp* will be the street (interstate in this case) name field. You can click the cell (Type) below Suffix T … and choose None. Click OK to exit the Network Directions Properties dialog, and click Next in the New Network Dataset dialog. A summary of the network dataset settings is displayed in the next window. Click Finish. Click Yes to build the network. Click No to add *interstates_ND.nd* to the map. Notice that *interstates_ND.nd,* a network dataset, and *interstates_ND_Junctions.shp,* a junction feature class, have been added to the Catalog tree.

4. Insert a data frame in ArcMap and rename it Task 3. Add *interstates_ND.nd* to Task 3. Click yes to add all feature classes that participate in *interstates_ND* to map. Add *uscities.shp* to Task 3. Choose Select By Attributes from the Selection menu. In the next dialog, make sure that *uscities* is the layer for selection and enter the following expression to select Helena, MT, and

Charlotte, NC: "City_Name" = 'Helena' Or "City-Name" = 'Charlotte'.

5. The Network Analyst toolbar should show *interstates_ND* in the Network Dataset box. Select New Route from the Network Analyst's dropdown menu. The Route analysis layer is added to the table of contents with its classes of Stops, Routes, and Barriers (Point, Line, and Polygon).

6. This step is to add Helena and Charlotte as stops for the shortest path analysis. Because the stops must be located on the network, you can use some help in locating them. Right-click *Route* in the table of contents and select Properties. On the Network Locations tab of the Layer Properties dialog, change the Search Tolerance to 1000 (meters). Click OK to exit the Layer Properties dialog. Zoom in on Helena, Montana. Click the Create Network Location tool on the Network Analyst toolbar and click a point on the interstate near Helena. The clicked point displays a symbol with 1. If the clicked point is not on the network, a question mark will show up next to the symbol. In that case, you can use the Select/Move Network Locations tool to move the point to be on the network. Repeat the same procedure to locate Charlotte on the network. The clicked point displays a symbol with 2. Click the Solve button on the Network Analyst toolbar to find the shortest path between the two stops.

7. The shortest route now appears in the map. Click the Directions Window on the Network Analyst toolbar. The Directions window shows the travel distance in miles, the travel time, and detailed driving directions of the shortest route from Helena to Charlotte.

Q5. What is the total travel distance in miles?

Q6. Approximately how many hours will it take to drive from Helena to Charlotte using the interstates?

Task 4   Build a Geodatabase Network Dataset

What you need: *moscowst.shp,* a line shapefile containing a street network in Moscow, Idaho; and *select_turns.dbf,* a dBASE file that lists selected turns in *moscowst.shp.*

moscowst.shp was compiled from the 2000 TIGER/Line files. *moscowst.shp* is projected onto a transverse Mercator coordinate system in meters. For Task 4, you will first examine the input data sets. Then build a personal geodatabase and a feature dataset. And then import *moscowst.shp* and *select_turns.dbf* as feature classes into the feature dataset. You will use the network dataset built in this task to run a closest facility analysis in Task 5.

1. Right-click *moscowst.shp* in the Catalog tree and select Item Description. *moscowst.shp* has the following attributes important for this task: MINUTES shows the travel time in minutes, ONEWAY identifies one-way streets as T, NAME shows the street name, and METERS shows the physical length in meters for each street segment.

Q7. How many one-way street segments (records) are in *moscowst.shp*?

2. Preview the table of *select_turns.dbf. select_turns.dbf* is a turn table originally created in ArcInfo Workstation. The table has the following attributes important for this task: ANGLE lists the turn angle, ARC1_ID shows the first arc for the turn, ARC2_ID shows the second arc for the turn, and MINUTES lists the turn impedance in minutes.

3. Now create a personal geodatabase. Right-click the Chapter 17 database in the Catalog tree, point to New, and select Personal Geodatabase. Rename the geodatabase *Network.mdb.*

4. Create a feature dataset. Right-click *Network .mdb,* point to New, and select Feature Dataset. In the next dialog, enter *MoscowNet* for the name. Then click Projected

Coordinate Systems and Import to import the coordinate system of *moscowst.shp* to be *MoscowNet*'s coordinate system. Select None for vertical coordinates. Take the default values for the tolerances. Then click Finish.

5. This step imports *moscowst.shp* to *MoscowNet*. Right-click *MoscowNet*, point to Import, and select Feature Class (single). In the next dialog, select *moscowst.shp* for the input features, check that the output location is *MoscowNet*, enter *MoscowSt* for the output feature class name, and click OK.

6. To add *select_turns.dbf* to *MoscowNet*, you need to use ArcToolbox. Double-click the Turn Table to Turn Feature Class tool in the Network Analyst Tools/Turn Feature Class toolset to open its dialog. Specify *select_turns.dbf* for the input turn table, specify *MoscowSt* in the *MoscowNet* feature dataset for the reference line features, enter *Select_Turns* for the output turn feature class name, and click OK.

7. Click the plus sign next to *MoscowNet* in the Catalog tree. *MoscowSt* and *Select_Turns* should be in the dataset.

8. With the input data ready, you can now build a network dataset. In the Catalog tree of ArcMap, right-click *MoscowNet*, point to New, and select Network Dataset. Do the following in the next four windows: take the default name (*MoscowNet_ND*) for the network dataset, select *MoscowSt* to participate in the network dataset, click yes to model turns and check the box next to *Select_Turns*, take the default connectivity settings, check None to model the elevation of your network features, make sure that Minutes and Oneway are the default attributes for the network dataset, and opt to establish driving directions. After reviewing the summary, click Finish. Click Yes to build the network. Click No

to add *MoscowNet_ND* to the map. Notice that *MoscowNet_ND*, a network dataset, and *MoscowNet_ND_ Junctions*, a junction feature class, have been added to the Catalog tree.

Task 5 Find Closest Facility

What you need: *MoscowNet*, a network dataset from Task 4; and *firestat.shp*, a point shapefile with two fire stations in Moscow, Idaho.

1. Insert a new data frame and rename it Task 5. Add the *MoscowNet* feature dataset and *firestat.shp* to Task 5. Turn off the *MoscowNet_ND_ Junctions* layer so that the map does not look too cluttered.

2. Make sure that the Network Analyst toolbar is available and *MoscowNet_ND* is the network dataset. Select New Closest Facility from the Network Analyst dropdown menu. The Closest Facility analysis layer is added to the table of contents with its analysis classes of Facilities, Incidents, Routes, and Barriers (Point, Line, and Polygon). Make sure that Closest Facility is checked to be visible.

3. Click Show/Hide Network Analyst Window on the Network Analyst toolbar to open the window. Right-click Facilities (0) in the Network Analyst window, and select Load Locations. In the next dialog, make sure that the locations will be loaded from *firestat*, before clicking OK.

4. Click the Closest Facility Properties button in the upper right of the Network Analyst window. On the Analysis Settings tab, opt to find 1 facility and to travel from Facility to Incident. Uncheck the box for Oneway in the Restrictions window. Click OK to dismiss the dialog. Click Incidents (0) to highlight it in the Network Analyst window. Then use the Create Network Location tool on the Network Analyst toolbar to

click an incident point of your choice on the network. Click the Solve button. The map should show the route connecting the closest facility to the incident. Click the Directions Window button on the Network Analyst toolbar. The window lists the route's distance and travel time and details the driving directions.

Q8. Suppose an incident occurs at the intersection of Orchard and F. How long will the ambulance from the closest fire station take to reach the incident?

Task 6   Find Service Area

What you need: *MoscowNet* and *firestat.shp,* same as Task 5.

1. Insert a new data frame and rename it Task 6. Add the *MoscowNet* feature dataset and *firestat.shp* to Task 6. Turn off the *MoscowNet_ND_Junctions* layer. Select New Service Area from the Network Analyst's dropdown menu. Click Show/Hide Network Analyst Window to open the window. The Network Analyst window opens with four empty lists of Facilities, Polygons, Lines, and Barriers (Point, Line, and Polygon).

2. Next add the fire stations as facilities. Right-click Facilities (0) in the Network Analyst window and select Load Locations. In the next dialog, make sure that the facilities are loaded from *firestat* and click OK. Location 1 and Location 2 should now be added as facilities in the Network Analyst window.

3. This step sets up the parameters for the service area analysis. Click the Service Area Properties button in the upper right of the Network Analyst window to open the dialog box. On the Analysis Settings tab, select Minutes for the impedance, enter "2 5" for default breaks of 2 and 5 minutes, check direction to be away from Facility, and uncheck Oneway restrictions. On the

Polygon Generation tab, check the box to generate polygons, opt for generalized polygon type and trim polygons, select not overlapping for multiple facilities options, and choose rings for the overlay type. Click OK to dismiss the Layer Properties dialog.

4. Click the Solve button on the Network Analyst toolbar to calculate the fire station service areas. The service area polygons now appear in the map as well as in the Network Analyst window under Polygon (4). Expand Polygons (4). Each fire station is associated with two service areas, one for 2 minutes and the other for 5 minutes. To see the boundary of a service area (e.g., 2 to 5 minutes from Location 1), you can simply click the service area under Polygon (4)

5. This step shows how to save a service area as a feature class. First select the service area (polygon) in the Network Analyst window. Then right-click the Polygon (4) layer in the window, and select Export Data. Save the data as a feature class in *MoscowNet.* The feature class attribute table contains the default fields of area and length.

Q9. What is the size of the 2-minute service area of Location 1 (fire station 1)?

Q10. What is the size of the 2-minute service area of Location 2 (fire station 2)?

Challenge Task

What you need: *uscities.shp* and *interstates.shp,* same as Task 3.

This challenge task asks you to find the shortest route by travel time from Grand Forks, North Dakota, to Houston, Texas.

Q1. What is the total travel distance in miles?

Q2. Approximately how many hours will it take to drive from Grand Forks to Houston using the interstates?

REFERENCES

Apparicio, P., M. Cloutier, and R. Shearmur. 2007. The Case of Montréal's Missing Food Deserts: Evaluation of Accessibility to Food Supermarkets. *International Journal of Health Geographics* 6:4 doi: 10.1186/1476-072X-6-4.

Atkinson, D. M., P. Deadman, D. Dudycha, and S. Traynor. 2005. Multi-Criteria Evaluation and Least Cost Path Analysis an Arctic All-Weather Road. *Applied Geography* 25: 287–307.

Bräysy, O., W. Dullaert, and P. Nakari. 2009. The Potential of Optimization in Communal Routing Problems: Case Studies from Finland. *Journal of Transport Geography* 17: 484–90.

Choi, Y., H. Park, C. Sunwoo, and K. C. Clarke. 2009. Multi-Criteria Evaluation and Least Cost Path Analysis for Optimal Haulage Routing of Dump Trucks in Large Scale Open-Pit Mines. *International Journal of Geographical Information Science* 23: 1541–1567.

Church, R. L. 1999. Location Modelling and GIS. In P. A. Longley, M. F. Goodchild, D. J. Maguire, and D. W. Rhind, eds., *Geographical Information Systems,* 2d ed., pp. 293–303. New York: Wiley.

Church, R. L., and C. S. ReVelle. 1974. The Maximal Covering Location Problem. *Papers of the Regional Science Association* 32: 101–18.

Collischonn, W., and J. V. Pilar. 2000. A Direction Dependent Least Costs Path Algorithm for Roads and Canals. *International Journal of Geographical Information Science* 14: 397–406.

Dijkstra, E. W. 1959. A Note on Two Problems in Connexion with Graphs. *Numerische Mathematik* 1: 269–71.

Falcucci, A., L. Maiorano, P. Ciucci, E. O. Garton, and L. Boitani. 2008. Land-Cover Change and the Future of the Apennine Brown Bear: A Perspective from the Past. *Journal of Mammalogy* 89: 1502–1511.

Farhan, B., and A. T. Murray. 2005. A GIS-Based Approach for Delineating Market Areas for Park and Ride Facilities. *Transactions in GIS* 9: 91–108.

Ghosh, A., S. McLafferty, and C. S. Craig. 1995. Multifacility Retail Networks. In Z. Drezner, ed., *Facility Location: A Survey of Applications and Methods,* pp. 301–30. New York: Springer.

Hakimi, S. L. 1964. Optimum Locations of Switching Centers and the Absolute Centers and Medians of a Graph. *Operations Research* 12: 450–59.

Hepner, G. H., and M. V. Finco. 1995. Modeling Dense Gas Contaminant Pathways over Complex Terrain Using a Geographic Information System. *Journal of Hazardous Materials* 42: 187–99.

Keenan, P. 2008. Modelling Vehicle Routing in GIS. 2008. *Operational Research* 8: 201–18.

Kindall, J. L., and F. T. van Manen. 2007. Identifying Habitat Linkages for American Black Bears in North Carolina, USA. *Journal of Wildlife Management* 71: 487–95.

Krizek, K. J., A. El-Geneidy, and K. Thompson. 2007. A Detailed Analysis of How an Urban Trail System Affects Cyclists' Travel. *Transportation* 34: 611–24.

Lin, S. 1965. Computer Solutions of the Travelling Salesman Problem. *Bell System Technical Journal* 44: 2245–69.

Lowe, J. C., and S. Moryadas. 1975. *The Geography of Movement.* Boston: Houghton Mifflin.

Malczewski, J. 2006. GIS-Based Multicriteria Decision Analysis: A Survey of the Literature. *International Journal of Geographical Information Science* 20: 703–26.

Miller, H. J., and S. Shaw. 2001. *Geographic Information Systems for Transportation: Principles and Applications.* New York: Oxford University Press.

Rasdorf, W., H. Cai, C. Tilley, S. Brun, and F. Robson. 2004. Accuracy Assessment of Interstate Highway Length Using Digital Elevation Model. *Journal of Surveying Engineering* 130: 142–50.

Rees, W. G. 2004. Least-Cost Paths in Mountainous Terrain. *Computers & Geosciences* 30: 203–9.

Schuurman, N., R. S. Fiedler, S. C. W. Grzybowski, and D. Grund. 2006. Defining Rational Hospital Catchments for Non-Urban Areas based on Travel-Time. *International Journal of Health Geographics*

5:43 doi:10.1186/1476-
072X-5-43.

Schuurman, N., M. Leight,
and M. Berube. 2008. A
Web-Based Graphical User
Interface for Evidence-Based
Decision Making for Health
Care Allocations in Rural
Areas. *International Journal
of Health Geographics* 7:49
doi:10.1186/1476-072X-7-49.

Snyder, S. A., J. H. Whitmore,
I. E. Schneider, and
D. R. Becker. 2008. Ecological
Criteria, Participant Preferences
and Location Models: A GIS
Approach toward ATV Trail
Planning. *Applied Geography*
28: 248–58.

Tarabanis, K., and I. Tsionas. 1999.
Using Network Analysis for

Emergency Planning in Case of
an Earthquake. *Transactions in
GIS* 3: 187–97.

Weigel, D., and B. Cao. 1999.
Applying GIS and OR
Techniques to Solve Sears
Technician-Dispatching and
Home-Delivery Problems.
Interfaces 29: 112–30.

Witten, K., D. Exeter, and
A. Field. 2003. The Quality
of Urban Environments:
Mapping Variation in Access to
Community Resources. *Urban
Studies* 40: 161–77.

Yu, C., J. Lee, and
M. J. Munro-Stasiuk. 2003.
Extensions to Least Cost
Path Algorithms for
Roadway Planning.
International Journal of

*Geographical Information
Science* 17: 361–76.

Zeng, W., and R. L. Church. 2009.
Finding Shortest Paths on Real
Road Networks: the Case For
A*. *International Journal of
Geographical Information
Science* 23: 531–43.

Zhan, F. B., and C. E. Noon. 1998.
Shortest Path Algorithms: An
Evaluation using Real Road
Networks. *Transportation
Science* 32: 65–73.

Ziliaskopoulos, A. K., and
H. S. Mahmassani. 1996.
A Note on Least Time Path
Computation Considering
Delays and Prohibitions for
Intersection Movements.
Transportation Research B
30: 359–67.

GIS MODELS AND MODELING

CHAPTER OUTLINE

18.1 Basic Elements of GIS Modeling

18.2 Binary Models

18.3 Index Models

18.4 Regression Models

18.5 Process Models

Previous chapters have presented tools for exploring, manipulating, and analyzing vector data and raster data. One of many uses of these tools is to build models. What is a model? A **model** is a simplified representation of a phenomenon or a system. Several types of models have already been covered in this book. A map is a model. So are the vector and raster data models for representing spatial features and the relational model for representing a database system. A model helps us better understand a phenomenon or a system by retaining the significant features and relationships of reality.

Chapter 18 discusses use of GIS for building models. Two points must be clarified at the start. First, Chapter 18 deals with models using geospatial data. Some researchers have used the term "spatially explicit models" to describe these models. Second, the emphasis is on the use of GIS in modeling rather than the models. Although Chapter 18 covers a number of models, the intent is simply to use them as examples. A basic requirement in modeling is the modeler's interest and knowledge of the system to be modeled (Hardisty, Taylor, and Metcalfe 1993). This is why many models are discipline specific. For example, models of the environment usually consist of atmospheric, hydrologic, land surface/subsurface, biogeochemical, and ecological models. It would be impossible for an introductory GIS book to discuss each of these environmental models, not to mention models from other disciplines.

Chapter 18 is divided into the following five sections. Section 18.1 discusses the basic elements of GIS modeling. Section 18.2 and 18.3 cover binary models and index models, respectively.

Section 18.4 deals with regression models, both linear and logistic. Section 18.5 introduces process models including soil erosion and other environmental models. Although these four types of models—binary, index, regression, and process—differ in degree of complexity, they share two common elements: a set of selected spatial variables, and the functional or mathematical relationship between the variables.

18.1 Basic Elements of GIS Modeling

Before building a GIS model, we must have a basic understanding of the type of model, the modeling process, and the role of GIS in the modeling process.

18.1.1 Classification of GIS Models

It is difficult to classify many models used by GIS users. DeMers (2002), for example, classifies models by purpose, methodology, and logic. But the boundary between the classification criteria is not always clear. Rather than proposing an exhaustive classification, some broad categories of models by purpose are covered here as an introduction to the models to be discussed later.

A model may be **descriptive** or **prescriptive.** A descriptive model describes the existing conditions of spatial data, and a prescriptive model offers a prediction of what the conditions could be or should be. If we use maps as analogies, a vegetation map would represent a descriptive model and a potential natural vegetation map would represent a prescriptive model. The vegetation map shows existing vegetation, whereas the potential natural vegetation map predicts the vegetation that could occupy a site without disturbance or climate change.

A model may be **deterministic** or **stochastic.** Both deterministic and stochastic models are mathematical models represented by equations with parameters and variables. A stochastic model considers the presence of some randomness in one or more of its parameters or variables, but a deterministic model does not. As a result of random

processes, the predictions of a stochastic model can have measures of errors or uncertainties, typically expressed in probabilistic terms. This is why a stochastic model is also called a *probabilistic* or *statistical* model. Among the local interpolation methods covered in Chapter 15, for instance, only kriging represents a stochastic model. Besides producing a prediction map, a kriging interpolator can also generate a standard error for each predicted value.

A model may be **dynamic** or **static.** A dynamic model emphasizes the changes of spatial data over time and the interactions between variables, whereas a static model deals with the state of geospatial data at a given time. Simulation is a technique that can generate different states of geospatial data over time. Many environmental models such as groundwater pollution and soil water distribution are best studied as dynamic models (Rogowski and Goyne 2002).

A model may be **deductive** or **inductive.** A deductive model represents the conclusion derived from a set of premises. These premises are often based on scientific theories or physical laws. An inductive model represents the conclusion derived from empirical data and observations. To assess the potential for a landslide, for example, one can use a deductive model based on laws in physics or use an inductive model based on recorded data from past landslides (Brimicombe 2003).

18.1.2 The Modeling Process

The development of a model follows a series of steps. The first step is to define the goals of the model. This is analogous to defining a research problem. What is the phenomenon to be modeled? Why is the model necessary? What spatial and temporal scales are appropriate for the model? The modeler can organize the essential structure of a model by using a sketch or a diagram.

The second step is to break down the model into elements and to define the properties of each element and the interactions between the elements in the form of a conceptual diagram (e.g., a flowchart). This is followed by a mathematical model in which the modeler gathers mathematical equations of the model and commands in a GIS (or the computer code) to carry out the computation.

The third step is the implementation and calibration of the model. The modeler needs data to run and calibrate the model. Model calibration is an iterative process, a process that repeatedly compares the output from the model to the observed data, adjusts the parameters, and reruns the model. Uncertainties in model prediction are a major problem in calibrating a deterministic model. Sensitivity analysis is a technique that can quantify these uncertainties by measuring the effects of input changes on the output (Lindsay 2006).

A calibrated model is a tool ready for prediction. But the model must be validated before it can be generally accepted. Model validation assesses the model's ability to predict under conditions that are different from those used in the calibration phase. A model that has not been validated is likely to be ignored by other researchers (Brooks 1997). Model validation requires a different set of data from those used for developing the model (Mulligan and Wainwright 2004). The modeler can split observed data into two subsets: one subset for developing the model and the other subset for model validation (e.g., Chang and Li 2000). But in many cases the required additional data set presents a problem and forces the modeler to, unwisely, forgo the validation step.

18.1.3 The Role of GIS in Modeling

GIS can assist the modeling process in several ways. First, a GIS is a tool that can process, display, and integrate different data sources including maps, digital elevation models (DEMs), GPS (global positioning system) data, images, and tables. These data are needed for the implementation, calibration, and validation of a model. A GIS can function as a database management tool and, at the same time, is useful for modeling-related tasks such as exploratory data analysis and data visualization.

Second, models built with a GIS can be vector-based or raster-based. The choice depends on the nature of the model, data sources, and the computing algorithm. A raster-based model is preferred if the spatial phenomenon to be modeled varies continuously over the space such as soil erosion and snow accumulation. A raster-based model is also preferred if satellite images and DEMs constitute a major portion of the input data, or if the modeling involves intense and complex computations. But raster-based models are not recommended for studies of travel demand, for example, because travel demand modeling requires the use of a topology-based road network (Chang, Chiang, and Hsu 2002). Vector-based models are generally recommended for spatial phenomena that involve well-defined locations and shapes.

Third, the distinction between raster-based and vector-based models does not preclude modelers from integrating both types of data in the modeling process. Algorithms for conversion between vector and raster data are easily available in GIS packages. The decision about which data format to use in analysis should be based on the efficiency and the expected result, rather than the format of the original data. For instance, if a vector-based model requires a precipitation layer (e.g., an isohyet layer) as the input, it would be easier to interpolate a precipitation raster from known points and then convert the raster to the precipitation layer.

Fourth, the process of modeling may take place in a GIS or require the linking of a GIS to other computer programs. Many GIS packages including ArcGIS, GRASS, IDRISI, ILWIS, MFworks, and PCRaster have extensive analytical functions for modeling. But a GIS package cannot accommodate statistical analysis as well as a statistical analysis package can. In those cases, the modeler would want to link a GIS to a statistical analysis package. Among the four types of GIS models to be discussed later, regression and process models usually require the coupling of a GIS with other programs. Binary and index models, on the other hand, can be built entirely in a GIS.

18.1.4 Integration of GIS and Other Modeling Programs

There are three scenarios for linking a GIS to other computer programs (Corwin, Vaughan, and Loague 1997; Brimicombe 2003). Modelers may encounter all three scenarios in the modeling process, depending on the tasks to be accomplished.

A **loose coupling** involves transfer of data files between the GIS and other programs. For example, one can export data to be run in a statistical analysis package from the GIS and import results from statistical analysis back to the GIS for data visualization or display. Under this scenario, the modeler must create and manipulate data files to be exported or imported unless the interface has already been established between the GIS and the target program. A **tight coupling** gives the GIS and other programs a common user interface. For instance, the GIS can have a menu selection to run a soil erosion program. An **embedded system** bundles the GIS and other programs with shared memory and a common interface. The Geostatistical Analyst extension to ArcGIS is an example of having geostatistical functions embedded into a GIS environment.

18.2 Binary Models

A **binary model** uses logical expressions to select target areas from a composite feature layer or multiple rasters. The output of a binary model is in binary format: 1 (true) for areas that meet the selection criteria and 0 (false) for areas that do not. We may consider a binary model an extension of data query (Chapter 10).

The choice of selection criteria is probably the most important step in building a binary model. This step is usually accomplished by conducting a thorough literature survey. And, if existing data for the phenomenon to be modeled are available, they can be used as references. Existing or historical data are also useful for model calibration and validation.

18.2.1 Vector-Based Method

To build a vector-based binary model, we can gather the input layers, overlay them, and perform data query from the composite feature layer (Figure 18.1). Suppose a county government wants to select potential industrial sites that meet the following criteria: at least 5 acres in size, commercial zones, not subject to flooding, not more than 1 mile from a heavy-duty road, and less than

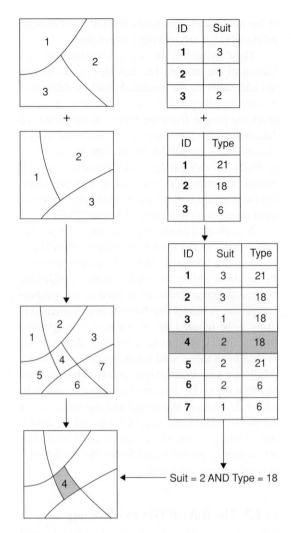

Figure 18.1
To build a vector-based binary model, first overlay the layers so that their geometries and attributes (Suit and Type) are combined. Then, use the query statement, Suit = 2 AND Type = 18, to select Polygon 4 and save it to the output layer.

10 percent slope. Operationally, the task involves the following five steps:

1. Gather all layers (land use, flood potential, road, and slope) relevant to the selection criteria.
   A DEM can be used to derive a slope raster, which can then be converted to a slope layer.

2. Select heavy-duty roads from the road layer, and create a 1-mile buffer zone around them.
3. Intersect the road buffer zone layer and other layers. Intersect, instead of other overlay operations, can limit the output to areas within 1 mile of heavy-duty roads.
4. Query the composite feature layer for potential industrial sites.
5. Select sites, which are equal to or larger than 5 acres.

18.2.2 Raster-Based Method

The raster-based method requires the input rasters, with each raster representing a criterion. A local operation with multiple rasters (Chapter 12) can then be used to derive the raster-based model from the input rasters (Figure 18.2).

To solve the same problem as in Section 18.2.1, the raster-based method proceeds by:

1. Derive a slope raster from a DEM, and a distance to heavy-duty road raster.
2. Convert the land use and flood potential layers into rasters with the same resolution as the slope raster.
3. Use a local operation with multiple rasters to query for potential industrial sites.
4. Use a zonal operation to select sites, which are equal to or larger than 5 acres. (The size of a site can be computed by multiplying its number of cells by the cell resolution.)

18.2.3 Applications of Binary Models

Siting analysis is probably the most common application of the binary model. A siting analysis determines if a unit area (i.e., a polygon or a cell) meets a set of selection criteria for locating a landfill, a ski resort, or a university campus. There are at least two approaches to conducting a siting analysis. One evaluates a set of nominated or preselected sites, and the other evaluates all potential sites. Although the two approaches may use different sets of selection criteria (e.g., more stringent

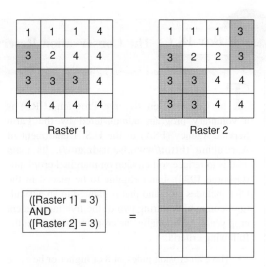

Figure 18.2
To build a raster-based binary model, use the query statement, [Raster 1] = 3 AND [Raster 2] = 3, to select three cells (shaded) and save them to the output raster.

criteria for evaluating preselected sites), they follow the same procedure for evaluation.

Another consideration in siting analysis is the threshold values for selection. Well-defined or "crisp" threshold values are used in the example in Section 18.2.1 to remove land from consideration: the road buffer is exactly 1 mile and the minimum parcel size is exactly 5 acres. These threshold values automatically exclude parcels that are more than 1 mile from heavy-duty roads or are smaller than 5 acres. Government programs such as the Conservation Reserve Program (Box 18.1) are often characterized by their detailed and explicit guidelines. Crisp threshold values simplify the process of siting analysis. But they can also become too restrictive or arbitrary in real-world applications. In an example cited by Steiner (1983), local residents questioned whether any land could meet the criteria of a county comprehensive plan on rural housing. To mitigate the suspicion, a study was made to show that there were indeed sites available. An alternative to crisp threshold values is to use the fuzzy set concept, which is covered in Section 18.3.2.

Box 18.1   The Conservation Reserve Program

The Conservation Reserve Program (CRP) is a voluntary program administered by the Farm Service Agency (FSA) of the U.S. Department of Agriculture **(http://www.fsa.usda.gov/).** Its main goal is to reduce soil erosion on marginal croplands (Osborne 1993). Land eligible to be placed in the CRP includes cropland that is planted to an agricultural commodity during two of the five most recent crop years. Additionally, the cropland must meet the following criteria:

- Have an erosion index of 8 or higher or be considered highly erodible land
- Be considered a cropped wetland

- Be devoted to any of a number of highly beneficial environmental practices, such as filter strips, riparian buffers, grass waterways, shelter belts, wellhead protection areas, and other similar practices
- Be subject to scour erosion
- Be located in a national or state CRP conservation priority area, or be cropland associated with or surrounding noncropped wetlands

The difficult part of implementing the CRP in a GIS is putting together the necessary map layers, unless they are already available in a statewide database (Wu et al. 2002).

Example 1 Silberman and Rees (2010) build a GIS model for identifying possible ski towns in the U.S. Rocky Mountains. They examine the characteristics of existing ski areas, before selecting location criteria that include annual snowfall, potential ski season, distance to national forests, and accessibility index. The accessibility index is defined as a combined measure of travel time and distance to settlement of 10,000, city of 50,000, and available airport. This set of selection criteria is then applied to all populated settlements in the Rocky Mountain region and to evaluate each settlement as a potential site for new ski resort development.

Example 2 Isaac et al. (2008) use a procedure similar to construction of a binary model for their predictive mapping of powerful owl breeding sites in urban Melbourne, Australia. To develop the selection criteria, they first buffer existing breeding sites with a distance of 1 kilometer and then intersect these buffer zones with data layers of tree density, hydrology, vegetation classes, land use zone, and slope. After analyzing ecological attributes within the buffer zones, they select distance to water (40 meters) and tree cover density (dense vegetation) as criteria for locating potential breeding sites of powerful owls.

18.3 INDEX MODELS

An **index model** calculates the index value for each unit area and produces a ranked map based on the index values. An index model is similar to a binary model in that both involve multicriteria evaluation (Malczewski, 2006) and both depend on overlay operations for data processing. But an index model produces for each unit area an index value rather than a simple yes or no.

18.3.1 The Weighted Linear Combination Method

The primary consideration in developing an index model, either vector- or raster-based, is the method for computing the index value. **Weighted linear combination** is a common method for computing the index value (Saaty 1980; Banai-Kashani 1989; Malczewski 2000). Following the analytic hierarchy process proposed by Saaty (1980), weighted linear combination involves evaluation at three levels (Figure 18.3).

First, the relative importance of each criterion, or factor, is evaluated against other criteria. Many studies have used expert-derived paired

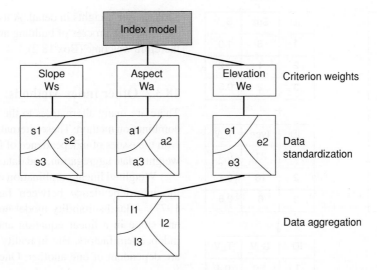

Figure 18.3

To build an index model with the selection criteria of slope, aspect, and elevation, the weighted linear combination method involves evaluation at three levels. First, determine the criterion weights (e.g., Ws for slope). Second, decide on the standardized values for each criterion (e.g., s1, s2, and s3 for slope). Third, compute the index (aggregate) value for each unit area.

comparison for evaluating criteria (Saaty 1980; Banai-Kashani 1989; Pereira and Duckstein 1993; Jiang and Eastman 2000). This method involves performing ratio estimates for each pair of criteria. For instance, if criterion A is considered to be three times more important than criterion B, then 3 is recorded for A/B and 1/3 is recorded for B/A. Using a criterion matrix of ratio estimates and their reciprocals as the input, the paired comparison method derives a weight for each criterion. The criterion weights are expressed in percentages, with the total equaling 100 percent or 1.0. Paired comparison is available in commercial software packages (e.g., Expert Choice).

Second, data for each criterion are standardized. A common method for data standardization is linear transformation. For example, the following formula can convert interval or ratio data into a standardized scale of 0.0 to 1.0:

(18.1)

$$S_i = \frac{X_i - X_{\min}}{X_{\max} - X_{\min}}$$

where S_i is the standardized value for the original value X_i, $X_{\min}$ is the lowest original value, and $X_{\max}$ is the highest original value. We cannot use Eq. (18.1) if the original data are nominal or ordinal data. In those cases, a ranking procedure based on expertise and knowledge can convert the data into a standardized range such as 0 to 1, 1 to 5, or 0 to 100.

Third, the index value is calculated for each unit area by summing the weighted criterion values and dividing the sum by the total of the weights:

(18.2)

$$I = \frac{\sum_{i=1}^{n} w_i x_i}{\sum_{i=1}^{n} w_i}$$

where I is the index value, n is the number of criteria, w_i is the weight for criterion i, and x_i is the standardized value for criterion i.

Figure 18.4 shows the procedure for developing a vector-based index model, and Figure 18.5 shows a raster-based index model. As long as

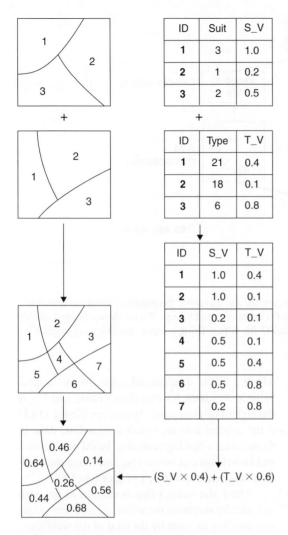

ID	Suit	S_V
1	3	1.0
2	1	0.2
3	2	0.5

+

ID	Type	T_V
1	21	0.4
2	18	0.1
3	6	0.8

ID	S_V	T_V
1	1.0	0.4
2	1.0	0.1
3	0.2	0.1
4	0.5	0.1
5	0.5	0.4
6	0.5	0.8
7	0.2	0.8

(S_V × 0.4) + (T_V × 0.6)

Figure 18.4

Building a vector-based index model requires several steps. First, standardize the Suit and Type values of the input layers into a scale of 0.0 to 1.0. Second, overlay the layers. Third, assign a weight of 0.4 to the layer with Suit and a weight of 0.6 to the layer with Type. Finally, calculate the index value for each polygon in the output by summing the weighted criterion values. For example, Polygon 4 has an index value of 0.26 ($0.5 \times 0.4 + 0.1 \times 0.6$).

criterion weighting and data standardization are well defined, it is not difficult to use the weighted linear combination method to build an index model in a GIS. But we must document standardized values

and criterion weights in detail. A user interface that simplifies the process of building an index model is therefore welcome (Box 18.2).

18.3.2 Other Index Methods

There are many alternatives to the weighted linear combination method. These alternatives mainly deal with the issues of independence of factors, criterion weights, data aggregation, and data standardization.

Weighted linear combination cannot deal with the interdependence between factors (Hopkins 1977). A land suitability model may include soils and slope in a linear equation and treat them as independent factors. But in reality soils and slope are dependent of one another. One solution to the interdependence problem is to use a nonlinear function and express the relationship between factors mathematically. But a nonlinear function is usually limited to two factors rather than multiple factors as required in an index model. Another solution is the rule of combination method proposed by Hopkins (1977). Using rules of combination, we would assign suitability values to sets of combinations of environmental factors and express them through verbal logic instead of numeric terms. The rule of combination method has been widely adopted in land suitability studies (e.g., Steiner 1983), but the method can become unwieldy given a large variety of criteria and data types (Pereira and Duckstein 1993).

Paired comparison for determining criterion weights is sometimes called direct assessment. An alternative to direct assessment is trade-off weighting (Hobbs and Meier 1994; Xiang 2001). Trade-off weighting determines the criterion weights by asking participants to state how much of one criterion they are willing to give up to obtain a given improvement in another criterion. In other words, trade-off weighting is based on the degree of compromise one is willing to make between two criteria when an ideal combination of the two criteria is not attainable. Although realistic in some real-world applications, trade-off weighting has shown to be more difficult to understand and use than direct assessment (Hobbs and Meier 1994).

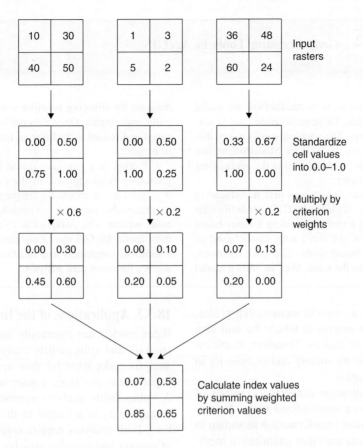

Input rasters

Standardize cell values into 0.0–1.0

Multiply by criterion weights

Calculate index values by summing weighted criterion values

Figure 18.5

Building a raster-based index model requires the following steps. First, standardize the cell values of each input raster into a scale of 0.0 to 1.0. Second, multiply each input raster by its criterion weight. Finally, calculate the index values in the output raster by summing the weighted criterion values. For example, the index value of 0.85 is calculated as follows: 0.45 + 0.20 + 0.20.

Data aggregation refers to the derivation of the index value. Weighted linear combination calculates the index value by summing the weighted criterion values. One alternative is to skip the computation entirely and assign the lowest value, the highest value, or the most frequent value among the criteria to the index value (Chrisman 2001). Another alternative is the ordered weighted averaging (OWA) operator, which uses ordered weights instead of criterion weights in computing the index value (Yager 1988; Jiang and Eastman 2000). Suppose that a set of weights is {0.6, 0.4}. If the ordered position of the criteria at location 1

is {A, B}, then criterion A has the weight of 0.6 and criterion B has the weight of 0.4. If the ordered position of the criteria at location 2 is {B, A}, then B has the weight of 0.6 and A has the weight of 0.4. Ordered weighted averaging is therefore more flexible than weighted linear combination. Flexibility in data aggregation can also be achieved by using fuzzy sets (Hall, Wang, and Subaryono 1992; Baja, Chapman, and Dragovich 2002; Hall and Arnberg 2002; Stoms, McDonald, and Davis 2002; Braimoh, Vlek, and Stein 2004). Fuzzy sets do not use sharp boundaries. Rather than being placed in a class (e.g., outside the road buffer), a unit area

Box 18.2 | **Geoprocessing Tools in ArcGIS**

Geoprocessing tools in ArcToolbox are useful for building models. These tools can be used in several different ways. The simplest way is to use dialogs. The other two methods are ModelBuilder and Python scripts, the methods used in the applications section of this chapter.

ModelBuilder, which was first introduced in ArcView 3.2, is back in ArcGIS. ModelBuilder is no longer just a tool for building a raster-based index model but can work with raster-based as well as vector-based tools. The user interface, however, remains the same. We can build a model diagram by stringing together a series of inputs, tools, and outputs. Once a model is built, it can be saved and reused with different model parameters or inputs.

Python is a general-purpose high-level programming language and, in the case of ArcGIS, it is used as an extension language to provide a programmable interface for modules, or blocks of code, written with ArcObjects. Python scripts are most useful for GIS operations that string together a series of commands and functions such as map algebra for raster data analysis.

is associated with a group of membership grades, which suggest the extents to which the unit area belongs to different classes. Therefore, fuzziness is a way to handle uncertainty and complexity in multicriteria evaluation.

Data standardization converts the values of each criterion into a standardized scale. A common method is linear transformation as shown in Eq. (18.1), but there are other quantitative methods. Habitat suitability studies may use expert-derived value functions, often nonlinear, for data standardization of each criterion (Pereira and Duckstein 1993; Lauver, Busby, and Whistler 2002). A fuzzy set membership function can also be used to transform data into fuzzy measures (Jiang and Eastman 2000).

Criterion weights, data aggregation, and data standardization are the same issues that a spatial decision support system (SDSS) must deal with. Designed to work with spatial data, an SDSS can assist the decision maker in making a choice from a set of alternatives according to given evaluation criteria (Densham 1991; Jankowski 1995; Malczewski 2006). Although its emphasis is on decision making, an SDSS shares the same methodology as an index model. An index model developer can therefore benefit from becoming familiar with the SDSS literature.

18.3.3 Applications of the Index Model

Index models are commonly used for suitability analysis and vulnerability analysis. A suitability analysis ranks areas for their appropriateness for a particular use (e.g., conservation protection). A vulnerability analysis assesses areas for their susceptibility to a hazard or disaster (e.g., forest fire). Both analyses require careful consideration of criteria and criterion weights. As an example, Rohde et al. (2006) choose a hierarchical filter process for their study of floodplain restoration. The process has three filters: filter 1 defines minimum pre-requisites, filter 2 evaluates ecological restoration suitability, and filter 3 introduces socioeconomic factors. For filter 2, Rohde et al. (2006) run a sensitivity analysis with four different weighting schemes to evaluate the relative influence of criteria weights on the ecological restoration suitability index. Development of an index model is therefore more involved than a binary model.

Example 1 In 1981 the Natural Resources Conservation Service (NRCS), known then as the Soil Conservation Service, proposed the Land Evaluation and Site Assessment (LESA) system, a tool intended to be used by state and local planners in determining the conditions that justify conversion of agricultural land to other uses (Wright et al. 1983).

LESA consists of two sets of factors: LE measures the inherent soil-based qualities of land for agricultural use, and SA measures demands for nonagricultural uses. Many state and local governments have developed, or are in the process of developing, their own LESA models. As an example, the California LESA model completed in 1997 (**http://www.consrv.ca.gov/DLRP/qh_lesa.htm**) uses the following factors and factor weights (in parentheses):

1. Land evaluation factors
   - Land capability classification (25 percent)
   - Storie index rating (25 percent)
2. Site assessment factors
   - Project size (15 percent)
   - Water resource availability (15 percent)
   - Surrounding agricultural lands (15 percent)
   - Surrounding protected resource lands (5 percent)

For a given location, one would first rate (standardize) each of the factors on a 100-point scale and then sum the weighted factor scores to derive a single index value.

Example 2   The U.S. Environmental Protection Agency developed the DRASTIC model for evaluating groundwater pollution potential (Aller et al. 1987). The acronym DRASTIC stands for the seven factors used in weighted linear combination: Depth to water, net Recharge, Aquifer media, Soil media, Topography, Impact of the vadose zone, and hydraulic Conductivity. The use of DRASTIC involves rating each parameter, multiplying the rating by a weight, and summing the total score by:

(18.3)

$$\text{Total Score} = \sum_{i=1}^{7} W_i P_i$$

where P_i is factor (input parameter) i and W_i is the weight applied to P_i. A critical review of the DRASTIC model in terms of the selection of factors and the interpretation of numeric scores and weights is available in Merchant (1994).

Example 3   Habitat Suitability Index (HSI) models typically evaluate habitat quality by using weighted linear combination and factors considered to be important to the wildlife species (Brooks 1997; Felix et al. 2004). An HSI model for pine marten developed by Kliskey et al. (1999) is as follows:

(18.4)

$$\text{HSI} = \text{sqrt} \left([(3SR_{BSZ} + SR_{SC} + SR_{DS})/6] \right.$$
$$\left. [(SR_{CC} + SR_{SS})/2] \right)$$

where SR_{BSZ}, SR_{SC}, SR_{DS}, SR_{CC}, and SR_{SS} are the ratings for biogeoclimatic zone, site class, dominant species, canopy closure, and seral stage, respectively. The model is scaled so that the HSI values range from 0 for unsuitable habitat to 1 for optimal habitat.

Example 4   Wildfire hazard and risk models are usually index models based on factors such as vegetation species (fuels), slope, aspect, and proximity to roads (Chuvieco and Congalton 1989, Chuvieco and Salas 1996). Lein and Stump (2009) use the weighted linear combination method to construct the following wildfire risk model in southeastern Ohio:

(18.5)

$$\text{Potential fire risk} = \text{fuel} + (2 \times \text{solar}) + \text{TWI}$$
$$+ \text{distance from roads} + \text{population density}$$

where fuel is based on vegetation species and canopy cover, solar is annual solar radiation calculated from a DEM and a slope raster, and TWI is topographic wetness index also calculated from a DEM and a slope raster. (TWI is defined as $\ln(a/\tan\beta)$, where a is local upslope contributing area and β is local slope.)

18.4 REGRESSION MODELS

A **regression model** relates a dependent variable to a number of independent (explanatory) variables in an equation, which can then be used for prediction or estimation (Rogerson 2001). Like an index model, a regression model can use overlay operations in a GIS to combine variables needed for the analysis. There are three types of regression model: linear regression, local regression, and logistic regression.

18.4.1 Linear Regression Models

A multiple linear regression model is defined by:

(18.6)

$$y = a + b_1 x_1 + b_2 x_2 + \cdots + b_n x_n$$

where y is the dependent variable, x_i is the independent variable i, and $b_1, \ldots, b_n$ are the regression coefficients. Typically all variables in the equation are numeric variables, although categorical variables, called dummy variables, may be used as independent variables. They can also be the transformation of some variables. Common transformations include square, square root, and logarithmic.

The primary purpose of linear regression is to predict values of y from values of x_i. But linear regression requires several assumptions about the error, or residual, between the predicted value and the actual value (Miles and Shevlin 2001):

- The errors have a normal distribution for each set of values of the independent variables.
- The errors have the expected (mean) value of zero.
- The variance of the errors is constant for all values of the independent variables.
- The errors are independent of one another.

Additionally, multiple linear regression assumes that the correlation among the independent variables is not high.

Example 1 A watershed-level regression model for snow accumulation developed by Chang and Li (2000) uses snow water equivalent (SWE) as the dependent variable and location and topographic variables as the independent variables. One of their models takes the form of:

(18.7)

$$\text{SWE} = b_0 + b_1 \text{EASTING} + b_2 \text{SOUTHING} + b_3 \text{ELEV} + b_4 \text{PLAN1000}$$

where EASTING and SOUTHING correspond to the column number and the row number in an elevation raster, ELEV is the elevation value, and PLAN1000 is a surface curvature measure. After the b_i coefficients in Eq. (18.7) are estimated using known values at snow courses, the model can be used to estimate SWE for all cells in the watershed and to produce a continuous SWE surface.

Example 2 A crime model developed by Ceccato, Haining, and Signoretta (2002) is expressed as:

(18.8)

$$y = 12.060 + 22.046x_1 + 275.707x_2$$

where y is the rate of vandalism, x_1 is percentage of unemployed inhabitants aged 25–64, and x_2 is a variable distinguishing between inner city and outer city areas. The model has a R^2 value of 0.412, meaning that the model explains more than 40 percent of variation in the rate of vandalism.

Other examples of regression models include wildlife home ranges (Anderson et al. 2005), non-point pollution risk (Potter et al. 2004), soil moisture (Lookingbill and Urban 2004), and residential burglaries (Malczewski and Poetz 2005).

18.4.2 Local Regression Models

Local regression analysis, also called geographically weighted regression analysis, uses the information for each known point to derive a local model (Fotheringham et al. 2002). The model's parameters can therefore vary in space, providing a basis to explore a relationship's spatial nonstationarity (i.e., the relationship between variables varies over space), as opposed to stationarity (i.e., the relationship between variables remains the same over space), which is assumed in a global regression model. Local regression has been applied to a wide variety of research topics: voter turnout (Calvo and Escolar 2003), species richness (Foody 2004), environmental degradation (Nelson et al. 2007), residential burglaries (Malczewski and Poetz 2005), housing attribute prices (Bitter et al. 2007), regional accessibility (Lim and Thill 2008), and the relationship between tourism/recreation and rural poverty (Deller 2009).

18.4.3 Logistic Regression Models

Logistic regression is used when the dependent variable is categorical (e.g., presence or absence) and the independent variables are categorical, numeric, or both (Menard 2002). A major advantage of using logistic regression is that it does not require the assumptions needed for linear regression. Logistic regression uses the logit of y as the dependent variable:

(18.9)

$$\text{logit}(y) = a + b_1 x_1 + b_2 x_2 + b_3 x_3 + \cdots$$

The logit of y is the natural logarithm of the odds (also called odds ratio):

(18.10)

$$\text{logit}(y) = \ln(p/(1 - p))$$

where ln is the natural logarithm, $p/(1 - p)$ is the odds, and p is the probability of the occurrence of y. To convert logit (y) back to the odds or the probability, Eq. (18.10) can be rewritten as:

(18.11)

$$p/(1 - p) = e^{(a + b_1 x_1 + b_2 x_2 + b_3 x_3 + \cdots)}$$

(18.12)

$$p = e^{(a + b_1 x_1 + b_2 x_2 + b_3 x_3 + \cdots)}/[1 + e^{(a + b_1 x_1 + b_2 x_2 + b_3 x_3 + \cdots)}]$$

or,

(18.13)

$$p = 1/[1 + e^{-(a + b_1 x_1 + b_2 x_2 + b_3 x_3 + \cdots)}]$$

where e is the exponent.

Example 1 A red squirrel habitat model developed by Pereira and Itami (1991) is based on the following logit model:

(18.14)

$$\text{logit}(y) = 0.002\ \text{elevation} - 0.228\ \text{slope} + 0.685\ \text{canopy1} + 0.443\ \text{canopy2} + 0.481\ \text{canopy3} + 0.009\ \text{aspectE–W}$$

where canopy1, canopy2, and canopy3 represent three categories of canopy.

Example 2 Chang, Chiang, and Hsu (2007) use logistic regression to develop a rainfall-triggered landslide model. The dependent variable is the pres-

ence or absence of landslide in a unit area. The independent variables include numerical variables of elevation, slope, aspect, distance to stream channel, distance to ridge line, topographic wetness index, and NDVI, and categorical variables of lithology and road buffer. NDVI stands for the normalized difference vegetation index, a measure of vegetation areas and their canopy condition that can be derived from satellite images.

Other logistic regression models have also been developed for predicting grassland bird habitat (Forman, Reineking, and Hersperger 2002), fish habitat (Eikaas, Kliskey, and McIntosh 2005), and visitors' awareness and attitude toward national park designation (Machairas and Hovardas 2005).

18.5 Process Models

A **process model** integrates existing knowledge about the environmental processes in the real world into a set of relationships and equations for quantifying the processes (Beck, Jakeman, and McAleer 1993). Modules or submodels are often needed to cover different components of a process model. Some of these modules may use mathematical equations derived from empirical data, whereas others may use equations derived from laws in physics. A process model offers both a predictive capability and an explanation that is inherent in the proposed processes (Hardisty, Taylor, and Metcalfe 1993). Therefore, process models are by definition predictive and dynamic models.

Environmental models are typically process models because they must deal with the interaction of many variables including physical variables such as climate, topography, vegetation, and soils as well as cultural variables such as land management (Brimicombe 2003). As would be expected, environmental models are complex and data-intensive and usually face issues of uncertainty to a greater extent than traditional natural science or social science models (Couclelis 2002; Mulligan and Wainwright 2004). Once built, an environmental model can improve our understanding of the physical and cultural variables, facilitate prediction, and perform simulations (Barnsley 2007).

18.5.1 Revised Universal Soil Loss Equation

Soil erosion is an environmental process that involves climate, soil properties, topography, soil surface conditions, and human activities. A well-known model of soil erosion is the Revised Universal Soil Loss Equation (RUSLE), the updated version of the Universal Soil Loss Equation (USLE) (Wischmeier and Smith 1965, 1978; Renard et al. 1997). RUSLE predicts the average soil loss carried by runoff from specific field slopes in specified cropping and management systems and from rangeland.

RUSLE is a multiplicative model with six factors:

(18.15)

$$A = R K L S C P$$

where A is the average soil loss, R is the rainfall–runoff erosivity factor, K is the soil erodibility factor, L is the slope length factor, S is the slope steepness factor, C is the crop management factor, and P is the support practice factor. L and S can be combined into a single topographic factor LS.

Among the six factors in RUSLE, the slope length factor L, and by extension LS, poses more questions than other factors (Renard et al. 1997). Slope length is defined as the horizontal distance from the point of origin of overland flow to the point where either the slope gradient decreases enough that deposition begins or the flow is concentrated in a defined channel (Wischmeier and Smith 1978). Previous studies have used GIS to estimate LS. For example, Moore and Burch (1986) have proposed a method based on the unit stream power theory for estimating LS:

(18.16)

$$LS = (A_s/22.13)^m(\sin \beta/0.0896)^n$$

where A_s is the upslope contributing area, β is the slope angle, m is the slope length exponent, and n is the slope steepness exponent. The exponents m and n are estimated to be 0.6 and 1.3, respectively.

RUSLE developers, however, recommend that the L and S components be separated in the computational procedure for the LS factor (Foster 1994; Renard et al. 1997). The equation for L is:

(18.17)

$$L = (\lambda/72.6)^m$$

where λ is the measured slope length, and m is the slope length exponent. The exponent m is calculated by:

$$m = \beta/(1+\beta)$$

$$\beta = (\sin\theta/0.0896)/[3.0(\sin\theta)^{0.8} + 0.56]$$

where β is the ratio of rill erosion (caused by flow) to interrill erosion (principally caused by raindrop impact), and θ is the slope angle. The equation for S is:

(18.18)

$S = 10.8 \sin \theta + 0.03$, for slopes of less than 9%

$S = 16.8 \sin \theta - 0.50$, for slopes of 9% or steeper

Both L and S also need to be adjusted for special conditions, such as the adjustment of the slope length exponent m for the erosion of thawing, cultivated soils by surface flow, and the use of a different equation than Eq. (18.17) for slopes shorter than 15 feet.

There are at least two methods that follow the same procedure and equations as proposed by RUSLE developers but still use a GIS to automate the estimation of L and S. One method proposed by Hickey, Smith, and Jankowski (1994) uses DEMs as the input and estimates for each cell the slope steepness by computing the maximum downhill slope and the slope length by iteratively calculating the cumulative downhill slope length. Another method proposed by Desmet and Govers (1996) also uses DEMs as the input and estimates for each cell the slope steepness by using a quadratic surface-fitting method and the slope length by calculating the unit-contributing area.

USLE and RUSLE have evolved over the past 50 years. This soil erosion model has gone through numerous cycles of model development, calibration, and validation. The process continues. An updated model called WEPP (Water

ArcGIS offers a number of looping tools, which allow a process to be repeated. Available in Arc-Toolbox and ModelBuilder, these tools include list, series, Boolean condition, count, and feedback. Both a list variable and a series variable can contain multiple values for looping, one value per iteration. A Boolean variable is similar to a *do ... loop* statement in computer programming: the iteration continues until the Boolean variable is false. A count variable is similar to *for ... next* statement in computer programming: the iteration runs for a fixed number of times. Feedback, as the term suggests, allows the output of a process to be used as an input to a previous process. Looping can be a useful tool for sensitivity analysis and model calibration and simulation.

Erosion Prediction Project) is expected to replace RUSLE (Laflen, Lane, and Foster 1991; Laflen et al. 1997; Covert et al. 2005; Zhang, Chang, and Wu 2008). In addition to modeling soil erosion on hillslopes, WEPP can model the deposition in the channel system. Integration of WEPP and GIS is still desirable because a GIS can be used to extract hillslopes and channels as well as identify the watershed (Cochrane and Flanagan 1999, 2003; Renschler 2003). A description and download site of WEPP is available at **http://topsoil.nserl .purdue.edu/nserlweb/weppmain/wepp.html**.

18.5.2 Critical Rainfall Model

Landslide is defined as a downslope movement of a mass of soil and rock material. A landslide hazard model measures the potential of landslide occurrence within a given area (Varnes 1984). For the past two decades, developments of landslide hazard models have taken advantage of GIS. There are two types of landslide models, physically-based and statistical models. An example of a statistical model is a logistic regression model, such as Example 2 in Section 18.4.3. This section introduces the critical rainfall model as a physically-based landslide model.

The infinite slope model defines slope stability as the ratio of the available shear strength (stabilizing forces), including soil and root cohesion,

to the shear stress (destabilizing forces). The critical rainfall model developed by Montgomery and Dietrich (1994) combines the infinite slope model with a steady-state hydrologic model to predict the critical rainfall Q_{cr} that can cause landslide. Q_{cr} can be computed by:

(18.19)

$$Q_{cr} = T \sin \theta \left(\frac{b}{a} \right) \left(\frac{\rho_s}{\rho_w} \right) \left[1 - \frac{(\sin \theta - C)}{(\cos \theta \tan \phi)} \right]$$

where T is saturated soil transmissivity, θ is local slope angle, a is the upslope contributing drainage area, b is the unit contour length (the raster resolution), ρ_s is wet soil density, ϕ is the internal friction angle of the soil, ρ_w is the density of water, and C is combined cohesion, which is calculated by:

(18.20)

$$(C_r + C_s)/(h \rho_s g)$$

where C_r is root cohesion, C_s is soil cohesion, h is soil depth, and g is the gravitational acceleration constant.

A DEM can be used to derive a, b, and θ in Eq. (18.19), whereas other parameters in Eq. (18.19) and Eq. (18.20) must be gathered from existing data, field work, or literature survey. The critical rainfall model is regularly used for predicting shallow landslides triggered by rainfall events (e.g., Chiang and Chang 2009).

18.5.3 GIS and Process Models

Process models are typically raster-based. The role of a GIS in building a process model depends on the complexity of the model. A simple process model may be prepared and run entirely within a GIS, especially if the GIS is capable of performing loops for model calibration and simulation (Box 18.3). But more often a GIS is delegated the role of performing modeling-related tasks such as data visualization, database management, and exploratory data analysis (Brimicombe 2003).

KEY CONCEPTS AND TERMS

Binary model: A GIS model that uses logical expressions to select features from a composite feature layer or multiple rasters.

Deductive model: A model that represents the conclusion derived from a set of premises.

Descriptive model: A model that describes the existing conditions of geospatial data.

Deterministic model: A mathematical model that does not involve randomness.

Dynamic model: A model that emphasizes the changes of geospatial data over time and the interactions between variables.

Embedded system: A GIS bundled with other computer programs in a system with shared memory and a common interface.

Index model: A GIS model that uses the index values calculated from a composite feature layer or multiple rasters to produce a ranked map.

Inductive model: A model that represents the conclusion derived from empirical data and observations.

Loose coupling: Linking a GIS to other computer programs through the transfer of data files.

Model: A simplified representation of a phenomenon or a system.

Prescriptive model: A model that offers a prediction of what the conditions of geospatial data could be or should be.

Process model: A GIS model that integrates existing knowledge into a set of relationships and equations for quantifying the physical processes.

Regression model: A GIS model that uses a dependent variable and a number of independent variables in a regression equation for prediction or estimation.

Static model: A model that deals with the state of geospatial data at a given time.

Stochastic model: A mathematical model that considers the presence of some randomness in one or more of its parameters or variables.

Tight coupling: Linking a GIS to other computer programs through a common user interface.

Weighted linear combination: A method that computes the index value for each unit area by summing the products of the standardized value and the weight for each criterion.

REVIEW QUESTIONS

1. Describe the difference between a descriptive model and a prescriptive model.

2. How does a static model differ from a dynamic model?

3. Describe the basic steps involved in the modeling process.

4. Suppose you use kriging to build an interpolation model. How do you calibrate the model?

5. In many instances, you can build a GIS model that is either vector-based or raster-based. What general guidelines should you use in deciding which model to build?

6. What does loose coupling mean in the context of linking a GIS to another software package?

7. Why is a binary model often considered an extension of data query?

8. Provide an example of a binary model from your discipline.

9. How does an index model differ from a binary model?

10. Many index models use the weighted linear combination method to calculate the index value. Explain the steps one follows in using the weighted linear combination method.

11. What are the general shortcomings of the weighted linear combination method?

12. Provide an example of an index model from your discipline.

13. What kinds of variables can be used in a logistic regression model?

14. How does a local regression model differ from a regression model?

15. What is an environmental model?

16. Provide an example of a process model from your discipline. Can the model be built entirely in a GIS?

APPLICATIONS: GIS MODELS AND MODELING

This applications section covers four tasks. Tasks 1 and 2 let you build binary models using vector data and raster data, respectively. Tasks 3 and 4 let you build index models using vector data and raster data, respectively. Different options for running the geoprocessing operations are covered in this section. Tasks 1 and 3 use ModelBuilder. Tasks 2 and 4 use Python scripts.

Task 1 Build a Vector-Based Binary Model

What you need: *elevzone.shp,* an elevation zone shapefile; and *stream.shp,* a stream shapefile.

Task 1 asks you to locate the potential habitats of a plant species. Both *elevzone.shp* and *stream .shp* are measured in meters and spatially registered. The field ZONE in *elevzone.shp* shows three elevation zones. The potential habitats must meet the following criteria: (1) in elevation zone 2 and (2) within 200 meters of streams. You will use ModelBuilder to complete the task.

1. Start ArcCatalog, and connect to the Chapter 18 database. Launch ArcMap.

Rename the data frame Task 1. Add *stream.shp* and *elevzone.shp* to Task 1. Open the ArcToolbox window. Set the Chapter 18 database for the current workspace. Right-click the Chapter 18 database in the Catalog tree, point to New, and select Toolbox. Rename the new toolbox Chap18.tbx.

2. Click the ModelBuilder window to open it. Select Model Properties from the Model menu in the Model window. On the General tab, change both the name and label to Task1 and click OK.

3. The first step is to buffer *streams* with a buffer distance of 200 meters. In ArcToolbox, drag the Buffer tool from the Analysis Tools/Proximity toolset and drop it in the Model window. Right-click Buffer and select Open. In the Buffer dialog, select *stream* from the dropdown list for the input features, name the output feature class *strmbuf.shp,* enter 200 (meters) for the distance, and select ALL for the dissolve type. Click OK.

4. The visual objects in the Model window are color-coded. The input is coded blue, the tool gold, and the output green. The model can be executed one tool (function) at a time or as an entire model. Run the Buffer tool first. Right-click Buffer and select Run. The tool turns red during processing. After processing, both the tool and the output have the added drop shadow. Right-click *strmbuf.shp,* and select Add to Display.

5. Next overlay *elevzone* and *strmbuf.* Drag the Intersect tool from the Analysis Tools/Overlay toolset and drop it in the Model window. Right-click Intersect and select Open. In the Intersect dialog, select *strmbuf.shp* and *elevzone* from the dropdown list for the input features, name the output feature class *pothab.shp,* and click OK.

6. Right-click Intersect and select Run. After the overlay operation is done, right-click *pothab.shp* and add it to display. Turn off all layers except *pothab* in ArcMap's table of contents.

7. The final step is to select areas from *pothab* that are in elevation zone 2. Drag the Select tool from the Analysis Tools/Extract toolset and drop it in the Model window. Right-click Select and select Open. In the Select dialog, select *pothab.shp* for the input features, name the output feature class *final.shp,* and click the SQL button for the expression. Enter the following SQL statement in the expression box: "ZONE" = 2. Click OK to dismiss the dialogs. Right-click Select and select Run. Add *final.shp* to display.

8. Select Auto Layout from the Model window's View menu and let ModelBuilder rearrange the model diagram. Finally, select Save from the Model menu and save it as Task1 in Chap18.tbx. To run the Task1 model next time, right-click Task1 in the Chap18 toolbox and select Edit.

Task 2 Build a Raster-Based Binary Model

What you need: *elevzone_gd,* an elevation zone grid; and *stream_gd,* a stream grid.

Task 2 tackles the same problem as Task 1 but uses raster data. Both *elevzone_gd* and *stream_gd* have the cell resolution of 30 meters. The cell value in *elevzone_gd* corresponds to the elevation zone. The cell value in *stream_gd* corresponds to the stream ID.

1. Insert a data frame in ArcMap and rename it Task 2. Add *stream_gd* and *elevzone_gd* to Task 2.

2. Click the Python window to open it. Enter the following statements (without >>>) in the Python window:

>>> import arcpy

>>> from arcpy import env

>>> from arcpy.sa import *

>>> env.workspace = "d:/chap18"

>>> arcpy.CheckExtension("Spatial")

>>> outEucDistance = EucDistance ("stream_gd", 200)

>>> outExtract = ExtractByAttributes ("elevzone_gd", "value = 2")

>>> outExtract2 = ExtractByMask ("outExtract", "outEucDistance")

>>> outExtract2.save("outExtract2")

The first five statements of the script import arcpy and Spatial Analyst tools, and define d:/chap18 as the workspace. This is followed by three statements using Spatial Analyst tools. The EucDistance tool creates a distance measure raster with a maximum distance of 200 meters from *stream_gd.* The ExtractByAttributes tool creates a raster (*outExtract*) by selecting zone 2 in *elevzone_gd.* The ExtractByMask tool uses *outEucDistance* as a mask to select areas from *outExtract* that fall within its boundary and saves the output into *outExtract2.* Finally *outExtract2* is saved in the workspace.

4. Compare *outExtract2* with *final* from Task 1. They should cover the same areas.

Q1. What is the difference between ExtractByAttributes and ExtractByMask?

Task 3 Build a Vector-Based Index Model

What you need: *soil.shp,* a soil shapefile; *landuse. shp,* a land-use shapefile; and *depwater .shp,* a depth to water shapefile.

Task 3 simulates a project on mapping groundwater vulnerability. The project assumes that groundwater vulnerability is related to three factors: soil characteristics, depth to water, and land use. Each factor has been rated on a standardized scale from 1 to 5. These standardized values are stored in SOILRATE in *soil.shp,* DWRATE in *depwater.shp,* and LURATE in *landuse.shp.* The score of 9.9 is assigned to areas such as urban and built-up areas in *landuse.shp,* which should not be included in the model. The project also assumes that the soil factor is more important than the other two factors and is therefore assigned a weight of 0.6 (60%), compared to 0.2 (20%) for the other two factors. The index model can be expressed as Index value = 0.6 × SOILRATE + 0.2 × LURATE + 0.2 × DWRATE. In Task 3, you will use a geodatabase and ModelBuilder to create the index model.

1. First create a new personal geodatabase and import the three input shapefiles as feature classes to the geodatabase. In the Catalog tree, right-click the Chapter 18 database, point to New, and select Personal Geodatabase. Rename the geodatabase *Task3.mdb.* Right-click *Task3.mdb,* point to Import, and select Feature Class (multiple). Use the browser in the next dialog to select *soil.shp, landuse.shp,* and *depwater.shp* for the input features. Make sure that *Task3.mdb* is the output geodatabase. Click OK to run the import operation.

2. Insert a new data frame in ArcMap and rename it Task 3. Add *soil, landuse,* and *depwater* from *Task3.mdb* to Task 3. You will use the ModelBuilder for the rest of the operation. Click the ModelBuilder window to open it. Select Model Properties from the Model menu and, on the General tab, change both the name and label to Task3. Drag the Intersect tool from the Analysis Tools/Overlay toolset to the ModelBuilder window. Click Intersect and select Open. In the next dialog, select *soil, landuse,* and *depwater* from *Task3.mdb* for the input features, save the output feature class as *vulner* in *Task3.mdb,* and click OK. Right-click Intersect in the ModelBuilder window and select Run. After the Intersect operation is finished, right-click *vulner* and add it to display.

3. The remainder of Task 3 consists of attribute data operations. Open the attribute table of *vulner.* The table has all three rates needed for computing the index value. But you must go through a couple of steps before computation: add a new field for the index value, and exclude areas with the LURATE value of 9.9 from computation.

4. Drag the Add Field Tool from the Data Management Tools/Fields toolset to the ModelBuilder window. Right-click Add Field and select Open. In the next dialog, select *vulner* for the input table, enter TOTAL for the field name, select DOUBLE for the field type, enter 11 for the field precision, and 3 for the field scale. Click OK. Right-click Add Field and select Run. Check the attribute table of *vulner* to make sure that TOTAL has been added with Nulls. vulner (2) in the ModelBuilder window is the same as *vulner.*

5. Drag the Calculate Field tool from the Data Management Tools/Fields toolset to the ModelBuilder window. Right-click Calculate Field and select Open. In the next dialog, select vulner (2) for the input table; select TOTAL for the field name; enter the expression, [SOILRATE]*0.6 + [LURATE]*0.2 + [DWRATE]*0.2;

and click OK. Right-click Calculate Field and select Run. When the operation is finished, you can save the model as Task3 in Chap18.tbx.

6. This final step in analysis is to assign a TOTAL value of −99 to urban areas. Open the attribute table of *vulner* in ArcMap. Click the Select By Attributes button. In the next dialog, enter the expression, [LURATE] = 9.9, and click Apply. Right-click the field Total and select Field Calculator. Enter −99 in the expression box, and click OK. Click the Clear Selection button before closing the attribute table.

Q2. Excluding −99 for urban areas, what is the value range of TOTAL?

7. This step is to display the index values of *vulner*. Select Properties from the context menu of *vulner* in ArcMap. On the Symbology tab, choose Quantities and Graduated colors in the Show box. Click the Value dropdown arrow and select TOTAL. Click Classify. In the Classification dialog, select 6 classes and enter 0, 3.0, 3.5, 4.0, 4.5, and 5.0 as Break Values. Then choose a color ramp like Red Light to Dark for the symbol. Double-click the default symbol for urban areas (range −99–0) in the Layer Properties dialog and change it to a Hollow symbol for areas not analyzed. Click OK to see the index value map.

8. Once the index value map is made, you can modify the classification so that the grouping of index values may represent a rank order such as very severe (5), severe (4), moderate (3), slight (2), very slight (1), and not applicable (−99). You can then convert the index value map into a ranked map by doing the following: save the rank of each class under a new field called RANK, and then use the Dissolve tool from the Data Management Tools/Generalization toolset to remove boundaries of polygons that fall within the same rank. The ranked map should look much simpler than the index value map.

Task 4 Build a Raster-Based Index Model

What you need: *soil,* a soils raster; *landuse,* a land-use raster; and *depwater,* a depth to water raster.

Task 4 performs the same analysis as Task 3 but uses raster data. All three rasters have the cell resolution of 90 meters. The cell value in *soil* corresponds to SOILRATE, the cell value in *landuse* corresponds to LURATE, and the cell value in *depwater* corresponds to DWRATE. The only difference is that urban areas in *landuse* are already classified as no data. In Task 4, you will use a Python script.

1. Insert a new frame in ArcMap and rename it Task 4. Add *soil, landuse*, and *depwater* to Task 4. Click the Python window to open it.

2. Assuming that the workspace is d:/chap18, type the following statements one at a time without >>>, in the Python window:

```
>>> import arcpy
>>> from arcpy import env
>>> from arcpy.sa import *
>>> env.workspace = "d:/chap18"
>>> arcpy.CheckExtension("Spatial")
>>> outsoil = Times("soil", 0.6)
>>> outlanduse = Times("landuse", 0.2)
>>> outdepwater = Times("depwater", 0.2)
>>> outsum = CellStatistics(["outsoil",
        "outlanduse", "outdepwater"], "SUM")
>>> outReclass = Reclassify("outsum",
        "value", RemapRange([[2,3,1],
        [3,3.5,2],[3.5,4,3],[4,4.5,4],[4.5,5,5]]))
>>> outReclass.save("reclass_vuln")
```

The first five statements of the script import arcpy and Spatial Analyst tools, and define the workspace. The next three statements multiply each of the input rasters by its weight. Then the script uses the Cell Statistics tool to sum the three weighted rasters to create *outsum,* uses the Reclassify tool to group the cell values of *outsum* into five classes, and save the classified output as *reclass_vuln* in the workspace. As you enter each of the analysis statements, you will see its output in ArcMap.

4. In ArcMap, *outReclass (or reclass-vuln)* has the following five classes: 1 for <= 3.00, 2 for 3.01 – 3.50, 3 for 3.51 – 4.00, 4 for 4.01 – 4.50, and 5 for 4.51 – 5.00.

5. Right-click *reclass_vuln* in the table of contents, and select Properties. On the Symbology tab, change the label of 1 to Very slight, 2 to Slight, 3 to Moderate, 4

to Severe, and 5 to Very severe. Click OK. Now the raster layer is shown with the proper labels.

Q3. What percentage of the study area is labeled "Very severe"?

Challenge Task

What you need: *soil.shp, landuse.shp,* and *depwater.shp,* same as Task 3. Write a Python script to complete the same operations as in Task 3.

REFERENCES

Aller, L., T. Bennett, J. H. Lehr, R. J. Petty, and G. Hackett 1987. *DRASTIC: A Standardized System for Evaluating Ground-water Pollution Potential Using Hydrogeologic Settings.* U.S. Environmental Protection Agency, EPA/600/2-87/035, pp. 622.

Anderson, D. P., J. D. Forester, M. G. Turner, J. L. Frair, E. H. Merrill, D. Fortin, J. S. Mao, and M. S. Boyce. 2005. Factors Influencing Female Home Range Sizes in Elk (*Cervus elaphus*) in North American Landscapes. *Landscape Ecology* 20: 257–71.

Baja, S., D. M. Chapman, and D. Dragovich. 2002. A Conceptual Model for Defining and Assessing Land Management Units Using a Fuzzy Modeling Approach in GIS Environment. *Environmental Management* 33: 226–38.

Banai-Kashani, R. 1989. A New Method for Site Suitability Analysis: The Analytic Hierarchy Process. *Environmental Management* 13: 685–93.

Barnsley, M. J. 2007. *Environmental Modeling: A Practical Introduction.* Boca Raton, FL: CRC Press.

Beck, M. B., A. J. Jakeman, and M. J. McAleer. 1993. Construction and Evaluation of Models of Environmental Systems. In A. J. Jakeman, M. B. Beck, and M. J. McAleer, eds., *Modelling Change in Environmental Systems,* pp. 3–35. Chichester, England: Wiley.

Braimoh, A. K., P. L. G. Vlek, and A. Stein. 2004. Land Evaluation for Maize Based on Fuzzy Set and Interpolation. *Environmental Management* 13: 685–93.

Brimicombe, A. 2003. *GIS, Environmental Modelling and Engineering.* London: Taylor & Francis.

Brooks, R. P. 1997. Improving Habitat Suitability Index Models. *Wildlife Society Bulletin* 25: 163–67.

Ceccato, V., R. Haining, and P. Signoretta, 2002. Exploring Offence Statistics in Stockholm City Using Spatial Analysis Tools. *Annals of the Association*

of American Geographers 92: 29–51.

Chang, K., S. Chiang, and M. Hsu. 2007. Modeling Typhoon- and Earthquake-Induced Landslides in a Mountainous Watershed Using Logistic Regression. *Geomorphology* 89: 335–47.

Chang, K., Z. Khatib, and Y. Ou. 2002. Effects of Zoning Structure and Network Detail on Traffic Demand Modeling. *Environment and Planning B* 29: 37–52.

Chang, K., and Z. Li. 2000. Modeling Snow Accumulation with a Geographic Information System. *International Journal of Geographical Information Science* 14: 693–707.

Chiang, S., and K. Chang. 2009. Application of Radar Data to Modeling Rainfall-Induced Landslides. *Geomorphology* 103: 299–309.

Chrisman, N. 2001. *Exploring Geographic Information Systems,* 2d ed. New York: Wiley.

Chuvieco, E., and R. G. Congalton. 1989. Application of Remote

Sensing and Geographic Information Systems to Forest Fire Hazard Mapping. *Remote Sensing of Environment* 29: 147–59.

Chuvieco, E., and J. Salas. 1996. Mapping the Spatial Distribution of Forest Fire Danger Using GIS. *International Journal of Geographical Information Systems* 10: 333–45.

Cochrane, T. A., and D. C. Flanagan. 1999. Assessing Water Erosion in Small Watersheds Using WEPP with GIS and Digital Elevation Models. *Journal of Soil and Water Conservation* 54: 678–85.

Cochrane, T. A., and D. C. Flanagan. 2003. Representative Hillslope Methods for Applying the WEPP Model with DEMs and GIS. *Transactions of the ASAE* 46: 1041–49.

Corwin, D. L., P. J. Vaughan, and K. Loague. 1997. Modeling Nonpoint Source Pollutants in the Vadose Zone with GIS. *Environmental Science & Technology* 31: 2157–75.

Couclelis, H. 2002. Modeling Frameworks, Paradigms and Approaches. In K. C. Clarke, B. O. Parks, and M. P. Crane, eds., *Geographic Information Systems and Environmental Modeling*, pp. 36–50. Upper Saddle River, NJ: Prentice Hall.

Covert, S. A., P. R. Robichaud, W. J. Elliot, and T. E. Link. 2005. Evaluation of Runoff Prediction from WEPP-Based Erosion Models for Harvested and Burned Forest Watersheds. *Transactions of the ASAE* 48: 1091–1100.

Deller, S. 2009. Rural Poverty, Tourism and Spatial Heterogeneity. *Annals of*

Tourism Research, doi:10.1016/j.annals.2009.09.001.

DeMers, M. N. 2002. *GIS Modeling in Raster.* New York: Wiley.

Densham, P. J. 1991. Spatial Decision Support Systems. In D. J. Maguire, M. F. Goodchild, and D. W. Rhind, eds., *Geographical Information Systems: Principles and Applications,* Vol 1., pp. 403–12. London: Longman.

Desmet, P. J. J., and G. Govers. 1996. Comparison of Routing Systems for DEMs and Their Implications for Predicting Ephemeral Gullies. *International Journal of Geographical Information Systems* 10: 311–31.

Eikaas, H. S., A. D. Kliskey, and A. R. McIntosh. 2005. Spatial Modeling and Habitat Quantification for Two Diadromous Fish in New Zealand Streams: A GIS-Based Approach with Application for Conservation Management. *Environmental Management* 36: 726–40.

Felix, A. B., H. Campa III, K. F. Millenbah, S. R. Winterstein, and W. E. Moritz. 2004. Development of Landscape-Scale Habitat-Potential Models for Forest Wildlife Planning and Management. *Wildlife Society Bulletin* 32: 795–806.

Forman, R. T., T. B. Reineking, and A. M. Hersperger. 2002. Road Traffic and Nearby Grassland Bird Patterns in a Suburbanizing Landscape. *Environmental Management* 29: 782–800.

Foster, G. R. 1994. Comments on "Length-Slope Factor for the Revised Universal Soil Loss Equation: Simplified Method of Estimation." *Journal of Soil and Water Conservation* 49: 171–73.

Hall, G. B., F. Wang, and Subaryono. 1992. Comparison of Boolean and Fuzzy Classification Methods in Land Suitability Analysis by Using Geographical Information Systems. *Environment and Planning A* 24: 497–516.

Hall, O., and W. Arnberg. 2002. A Method for Landscape Regionalization Based On Fuzzy Membership Signatures. *Landscape and Urban Planning* 59: 227–40.

Hardisty, J., D. M. Taylor, and S. E. Metcalfe. 1993. *Computerized Environmental Modelling.* Chichester, England: Wiley.

Hickey, R., A. Smith, and P. Jankowski. 1994. Slope Length Calculations from a DEM Within ARC/INFO GRID. *Computers, Environment, and Urban Systems* 18: 365–80.

Hobbs, B. F., and P. M. Meier. 1994. Multicriteria Methods for Resource Planning: An Experimental Comparison. *IEEE Transactions on Power Systems* 9 (4): 1811–17.

Hopkins, L. D. 1977. Methods for Generating Land Suitability Maps: A Comparative Evaluation. *Journal of the American Institute of Planners* 43: 386–400.

Isaac, B., R. Cooke, D. Simmons, and F. Hogan. 2008. Predictive Mapping of Powerful Owl *(Ninox strenua)* Breeding Sites using Geographical Information Systems (GIS) in Urban Melbourne, Australia. *Landscape and Urban Planning* 84: 212–18.

Jankowski, J. 1995. Integrating Geographic Information Systems and Multiple Criteria Decision-Making Methods.

International Journal of Geographical Information Systems 9: 251–73.

Jiang, H., and J. R. Eastman. 2000. Application of Fuzzy Measures in Multi-Criteria Evaluation in GIS. *International Journal of Geographical Information Science* 14: 173–84.

Kliskey, A. D., E. C. Lofroth, W. A. Thompson, S. Brown, and H. Schreier. 1999. Simulating and Evaluating Alternative Resource-Use Strategies Using GIS-Based Habitat Suitability Indices. *Landscape and Urban Planning* 45: 163–75.

Laflen, J. M., W. J. Elliot, D. C. Flanagan, C. R. Meyer, and M. A. Nearing. 1997. WEPP—Predicting Water Erosion Using a Process-Based Model. *Journal of Soil and Water Conservation* 52: 96–102.

Laflen, J. M., L. J. Lane, and G. R. Foster. 1991. WEPP: A New Generation of Erosion Prediction Technology. *Journal of Soil and Water Conservation* 46: 34–38.

Lauver, C. L., W. H. Busby, and J. L. Whistler. 2002. Testing a GIS Model of Habitat Suitability for a Declining Grassland Bird. *Environmental Management* 30: 88–97.

Lein, J. K., and N. I. Stump. 2009. Assessing Wildfire Potential within the Wildland-Urban Interface: A Southeastern Ohio Example. *Applied Geography* 29: 21–34.

Lindsay, J. B. 2006. Sensitivity of Channel Mapping Techniques to Uncertainty in Digital Elevation Data. *International Journal of Geographical Information Science* 20: 669–92.

Lookingbill, T., and D. Urban. 2004. An Empirical Approach

Towards Improved Spatial Estimates of Soil Moisture for Vegetation Analysis. *Landscape Ecology* 19: 417–33.

Machairas, I., and T. Hovardas. 2005. Determining Visitor's Dispositions Toward the Designation of a Greek National Park. *Environmental Management* 36: 73–88.

Malczewski, J. 2000. On the Use of Weighted Linear Combination Method in GIS: Common and Best Practice Approaches. *Transactions in GIS* 4: 5–22.

Malczewski, J. 2006. GIS-Based Multicriteria Decision Analysis: A Survey of the Literature. *International Journal of Geographical Information Science* 20: 703–26.

Malczewski, J., and A. Poetz. 2005. Residential Burglaries and Neighborhood Socioeconomic Context in London, Ontario: Global and Local Regression Analysis. *The Professional Geographer* 57: 516–29.

Menard, S. 2002. *Applied Logistic Regression Analysis,* 2d ed. Thousand Oaks, CA: Sage.

Merchant, J. W. 1994. GIS-Based Groundwater Pollution Hazard Assessment: A Critical Review of the DRASTIC Model. *Photogrammetric Engineering & Remote Sensing* 60: 1117–27.

Miles, J., and Shevlin, M. 2001. *Applying Regression & Correlation: A Guide for Students and Researchers.* London: Sage.

Montgomery, D. R., and W. E. Dietrich. 1994. A Physically Based Model for Topographic Control on Shallow Landsliding. *Water Resources Research* 30: 1153–1171.

Moore, I. D., and G. J. Burch. 1986. Physical Basis of the Length-Slope Factor in the Universal Soil Loss Equation. *Soil Science Society of America Journal* 50: 1294–98.

Pereira, J. M. C., and L. Duckstein. 1993. A Multiple Criteria Decision-Making Approach to GIS-Based Land Suitability Evaluation. *International Journal of Geographical Information Systems* 7: 407–24.

Pereira, J. M. C., and R. M. Itami. 1991. GIS-Based Habitat Modeling Using Logistic Multiple Regression: A Study of the Mt. Graham Red Squirrel. *Photogrammetric Engineering & Remote Sensing* 57: 1475–86.

Potter, K. M., F. W. Cubbage, G. B. Blank, and R. H. Schaberg. 2004. A Watershed-Scale Model for Predicting Nonpoint Pollution Risk in North Carolina. *Environmental Management* 34: 62–74.

Renard, K. G., G. R. Foster, G. A. Weesies, D. K. McCool, and D. C. Yoder, coordinators. 1997. Predicting Soil Erosion by Water: A Guide to Conservation Planning with the Revised Universal Soil Loss Equation (RUSLE). *Agricultural Handbook 703*. Washington, DC: U.S. Department of Agriculture.

Renschler, C. S. 2003. Designing Geo-Spatial Interfaces to Scale Process Models: The GeoWEPP Approach. *Hydrological Processes* 17: 1005–17.

Rogerson, P. A. 2001. *Statistical Methods for Geography.* London: Sage.

Rogowski, A., and J. Goyne. 2002. Dynamic Systems Modeling and Four Dimensional Geographic Information Systems. In

K. C. Clarke, B. O. Parks, and M. P. Crane, eds., *Geographic Information Systems and Environmental Modeling,* pp. 122–59. Upper Saddle River, NJ: Prentice Hall.

Saaty, T. L. 1980. *The Analytic Hierarchy Process.* New York: McGraw-Hill.

Silberman, J. A., and P. W. Rees. 2010. Reinventing Mountain Settlements: A GIS Model for Identifying Possible Ski Towns in the U.S. Rocky Mountains. *Applied Geography* 30: 36–49.

Steiner, F. 1983. Resource Suitability: Methods for Analyses. *Environmental Management* 7: 401–20.

Stoms, D. M., J. M. McDonald, and F. W. Davis. 2002. Environmental Assessment: Fuzzy Assessment of Land Suitability for Scientific Research Reserves. *Environmental Management* 29: 545–58.

Varnes, D. J. 1984. Landslide Hazard Zonation: *A Review of*

Principles and Practice. Paris: UNESCO Press.

Wischmeier, W. H., and D. D. Smith. 1965. Predicting Rainfall-Erosion Losses from Cropland East of the Rocky Mountains: Guide for Selection of Practices for Soil and Water Conservation. *Agricultural Handbook 282.* Washington, DC: U.S. Department of Agriculture.

Wischmeier, W. H., and D. D. Smith. 1978. Predicting Rainfall Erosion Losses: A Guide to Conservation Planning. *Agricultural Handbook 537.* Washington, DC: U.S. Department of Agriculture.

Wright, L. E., W. Zitzmann, K. Young, and R. Googins. 1983. LESA—Agricultural Land Evaluation and Site Assessment. *Journal of Soil and Water Conservation* 38: 82–89.

Wu, J., M. D. Random, M. D. Nellis, G. J. Kluitenberg,

H. L. Seyler, and B. C. Rundquist. 2002. Using GIS to Assess and Manage the Conservation Reserve Program in Finney County, Kansas. *Photogrammetric Survey and Remote Sensing* 68: 735–44.

Xiang, W. 2001. Weighting-by-Choosing: A Weight Elicitation Method for Map Overlays. *Landscape and Urban Planning* 56: 61–73.

Yager, R. 1988. On Ordered Weighted Averaging Aggregation Operators in Multicriteria Decision Making. *IEEE Transactions on Systems, Man, and Cybernetics* 18: 183–90.

Zhang, J. X., K. Chang, and J. Q. Wu. 2008. Effects of DEM Resolution and Source on Soil Erosion Modelling: A Case Study using the WEPP Model. *International Journal of Geographical Information Science* 22: 925–42.

INDEX

U.S. STATE PLANE COORDINATE SYSTEMS OF 1983

State	Zone (Projection)
Alabama	East, West (Transverse Mercator)
Alaska	10 zones[a]
Arizona	East, Central, West (Transverse Mercator)
Arkansas	North, South (Lambert)
California	Zones 1–6 (Lambert)
Colorado	North, Central, South (Lambert)
Connecticut	1 zone (Lambert)
Delaware	1 zone (Transverse Mercator)
Florida	East, West (Transverse Mercator), North (Lambert)
Georgia	East, West (Transverse Mercator)
Hawaii	Zones 1–5 (Transverse Mercator)
Idaho	East, Central, West (Transverse Mercator)
Illinois	East, West (Transverse Mercator)
Indiana	East, West (Transverse Mercator)
Iowa	North, South (Lambert)
Kansas	North, South (Lambert)
Kentucky	North, South (Lambert)
Louisiana	North, South, Offshore (Lambert)
Maine	East, Central, West (Transverse Mercator)
Maryland	1 zone (Lambert)
Massachusetts	Mainland, Island (Lambert)
Michigan	North, Central, South (Lambert)
Minnesota	North, Central, South (Lambert)
Mississippi	East, West (Transverse Mercator)
Missouri	East, Central, West (Transverse Mercator)
Montana	1 zone (Lambert)
Nebraska	1 zone (Lambert)
Nevada	East, Central, West (Transverse Mercator)
New Hampshire	1 zone (Transverse Mercator)
New Jersey	1 zone (Transverse Mercator)
New Mexico	East, Central, West (Transverse Mercator)
New York	East, Central, West (Transverse Mercator), Long Island (Lambert)
North Carolina	1 zone (Lambert)
North Dakota	North, South (Lambert)
Ohio	North, South (Lambert)
Oklahoma	North, South (Lambert)
Oregon	North, South (Lambert)
Pennsylvania	North, South (Lambert)
Rhode Island	1 zone (Transverse Mercator)
South Carolina	1 zone (Lambert)
South Dakota	North, South (Lambert)
Tennessee	1 zone (Lambert)
Texas	North, North Central, Central, South Central, South (Lambert)
Utah	North, Central, South (Lambert)
Vermont	1 zone (Transverse Mercator)
Virginia	North, South (Lambert)
Washington	North, South (Lambert)
West Virginia	North, South (Lambert)
Wisconsin	North, Central, South (Lambert)
Wyoming	East, East Central, West Central, West (Transverse Mercator)

[a]Zone 1 (oblique Mercator), Zones 2–9 (Transverse Mercator), Zone 10 (Lambert)

Useful Conversion Factors

Length

1 statute mile = 5280 feet.
1 foot = 12 inches.
1 kilometer = 1000 meters.
1 meter = 100 centimeters.

To convert
  miles to kilometers, multiply miles by 1.6903.
  feet to meters, multiply feet by 0.3048.
  inches to centimeters, multiply inches by 2.54.

Angle

To convert
  radians to degrees, multiply radians by 57.2956.
  degrees to radians, multiply degrees by 0.0175.

Area

1 square mile = 640 acres.
1 hectare = 10,000 square meters.
1 square kilometer = 100 hectares.

To convert from square kilometers to square miles, multiply square kilometers by 0.3861.

To convert from square meters to square feet, multiply square meters by 10.7636.

To convert from hectares to acres, multiply hectares by 2.4711.

To convert from acres to square feet, multiply acres by 43,560.

To convert from acres to square meters, multiply acres by 4046.7808.

UTM (Universal Transverse Mercator) Zones

The following shows the UTM zone numbers and their longitude ranges (in parentheses). All longitude values are in degrees east (E) and west (W) of the prime meridian (0°).

1	2	3	4	5
(180W–174W)	(174W–168W)	(168W–162W)	(162W–156W)	(156W–150W)
6	7	8	9	10
(150W–144W)	(144W–138W)	(138W–132W)	(132W–126W)	(126W–120W)
11	12	13	14	15
(120W–114W)	(114W–108W)	(108W–102W)	(102W–96W)	(96W–90W)
16	17	18	19	20
(90W–84W)	(84W–78W)	(78W–72W)	(72W–66W)	(66W–60W)
21	22	23	24	25
(60W–54W)	(54W–48W)	(48W–42W)	(42W–36W)	(36W–30W)
26	27	28	29	30
(30W–24W)	(24W–18W)	(18W–12W)	(12W–6W)	(6W–0)
31	32	33	34	35
(0–6E)	(6E–12E)	(12E–18E)	(18E–24E)	(24E–30E)
36	37	38	39	40
(30E–36E)	(36E–42E)	(42E–48E)	(48E–54E)	(54E–60E)
41	42	43	44	45
(60E–66E)	(66E–72E)	(72E–78E)	(78E–84E)	(84E–90E)
46	47	48	49	50
(90E–96E)	(96E–102E)	(102E–108E)	(108E–114E)	(114E–120E)
51	52	53	54	55
(120E–126E)	(126E–132E)	(132E–138E)	(138E–144E)	(144E–150E)
56	57	58	59	60
(150E–156E)	(156E–162E)	(162E–168E)	(168E–174E)	(174E–180E)